A Primer of Ecological Statistics

SECOND EDITION

A Primer of Ecological Statistics

Second Edition

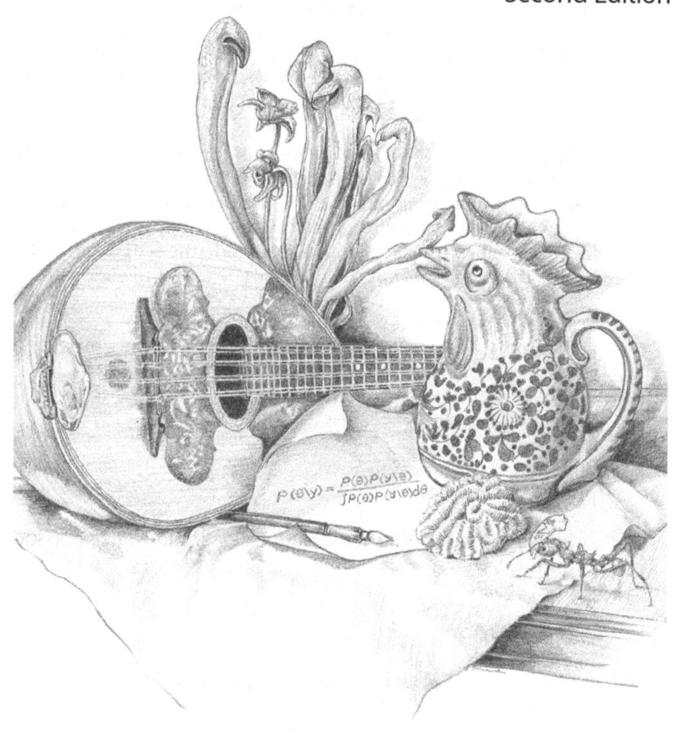

Nicholas J. Gotelli *University of Vermont*

Aaron M. Ellison *Harvard Forest*

Sinauer Associates, Inc. *Publishers*
Sunderland, Massachusetts U.S.A.

Oxford University Press is a department of the University of Oxford.
It furthers the University's objective of excellence in research,
scholarship, and education by publishing worldwide. Oxford is a
registered trade mark of Oxford University Press in the UK and certain
other countries.

Published in the United States of America by Oxford University Press
198 Madison Avenue, New York, NY 10016, United States of America

Sinauer Associates is an imprint of Oxford University Press.

For titles covered by Section 112 of the US Higher Education
Opportunity Act, please visit www.oup.com/us/he for the latest
information about pricing and alternate formats.

Address editorial correspondence to:
Sinauer Associates
23 Plumtree Road
Sunderland, MA 01375 U.S.A.
publish@sinauer.com

Address orders, sales, license, permissions, and translation inquiries to:
Oxford University Press U.S.A.
2001 Evans Road
Cary, NC 27513 U.S.A.
Orders: 1-800-445-9714

Library of Congress Cataloging-in-Publication Data
Gotelli, Nicholas J., 1959–
A primer of ecological statistics / Nicholas J. Gotelli,University of Vermont,Aaron M. Ellison, Harvard University.—Second edition.
pages ; cm
Includes bibliographical references and index.
ISBN 978-1-60535-064-6
1. Ecology—Statistical methods. I. Ellison, Aaron M., 1960- II. Title.
QH541.15.S72G68 2013
577.072—dc23

2012037594

Printed in the United States of America

For
Maryanne
&
Elizabeth

Who measures heaven, earth, sea, and sky
Thus seeking lore or gaiety
Let him beware a fool to be.

SEBASTIAN BRANT, *Ship of Fools*, 1494. Basel, Switzerland.

[N]umbers are the words without which exact description of any natural phenomenon is impossible.... Assuredly, every objective phenomenon, of whatever kind, is quantitative as well as qualitative; and to ignore the former, or to brush it aside as inconsequential, virtually replaces objective nature by abstract toys wholly devoid of dimensions — toys that neither exist nor can be conceived to exist.

E. L. MICHAEL, "Marine ecology and the coefficient of association: A plea in behalf of quantitative biology," 1920. *Journal of Ecology* 8: 54–59.

[W]e now know that what we term natural laws are merely statistical truths and thus must necessarily allow for exceptions. ...[W]e need the laboratory with its incisive restrictions in order to demonstrate the invariable validity of natural law. If we leave things to nature, we see a very different picture: every process is partially or totally interfered with by chance, so much so that under natural circumstances a course of events absolutely conforming to specific laws is almost an exception.

CARL JUNG, Foreword to *The I Ching or Book of Changes.*Third Edition, 1950, translated by R. Wilhelm and C. F. Baynes. Bollingen Series XIX, Princeton University Press.

Brief Contents

Contents

PART I

Fundamentals of Probability and Statistical Thinking

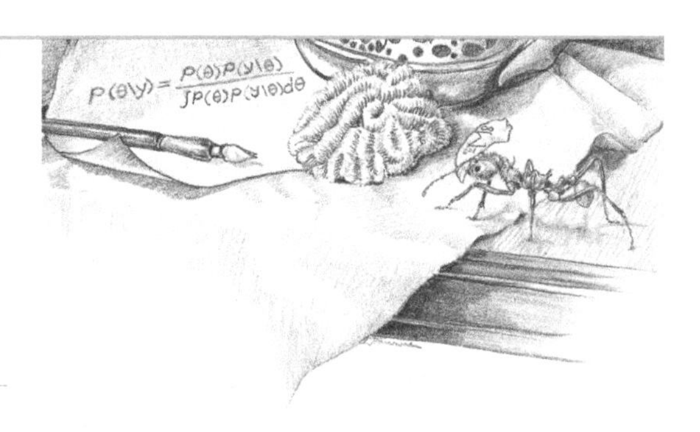

PART II

Designing Experiments

CHAPTER 6

Designing Successful Field Studies 137

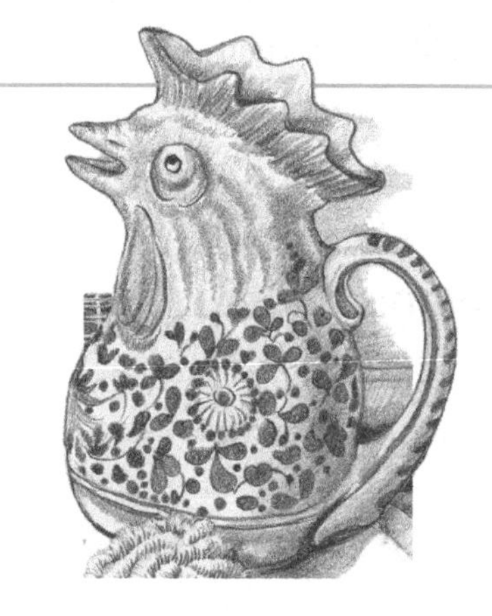

$P(\theta|y) = \frac{P(\theta)P(y|\theta)}{\int P(\theta)P(y|\theta)d\theta}$

Preface

What is the "added value" for readers in purchasing this Second Edition? For this Second Edition, we have added a new Part 4, on "Estimation," which includes two chapters: "The Measurement of Biodiversity" (Chapter 13), and "Detecting Populations and Estimating their Size" (Chapter 14). These two new chapters describe measurements and methods that are beginning to be widely used and that address questions that are central to the disciplines of community ecology and demography. Some of the methods we describe have been developed only in the last 10 years, and many of them continue to evolve rapidly.

The title of this new section, "Estimation," reflects an ongoing shift in ecology from testing hypotheses to estimating parameters. Although Chapter 3 describes summary measures of location and spread, most of the focus of the First Edition—and of many ecological publications—is on hypothesis testing (Chapter 4). In the two new chapters, we are more concerned with the process of estimation itself, although the resulting estimators of biodiversity and population size certainly can be used in conventional statistical tests once they are calculated. It is useful to think of biodiversity and population sizes as latent or unknown "state variables." We want to measure these things, but it is impossible to find and tag all of the trout in a stream or discover all of the species of ants that lurk in a bog. Instead, we work only with small samples of diversity, or limited collections of marked animals, and we use these samples to estimate the "true" value of the underlying state variables. The two new chapters are united not only by their focus on estimation, but also by their common underlying statistical framework: methods for asymptotic estimators of species richness discussed in Chapter 13 were derived from methods developed for mark-recapture studies discussed in Chapter 14. We hope the methods described in these two chapters will provide some fresh statistical tools for ecologists to complement the more classical topics we cover in Chapters 1–12.

Chapter 13 begins with what seems like a relatively simple question: how many species are there in an assemblage? The central problem is that nature is (still) vast, our sample of its diversity is tiny, and there are always many rare and undetected species. Moreover, because the number of species we count is heavily influenced by the number of individuals and samples we have collected, the data must be standardized for meaningful comparisons. We describe methods for both interpolation (rarefaction) and extrapolation (asymptotic species richness estimators) to effectively standardize biodiversity comparisons. We also advocate the use of Hill numbers as a class of diversity indices that have useful statistical properties, although they too are subject to sampling effects. We thank Anne Chao, Rob Colwell, and Lou Jost for their ongoing collaboration, as well as their insights, extended correspondence, and important contributions to the literature on biodiversity estimation.

In Chapter 14, we delve deeper into the problem of incomplete detection, addressing two main questions. First, how do we estimate the probability that we can detect a species, given that it is actually present at a site? Second, for a species that is present at a site, how large is its population? Because we have a limited sample of individuals that we can see and count, we must first estimate the detection probability of an individual species before we can estimate the size of its population. Estimates of occupancy probability and population size are used in quantitative models for the management of populations of both rare and exploited species. And, just as we use species richness estimators to extrapolate beyond our sample, we use estimates from occupancy models and mark-recapture studies to extrapolate and forecast what will happen to these populations as, for example, habitats are fragmented, fishing pressure increases, or the climate changes. We thank Elizabeth Crone, Dave Orwig, Evan Preisser, and Rui Zhang for letting us use some of their unpublished data in this chapter and for discussing their analyses with us. Bob Dorazio and Andy Royle discussed these models with us; they, along with Evan Cooch, have developed state-of-the-art open-source software for occupancy modeling and mark-recapture analysis.

In this edition, we have also corrected numerous minor errors that astute readers have pointed out to us over the past 8 years. We are especially grateful for the careful reading of Victor Lemes Landeiro, who, together with his colleagues Fabricio Beggiato Baccaro, Helder Mateus Viana Espirito Santo, Miriam Plaza Pinto, and Murilo Sversut Dias, translated the entire book into Portuguese. Specific contributions of these individuals, and all the others who sent us comments and pointed out errors, are noted in the Errata section of the book's website (harvardforest.fas.harvard.edu/ellison/publications/primer/errata). Although we have corrected all errors identified in the first edition, others undoubtedly remain (another problem in detection probability); please contact us if you find any.

All code posted on the Data and Code section of the book's website (harvardforest.fas.harvard.edu/ellison/publications/primer/datafiles) that we used for the analyses in the book has been updated and reworked in the R programming language (r-project.org). In the time since we wrote the first edition, R has become the de facto standard software for statistical analysis and is used by most ecologists and statisticians for their day-to-day analytical work. We have provided some R scripts for our figures and analyses, but they have minimal annotation, and there are already plenty of excellent resources available for ecological modeling and graphics with R.

Last but not least, we thank Andy Sinauer, Azelie Aquadro, Joan Gemme, Chris Small, Randy Burgess, and the entire staff at Sinauer Associates for carefully editing our manuscript pages and turning them into an attractive book.[1]

Preface to the First Edition

Why another book on statistics? The field is already crowded with texts on biometry, probability theory, experimental design, and analysis of variance. The short answer is that we have yet to find a single text that meets the two specific needs of ecologists and environmental scientists. The first need is a general introduction to probability theory including the assumptions of statistical inference and hypothesis testing. The second need is a detailed discussion of specific designs and analyses that are typically encountered in ecology.

This book reflects several collective decades of teaching statistics to undergraduate and graduate students at the University of Oklahoma, the University of Vermont, Mount Holyoke College, and the Harvard Forest. The book represents our personal perspective on what is important at the interface of ecology and statistics and the most effective way to teach it. It is the book we both wish we had in our hands when we started studying ecology in college.

A "primer" suggests a short text, with simple messages and safe recommendations for users. If only statistics were so simple! The subject is vast, and new methods and software tools for ecologists appear almost monthly. In spite of the book's length (which stunned even us), we still view it as a distillation of statistical knowledge that is important for ecologists. We have included both classic expositions as well as discussions of more recent methods and techniques. We hope there will be material that will appeal to both the neophyte and the experienced researcher. As in art and music, there is a strong element of style and personal preference in the choice and application of statistics. We have explained our own preferences for certain kinds of analyses and designs, but readers and users of statistics will ultimately have to choose their own set of statistical tools.

[1] And for dealing with all of the footnotes!

How To Use This Book

In contrast to many statistics books aimed at biologists or ecologists—"biometry" texts—this is not a book of "recipes," nor is it a set of exercises or problems tied to a particular software package that is probably already out of date. Although the book contains equations and derivations, it is not a formal statistical text with detailed mathematical proofs. Instead, it is an exposition of basic and advanced topics in statistics that are important specifically for ecologists and environmental scientists. It is a book that is meant to be read and used, perhaps as a supplement to a more traditional text, or as a stand-alone text for students who have had at least a minimal introduction to statistics. We hope this book will also find a place on the shelf (or floor) of environmental professionals who need to use and interpret statistics daily but who may not have had formal training in statistics or have ready access to helpful statisticians.

Throughout the book we make extensive use of footnotes. Footnotes give us a chance to greatly expand the material and to talk about more complex issues that would bog down the flow of the main text. Some footnotes are purely historical, others cover mathematical and statistical proofs or details, and others are brief essays on topics in the ecological literature. As undergraduates, we both were independently enamored of Hutchinson's classic *An Introduction to Population Biology* (Yale University Press 1977). Hutchinson's liberal use of footnotes and his frequent forays into history and philosophy served as our stylistic model.

We have tried to strike a balance between being concise and being thorough. Many topics recur in different chapters, although sometimes in slightly different contexts. Figure and table legends are somewhat expansive because they are meant to be readable without reference to the text. Because Chapter 12 requires matrix algebra, we provide a brief Appendix that covers the essentials of matrix notation and manipulation. Finally, we have included a comprehensive glossary of terms used in statistics and probability theory. This glossary may be useful not only for reading this book, but also for deciphering statistical methods presented in journal articles.

Mathematical Content

Although we have not shied away from equations when they are needed, there is considerably less math here than in many intermediate level texts. We also try to illustrate all methods with empirical analyses. In almost all cases, we have used data from our own studies so that we can illustrate the progression from raw data to statistical output.

Although the chapters are meant to stand alone, there is considerable cross-referencing. Chapters 9–12 cover the traditional topics of many biometry texts and contain the heaviest concentration of formulas and equations. Chapter 12, on multivariate methods, is the most advanced in the book. Although we have tried to explain multivariate methods in plain English (no easy task!), there is no way to sugar-coat the heavy dose of matrix algebra and new vocabulary necessary for multivariate analysis.

Coverage and Topics

Statistics is an evolving field, and new methods are always being developed to answer ecological questions. This text, however, covers core material that we believe is foundational for any statistical analysis. The book is organized around design-based statistics—statistics that are to be used with designed observational and experimental studies. An alternative framework is model-based statistics, including model selection criteria, likelihood analysis, inverse modeling, and other such methods.

What characterizes the difference between these two approaches? In brief, design-based analyses primarily address the problem of $P(\text{data} \mid \text{model})$: assessing the probability of the data given a model that we have specified. An important element of design-based analysis is being explicit about the underlying model. In contrast, model-based statistics address the problem of $P(\text{model} \mid \text{data})$—that is, assessing the probability of a model given the data that are present. Model-based methods are often most appropriate for large datasets that may not have been collected according to an explicit sampling strategy.

We also discuss Bayesian methods, which somewhat straddle the line between these two approaches. Bayesian methods address $P(\text{model} \mid \text{data})$, but can be used with data from designed experiments and observational studies.

The book is divided into three parts. Part I discusses the fundamentals of probability and statistical thinking. It introduces the logic and language of probability (Chapter 1), explains common statistical distributions used in ecology (Chapter 2), and introduces important measures of location and spread (Chapter 3). Chapter 4 is more of a philosophical discussion about hypothesis testing, with careful explanations of Type I and Type II errors and of the ubiquitous statement that "$P < 0.05$." Chapter 5 closes this section by introducing the three major paradigms of statistical analysis (frequentist, Monte Carlo, and Bayesian), illustrating their use through the analysis of a single dataset.

Part II discusses how to successfully design and execute observational studies and field experiments. These chapters contain advice and information that is to be used before any data are collected in the field. Chapter 6 addresses the practical problems of articulating the reason for the study, setting the sample size, and dealing with independence, replication, randomization, impact studies, and environmental heterogeneity. Chapter 7 is a bestiary of experimental designs. However, in contrast to almost all other texts, there are almost no equations in this chapter. We discuss the strengths and weaknesses of different designs without getting bogged down in equations or analysis of variance (ANOVA) tables, which have mostly been quarantined in Part III. Chapter 8 addresses the problem of how to curate and manage the data once it is collected. Transformations are introduced in this chapter as a way to help screen for outliers and errors in data transcription. We also stress the importance of creating metadata—documentation that accompanies the raw data into long-term archival storage.

Part III discusses specific analyses and covers the material that is the main core of most statistics texts. These chapters carefully develop many basic models, but also try to introduce some more current statistical tools that ecologists and environmental scientists find useful. Chapter 9 develops the basic linear regression model and also introduces more advanced topics such as non-linear, quantile, and robust regression, and path analysis. Chapter 10 discusses ANOVA (analysis of variance) and ANCOVA (analysis of covariance) models, highlighting which ANOVA model is appropriate for particular sampling or experimental designs. This chapter also explains how to use a priori contrasts and emphasizes plotting ANOVA results in a meaningful way to understand main effects and interaction terms. Chapter 11 discusses categorical data analysis, with an emphasis on chi-square analysis of one- and two-way contingency tables and log-linear modeling of multi-way contingency tables. The chapter also introduces goodness-of-fit tests for discrete and continuous variables. Chapter 12 introduces a variety of multivariate methods for both ordination and classification. The chapter closes with an introduction to redundancy analysis, a multivariate analog of multiple regression.

About the Cover

The cover is an original still life by the ecologist Elizabeth Farnsworth, inspired by seventeenth-century Dutch still-life paintings. Netherlands painters, including De Heem, Claesz, and Kalf, displayed their skill by arranging and painting a wide range of natural and manmade objects, often with rich allusions and hidden symbolism. Elizabeth's beautiful and biologically accurate renderings of plants and animals follow a long tradition of biological illustrators such as Dürer and Escher. In Elizabeth's drawing, the centerpiece is a scrolled pen-and-paper illustration of Bayes' Theorem, an eighteenth-century mathematical exposition that lies at the core of probability theory. Although modern Bayesian methods are computationally intensive, an understanding of Bayes' Theorem still requires pen-and-paper contemplation.

Dutch still lifes often included musical instruments, perhaps symbolizing the Pythagorean beauty of music and numbers. Elizabeth's drawing incorporates Aaron's bouzouki, as the writing of this book was interwoven with many sessions of acoustic music with Aaron and Elizabeth.

The human skull is the traditional symbol of mortality in still life paintings. Elizabeth has substituted a small brain coral, in honor of Nick's early dissertation work on the ecology of subtropical gorgonians. In light of the current worldwide collapse of coral reefs, the brain coral seems an appropriate symbol of mortality. In the foreground, an *Atta* ant carries a scrap of text with a summation sign, reminding us of the collective power of individual ant workers whose summed activities benefit the entire colony. As in early medieval paintings, the ant is drawn disproportionately larger than the other objects, reflecting its dominance and importance in terrestrial ecosystems. In the background is a specimen of *Sarracenia minor*, a carnivorous plant from the southeastern U.S. that we grow in the greenhouse. Although *Sarracenia minor* is too rare for experimental manipulations in the field, our ongoing studies of the more common *Sarracenia purpurea* in the northeastern U.S. may help us to develop effective conservation strategies for all of the species in this genus.

The final element in this drawing is a traditional Italian Renaissance chicken carafe. In the summer of 2001, we were surveying pitcher plants in remote bogs of the Adirondack Mountains. At the end of a long field day, we found ourselves in a small Italian restaurant in southern Vermont. Somewhat intemperately, we ordered and consumed an entire liter-sized chicken carafe of Chianti. By the end of the evening, we had committed ourselves to writing this book, and "The Chicken"—as we have always called it—was officially hatched.

Acknowledgments

We thank several colleagues for their detailed suggestions on different chapters: Marti Anderson (11, 12), Jim Brunt (8), George Cobb (9, 10), Elizabeth Farnsworth (1–5), Brian Inouye (6, 7), Pedro Peres-Neto (11,12), Catherine Potvin (9, 10), Robert Rockwell (1–5), Derek Roff (original book proposal), David Skelly (original book proposal), and Steve Tilley (1–5). Laboratory discussion groups at the Harvard Forest and University of Vermont also gave us feedback on many parts of the book. Special kudos go to Henry Horn, who read and thoroughly critiqued the entire manuscript. We thank Andy Sinauer, Chris Small, Carol Wigg, Bobbie Lewis, and Susan McGlew of Sinauer Associates, and, especially, Michele Ruschhaupt of The Format Group for transforming pages of crude word-processed text into an elegant and pleasing book.

The National Science Foundation supports our research on pitcher plants, ants, null models, and mangroves, all of which appear in many of the examples in this book. The University of Vermont and the Harvard Forest, our respective home institutions, provide supportive environments for book writing.

PART I

Fundamentals of Probability and Statistical Thinking

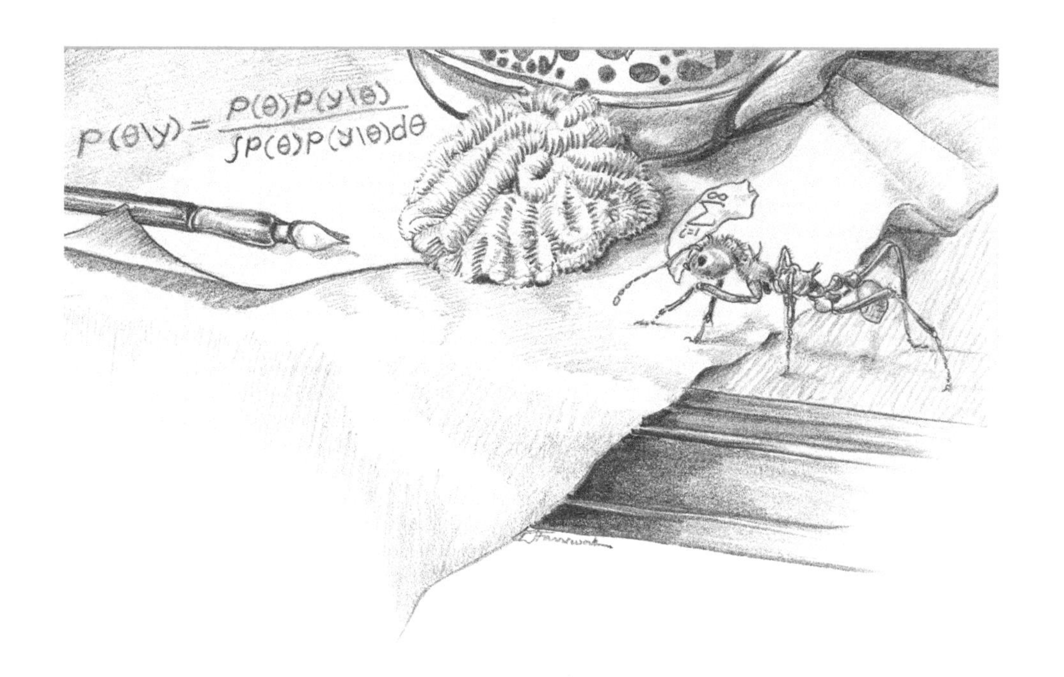

CHAPTER 1

An Introduction to Probability

In this chapter, we develop basic concepts and definitions required to understand probability and sampling. The details of probability calculations lie behind the computation of all statistical significance tests. Developing an appreciation for probability will help you design better experiments and interpret your results more clearly.

The concepts in this chapter lay the foundations for the use and understanding of statistics. They are far more important, actually, than some of the detailed topics that will follow. This chapter provides the necessary background to address the critical questions such as "What does it mean when you read in a scientific paper that the means of two samples were significantly different at $P = 0.003$?" or "What is the difference between a Type I and a Type II statistical error?" You would be surprised how often scientists with considerable experience in statistics are unable to clearly answer these questions. If you can understand the material in this introductory chapter, you will have a strong basis for your study of statistics, and you will always be able to correctly interpret statistical material in the scientific literature—even if you are not familiar with the details of the particular test being used.

In this chapter, we also introduce you to the problem of measurement and quantification, processes that are essential to all sciences. We cannot begin to investigate phenomena scientifically unless we can quantify processes and agree on a common language with which to interpret our measurements. Of course, the mere act of quantification does not by itself make something a science: astrology and stock market forecasting use a lot of numbers, but they don't qualify as sciences.

One conceptual challenge for students of ecology is translating their "love of nature" into a "love of pattern." For example, how do we quantify patterns of plant and animal abundance? When we take a walk in the woods, we find ourselves asking a stream of questions such as: "What is our best estimate for the density of *Myrmica* (a common forest ant) colonies in this forest? Is it 1 colony/m^2? 10 colonies/m^2? What would be the best way to measure *Myrmica* density? How does *Myrmica* density vary in different parts of the woods? What mechanisms or hypotheses might account for such variation?" And, finally, "What experiments and observations could we make to try and test or falsify these hypotheses?"

But once we have "quantified nature," we still have to summarize, synthesize, and interpret the data we have collected. Statistics is the common language used in all sciences to interpret our measurements and to test and discriminate among our hypotheses (e.g., Ellison and Dennis 2010). Probability is the foundation of statistics, and therefore is the starting point of this book.

What Is Probability?

If the weather forecast calls for a 70% chance of rain, we all have an intuitive idea of what that means. Such a statement quantifies the **probability**, or the likely outcome, of an event that we are unsure of. Uncertainty exists because there is **variation** in the world, and that variation is not always predictable. Variation in biological systems is especially important; it is impossible to understand basic concepts in ecology, evolution, and environmental science without an appreciation of natural variation.[1]

Although we all have a general understanding of probability, defining it precisely is a different matter. The problem of defining probability comes into sharper focus when we actually try to measure probabilities of real events.

Measuring Probability

The Probability of a Single Event: Prey Capture by Carnivorous Plants

Carnivorous plants are a good system for thinking about the definition of probability. The northern pitcher plant *Sarracenia purpurea* captures insect prey in leaves that fill with rainwater.[2] Some of the insects that visit a pitcher plant fall into

Charles Darwin

[1] Variation in traits among individuals is one of the key elements of the theory of evolution by natural selection. One of Charles Darwin's (1809–1882) great intellectual contributions was to emphasize the significance of such variation and to break free from the conceptual straightjacket of the typological "species concept," which treated species as fixed, static entities with well-defined, unchanging boundaries.

*Pitcher plant (*Sarracenia purpurea*)*

[2] Carnivorous plants are some of our favorite study organisms. They have fascinated biologists since Darwin first proved that they can absorb nutrients from insects in 1875. Carnivorous plants have many attributes that make them model ecological systems (Ellison and Gotelli 2001; Ellison et al. 2003).

the pitcher and drown; the plant extracts nutrients from the decomposing prey. Although the trap is a marvelous evolutionary adaptation for life as a carnivore, it is not terribly efficient, and most of the insects that visit a pitcher plant escape.

How could we estimate the probability that an insect visit will result in a successful capture? The most straightforward way would be to keep careful track of the number of insects that visit a plant and the number that are captured. *Visiting a plant* is an example of what statisticians call an **event**. An event is a simple process with a well-recognized beginning and end.

In this simple universe, visiting a plant is an event that can result in two **outcomes**: *escape* or *capture*. Prey capture is an example of a **discrete outcome** because it can be assigned a positive integer. For example, you could assign a 1 for a capture and a 2 for an escape. The **set** formed from all the possible outcomes of an event is called a **sample space.**[3] Sample spaces or sets made up of discrete outcomes are called **discrete sets** because their outcomes are all countable.[4]

[3] Even in this simple example, the definitions are slippery and won't cover all possibilities. For example, some flying insects may hover above the plant but never touch it, and some crawling insects may explore the outer surface of the plant but not cross the lip and enter the pitcher. Depending on how precise we care to be, these possibilities may or may not be included in the sample space of observations we use to determine the probability of being captured. The sample space establishes the domain of inference from which we can draw conclusions.

G. F. L. P. Cantor

[4] The term *countable* has a very specific meaning in mathematics. A set is considered countable if every element (or outcome) of that set can be assigned a positive integer value. For example, we can assign to the elements of the set *Visit* the integers 1 (= capture) and 2 (= escape). The set of integers itself is countable, even though there are infinitely many integers. In the late 1800s, the mathematician Georg Ferdinand Ludwig Philipp Cantor (1845–1918), one of the founders of what has come to be called set theory, developed the notion of cardinality to describe such countability. Two sets are considered to have the same cardinality if they can be put in one-to-one correspondence with each other. For example, the elements of the set of all even integers {2, 4, 6, ...} can each be assigned an element of the set of all integers (and vice versa). Sets that have the same cardinality as the integers are said to have cardinality 0, denoted as $\aleph_0$. Cantor went on to prove that the set of rational numbers (numbers that can be written as the quotient of any two integers) also has cardinality $\aleph_0$, but the set of irrational numbers (numbers that cannot be written as the quotient of any two integers, such as $\sqrt{2}$) is not countable. The set of irrational numbers has cardinality 1, and is denoted $\aleph_1$. One of Cantor's most famous results is that there is a one-to-one correspondence between the set of all points in the small interval [0,1] and the set of all points in *n*-dimensional space (both have cardinality 1). This result, published in 1877, led him to write to the mathematician Julius Wilhelm Richard Dedekind, "I see it, but I don't believe it!" Curiously, there are types of sets that have higher cardinality than 1, and in fact there are infinitely many cardinalities. Still weirder, the set of cardinal numbers $\{\aleph_0, \aleph_1, \aleph_2, \ldots\}$ is itself a countably infinite set (its cardinality = 0)!

Each insect that visits a plant is counted as a **trial.** Statisticians usually refer to each trial as an individual **replicate,** and refer to a set of trials as an **experiment.**[5] We define the probability of an outcome as the number of times that outcome occurs divided by the number of trials. If we watched a single plant and recorded 30 prey captures out of 3000 visits, we would calculate the probability as

$$\frac{\text{number of captures}}{\text{number of visits}}$$

or 30 in 3000, which we can write as 30/3000, or 0.01. In more general terms, we calculate the probability P that an outcome occurs to be

$$P = \frac{\text{number of outcomes}}{\text{number of trials}} \tag{1.1}$$

By definition, there can never be more outcomes than there are trials, so the numerator of this fraction is never larger than the denominator, and therefore

$$0.0 \leq P \leq 1.0$$

In other words, probabilities are always bounded by a minimum of 0.0 and a maximum of 1.0. A probability of 0.0 represents an event that will never happen, and a probability of 1.0 represents an event that will always happen.

However, even this estimate of probability is problematic. Most people would say that the probability that the sun will rise tomorrow is a sure thing (P = 1.0). Certainly if we assiduously recorded the rising of the sun each morning, we would find that it always does, and our measurements would, in fact, yield a probability estimate of 1.0. But our sun is an ordinary star, a nuclear reactor that releases energy when hydrogen atoms fuse to form helium. After about 10 billion years, all the hydrogen fuel is used up, and the star dies. Our sun is a middle-aged star, so if you could keep observing for another 5 billion years, one morning the sun would no longer rise. And if you started your observations at that point, your estimate of the probability of a sunrise would be 0.0. So, is the probability that the sun rises each day really 1.0? Something less than 1.0? 1.0 for now, but 0.0 in 5 billion years?

[5] This statistical definition of an experiment is less restrictive than the conventional definition, which describes a set of manipulated subjects (the treatment) and an appropriate comparison group (the control). However, many ecologists use the term **natural experiment** to refer to comparisons of replicates that have not been manipulated by the investigator, but that differ naturally in the quantity of interest (e.g., islands with and without predators; see Chapter 6).

This example illustrates that any estimate or measure of probability is completely contingent on how we define the sample space—the set of all possible events that we use for comparison. In general, we all have intuitive estimates or guesses for probabilities for all kinds of events in daily life. However, to quantify those guesses, we have to decide on a sample space, take samples, and count the frequency with which certain events occur.

Estimating Probabilities by Sampling

In our first experiment, we watched 3000 insects visit a plant; 30 of them were captured, and we calculated the probability of capture as 0.01. Is this number reasonable, or was our experiment conducted on a particularly good day for the plants (or a bad day for the insects)? There will always be variability in these sorts of numbers. Some days, no prey are captured; other days, there are several captures. We could determine the true probability of capture precisely if we could watch every insect visiting a plant at all times of day or night, every day of the year. Each time there was a visit, we would determine whether the outcome *Capture* occurred and accordingly update our estimate of the probability of capture. But life is too short for this. How can we estimate the probability of capture without constantly monitoring plants?

We can efficiently estimate the probability of an event by taking a **sample** of the population of interest. For example, every week for an entire year we could watch 1000 insects visit plants on a single day. The result is a set of 52 samples of 1000 trials each, and the number of insects captured in each sample. The first three rows of this 52-row dataset are shown in Table 1.1. The probabilities of capture differ on the different days, but not by very much; it appears to be relatively uncommon for insects to be captured while visiting pitcher plants.

TABLE 1.1 **Sample data for number of insect captures per 1000 visits to the carnivorous plant *Sarracenia purpurea***

ID number	Observation date	Number of insect captures
1	June 1, 1998	10
2	June 8, 1998	13
3	June 15, 1998	12
⋮	⋮	⋮
52	May 24, 1999	11

In this hypothetical dataset, each row represents a different week in which a sample was collected. If the sampling was conducted once a week for an entire year, the dataset would have exactly 52 rows. The first column indicates the ID number, a unique consecutive integer (1 to 52) assigned to each row of the dataset. The second column gives the sample date, and the third column gives the number of insect captures out of 1000 insect visits that were observed. From this dataset, the probability of capture for any single sample can be estimated as the number of insect captures/1000. The data from all 52 capture numbers can be plotted and summarized as a histogram that illustrates the variability among the samples (see Figure 1.1).

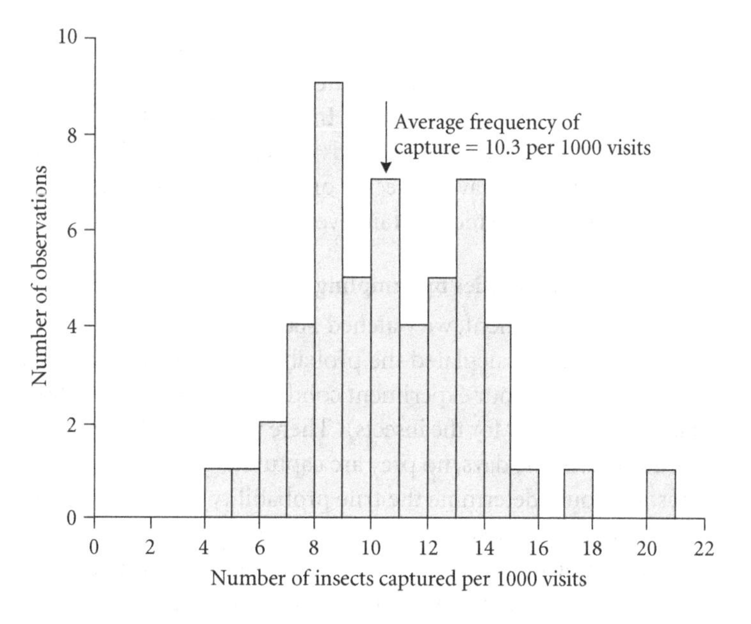

Figure 1.1 A histogram of capture frequencies. Each week for an entire year, an investigator might observe 1000 insect visits to a carnivorous pitcher plant and count how many times out of those visits the plant captures an insect (see Table 1.1). The collected data from all 52 observations (one per week) can be plotted in the form of a histogram. The number of captures observed ranges from as few as 4 one week (bar at left end) to as many as 20 another week (bar at right end); the average of all 52 observations is 10.3 captures per 1000 visits.

In Figure 1.1, we concisely summarize the results of one year's worth of samples[6] in a graph called a **histogram**. In this particular histogram, the numbers on the horizontal axis, or ***x*-axis**, indicate how many insects were captured while visiting plants. The numbers range from 4 to 20, because in the 52 samples, there were days when only 4 insects were captured, days when 20 insects were captured, and some days in between. The numbers on the vertical axis, or ***y*-axis** indicate the **frequency**: the number of trials that resulted in a particular outcome. For example, there was only one day in which there were 4 captures recorded, two days in which there were 6 captures recorded, and five days in which 9 captures were recorded. One way we can estimate the probability of capture is to take the average of all the samples. In other words, we would calculate the average number of captures in each of our 52 samples of 1000 visits. In this example, the average number of captures out of 1000 visits is 10.3. Thus, the probability of being captured is 10.3/1000 = 0.0103, or just over one in a hun-

[6] Although we didn't actually conduct this experiment, these sorts of data were collected by Newell and Nastase (1998), who videotaped insect visits to pitcher plants and recorded 27 captures out of 3308 visits, with 74 visits having an unknown fate. For the analyses in this chapter, we simulated data using a random-number generator in a computer spreadsheet. For this simulation, we set the "true" probability of capture at 0.01, the sample size at 1000 visits, and the number of samples at 52. We then used the resulting data to generate the histogram shown in Figure 1.1. If you conduct the same simulation experiment on your computer, you should get similar results, but there will be some variability depending on how your computer generates random data. We discuss simulation experiments in more detail in Chapters 3 and 5.

dred. We call this average the *expected value of the probability*, or **expectation**, and denote it as $E(P)$. Distributions like the one shown in Figure 1.1 are often used to describe the results of experiments in probability. We return to them in greater detail in Chapter 3.

Problems in the Definition of Probability

Most statistics textbooks very briefly define probability just as we have done: the (expected) frequency with which events occur. The standard textbook examples are that the toss of a fair coin has a 0.50 probability (equals a 50% chance) of landing heads,[7] or that each face of a six-sided fair die has a 1 in 6 chance of turning up. But let us consider these examples more carefully. Why do we accept these values as the correct probabilities?

We accept these answers because of our view of how coins are tossed and dice are rolled. For example, if we were to balance the coin on a tabletop with "heads" facing toward us, and then tip the coin gently forward, we could all agree that the probability of obtaining heads would no longer be 0.50. Instead, the probability of 0.50 only applies to a coin that is flipped vigorously into the air.[8] Even here, however, we should recognize that the 0.50 probability really represents an estimate based on a minimal amount of data.

For example, suppose we had a vast array of high-tech microsensors attached to the surface of the coin, the muscles of our hand, and the walls of the room we are in. These sensors detect and quantify the exact amount of torque in our toss, the temperature and air turbulence in the room, the micro-irregularities on the surface of the coin, and the microeddies of air turbulence that are generated as the coin flips. If all these data were instantaneously streamed into a fast

[7] In a curious twist of fate, one of the Euro coins may not be fair. The Euro coin, introduced in 2002 by 12 European states, has a map of Europe on the "tails" side, but each country has its own design on the "heads" side. Polish statisticians Tomasz Gliszcynski and Waclaw Zawadowski flipped (actually, spun) the Belgian Euro 250 times, and it came up heads 56% of the time (140 heads). They attributed this result to a heavier embossed image on the "heads" side, but Howard Grubb, a statistician at the University of Reading, points out that for 250 trials, 56% is "not significantly different" from 50%. Who's right? (as reported by *New Scientist:* www.newscientist.com/, January 2, 2002). We return to this example in Chapter 11.

[8] Gamblers and cardsharps try to thwart this model by using loaded coins or dice, or by carefully tossing coins and dice in a way that favors one of the faces but gives the appearance of randomness. Casinos work vigilantly to ensure that the random model is enforced; dice must be shaken vigorously, roulette wheels are spun very rapidly, and blackjack decks are frequently reshuffled. It isn't necessary for the casino to cheat. The payoff schedules are calculated to yield a handsome profit, and all that the casino must do is to ensure that customers have no way to influence or predict the outcome of each game.

computer, we could develop a very complex model that describes the trajectory of the coin. With such information, we could predict, much more accurately, which face of the coin would turn up. In fact, if we had an infinite amount of data available, perhaps there would be no uncertainty in the outcome of the coin toss.[9] The trials that we use to estimate the probability of an event may not be so similar after all. Ultimately, each trial represents a completely unique set of conditions; a particular chain of cause and effect that determines entirely whether the coin will turn up heads or tails. If we could perfectly duplicate that set of conditions; there would be no uncertainty in the outcome of the event, and no need to use probability or statistics at all![10]

Werner Heisenberg

Erwin Schrödinger

[9] This last statement is highly debatable. It assumes that we have measured all the right variables, and that the interactions between those variables are simple enough that we can map out all of the contingencies or express them in a mathematical relationship. Most scientists believe that complex models will be more accurate than simple ones, but that it may not be possible to eliminate all uncertainty (Lavine 2010). Note also that complex models don't help us if we cannot measure the variables they include; the technology needed to monitor our coin tosses isn't likely to exist anytime soon. Finally, there is the more subtle problem that the very act of measuring things in nature may alter the processes we are trying to study. For example, suppose we want to quantify the relative abundance of fish species near the deep ocean floor, where sunlight never penetrates. We can use underwater cameras to photograph fish and then count the number of fish in each photograph. But if the lights of the camera attract some fish species and repel others, what, exactly, have we measured? The Heisenberg Uncertainty Principle (named after the German physicist Werner Heisenberg, 1901–1976) in physics says that it is impossible to simultaneously measure the position and momentum of an electron: the more accurately you measure one of those quantities, the less accurately you can measure the other. If the very act of observing fundamentally changes processes in nature, then the chain of cause and effect is broken (or at least badly warped). This concept was elaborated by Erwin Schrödinger (1887–1961), who placed a (theoretical) cat and a vial of cyanide in a quantum box. In the box, the cat can be both alive and dead simultaneously, but once the box is opened and the cat observed, it is either alive or eternally dead.

[10] Many scientists would embrace this lack of uncertainty. Some of our molecular biologist colleagues would be quite happy in a world in which statistics are unnecessary. When there is uncertainty in the outcome of their experiments, they often assume that it is due to bad experimental technique, and that eliminating measurement error and contamination will lead to clean and repeatable data that are correct. Ecologists and field biologists usually are more sanguine about variation in their systems. This is not necessarily because ecological systems are inherently more noisy than molecular ones; rather, ecological data may be more variable than molecular data because they are often collected at different spatial and temporal scales.

Returning to the pitcher plants, it is obvious that our estimates of capture probabilities will depend very much on the details of the trials. The chances of capture will differ between large and small plants, between ant prey and fly prey, and between plants in sun and plants in shade. And, in each of these groups, we could make further subdivisions based on still finer details about the conditions of the visit. Our estimates of probabilities will depend on how broadly or how narrowly we limit the kinds of trials we will consider.

To summarize, when we say that events are random, stochastic, probabilistic, or due to chance, what we really mean is that their outcomes are determined in part by a complex set of processes that we are unable or unwilling to measure and will instead treat as random. The strength of other processes that we can measure, manipulate, and model represent deterministic or mechanistic forces. We can think of patterns in our data as being determined by mixtures of such deterministic and random forces. As observers, we impose the distinction between deterministic and random. It reflects our implicit conceptual model about the forces in operation, and the availability of data and measurements that are relevant to these forces.[11]

The Mathematics of Probability

In this section, we present a brief mathematical treatment of how to calculate probabilities. Although the material in this section may seem tangential to your quest (*Are the data "significant"?*), the correct interpretation of statistical results hinges on these operations.

Defining the Sample Space

As we illustrated in the example of prey capture by carnivorous plants, the probability P of being captured (the outcome) while visiting a plant (the event) is

---

[11] If we reject the notion of any randomness in nature, we quickly move into the philosophic realm of mysticism: "What, then, is this Law of Karma? *The* Law, without exception, which rules the whole universe, from the invisible, imponderable atom to the suns; ... and this law is that *every cause produces its effect,* without any possibility of delaying or annulling that effect, once the cause begins to operate. The law of causation is everywhere supreme. This is the Law of Karma; Karma is the inevitable link between cause and effect" (Arnould 1895).

Western science doesn't progress well by embracing this sort of all-encompassing complexity. It is true that many scientific explanations for natural phenomena are complicated. However, those explanations are reached by first eschewing all that complexity and posing a **null hypothesis.** A null hypothesis tries to account for patterns in the data in the simplest way possible, which often means initially attributing variation in the data to randomness (or measurement error). If that simple null hypothesis can be rejected, we can move on to entertain more complex hypotheses. See Chapter 4 for more details on null hypotheses and hypothesis testing, and Beltrami (1999) for a discussion of randomness.

defined simply as the number of captures divided by the number of visits (the number of trials) (Equation 1.1). Let's examine in detail each of these terms: outcome, event, trial, and probability.

First, we need to define our universe of possible events, or the sample space of interest. In our first example, an insect could successfully visit a plant and leave, or it could be captured. These two possible outcomes form the sample space (or set), which we will call *Visit*:

$$Visit = \{(Capture), (Escape)\}$$

We use curly braces {} to denote the set and parentheses () to denote the events of a set. The objects within the parentheses are the outcome(s) of the event. Because there are only two possible outcomes to a visit, if the probability of the outcome *Capture* is 1 in 1000, or 0.001, then the probability of *Escape* is 999 in 1000, or 0.999. This simple example can be generalized to the **First Axiom of Probability**:

Axiom 1: The sum of all the probabilities of outcomes within a single sample space = 1.0.

We can write this axiom as

$$\sum_{i=1}^{n} P(A_i) = 1.0$$

which we read as "the sum of the probabilities of all outcomes A_i equals 1.0." The "sideways W" (actually the capital Greek letter sigma) is a summation sign, the shorthand notation for adding up the elements that follow. The subscript i indicates the particular element and says that we are to sum up the elements from $i = 1$ to n, where n is the total number of elements in the set. In a properly defined sample space, we say that the outcomes are **mutually exclusive** (an individual is either captured or it escapes), and that the outcomes are **exhaustive** (*Capture* or *Escape* are the only possible outcomes of the event). If the events are not mutually exclusive and exhaustive, the probabilities of events in the sample space will not sum to 1.0.

In many cases, there will be more than just two possible outcomes of an event. For example, in a study of reproduction by the imaginary orange-spotted whirligig beetle, we found that each of these beasts always produces exactly 2 litters, with between 2 and 4 offspring per litter. The lifetime reproductive success of an orange-spotted whirligig can be described as an outcome (a,b), where a represents the number of offspring in the first litter and b the number of offspring in the second litter. The sample space *Fitness* consists of all the possible lifetime reproductive outcomes that an individual could achieve:

$$Fitness = \{(2,2), (2,3), (2,4), (3,2), (3,3), (3,4), (4,2), (4,3), (4,4)\}$$

Because each whirligig can give birth to only 2, 3, or 4 offspring in a litter, these 9 pairs of integers are the only possible outcomes. In the absence of any other information, we initially make the simplifying assumption that the probabilities of each of these different reproductive outcomes are equal. We use the definition of probability from Equation 1.1 (number of outcomes/number of trials) to determine this value. Because there are 9 possible outcomes in this set, $P(2,2) = P(2,3) = \ldots = P(4,4) = 1/9$. Notice also that these probabilities obey Axiom 1: the sum of the probabilities of all outcomes = 1/9 + 1/9 + 1/9 + 1/9 + 1/9 + 1/9 + 1/9 + 1/9 + 1/9 = 1.0.

Complex and Shared Events: Combining Simple Probabilities

Once probabilities of simple events are known or have been estimated, we can use them to measure probabilities of more complicated events. **Complex events** are composites of simple events in the sample space. **Shared events** are multiple simultaneous occurrences of simple events in the sample space. The probabilities of complex and shared events can be decomposed into the sums or products of probabilities of simple events. However, it can be difficult to decide when probabilities are to be added and when they are to be multiplied. The answer can be found by determining whether the new event can be achieved by one of several different pathways (a complex event), or whether it requires the simultaneous occurrence of two or more simple events (a shared event).

If the new event can occur via different pathways, it is a complex event and can be represented as an ***or*** **statement**: Event *A* *or* Event *B* *or* Event *C*. Complex events thus are said to represent the **union** of simple events. Probabilities of complex events are determined by summing the probabilities of simple events.

In contrast, if the new event requires the simultaneous occurrence of several simple events, it is a shared event and can be represented as an ***and*** **statement**: Event *A* *and* Event *B* *and* Event *C*. Shared events therefore are said to represent the **intersection** of simple events. Probabilities of shared events are determined by multiplying probabilities together.

COMPLEX EVENTS: SUMMING PROBABILITIES The whirligig example can be used to illustrate the calculation of probabilities of complex events. Suppose we wish to measure the lifetime reproductive output of a whirligig beetle. We would count the total number of offspring that a whirligig produces in its lifetime. This number is the sum of the offspring of the two litters, which results in an integer number between 4 and 8, inclusive. How would we determine the probability that an orange-spotted whirligig produces 6 offspring? First, note that the event *produces 6 offspring* can occur in three ways:

$$6\ offspring = \{(2,4), (3,3), (4,2)\}$$

and this complex event itself is a set.

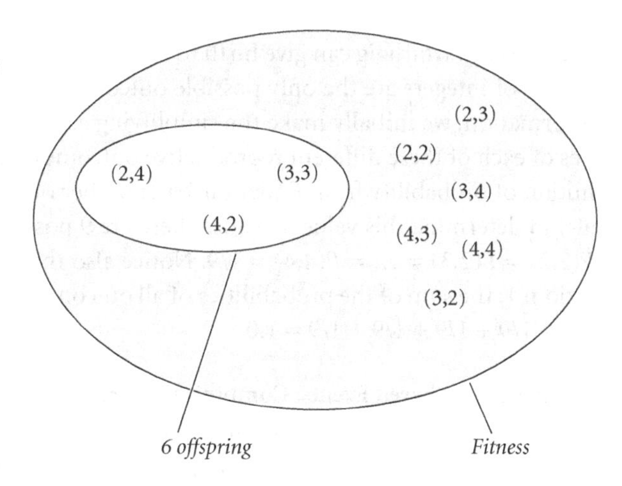

Figure 1.2 Venn diagram illustrating the concept of sets. Each pair of numbers represents the number of offspring produced in two consecutive litters by an imaginary species of whirligig beetle. We assume that this beetle produces exactly 2, 3, or 4 offspring each time it reproduces, so these are the only integers represented in the diagram. A set in a Venn diagram is a ring that encompasses certain elements. The set *Fitness* contains all of the possible reproductive outcomes for two consecutive litters. Within the set *Fitness* is a smaller set *6 offspring*, consisting of those litters that produce a total of exactly 6 offspring (i.e., each pair of integers adds up to 6). We say that *6 offspring* is a proper subset of *Fitness* because the elements of the former are completely contained within the latter.

We can illustrate these two sets with a **Venn diagram.**[12] Figure 1.2 illustrates Venn diagrams for our sets *Fitness* and *6 offspring*. Graphically, you can see that *6 offspring* is a **proper subset** of the larger set *Fitness* (that is, all of the elements of the former are elements of the latter). We indicate that one set is a subset of another set with the symbol ⊂, and in this case we would write

$$6\ offspring \subset Fitness$$

Three of the 9 possible outcomes of *Fitness* give rise to 6 offspring, and so we would estimate the probability of having 6 offspring to be 1/9 + 1/9 + 1/9 = 3/9 (recalling our assumption that each of the outcomes is equally likely). This result is generalized in the **Second Axiom of Probability:**

Axiom 2: The probability of a complex event equals the sum of the probabilities of the outcomes that make up that event.

John Venn

[12] John Venn (1834–1923) studied at Gonville and Caius College, Cambridge University, England, from which he graduated in 1857. He was ordained as a priest two years later and returned to Cambridge in 1862 as a lecturer in Moral Science. He also studied and taught logic and probability theory. He is best known for the diagrams that represent sets, their unions and intersections, and these diagrams now bear his name. Venn is also remembered for building a machine for bowling cricket balls.

You can think of a complex event as an *or* statement in a computer program: if the simple events are *A*, *B*, and C, the complex event is (*A* or *B* or *C*), because any one of these outcomes will represent the complex event. Thus

$$P(A \text{ or } B \text{ or } C) = P(A) + P(B) + P(C) \quad (1.2)$$

For another simple example, consider the probability of drawing a single card from a well-shuffled 52-card deck, and obtaining an ace. We know there are 4 aces in the deck, and the probability of drawing each particular ace is 1/52. Therefore, the probability of the complex event of drawing any 1 of the 4 aces is:

$$P(ace) = 1/52 + 1/52 + 1/52 + 1/52 = 4/52 = 1/13$$

SHARED EVENTS: MULTIPLYING PROBABILITIES In the whirligig example, we calculated the probability that a whirligig produces exactly 6 offspring. This complex event could occur by any 1 of 3 different litter pairs {(2,4), (3,3), (4,2)} and we determined the probability of this complex event by adding the simple probabilities.

Now we will turn to the calculation of the probability of a shared event—the simultaneous occurrence of two simple events. We assume that the number of offspring produced in the second litter is independent of the number produced in the first litter. **Independence** is a critical, simplifying assumption for many statistical analyses. When we say that two events are independent of one another, we mean that the outcome of one event is not affected by the outcome of the other. If two events are independent of one another, then the probability that both events occur (a shared event) equals the *product* of their individual probabilities:

$$P(A \cap B) = P(A) \times P(B) \text{ (if } A \text{ and } B \text{ are independent)} \quad (1.3)$$

The symbol $\cap$ indicates the **intersection** of two independent events—that is, both events occurring simultaneously. For example, in the litter size example, suppose that an individual can produce 2, 3, or 4 offspring in the first litter, and that the chances of each of these events are 1/3. If the same rules hold for the production of the second litter, then the probability of obtaining the pair of litters (2,4) equals 1/3 × 1/3 = 1/9. Notice that this is the same number we arrived at by treating each of the different litter pairs as independent, equiprobable events.

Probability Calculations: Milkweeds and Caterpillars

Here is a simple example that incorporates both complex and shared events. Imagine a set of serpentine rock outcroppings in which we can find populations of milkweed, and in which we also can find herbivorous caterpillars. Two kinds of milkweed populations are present: those that have evolved secondary chemicals that make them resistant (R) to the herbivore, and those that have not. Suppose you census a number of milkweed populations and determine that $P(R) = 0.20$. In other words, 20% of the milkweed populations are resistant to the herbivore. The remaining 80% of the populations represent the complement of this set. The complement includes all the other elements of the set, which we can write in short hand as *not R*. Thus

$$P(R) = 0.20 \qquad P(not\ R) = 1 - P(R) = 0.80$$

Similarly, suppose the probability that the caterpillar (C) occurs in a patch is 0.7:

$$P(C) = 0.7 \qquad P(not\ C) = 1 - P(C) = 0.30$$

Next we will specify the ecological rules that determine the interaction of milkweeds and caterpillars and then use probability theory to determine the chances of finding either caterpillars, milkweeds, or both in these patches. The rules are simple. First, all milkweeds and all caterpillars can disperse and reach all of the serpentine patches. Milkweed populations can always persist when the caterpillar is absent, but when the caterpillar is present, only resistant milkweed populations can persist. As before, we assume that milkweeds and caterpillars initially colonize patches independently of one another.[13]

Let's first consider the different combinations of resistant and non-resistant populations occurring with and without herbivores. These are two simultaneous events, so we will multiply probabilities to generate the 4 possible shared events (Table 1.2). There are a few important things to notice in Table 1.2. First, the sum of the resulting probabilities of the shared events (0.24 + 0.56 + 0.06 + 0.14) = 1.0, and these 4 shared events together form a proper set. Second, we can add some of these probabilities together to define new complex events and also recover some of the underlying simple probabilities. For example, what is the probability that milkweed populations will be resistant? This can be estimated as the probability of finding resistant populations with caterpillars

[13] Lots of interesting biology occurs when the assumption of independence is violated. For example, many species of adult butterflies and moths are quite selective and seek out patches with appropriate host plants for laying their eggs. Consequently, the occurrence of caterpillars may not be independent of the host plant. In another example, the presence of herbivores increases the selective pressure for the evolution of host resistance. Moreover, many plant species have so-called facultative chemical defenses that are switched on only when herbivores show up. Consequently, the occurrence of resistant populations may not be independent of the presence of herbivores. Later in this chapter, we will learn some methods for incorporating non-independent probabilities into our calculations of complex events.

TABLE 1.2 **Probability calculations for shared events**

Shared event	Probability calculation	Outcome: Milkweed present?	Outcome: Caterpillar present?
Susceptible population and no caterpillar	$[1 - P(R)] \times [1 - P(C)] =$ $(1.0 - 0.2) \times (1.0 - 0.7) = 0.24$	Yes	No
Susceptible population and caterpillar	$[1 - P(R)] \times [P(C)] =$ $(1.0 - 0.2) \times (0.7) = 0.56$	No	Yes
Resistant population and no caterpillar	$[P(R)] \times [1 - P(C)] =$ $(0.2) \times (1.0 - 0.7) = 0.06$	Yes	No
Resistant population and caterpillar	$[P(R)] \times [P(C)] =$ $(0.2) \times (0.7) = 0.14$	Yes	Yes

The joint occurrence of independent events is a shared event, and its probability can be calculated as the product of the probabilities of the individual events. In this hypothetical example, milkweed populations that are resistant to herbivorous caterpillars occur with probability $P(R)$, and milkweed populations that are susceptible to caterpillars occur with probability $1 - P(R)$. The probability that a milkweed patch is colonized by caterpillars is $P(C)$ and the probability that a milkweed patch is not colonized by caterpillars is $1 - P(C)$. The simple events are the occurrence of caterpillars (C) and of resistant milkweed populations (R). The first column lists the four complex events, defined by resistant or susceptible milkweed populations and the presence or absence of caterpillars. The second column illustrates the probability calculation of the complex event. The third and fourth columns indicate the ecological outcome. Notice that the milkweed population goes extinct if it is susceptible and is colonized by caterpillars. The probability of this event occurring is 0.56, so we expect this outcome 56% of the time. The most unlikely outcome is a resistant milkweed population that does not contain caterpillars ($P = 0.06$). Notice also that the four shared events form a proper set, so their probabilities sum to 1.0 ($0.24 + 0.56 + 0.06 + 0.14 = 1.0$).

($P = 0.14$) plus the probability of finding resistant populations without caterpillars ($P = 0.06$). The sum (0.20) indeed matches the original probability of resistance [$P(R) = 0.20$]. The independence of the two events ensures that we can recover the original values in this way.

However, we have also learned something new from this exercise. The milkweeds will disappear if a susceptible population encounters the caterpillar, and this should happen with probability 0.56. The complement of this event, $(1 - 0.56) = 0.44$, is the probability that a site will contain a milkweed population. Equivalently, the probability of milkweed persistence can be calculated as $P(\text{milkweed present}) = 0.24 + 0.06 + 0.14 = 0.44$, adding together the probabilities for the different combinations that result in milkweeds. Thus, although the probability of resistance is only 0.20, we expect to find milkweed populations occurring in 44% of the sampled patches because not all susceptible populations are hit by caterpillars. Again, we emphasize that these calculations are correct only if the initial colonization events are independent of one another.

Complex and Shared Events: Rules for Combining Sets

Many events are not independent of one another, however, and we need methods to take account of that non-independence. Returning to our whirligig example, what if the number of offspring produced in the second litter was somehow related to the number produced in the first litter? This might happen because organisms have a limited amount of energy available for producing offspring, so that energy invested in the first litter of offspring is not available for investment in the second litter. Would that change our estimate of the probability of producing 6 offspring? Before we can answer this question, we need a few more tools in our probabilistic toolkit. These tools tell us how to combine events or sets, and allow us to calculate the probabilities of combinations of events.

Suppose in our sample space there are two identifiable events, each of which consists of a group of outcomes. For example, in the sample space *Fitness*, we could describe one event as a whirligig that produces exactly 2 offspring in its first litter. We will call this event *First litter 2*, and abbreviate it as *F*. The second event is a whirligig that produces exactly 4 offspring in its second litter. We will call this event *Second litter 4* and abbreviate it as *S*:

$$Fitness = \{(2,2), (2,3), (2,4), (3,2), (3,3), (3,4), (4,2), (4,3), (4,4)\}$$
$$F = \{(2,2), (2,3), (2,4)\}$$
$$S = \{(2,4), (3,4), (4,4)\}$$

We can construct two new sets from *F* and *S*. The first is the new set of outcomes that equals all the outcomes that are in either *F* or *S* alone. We indicate this new set using the notation $F \cup S$, and we call this new set the **union** of these two sets:

$$F \cup S = \{(2,2), (2,3), (2,4), (3,4), (4,4)\}$$

Note that the outcome (2,4) occurs in both *F* and *S*, but it is counted only once in $F \cup S$. Also notice that the union of *F* and *S* is a set that contains more elements than are contained in either *F* or *S*, because these sets are "added" together to create the union.

The second new set equals the outcomes that are in both *F* and *S*. We indicate this set with the notation $F \cap S$, and call this new set the **intersection** of the two sets:

$$F \cap S = \{(2,4)\}$$

Notice that the intersection of *F* and *S* is a set that contains fewer elements than are contained in either *F* or *S* alone, because now we are considering only the elements that are common to both sets. The Venn diagram in Figure 1.3 illustrates these operations of union and intersection.

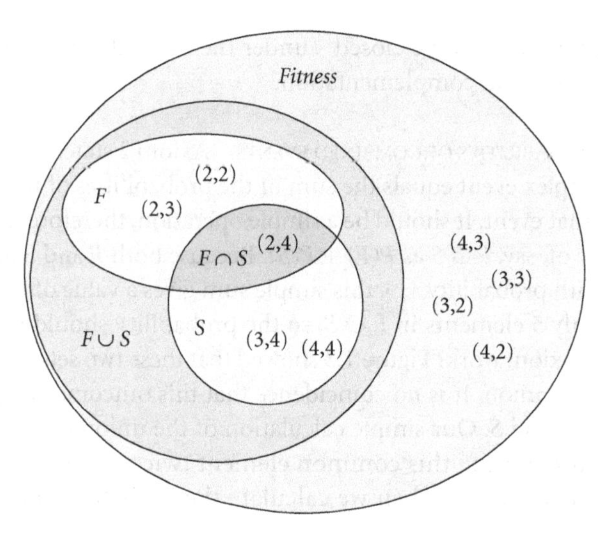

Figure 1.3 Venn diagram illustrating unions and intersections of sets. Each ring represents a different set of the numbers of offspring produced in a pair of litters by an imaginary whirligig beetle species, as in Figure 1.2. The largest ring is the set *Fitness*, which encompasses all of the possible reproductive outputs of the beetle. The small ring F is the set of all pairs of litters in which there are exactly 2 offspring in the first litter. The small ring S is the set of all pairs of litters in which there are exactly 4 offspring in the second litter. The area in which the rings overlap represents the intersection of F and S ($F \cap S$) and contains only those elements common to both. The ring that encompasses F and S represents the union of F and S ($F \cup S$) and contains all elements found in either set. Notice that the union of F and S does not double count their common element, (2,4). In other words, the union of two sets is the sum of the elements in both sets, minus their common elements. Thus, $F \cup S = (F + S) - (F \cap S)$.

We can construct a third useful set by considering the set F^c, called the complement of F, which is the set of objects in the remaining sample space (in this example, *Fitness*) that are not in the set F:

$$F^c = \{(3,2), (3,3), (3,4), (4,2), (4,3), (4,4)\}$$

From Axioms 1 and 2, you should see that

$$P(F) + P(F^c) = 1.0$$

In other words, because F and F^c collectively include all possible outcomes (by definition, they comprise the entire set), the sum of the probabilities associated with them must add to 1.0.

Finally, we need to introduce the **empty set**. The empty set contains no elements and is written as $\{\varnothing\}$. Why is this set important? Consider the set consisting of the intersection of F and F^c. Because they have no elements in common, if we did not have an empty set, then $F \cap F^c$ would be undefined. Having an empty set allows sets to be closed[14] under the three allowable operations: union, intersection, and complementation.

CALCULATING PROBABILITIES OF COMBINED EVENTS Axiom 2 stated that the probability of a complex event equals the sum of the probabilities of the outcomes that make up that event. It should be a simple operation, therefore, to calculate the probability of, say, $F \cup S$ as $P(F) + P(S)$. Because both F and S have 3 outcomes, each with probability 1/9, this simple sum gives a value of 6/9. However, there are only 5 elements in $F \cup S$, so the probability should be only 5/9. Why didn't the axiom work? Figure 1.3 showed that these two sets have the outcome (2,4) in common. It is no coincidence that this outcome is equal to the intersection of F and S. Our simple calculation of the union of these two sets resulted in our counting this common element twice. In general, we need to avoid double counting when we calculate the probability of a union.[15]

---

[14] **Closure** is a mathematical property, not a psychological state. A collection of objects *G* (technically, a **group**) is closed under an operation $\oplus$ (such as union, intersection, or complementation) if, for all elements *A* and *B* of *G*, $A \oplus B$ is also an element of *G*.

[15] Determining whether events have been double-counted or not can be tricky. For example, in basic population genetics, the familiar Hardy-Weinberg equation gives the frequencies of different genotypes present in a randomly mating population. For a single-gene, two-allele system, the Hardy-Weinberg equation predicts the frequency of each genotype (*RR*, *Rr*, and *rr*) in a population in which *p* is the initial allele frequency of *R* and *q* is the initial allele frequency of *r*. Because the genotypes *R* and *r* form a closed set, $p + q = 1.0$. Each parent contributes a single allele in its gametes, so the formation of offspring represents a shared event, with two gametes combining at fertilization and contributing their alleles to determine the genotype of the offspring. The Hardy-Weinberg equation predicts the probability (or frequency) of the different genotype combinations in the offspring of a randomly mating population as

$$P(RR) = p^2$$
$$P(Rr) = 2pq$$
$$P(rr) = q^2$$

The production of an *RR* genotype requires that both gametes contain the *R* allele. Because the frequency of the *R* allele in the population is *p*, the probability of this complex event $= p \times p = p^2$. But why does the heterozygote genotype (*Rr*) occur with frequency 2*pq* and not just *pq*? Isn't this a case of double counting? The answer is no, because there are two ways to create a heterozygous individual: we can combine a male *R* gamete with a female *r* gamete, *or* we can combine a male *r* gamete with a female *R* gamete. These are two distinct events, even though the resulting zygote has an identical genotype (*Rr*) in both cases. Therefore, the probability of heterozygote formation is a complex event, whose elements are two shared events:

$$P(Rr) = P(\text{male } R \text{ and female } r) \textit{ or } P(\text{male } r \text{ and female } R)$$
$$P(Rr) = pq + qp = 2pq$$

We therefore calculate the probability of a union of any two sets or events A and B as

$$P(A \cup B) = P(A) + P(B) - P(A \cap B) \tag{1.4}$$

The probability of an intersection is the probability of both events happening. In the example shown in Figure 1.3, $F \cap S = \{(2,4)\}$, which has probability 1/9, the product of the probabilities of F and S ($(1/3) \times (1/3) = 1/9$). So, subtracting 1/9 from 6/9 gives us the desired probability of 5/9 for $P(F \cup S)$.

If there is no overlap between A and B, then they are **mutually exclusive events**, meaning they have no outcomes in common. The intersection of two mutually exclusive events ($A \cap B$) is therefore the empty set $\{\varnothing\}$. Because the empty set has no outcomes, it has probability = 0, and we can simply add the probabilities of two mutually exclusive events to obtain the probability of their union:

$$P(A \cup B) = P(A) + P(B) \quad (\text{if } A \cap B = \{\varnothing\}) \tag{1.5}$$

We are now ready to return to our question of estimating the probability that an orange-spotted whirligig producing 6 offspring, if the number of offspring produced in the second litter depends on the number of offspring produced in the first litter. Recall our complex event *6 offspring*, which consisted of the three outcomes (2,4), (3,3), and (4,2); its probability is $P(6\ offspring) = 3/9$ (or 1/3). We also found another set F, the set for which the number of offspring in the first litter was equal to 2, and we found that $P(F) = 3/9$ (or 1/3). If you observed that the first litter was 2 offspring, what is the probability that the whirligig will produce 4 offspring the next time (for a total of 6)? Intuitively, it seems that this probability should be 1/3 as well, because there are 3 outcomes in F, and only one of them (2,4) results in a total of 6 offspring. This is the correct answer. But why is this probability not equal to 1/9, which is the probability of getting (2,4) in *Fitness*? The short answer is that the additional information of knowing the size of the first litter influenced the probability of the total reproductive output. Conditional probabilities resolve this puzzle.

Conditional Probabilities

If we are calculating the probability of a complex event, and we have information about an outcome in that event, we should modify our estimates of the probabilities of other outcomes accordingly. We refer to these updated estimates as **conditional probabilities**. We write this as

$$P(A \mid B)$$

or the probability of event or outcome A *given* event or outcome B. The vertical slash ($|$) indicates that the probability of A is calculated assuming that the event or outcome B has already occurred. This conditional probability is defined as

$$P(A \mid B) = \frac{P(A \cap B)}{P(B)} \quad (1.6)$$

This definition should make intuitive sense. If outcome B has already occurred, then any outcomes in the original sample space not in B (i.e., B^c) cannot occur, so we restrict the outcomes of A that we could observe to those that also occur in B. So, $P(A \mid B)$ should somehow be related to $P(A \cap B)$, and we have defined it to be proportional to that intersection. The denominator $P(B)$ is the restricted sample space of events for which B has already occurred. In our whirligig example, $P(F \cap S) = 1/9$, and $P(F) = 1/3$. Dividing the former by the latter gives the value for $P(S \mid F) = (1/9)/(1/3) = 1/3$, as suggested by our intuition.

Rearranging the formula in Equation 1.6 for calculating conditional probabilities gives us a general formula for calculating the probability of an intersection:

$$P(A \cap B) = P(A \mid B) \times P(B) = P(B \mid A) \times P(A) \quad (1.7)$$

You should recall that earlier in the chapter we defined the probability of the intersection of two *independent* events to be equal to $P(A) \times P(B)$. This is a special case of the formula for calculating the probability of intersection using the conditional probability $P(A \mid B)$. Simply note that if two events A and B are independent, then $P(A \mid B) = P(A)$, so that $P(A \mid B) \times P(B) = P(A) \times P(B)$. By the same reasoning, $P(A \cap B) = P(B \mid A) \times P(A)$.

Bayes' Theorem

Until now, we have discussed probability using what is known as the **frequentist paradigm**, in which probabilities are estimated as the relative frequencies of outcomes based on an infinitely large set of trials. Each time scientists want to estimate the probability of a phenomenon, they start by assuming no prior knowledge of the probability of an event, and re-estimate the probability based on a large number of trials. In contrast, the **Bayesian paradigm**, based on a formula for conditional probability developed by Thomas Bayes,[16] builds on the

Thomas Bayes

[16] Thomas Bayes (1702–1761) was a Nonconformist minister in the Presbyterian Chapel in Tunbridge Wells, south of London, England. He is best known for his "Essay towards solving a problem in the doctrine of chances," which was published two years after his death in *Philosophical Transactions* 53: 370–418 (1763). Bayes was elected a Fellow of the Royal Society of London in 1742. Although he was cited for his contributions to mathematics, he never published a mathematical paper in his lifetime.

idea that investigators may already have a belief of the probability of an event, before the trials are conducted. These **prior probabilities** may be based on previous experience (which itself could be a frequentist estimate from earlier studies), intuition, or model predictions. These prior probabilities are then modified by the data from the current trials to yield **posterior probabilities** (discussed in more detail in Chapter 5). However, even in the Bayesian paradigm, quantitative estimates of prior probabilities will ultimately have to come from trials and experiments.

Bayes' formula for calculating conditional probabilities, now known as **Bayes' Theorem**, is

$$P(A \mid B) = \frac{P(B \mid A)P(A)}{P(B \mid A)P(A) + P(B \mid A^c)P(A^c)} \quad \textbf{(1.8)}$$

This formula is obtained by simple substitution. From Equation 1.6, we can write the conditional probability $P(A \mid B)$ on the right side of the equation as

$$P(A \mid B) = \frac{P(A \cap B)}{P(B)}$$

From Equation 1.7, the numerator of this second equation can be rewritten as $P(B \mid A) \times P(A)$, yielding the numerator of Bayes' Theorem. By Axiom 1, the denominator, $P(B)$, can be rewritten as $P(B \cap A) + P(B \cap A^c)$, as these two terms sum to $P(B)$. Again using our formula for the probability of an intersection (Equation 1.7), this sum can be rewritten as $P(B \mid A) \times P(A) + P(B \mid A^c) \times P(A^c)$, which is the denominator of Bayes' Theorem.

Although Bayes' Theorem is simply an expansion of the definition of conditional probability, it contains a very powerful idea. That idea is that the probability of an event or outcome A conditional on another event B can be determined if you know the probability of the event B conditional on the event A and you know the complement of A, A^c (which is why Bayes' Theorem is often called a theorem of inverse probability). We will return to a detailed exploration of Bayes' Theorem and its use in statistical inference in Chapter 5.

For now, we conclude by highlighting the very important distinction between $P(A \mid B)$ and $P(B \mid A)$. Although these two conditional probabilities look similar, they are measuring completely different things. As an example, let's return to the susceptible and resistant milkweeds and the caterpillars.

First consider the conditional probability

$$P(C \mid R)$$

This expression indicates the probability that caterpillars (C) are found *given* a resistant population of milkweeds (R). To estimate $P(C \mid R)$, we would need to examine a random sample of resistant milkweed populations and to determine the frequency with which these populations were hosting caterpillars. Now consider this conditional probability:

$$P(R \mid C)$$

In contrast to the previous expression, $P(R \mid C)$ is the probability that a milkweed population is resistant (R), given that it is being eaten by caterpillars (C). To estimate $P(R \mid C)$, we would need to examine a random sample of caterpillars and determine the frequency with which they were actually feeding on resistant milkweed populations.

These are two quite different quantities and their probabilities are determined by different factors. The first conditional probability $P(C \mid R)$ (the probability of the occurrence of caterpillars given resistant milkweeds) will depend, among other things, on the extent to which resistance to herbivory is directly or indirectly responsible for the occurrence of caterpillars. In contrast, the second conditional probability $P(R \mid C)$ (the probability of resistant milkweed populations given the presence of caterpillars) will depend, in part, on the incidence of caterpillars on resistant plants versus other situations that could also lead to the presence of caterpillars (e.g., caterpillars feeding on other plant species or caterpillars feeding on susceptible plants that were not yet dead).

Finally, note that the conditional probabilities are also distinct from the simple probabilities: $P(C)$ is just the probability that a randomly chosen individual is a caterpillar, and $P(R)$ is the probability that a randomly chosen milkweed population is resistant to caterpillars. In Chapter 5, we will see how to use Bayes' Theorem to calculate conditional probabilities when we cannot directly measure $P(A \mid B)$.

Summary

The probability of an outcome is simply the number of times that the outcome occurs divided by the total number of trials. If simple probabilities are known or estimated, probabilities can be determined for complex events (Event A *or* Event B) through summation, and probabilities can be determined for shared events (Event A *and* Event B) through multiplication. The definition of probability, together with axioms for the additivity of probabilities and three operations on sets (union, intersection, complementation), form the fundamentals of the **probability calculus**.

CHAPTER 2

Random Variables and Probability Distributions

In Chapter 1, we explored the notion of probability and the idea that the outcome of a single trial is an uncertain event. However, when we accumulate data on many trials, we may begin to see regular patterns in the frequency distribution of events (e.g., Figure 1.1). In this chapter, we will explore some useful mathematical functions that can generate these frequency distributions.

Certain probability distributions are assumed by many of our common statistical tests. For example, parametric analysis of variance (ANOVA; see Chapter 10) assumes that random samples of measured values fit a normal, or bell-shaped, distribution, and that the variance of that distribution is similar among different groups. If the data meet those assumptions, ANOVA can be used to test for differences among group means. Probability distributions also can be used to build models and make predictions. Finally, probability distributions can be fit to real datasets without specifying a particular mechanistic model. We use probablility distributions because they work—they fit lots of data in the real world.

Our first foray into probability theory involved estimating the probabilities of individual outcomes, such as prey capture by pitcher plants or reproduction of whirligig beetles. We found numerical values for each of these probabilities by using simple rules, or **functions**. More formally, the mathematical rule (or function) that assigns a given numerical value to each possible outcome of an experiment in the sample space of interest is called a **random variable**. We use the term "random variable" in this mathematical sense, not in the colloquial sense of a random event.

Random variables come in two forms: **discrete random variables** and **continuous random variables**. Discrete random variables are those that take on finite or countable values (such as integers; see Footnote 4 in Chapter 1). Common examples include presence or absence of a given species (which takes the value of 1 or 0), or number of offspring, leaves, or legs (integer values). Continuous random variables, on the other hand, are those that can take on any value within a smooth interval. Examples include the biomass of a starling, the leaf area consumed by an herbivore, or the dissolved oxygen content of a water sample. In one sense, all values of a continuous variable that we measure in the real world are discrete: our instrumentation only allows us to measure things with a finite level of precision.[1] But there is no theoretical limit to the precision we could obtain in the measurement of a continuous variable. In Chapter 7, we will return to the distinction between discrete and continuous variables as a key consideration in the design of field experiments and sampling designs.

Discrete Random Variables

Bernoulli Random Variables

The simplest experiment has only two outcomes, such as organisms being present or absent, coins landing heads or tails, or whirligigs reproducing or not. The random variable describing the outcome of such an experiment is a **Bernoulli random variable**, and an experiment of independent trials in which there are only two possible outcomes for each trial is a **Bernoulli trial**.[2] We use the notation

[1] It is important to understand the distinction between **precision** and **accuracy** in measurements. Accuracy refers to how close the measurement is to the true value. Accurate measurements are **unbiased**, meaning they are neither consistently above nor below the true value. Precision refers to the agreement among a series of measurements and the degree to which these measurements can be discriminated. For example, a measurement can be precise to 3 decimal places, meaning we can discriminate among different measurements out to 3 decimal places. Accuracy is more important than precision. It would be much better to use an accurate balance that was precise to only 1 decimal place than an inaccurate balance that was precise to 5 decimal places. The extra decimal places don't bring you any closer to the true value if your instrument is flawed or biased.

Jacob Bernoulli

[2] In our continuing series of profiles of dead white males in wigs, Jacob (aka Jaques or James) Bernoulli (1654–1705) was one of a family of famous physicists and mathematicians. He and his brother Johann are considered to be second only to Newton in their development of calculus, but they argued constantly about the relative merits of each other's work. Jacob's major mathematical work, *Ars conjectandi* (published in 1713, 8 years after his death), was the first major text on probability. It included the first exposition of many of the topics discussed in Chapters 2 and 3, such as general theories of permutations and combinations, the first proofs of the binomial theorem, and the Law of Large Numbers. Jacob Bernoulli also made substantial contributions to astronomy and mechanics (physics).

$$X \sim \text{Bernoulli}(p) \tag{2.1}$$

to indicate that the random variable X is a Bernoulli random variable. The symbol ~ is read "is distributed as." X takes on the values of the number of "successes" in the trial (e.g., present, captured, reproducing), and p is the probability of a successful outcome. The most common example of a Bernoulli trial is a flip of a single fair coin (perhaps not a Belgian Euro; see Footnote 7 in Chapter 1, and Chapter 11), where the probability of heads = the probability of tails = 0.5. However, even a variable with a large number of outcomes can be redefined as a Bernoulli trial if we can collapse the range of responses to two outcomes. For example, a set of 10 reproductive events of whirligig beetles can be analyzed as a single Bernoulli trial, where success is defined as *having exactly 6 offspring* and failure is defined as *having more than or fewer than 6 offspring.*

An Example of a Bernoulli Trial

A one-time census of all the towns in Massachusetts for the rare plant species *Rhexia mariana* (meadow beauty) provides an example of a Bernoulli trial. The occurrence of *Rhexia* is a Bernoulli random variable X, which takes on two outcomes: $X = 1$ (*Rhexia* present) or $X = 0$ (*Rhexia* absent). There are 349 towns in Massachusetts, so our single Bernoulli trial is the one search of all of these towns for occurrences of *Rhexia*. Because *Rhexia* is a rare plant, let us imagine that the probability that *Rhexia* is present (i.e., $X = 1$) is low: $P(X = 1) = p = 0.02$. Thus, if we census a single town, there is only a 2% chance ($p = 0.02$) of finding *Rhexia*. But what is the expected probability that *Rhexia* occurs in any 10 towns, and not in any of the remaining 339 towns?

Because the probability of *Rhexia* being present in any given town $p = 0.02$, we know from the First Axiom of Probability (see Chapter 1) that the probability of *Rhexia* being absent from any single town = $(1 - p) = 0.98$. By definition, each event (occurrence in a single town) in our Bernoulli trial must be independent. For this example, we assume the presence or absence of *Rhexia* in a given town to be independent of its presence or absence in any other town. Because the probability that *Rhexia* occurs in one town is 0.02, and the probability that it occurs in another town is also 0.02, the probability that it occurs in both of these specific towns is $0.02 \times 0.02 = 0.0004$. By extension, the probability that *Rhexia* occurs in 10 specified towns is $0.02^{10} = 1.024 \times 10^{-17}$ (or 0.00000000000000001024, which is a very small number).

However, this calculation does not give us the exact answer we are looking for. More precisely, we want the probability that *Rhexia* occurs in exactly 10 towns *and* does *not* occur in the remaining 339 towns. Imagine that the first town surveyed has no *Rhexia*, which should occur with probability $(1 - p) = 0.98$. The second town has *Rhexia*, which should occur with probability $p = 0.02$. Because the presence or absence of *Rhexia* in each town is independent, the probability of getting one town with *Rhexia* and one without by picking any two towns at random should be the product of p and $(1 - p) = 0.02 \times 0.98 = 0.0196$ (see Chapter 1). By extension, then, the probability of *Rhexia* occurring in a given set of 10 towns in Massachusetts should be the probability of occurrence in 10 towns (0.02^{10}) multiplied by the probability that it does *not* occur in the remaining 339 towns (0.98^{339}) $= 1.11 \times 10^{-20}$. Once again this is an astronomically small number.

But we are not finished yet! Our calculation so far gives us the probability that *Rhexia* occurs in exactly 10 particular towns (and in no others). But we are actually interested in the probability that *Rhexia* occurs in *any* 10 towns in Massachusetts, not just a specific list of 10 particular towns. From the Second Axiom of Probability (see Chapter 1), we learned that probabilities of complex events that can occur by different pathways can be calculated by adding the probabilities for each of the pathways.

So, how many different sets of 10 town each are possible from a list of 349? Lots! In fact, 6.5×10^{18} different **combinations** of 10 are possible in a set of 349 (we explain this calculation in the next section). Therefore, the probability that *Rhexia* occurs in *any* 10 towns equals the probability that it occurs in one set of 10 towns (0.02^{10}) times the probability that it does not occur in the remaining set of 339 towns (0.98^{339}) times the number of unique combinations of 10 towns that can be produced from a list of 349 (6.5×10^{18}). This final product = 0.07; there is a 7% chance of finding exactly 10 towns with *Rhexia*. This is also a small number, but not nearly as small as the first two probability values we calculated.

Many Bernoulli Trials = A Binomial Random Variable

Because a central feature of experimental science is replication, we would rarely conduct a single Bernoulli trial. Rather, we would conduct replicate, independent Bernoulli trials in a single experiment. We define a **binomial random variable** X to be the number of successful results in n independent Bernoulli trials. Our shorthand notation for a binomial random variable is

$$X \sim \text{Bin}(n,p) \tag{2.2}$$

to indicate the probability of obtaining X successful outcomes in n independent Bernoulli trials, where the probability of a successful outcome of any given event is p. Note that if $n = 1$, then the binomial random variable X is equivalent to a Bernoulli random variable. Binomial random variables are one of the most common types of random variables encountered in ecological and environmental studies. The probability of obtaining X successes for a binomial random variable is

$$P(X) = \frac{n!}{X!(n-X)!} p^X (1-p)^{n-X} \tag{2.3}$$

where n is the number of trials, X is the number of successful outcomes ($X \leq n$), and $n!$ means n **factorial**,[3] which is calculated as

$$n \times (n-1) \times (n-2) \times \ldots \times (3) \times (2) \times (1)$$

Equation 2.3 has three components, and two of them should look familiar from our analysis of the *Rhexia* problem. The component p^X is the probability of obtaining X independent successes, each with probability p. The component $(1 - p)^{(n-X)}$ is the probability of obtaining $(n - X)$ failures, each with probability $(1 - p)$. Note that the sum of the successes (X) and the failures $(n - X)$ is just n, the total number of Bernoulli trials. As we saw in the *Rhexia* example, the probability of obtaining X successes with probability p and $(n - X)$ failures with probability $(1 - p)$ is the product of these two independent events $p^X(1 - p)^{(n-X)}$.

But then why do we need the term

$$\frac{n!}{X!(n-X)!}$$

and where does it come from? The equivalent notation for this term is

$$\binom{n}{X}$$

(read as "n choose X"), which is known as the **binomial coefficient**. The binomial coefficient is needed because there is more than one way to obtain most combinations of successes and failures (the combinations of 10 towns we described in the *Rhexia* example above). For example, the outcome *one success* in a set of two Bernoulli trials can actually occur in two ways: (1,0) or (0,1). So the probability of getting one success in a set of two Bernoulli trials equals the probability of an outcome of one success $[= p(1 - p)]$ times the number of possible outcomes of one success (= 2). We could write out all of the different outcomes with X successes and count them, but as n gets large, so does X; there are 2^n possible outcomes for n trials.

It is more straightforward to compute directly the number of outcomes of X (successes), and this is what the binomial coefficient does. Returning to the *Rhexia* example, we have 10 occurrences of this plant ($X = 10$) and 349 towns ($n = 349$), but we don't know which particular towns they are in. Beginning our

[3] The factorial operation can be applied only to non-negative integers. By definition, $0! = 1$.

search, there initially are 349 towns in which *Rhexia* could occur. Once *Rhexia* is found in a given town, there are only 348 other towns in which it could occur. So there are 349 combinations of towns in which you could have only one occurrence, and 348 combinations in which you could have the second occurrence. By extension, for X occurrences, the total number of ways that 10 *Rhexia* occurrences could be distributed among the 349 towns would be

$$349 \times 348 \times 347 \times 346 \times 345 \times 344 \times 343 \times 342 \times 341 \times 340 = 2.35 \times 10^{25}$$

In general, the total number of ways to obtain X successes in n trials is

$$n \times (n-1) \times (n-2) \times \ldots \times (n-X+1)$$

which looks a lot like the formula for $n!$. The terms in the above equation that are missing from $n!$ are all the remaining ones below $(n - X + 1)$, or

$$(n-X) \times (n-X-1) \times (n-X-2) \times \ldots \times 1$$

which simply equals $(n - X)!$. Thus, if we divide $n!$ by $(n - X)!$, we're left with the total number of ways of obtaining X successes in n trials. But this is not quite the same as the binomial coefficient described above, which further divides our result by $X!$:

$$\frac{n!}{(n-X)!X!}$$

The reason for the further division is that we don't want to double-count identical patterns of successes that simply occurred in a different order. Returning to *Rhexia*, if we found populations first in the town of Barnstable and second in Chatham, we would not want to count that as a different outcome from finding it first in Chatham and second in Barnstable. By the same reasoning as we used above, there are exactly $X!$ ways of reordering a given outcome. Thus, we want to "discount" (divide by) our result by $X!$. The utility of such different **permutations** will become more apparent in Chapter 5, when we discuss Monte Carlo methods for hypothesis testing.[4]

[4] One more example to persuade yourself that the binomial coefficient works. Consider the following set of five marine fishes:

{(wrasse), (blenny), (goby), (eel), (damselfish)}

How many unique pairs of fishes can be formed? If we list all of them, we find 10 pairs:

(wrasse), (blenny)
(wrasse), (goby)
(wrasse), (eel)
(wrasse), (damselfish)
(blenny), (goby)

The Binomial Distribution

Now that we have a simple function for calculating the probability of a binomial random variable, what do we do with it? In Chapter 1, we introduced the histogram, a type of graph used to summarize concisely the number of trials that resulted in a particular outcome. Similarly, we can plot a histogram of the number of binomial random variables from a series of Bernoulli trials that resulted in each possible outcome. Such a histogram is called a **binomial distribution**, and it is generated from a binomial random variable (Equation 2.3).

For example, consider the Bernoulli random variable for which the probability of success = the probability of failure = 0.5 (such as flipping fair coins). Our experiment will consist of flipping a fair coin 25 times ($n = 25$), and our possible outcomes are given by the sample space *number of heads* = {0, 1, . . ., 24, 25}. Each outcome X_i is a binomial random variable, and the probability of each outcome is given by the binomial formula, which we will refer to as a **probability distribution function**, because it is the function (or rule) that provides the numerical value (the probability) of each outcome in the sample space.

Using our formula for a binomial random variable, we can tabulate the probabilities of every possible outcome from hunting for *Rhexia* 25 times (Table 2.1). We can draw this table as a histogram (Figure 2.1), and the histogram can be interpreted in two different ways. First, the values on the y-axis can be read as the probability of obtaining a given random variable X (e.g., the number of occurrences of *Rhexia*) in 25 trials where the probability of an occurrence is 0.5. In this interpretation, we refer to Figure 2.1 as a **probability distribution**. Accordingly, if we add up the values in Table 2.1, they sum to exactly 1.0 (with rounding error), because they define the entire sample space (from the First Axiom of Probability). Second, we can interpret the values on the y-axis as the expected relative frequency of each random variable in a large number of experiments, each of which had 25 replicates. This definition of relative frequency matches the formal definition of probability given in Chapter 1. It is also the basis of the term **frequentist** to describe statistics based on the expected frequency of an event based on an infinitely large number of trials.

(blenny), (eel)
(blenny), (damselfish)
(goby), (eel)
(goby), (damselfish)
(eel), (damselfish)

Using the binomial coefficient, we would set $n = 5$ and $X = 2$ and arrive at the same number:

$$\binom{5}{2} = \frac{5!}{3!2!} = \frac{120}{6 \times 2} = 10$$

The first column gives the full range of possible numbers of successes out of 25 trials (i.e., from 0 to 25). The second column gives the probability of obtaining exactly that number of successes, calculated from Equation 2.3. Allowing for rounding error, these probabilities sum to 1.0, the total area under the probability curve. Notice that a binomial with $p = 0.5$ gives a perfectly symmetric distribution of successes, centered around 12.5, the expectation of the distribution (Figure 2.1).

TABLE 2.1 **Binomial probabilities for $p = 0.5$, 25 trials**

Number of successes (X)	**Probability of X in 25 trials ($P(X)$)**
0	0.00000003
1	0.00000075
2	0.00000894
3	0.00006855
4	0.00037700
5	0.00158340
6	0.00527799
7	0.01432598
8	0.03223345
9	0.06088540
10	0.09741664
11	0.13284087
12	0.15498102
13	0.15498102
14	0.13284087
15	0.09741664
16	0.06088540
17	0.03223345
18	0.01432598
19	0.00527799
20	0.00158340
21	0.00037700
22	0.00006855
23	0.00000894
24	0.00000075
25	0.00000003
	$\Sigma = 1.00000000$

Figure 2.1 is **symmetric**—the left- and right-hand sides of the distribution are mirror images of one another.[5] However, that result is a special property of a binomial distribution for which $p = (1 - p) = 0.50$. Alternatives are possible,

[5] If you ask most people what the chances are of obtaining 12 or 13 heads out of 25 tosses of a fair coin, they will say about 50%. However, the actual probability of obtaining 12 heads is only 0.155 (Table 2.1), about 15.5%. Why is this number so small? The answer is that the binomial equation gives the **exact probability**: the value of 0.155 is the probability of obtaining *exactly* 12 heads—no more or less. However, scientists are

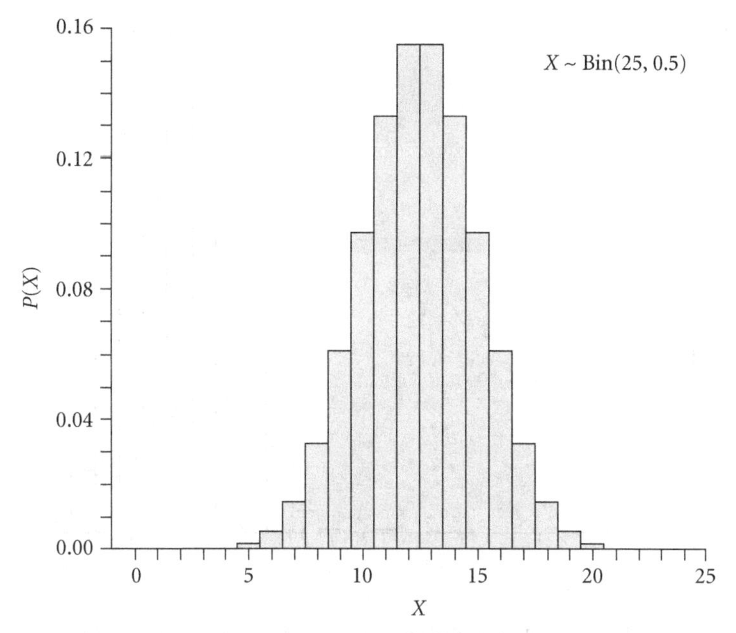

Figure 2.1 Probability distribution for a binomial random variable. A binomial random variable has only two possible outcomes (e.g., *Yes, No*) in a single trial. The variable X represents the number of positive outcomes of 25 trials. The variable $P(X)$ represents the probability of obtaining that number of positive outcomes, calculated from Equation 2.3. Because the probability of a successful outcome was set at $p = 0.5$, this binomial distribution is symmetric, and the midpoint of the distribution is 12.5, or half of the 25 trials. A binomial distribution is specified by two parameters: n, the number trials; and p, the probability of a positive outcome.

such as with biased coins, which come up heads more frequently than 50% of the time. For example, setting $p = 0.80$ and 25 trials, a different shape of the binomial distribution is obtained, as shown in Figure 2.2. This distribution is asymmetric and shifted toward the right; samples with many successes are more probable than they are in Figure 2.1. Thus, the exact shape of the binomial distribution depends both on the total number of trials n, and the probability of suc-

often more interested in knowing the extreme or **tail probability**. The tail probability is the chance of obtaining 12 *or fewer* heads in 25 trials. The tail probability is calculated by adding the probabilities for each of the outcomes from 0 to 12 heads. This sum is indeed 0.50, and it corresponds to the area under the left half of the binomial distribution in Figure 2.1.

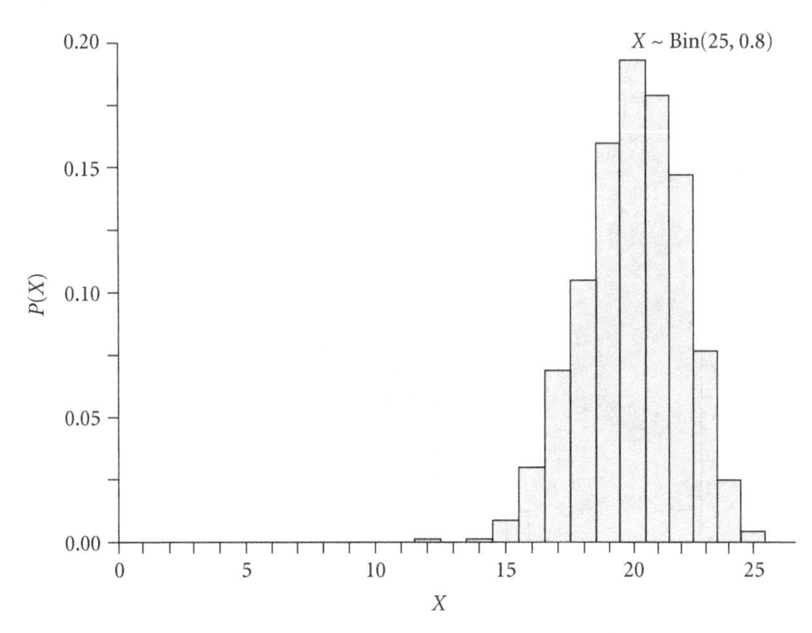

Figure 2.2 Probability distribution for a binomial random variable $X \sim \text{Bin}(25, 0.8)$—that is, there are 25 trials, and the probability of a successful outcome in each trial is $p = 0.8$. This figure follows the same layout and notation as Figure 2.1. However, here the probability of a positive outcome is $p = 0.8$ (instead of $p = 0.5$), and the distribution is no longer symmetric. Notice the expectation for this binomial distribution is $25 \times 0.8 = 20$, which corresponds to the mode (highest peak) of the histogram.

cess, p. You should try generating your own binomial distributions[6] using different values for n and p.

Poisson Random Variables

The binomial distribution is appropriate for cases in which there is a fixed number of trials (n) and the probability of success is not too small. However, the formula quickly becomes cumbersome when n becomes large and p becomes small, such as the occurrences of rare plants or animals. Moreover, the binomial is only

[6] In Chapter 1 (see Footnote 7), we discussed a coin-spinning experiment for the new Euro coin, which yielded an estimate of $p = 0.56$ for $n = 250$ trials. Use spreadsheet or statistical software to generate the binomial distribution with these parameters, and compare it to the binomial distribution for a fair coin ($p = 0.50$, $n = 250$). You will see that these probability distributions do, indeed differ, and that the $p = 0.56$ is shifted very slightly to the right. Now try the same exercise with $n = 25$, as in Figure 2.1. With this relatively small sample size, it will be virtually impossible to distinguish the two distributions. In general, the closer two distributions are in their expectation (see Chapter 3), the larger a sample we will need to distinguish between them.

useful when we can directly count the trials themselves. However, in many cases, we do not count individual trials, but count events that occur within a sample. For example, suppose the seeds from an orchid (a flowering plant) are randomly scattered in a region, and we count the number of seeds in sampling **quadrats** of fixed size. In this case, the occurrence of each seed represents a successful outcome of an unobserved dispersal trial, but we really cannot say how many trials have taken place to generate this distribution. Samples in time can also be treated this way, such as the counts of the number of birds visiting a feeder over a 30-minute period. For such distributions, we use the **Poisson distribution** rather than the binomial distribution.

A **Poisson random variable**[7] X is the number of occurrences of an event recorded in a sample of fixed area or during a fixed interval of time. Poisson random variables are used when such occurrences are rare—that is, when the most common number of counts in any sample is 0. The Poisson random variable itself is the number of events in each sample. As always, we assume that the occurrences of each event are independent of one another.

Poisson random variables are described by a single parameter λ, sometimes referred to as a **rate parameter** because Poisson random variables can describe the frequency of rare events in time. The parameter λ is the average value of the number of occurrences of the event in each sample (or over each time interval). Estimates of λ can be obtained either from data collection or from prior knowledge. Our shorthand notation for a Poisson random variable is

$$X \sim \text{Poisson}(\lambda) \tag{2.4}$$

Siméon-Denis Poisson

[7] Poisson random variables are named for Siméon-Denis Poisson (1781–1840), who claimed that "life was only good for two things: to do mathematics and to teach it" (*fide* Boyer 1968). He applied mathematics to physics (in his two-volume work *Traité de mécanique*, published in 1811 and 1833), and provided an early derivation of the Law of Large Numbers (see Chapter 3). His treatise on probability, *Recherches sur la probabilité des jugements*, was published in 1837.

The most famous literary reference to the Poisson distribution occurs in Thomas Pynchon's novel *Gravity's Rainbow* (1972). One of the lead characters of the novel, Roger Mexico, works for The White Visitation in PISCES (Psychological Intelligence Scheme for Expediting Surrender) during World War II. He plots on a map of London the occurrence of points hit by Nazi bombs, and fits the data to a Poisson distribution. *Gravity's Rainbow*, which won the National Book Award in 1974, is filled with other references to statistics, mathematics, and science. See Simberloff (1978) for a discussion of entropy and biophysics in Pynchon's novels.

and we calculate the probability of any observation x as

$$P(x) = \frac{\lambda^x}{x!} e^{-\lambda} \tag{2.5}$$

where e is a constant, the base of the natural logarithm ($e \sim 2.71828$). For example, suppose the average number of orchid seedlings found in a 1-m^2 quadrat is 0.75. What are the chances that a single quadrat will contain 4 seedlings? In this case, $\lambda = 0.75$ and $x = 4$. Using Equation 2.5,

$$P(4\ seedlings) = \frac{0.75^4}{4!} e^{-0.75} = 0.0062$$

A much more likely event is that a quadrat will contain no seedlings:

$$P(0\ seedlings) = \frac{0.75^0}{0!} e^{-0.75} = 0.4724$$

There is a close relationship between the Poisson and the binomial distributions. For a binomial random variable $X \sim \text{Bin}(n, p)$, where the number of successes X is very small relative to the sample size (or number of trials) n, we can approximate X by using the Poisson distribution, and by estimating $P(X)$ as Poisson(λ). However, the binomial distribution depends on both the probability of success p and the number of trials n, whereas the Poisson distribution depends only on the average number of events per sample, λ.

An entire family of Poisson distributions is possible, depending on the value of λ (Figure 2.3). When λ is small, the distribution has a strong "reverse-J" shape, with the most likely events being 0 or 1 occurrences per sample, but with a long probability tail extending to the right, corresponding to very rare samples that contain many events. As λ increases, the center of the Poisson distribution shifts to the right and becomes more symmetric, resembling a normal or binomial distribution. The binomial and the Poisson are both discrete distributions that take on only integer values, with a minimum value of 0. However, the binomial is always bounded between 0 and n (the number of trials). In contrast, the right-hand tail of the Poisson is not bounded and extends to infinity, although the probabilities quickly become vanishingly small for large numbers of events in a single sample.

An Example of a Poisson Random Variable: Distribution of a Rare Plant

To close this discussion, we describe the application of the Poisson distribution to the distribution of the rare plant *Rhexia mariana* in Massachusetts. The data consist of the number of populations of *Rhexia* recorded in each of 349 Massachusetts towns between 1913 and 2001 (Craine 2002). Although each population can be thought of as an independent Bernoulli trial, we know only the num-

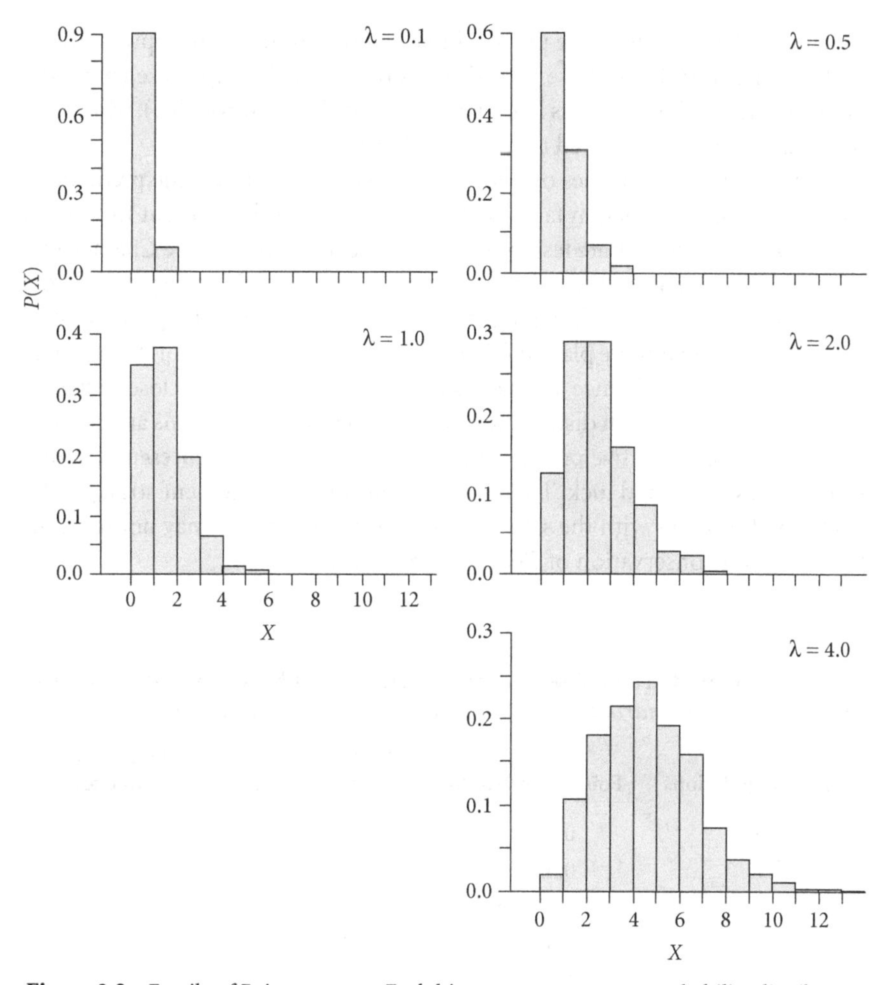

Figure 2.3 Family of Poisson curves. Each histogram represents a probability distribution with a different rate parameter λ. The value of X indicates the number of positive outcomes, and $P(X)$ is the probability of that outcome, calculated from Equation 2.5. In contrast to the binomial distributions shown in Figures 2.1 and 2.2, the Poisson distribution is specified by only a single parameter, λ, which is the rate of occurrence of independent events (or the number that occur during some specified time interval). As the value of λ increases, the Poisson distribution becomes more symmetric. When λ is very large, the Poisson distribution is almost indistinguishable from a normal distribution.

ber of populations, not the number of individual plants; the data from each town is our sampling unit. For these 349 sampling units (towns), most samples (344) had 0 populations in any year, but one town has had as many as 5 populations. The average number of populations is 12/349, or 0.03438 populations/town, which we will use as the rate parameter λ in a Poisson distribution. Next, we use

Equation 2.5 to calculate the probability of a given number of populations in each town. Multiplying those probabilities by the total sample size ($n = 349$) gives the expected frequencies in each population class (0 through 5), which can be compared to the observed numbers (Table 2.2).

The observed frequencies of populations match very closely the predictions from the Poisson model; in fact, they are not significantly different from each other (we used a chi-square test to compare these distributions; see Chapter 11). We can imagine that the town that has five populations of this rare plant is a key target for conservation and management activities—is the soil especially good there, or are there unique plant habitats in that town? Regardless of the ultimate cause of *Rhexia* occurrence and persistence in these towns, the close match of the data with the Poisson distribution suggests that the populations are random and independent; the five populations in a single town may represent nothing more or less than good luck. Therefore, establishing management strategies by searching for towns with the same set of physical conditions may not result in the successful conservation of *Rhexia* in those towns.

TABLE 2.2 **Expected and observed frequencies of numbers of populations of a rare plant (*Rhexia mariana*, meadow beauty) in Massachusetts towns**

Number of *Rhexia* populations	Poisson probability	Poisson expected frequency	Observed frequency
0	0.96	337.2	344
1	0.033	11.6	3
2	5.7×10^{-4}	0.19	0
3	6.5×10^{-6}	2.3×10^{-3}	0
4	5.6×10^{-8}	1.9×10^{-5}	1
5	3.8×10^{-10}	1.4×10^{-7}	1
Total	1.0000	349	349

The first column gives the number of populations found in a particular number of towns. Although there is no theoretical maximum for this number, the observed number was never greater than 5, so only the values for 0 through 5 are shown. The second column gives the Poisson probability for obtaining the observed number, assuming a Poisson distribution with a mean rate parameter $\lambda = 0.03438$ populations/town. This value corresponds to the observed frequency in the data (12 observed populations divided by 349 censused towns = 0.03438). The third column gives the expected Poisson frequency, which is simply the Poisson probability multiplied by the total sample size of 349. The final column gives the observed frequency, that is, the number of towns that contained a certain number of *Rhexia* populations. Notice the close match between the observed and expected Poisson frequencies, suggesting that the occurrence of *Rhexia* populations in a town is a random, independent event. (Data from Craine 2002.)

The Expected Value of a Discrete Random Variable

Random variables take on many different values, but the entire distribution can be summarized by determining the typical, or average, value. You are familiar with the arithmetic mean as the measure of central tendency for a set of numbers, and in Chapter 3 we will discuss it and other useful summary statistics that can be calculated from data. However, when we are dealing with probability distributions, a simple average is misleading. For example, consider a binomial random variable X that can take on the values 0 and 50 with probabilities 0.1 and 0.9, respectively. The arithmetic average of these two values = (0 + 50)/2 = 25, but the most probable value of this random variable is 50, which will occur 90% of the time. To get the average probable value, we take the average of the values weighted by their probabilities. In this way, values with large probabilities count more than values with small probabilities.

Formally, consider a discrete random variable X, which can take on values a_1, $a_2, \ldots, a_n$, with probabilities $p_1, p_2, \ldots, p_n$, respectively. We define the **expected value** of X, which we write as $E(X)$ (read "the **expectation** of X") to be

$$E(X) = \sum_{i=1}^{n} a_i p_i = a_1 p_1 + a_2 p_2 + \ldots + a_n p_n \qquad (2.6)$$

Three points regarding $E(X)$ are worth noting. First, the series

$$\sum_{i=1}^{n} a_i p_i$$

can have either a finite or an infinite number of terms, depending on the range of values that X can take on. Second, unlike an arithmetic average, the sum is not divided by the number of terms. Because probabilities are relative weights (probabilities are all scaled between 0 and 1), the division has already been done implicitly. And finally, except when all the p_i are equal, this sum is not the same as the arithmetic average (see Chapter 3 for further discussion). For the three discrete random variables we have introduced—Bernoulli, binomial, and Poisson—the expected values are p, np, and λ, respectively.

The Variance of a Discrete Random Variable

The expectation of a random variable describes the average, or central tendency, of the values. However, this number (like all averages) gives no insight into the spread, or **variation**, among the values. And, like the average family with 2.2 children, the expectation will not necessarily give an accurate depiction of an individual datum. In fact, for some discrete distributions such as the binomial or Poisson, none of the random variables may ever equal the expectation.

For example, a binomial random variable that can take on the values –10 and +10, each value with probability 0.5, has the same expectation as a binomial random variable that can take on the values –1000 and +1000, each with probability = 0.5. Both have $E(X) = 0$, but in neither distribution will a random value of 0 ever be generated. Moreover, the observed values of the first distribution are much closer to the expected value than are the observed values of the second. Although the expectation accurately describes the midpoint of a distribution, we need some way to quantify the spread of the values around that midpoint.

The **variance** of a random variable, which we write as $\sigma^2(X)$, is a measurement of how far the actual values of a random variable differ from the expected value. The variance of a random variable X is defined as

$$\begin{aligned}\sigma^2(X) &= E[X - E(X)]^2 \\ &= \sum_{i=1}^{n} p_i \left(a_i - \sum_{i=1}^{n} a_i p_i \right)^2\end{aligned} \tag{2.7}$$

As in Equation 2.6, $E(X)$ is the expected value of X, and the a_i's are the different possible values of the variable X, each of which occurs with probability p_i.

To calculate $\sigma^2(X)$, we first calculate $E(X)$, subtract this value from X, then square this difference. This is a basic measure of how much each value X differs from the expectation $E(X)$.[8] Because there are many possible values of X for random variables (e.g., two possible values for a Bernoulli random variable, n possible trial values for a binomial random variable, and infinitely many for a Poisson random variable), we repeat this subtraction and squaring for each possible value of X. Finally, we calculate the expectation of these squared deviates by following the same procedure as in Equation 2.6: each squared deviate is weighted by its probability of occurrence (p_i) and then summed.

Thus, in our example above, if Y is the binomial random variable that can take on values –10 and +10, each with $P(Y) = 0.5$, then $\sigma^2(Y) = 0.5(-10 - 0)^2 + 0.5(10 - 0)^2 = 100$. Similarly, if Z is the binomial random variable that can take

[8] You may be wondering why we bother to square the deviation $X - E(X)$ once it is calculated. If you simply add up the unsquared deviations you will discover that the sum is always 0 because $E(X)$ sits at the midpoint of all the values of X. We are interested in the magnitude of the deviation, rather than its sign, so we could use the sum of absolute values of the deviations $\Sigma|X - E(X)|$. However, the algebra of absolute values is not as simple as that of squared terms, which also have better mathematical properties. The sum of the squared deviations, $\Sigma([X - E(X)]^2)$, forms the basis for analysis of variance (see Footnote 1 in Chapter 10).

on values –1000 and +1000, then $\sigma^2(Z) = 100{,}000$. This result supports our intuition that Z has greater "spread" than Y. Finally, notice that if all of the values of the random variable are identical [i.e., $X = E(X)$], then $\sigma^2(X) = 0$, and there is no variation in the data.

For the three discrete random variables we have introduced—Bernoulli, binomial, and Poisson—the variances are $p(1 - p)$, $np(1 - p)$, and λ, respectively (Table 2.3). The expectation and the variance are the two most important descriptors of a random variable or probability distribution; Chapter 3 will discuss other summary measures of random variables. For now, we turn our attention to continuous variables.

Continuous Random Variables

Many ecological and environmental variables cannot be described by discrete variables. For example, wing lengths of birds or pesticide concentrations in fish tissues can take on any value (bounded by an **interval** with appropriate upper and lower limits), and the **precision** of the measured value is limited only by available instrumentation. When we work with discrete random variables,

TABLE 2.3 **Three discrete statistical distributions**

Distribution	Probability value	E(X)	$\sigma^2(X)$	Comments	Ecological example
Bernoulli	$P(X) = p$	p	$p(1-p)$	Use for dichotomous outcomes	To reproduce or not, that is the question
Binomial	$P(X) = \binom{n}{X} p^X (1-p)^{n-X}$	np	$np(1-p)$	Use for number of successes in n independent trials	Presence or absence of species
Poisson	$P(x) = \frac{\lambda^x}{x!} e^{-\lambda}$	λ	λ	Use for independent rare events where λ is the rate at which events occur in time or space	Distribution of rare species across a landscape

The probability value equation determines the probability of obtaining a particular value X for each distribution. The expectation $E(X)$ of the distribution of values is estimated by the mean or average of a sample. The variance $\sigma^2(X)$ is a measure of the spread or deviation of the observations from $E(X)$. These distributions are for discrete variables, which are measured as integers or counts.

we are able to define the total sample space as a set of possible discrete outcomes. However, when we work with continuous variables, we cannot identify all the possible events or outcomes as there are infinitely many of them (often uncountably infinitely many; see Footnote 4 in Chapter 1). Similarly, because observations can take on any value within the defined interval, it is difficult to define the probability of obtaining a particular value. We illustrate these issues as we describe our first type of continuous random variable, the uniform random variable.

Uniform Random Variables

Both of the problems mentioned above—defining the appropriate sample space and obtaining the probability of any given value—are solvable. First, we recognize that our sample space is no longer discrete, but continuous. In a continuous sample space, we no longer consider discrete outcomes (such as $X = 2$), but instead focus on events that occur within a given subinterval (such as $1.5 < X < 2.5$). The probability that an event occurs within a subinterval can itself be treated as an event, and our rules of probability continue to hold for such events. We start with a theoretical example: the **closed unit interval**, which contains all numbers between 0 and 1, including the two endpoints 0 and 1, and which we write as [0,1]. In this closed-unit interval, suppose that the probability of an event X occurring between 0 and $1/4 = p_1$, and the probability of this same event occurring between 1/2 and $3/4 = p_2$. By the rule that the probability of the union of two independent events equals the sum of their probabilities (see Chapter 1), the probability that X occurs in either of these two intervals (0 to 1/4 or 1/2 to $3/4) = p_1 + p_2$.

The second important rule is that all the probabilities of an event X in a continuous sample space must sum to 1 (this is the First Axiom of Probability). Suppose that, within the closed unit interval, all possible outcomes have equal probability (imagine, for example, a die with an infinite number of sides). Although we cannot define precisely the probability of obtaining the value 0.1 on a roll of this infinite die, we could divide the interval into 10 half-**open intervals** *of equal length* {[0,1/10), [1/10, 2/10),...,[9/10,1]} and calculate the probability of any roll being within one of these subintervals. As you might guess, the probability of a roll being within any one of these ten subintervals would be 0.1, and the sum of all the probabilities in this set = $10 \times 0.1 = 1.0$.

Figure 2.4 illustrates this principle. In this example, the interval ranges from 0 to 10 (on the x-axis). Draw a line L parallel to the x-axis with a y-intercept of 0.1. For any subinterval U on this interval, the probability of a single roll of the die falling in that interval can be found by finding the *area* of the rectangle bounding the subinterval. This rectangle has length equal to the size of the subin-

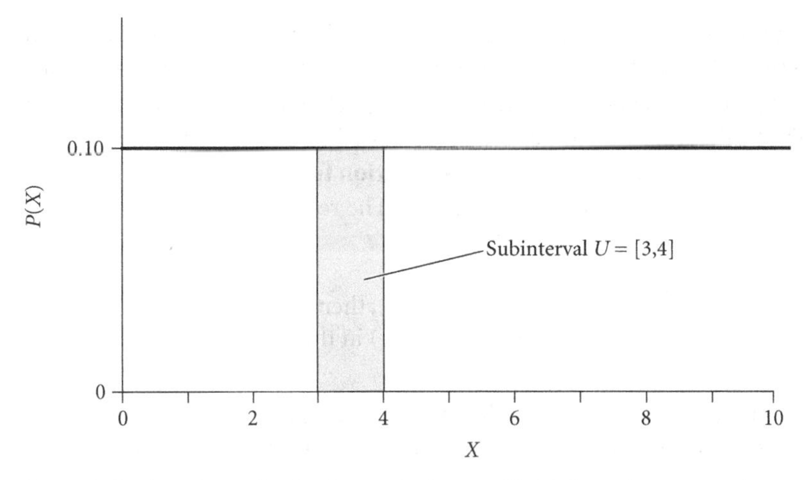

Figure 2.4 A uniform distribution on the interval [0,10]. In a continuous uniform distribution, the probability of an event occurring in a particular subinterval depends on the relative area in the subinterval; it is the same regardless of where that subinterval occurs within the bounded limits of the distribution. For example, if the distribution is bounded between 0 and 10, the probability that an event occurs in the subinterval [3,4] is the relative area that is bounded by that subinterval, which in this case is 0.10. The probability is the same for any other subinterval of the same size, such as [1,2], or [4,5]. If the subinterval chosen is larger, the probability of an event occurring in that subinterval is proportionally larger. For example the probability of an event occurring in the subinterval [3,5] is 0.20 (since 2 interval units of 10 are traversed), and is 0.6 in the subinterval [2,8].

terval and height = 0.1. If we divide the interval into u equal subintervals, the sum of all the areas of these subintervals would equal the sum of the area of the whole interval: 10 (length) × 0.1 (height) = 1.0. This graph also illustrates that the probability of any *particular* outcome a within a continuous sample space is 0, because the subinterval that contains only a is infinitely small, and an infinitely small number divided by a larger number = 0.

We can now define a **uniform random variable** X with respect to any particular interval I. The probability that this uniform random variable X occurs in any subinterval U equals the product $U \times I$. In the example illustrated in Figure 2.4, we define the following function to describe this uniform random variable:

$$f(x) = \begin{cases} 1/10, & 0 \le x \le 10 \\ 0 \text{ otherwise} \end{cases}$$

This function $f(x)$ is called the **probability density function** (**PDF**) for this uniform distribution. In general, the PDF of a continuous random variable is found by assigning probabilities that a continuous random variable X occurs within an interval I. The probability of X occurring within the interval I equals the area of the region bounded by I on the x-axis and $f(x)$ on the y-axis. By the rules of probability, the total area under the curve described by the PDF = 1.

We can also define a **cumulative distribution function** (**CDF**) of a random variable X as the function $F(y) = P(X < y)$. The relationship between the PDF and the CDF is as follows:

If X is a random variable with PDF $f(x)$, then the CDF $F(y) = P(X < Y)$ is equal to the area under $f(x)$ in the interval $x < y$

The CDF represents the **tail probability**—that is, the probability that a random variable X is less than or equal to some value y, $[P(X) < y]$—and is the same as the familiar P-value that we will discuss in Chapter 4. Figure 2.5 illustrates the PDF and the CDF for a uniform random variable on the closed unit interval.

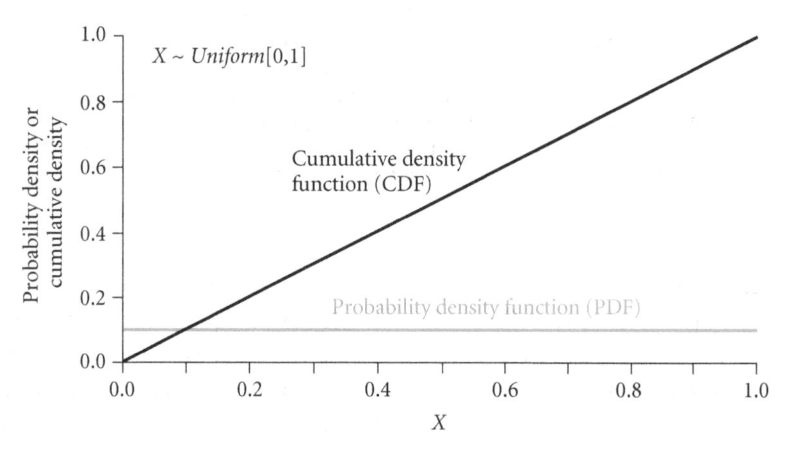

Figure 2.5 Probability density function and cumulative distribution function for the uniform distribution measured over the closed unit interval [0,1]. The probability density function (PDF) shows the probability $P(X)$ for any value X. In this continuous distribution, the exact probability of value X is technically 0, because the area under the curve is zero when measured at a single point. However, the area under the curve of any measurable subinterval is just the proportion of the total area under the curve, which by definition equals 1.0. In the uniform distribution, the probability of an event in any subinterval is the same regardless of where the subinterval is located. The cumulative density function (CDF) for this same distribution illustrates the cumulative area under the curve for the subinterval that is bounded at the low end by 0 and at the high end by 1.0. Because this is a uniform distribution, these probabilities accumulate in a linear fashion. When the end of the interval 1.0 is reached, CDF = 1.0, because the entire area under the curve has been included.

The Expected Value of a Continuous Random Variable

We introduced the expected value of a random variable in the context of a discrete distribution. For a discrete random variable,

$$E(X) = \sum_{i=1}^{n} a_i p_i$$

This calculation makes sense because each value of a_i of X has an associated probability p_i. However, for continuous distributions such as the uniform distribution, the probability of any particular observation = 0, and so we use the probabilities of events occurring within subintervals of the sample space. We take the same approach to find the expected value of a continuous random variable.

To find the expected value of a continuous random variable, we will use very small subintervals of the sample space, which we will denote as Δx. For a probability density function $f(x)$, the product of $f(x_i)$ and Δx gives the probability of an event occurring in the subinterval Δx, written as $P(X = x_i) = f(x_i)\,\Delta x$. This probability is the same as p_i in the discrete case. As in Figure 2.4, the product of $f(x_i)$ and Δx describes the area of a very narrow rectangle. In the discrete case, we found the expected value by summing the product of each x_i by its associated probability p_i. In the continuous case, we will also find the expected value of a continuous random variable by summing the products of each x_i and its associated probability $f(x_i)\Delta x$. Obviously, the value of this sum will depend on the size of the small subinterval Δx. But it turns out that if our PDF $f(x)$ has "reasonable" mathematical properties, and if we let Δx get smaller and smaller, then the sum

$$\sum_{i=1}^{n} x_i f(x_i)\Delta x$$

will approach a unique, limiting value. This limiting value = $E(X)$ for a continuous random variable.[9] For a uniform random variable X, where $f(x)$ is defined on the interval $[a,b]$, and where $a < b$,

$$E(X) = (b + a)/2$$

[9] If you've studied calculus, you will recognize this approach. For a continuous random variable X, where $f(x)$ is differentiable within the sample space,

$$E(X) = \int x f(x) dx$$

The integral represents the sum of the product $x \times f(x)$, where x becomes infinitely small in the limit.

The variance of a uniform random variable is

$$\sigma^2(X) = \frac{(b-a)^2}{12}$$

Normal Random Variables

Perhaps the most familiar probability distribution is the "bell curve"—the **normal** (or **Gaussian**) **probability distribution**. This distribution forms the theoretical basis for linear regression and analysis of variance (see Chapters 9 and 10), and it fits many empirical datasets. We will introduce the normal distribution with an example from the spider family Linyphiidae. Members of the different linyphiid genera can be distinguished by the length of their tibial

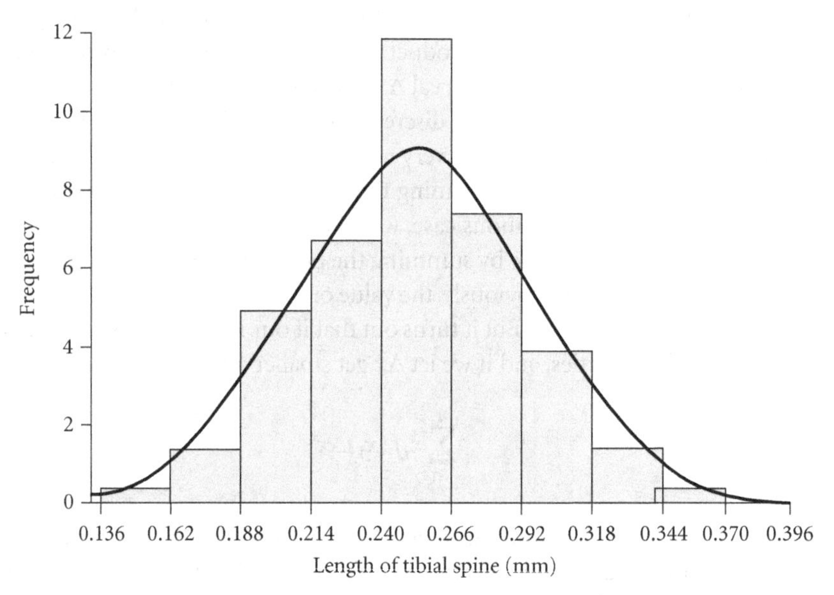

Figure 2.6 Normal distribution of a set of morphological measurements. Each observation in this histogram represents one of 50 measurements of tibial spine length in a sample of spiders (raw data in Table 3.1). The observations are grouped in "bins" that span 0.026-mm intervals; the height of each bar is the frequency (the number of observations that fell in that bin). Superimposed on the histogram is the normal distribution, with a mean of 0.253 and a standard deviation of 0.0039. Although the histogram does not conform perfectly to the normal distribution, the overall fit of the data is very good: the histogram exhibits a single central peak, an approximately symmetric distribution, and a steady decrease in the frequency of very large and very small measurements in the tails of the distribution.

spines. An arachnologist measured 50 such spines and obtained the distribution illustrated in Figure 2.6. The histogram of these measurements illustrates several features that are characteristic of the normal distribution.

First, notice that most of the observations are clustered around a central, or average tibial spine length. However, there are long tails of the histogram that extend to both the left and the right of the center. In a true normal distribution (in which we measured an infinite number of unfortunate spiders), these tails extend indefinitely in both directions, although the probability density quickly becomes vanishingly small as we move away from the center. In real datasets, the tails do not extend forever because we always have a limited amount of data, and because most measured variables cannot take on negative values. Finally, notice that the distribution is approximately symmetric: the left- and right-hand sides of the histogram are almost mirror images of one another.

If we consider spine length to be a random variable X, we can use the normal probability density function to approximate this distribution. The normal distribution is described by two parameters, which we will call μ and σ, and so $f(x) = f(\mu,\sigma)$. The exact form of this function is not important here,[10] but it has the properties that $E(X) = \mu$, $\sigma^2(X) = \sigma^2$, and the distribution is symmetrical around μ. $E(X)$ is the expectation and represents the central tendency of the data; $\sigma^2(X)$ is the variance and represents the spread of the obser-

[10] If you're craving the details, the PDF for the normal distribution is

$$f(x) = \frac{1}{\sigma\sqrt{2\pi}} e^{-\frac{1}{2}\left(\frac{X-\mu}{\sigma}\right)^2}$$

where π is 3.14159..., e is the base of the natural logarithm (2.71828...), and μ and σ are the **parameters** defining the distribution. You can see from this formula that the more distant X is from μ, the larger the negative exponent of e, and hence the smaller the calculated probability for X. The CDF of this distribution,

$$F(X) = \int_{-\infty}^{X} f(x)dx$$

does not have an analytic solution. Most statistics texts provide look-up tables for the CDF of the normal distribution. It also can be approximated using numerical integration techniques in standard software packages such as MatLab or R.

vations around the expectation. A random variable X described by this distribution is called a **normal random variable**, or a **Gaussian random variable**,[11] and is written as

$$X \sim N(\mu,\sigma) \tag{2.8}$$

Many different normal distributions can be created by specifying different values for μ and σ. However, statisticians often use a **standard normal distribution**, where $\mu = 0$ and $\sigma = 1$. The associated **standard normal random variable** is usually referred to simply as Z. $E(Z) = 0$, and $\sigma^2(Z) = 1$.

Useful Properties of the Normal Distribution

The normal distribution has three useful properties. First, normal distributions can be added. If you have two independent normal random variables X and Y, their sum also is a normal random variable with $E(X + Y) = E(X) + E(Y)$ and $\sigma^2(X + Y) = \sigma^2(X) + \sigma^2(Y)$.

Second, normal distributions can be easily transformed with **shift** and **change of scale** operations. Consider two random variables X and Y. Let $X \sim N(\mu,\sigma)$, and let $Y = aX + b$, where a and b are constants. We refer to the operation of

Karl Friedrich Gauss

[11] The Gaussian distribution is named for Karl Friedrich Gauss (1777–1855), one of the most important mathematicians in history. A child prodigy, he is reputed to have corrected an arithmetic error in his father's payroll calculations at the age of 3. Gauss also proved the Fundamental Theorem of Algebra (every polynomial has a root of the form $a + bi$, where $i = \sqrt{-1}$); the Fundamental Theorem of Arithmetic (every natural number can be represented as a unique product of prime numbers); formalized number theory; and, in 1801, developed the method of fitting a line to a set of points using least squares (see Chapter 9). Unfortunately, Gauss did not publish his method of least squares, and it is generally credited to Legendre, who published it 10 years later. Gauss found that the distribution of errors in the lines fit by least squares approximated what we now call the normal distribution, which was introduced nearly 100 years earlier by the American mathematician Abraham de Moivre (1667–1754) in his book *The Doctrine of Chances*. Because Gauss was the more famous of the two (at the time, the United States was a mathematical backwater), the normal probability distribution was originally referred to as the Gaussian distribution. De Moivre himself identified the normal distribution through his studies of the binomial distribution described earlier in this chapter. However, the modern name "normal" was not given to this distribution until the end of the nineteenth century (by the mathematician Poincaré), when the statistician Karl Pearson (1857–1936) rediscovered de Moivre's work and showed that his discovery of this distribution predated Gauss'.

multiplying X by a constant a as a change of scale operation because one unit of X becomes a units of Y; hence Y increases as a function of a. In contrast, the addition of a constant b to X is referred to as a shift operation because we simply move (shift) our random variable over b units along the x axis by adding b to X. Shift and change of scale operations are illustrated in Figure 2.7.

If X is a normal random variable, then a new random variable Y, created by either a change of scale operation, a shift operation, or both on the normal random variable X is also a normal random variable. Conveniently, the expectation and variance of the new random variable is a simple function of the shift and scale constants. For two random variables $X \sim N(\mu,\sigma)$ and $Y = aX + b$, we calculate $E(Y) = a\mu + b$ and $\sigma^2(Y) = a^2\sigma^2$. Notice that the expectation of the new

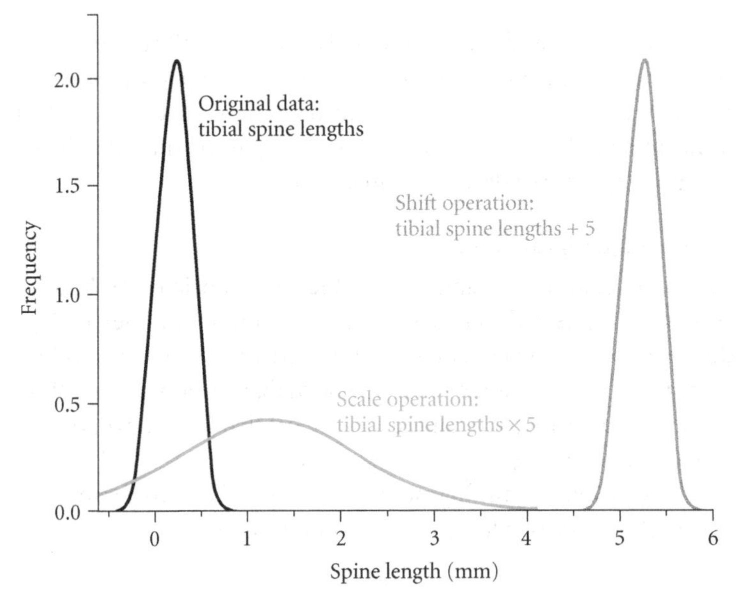

Figure 2.7 Shift and scale operations on a normal distribution. The normal distribution has two convenient algebraic properties. The first is a shift operation: if a constant b is added to a set of measurements with a mean μ, the mean of the new distribution is shifted to $\mu + b$, but the variance is unaffected. The black curve is the normal distribution fit to a set of 200 measurements of spider tibial spine length (see Figure 2.6). The gray curve shows the shifted normal distribution after a value of 5 was added to each one of the original observations. The mean has shifted 5 units to the right, but the variance is unaltered. In a scale operation (blue curve), multiplying each observation by a constant a causes the mean to be increased by a factor of a, but the variance to be increased by a factor of a^2. This curve is the normal distribution fit to the data after they have been multiplied by 5. The mean is shifted to a value of 5 times the original, and the variance has increased by a factor of $5^2 = 25$.

random variable $E(Y)$ is created by directly applying the shift and scale operations to $E(X)$. However, the variance of the new random variable $\sigma^2(Y)$ is changed only by the scale operation. Intuitively, you can see that if all of the elements of a dataset are shifted by adding a constant, the variance should not change because the relative spread of the data has not been affected (see Figure 2.7). However, if those data are multiplied by a constant a (change of scale), the variance will be increased by the quantity a^2; because the relative distance of each element from the expectation is now increased by a factor of a (Figure 2.7), this quantity is squared in our variance calculation.

A final (and convenient) property of the normal distribution is the special case of a change of scale and shift operation in which $a = 1/\sigma$ and $b = -1(\mu/\sigma)$:

For $X \sim N(\mu, \sigma)$, $Y = (1/\sigma)X - \mu/\sigma = (X - \mu)/\sigma$ gives

$$E(Y) = 0 \text{ and } \sigma^2(Y) = 1$$

which is a standard normal random variable. This is an especially useful result, because it means that any normal random variable can be transformed into a standard normal random variable. Moreover, any operation that can be applied to a standard normal random variable will apply to any normal random variable after it has been appropriately scaled and shifted.

Other Continuous Random Variables

There are many other continuous random variables and associated probability distributions that ecologists and statisticians use. Two important ones are log-normal random variables and exponential random variables (Figure 2.8). A **log-normal random variable** X is a random variable such that its natural logarithm, $\ln(X)$, is a normal random variable. Many key ecological characteristics of organisms, such as body mass, are log-normally distributed.[12]

Like the normal distribution, the log-normal is described by two parameters, μ and σ. The expected value of a log-normal random variable is

$$E(X) = e^{\frac{2\mu+\sigma^2}{2}}$$

[12] The most familiar ecological example of a log-normal distribution is the distribution of relative abundances of species in a community. If you take a large, random sample of individuals from a community, sort them according to species, and construct a histogram of the frequency of species represented in different abundance classes, the data will often resemble a normal distribution when the abundance classes are plotted on a logarithmic scale (Preston 1948). What is the explanation for this pattern? On the one hand, many non-biological datasets (such as the distribution of economic wealth among countries, or the distribution of "survival times" of drinking glasses in a busy restaurant) also follow a log-normal distribution. Therefore, the pattern may reflect a generic statistical response of exponentially increasing populations (a loga-

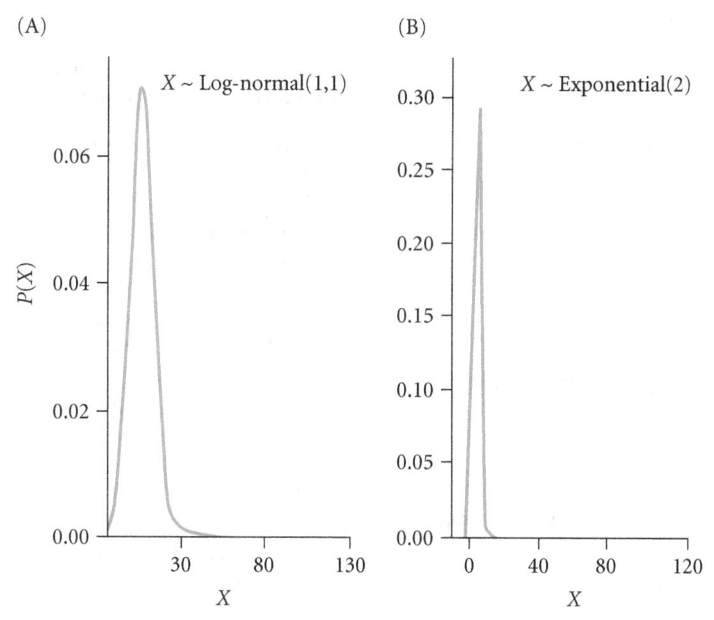

Figure 2.8 Log-normal and exponential distributions fit certain kinds of ecological data, such as species abundance distributions and seed dispersal distances. (A) The log-normal distribution is described by two parameters, a mean and a variance, both set to 1.0 in this example. (B) The exponential distribution is described by a single parameter *b*, set to 2.0 in this example. See Table 2.4 for the equations used with the log-normal and exponential distributions. Both the log-normal and the exponential distributions are asymmetric, with long right-hand tails that skew the distribution to the right.

and the variance of a log-normal distribution is

$$\sigma^2(X) = \left[e^{\frac{\mu+\sigma^2}{2}}\right]^2 \times \left[e^{\sigma^2} - 1\right]$$

When plotted on a logarithmic scale, the log-normal distribution shows a characteristic bell-shaped curve. However, if these same data are back-transformed to their original values (by applying the transformation e^X), the resulting dis-

rithmic phenomenon) to many independent factors (May 1975). On the other hand, specific biological mechanisms may be at work, including patch dynamics (Ugland and Gray 1982) or hierarchical niche partitioning (Sugihara 1980). The study of log-normal distributions is also complicated by sampling problems. Because rare species in a community may be missed in small samples, the shape of the species-abundance distribution changes with the intensity of sampling (Wilson 1993). Moreover, even in well-sampled communities, the tail of the frequency histogram may not fit a true log-normal distribution very well (Preston 1981). This lack of fit may arise because a large sample of a community will usually contain mixtures of resident and transient species (Magurran and Henderson 2003).

TABLE 2.4 **Four continuous distributions**

Distribution	Probability value	E(X)	$\sigma^2(X)$	Comments	Ecological example
Uniform	$P(a < X < b) = 1.0$	$\frac{b+a}{2}$	$\frac{(b-a)^2}{12}$	Use for equiprobable outcomes over interval $[a,b]$	Even distribution of resources
Normal	$\frac{1}{\sigma\sqrt{2\pi}}e^{-\frac{1}{2}\left(\frac{X-\mu}{\sigma}\right)^2}$	μ	σ^2	Generates a symmetric "bell curve" for continuous data	Distribution of tibial spine lengths or other continuous size variables
Log-normal	$\frac{1}{\sigma X\sqrt{2\pi}}e^{-\frac{1}{2}\left(\frac{\ln(X)-\mu}{\sigma}\right)^2}$	$e^{\frac{2\mu+\sigma^2}{2}}$	$\left[e^{\frac{\mu+\sigma^2}{2}}\right]^2 \times \left[e^{\sigma^2}-1\right]$	Log-transformed data of right-skewed data are often fit by a normal distribution	Distribution of species abundance classes
Exponential	$P(X) = \beta e^{-\beta X}$	$1/\beta$	$1/\beta^2$	Continuous distribution analog of Poisson	Seed dispersal distance

The Probability value equation determines the probability of obtaining a particular value X for each distribution. The expectation $E(X)$ of the distribution of values is estimated by the mean or average of a sample. The variance $\sigma^2(X)$ is a measure of the spread or deviation of the observations from $E(X)$. These distributions are for variables measured on a continuous scale that can take on any real number value, although the exponential distribution is limited to positive values.

tribution is skewed, with a long probability tail extending to the right. See Chapter 3 for a further discussion of measures of skewness, and Chapter 8 for more examples of data transformations.

Exponential random variables are related to Poisson random variables. Recall that Poisson random variables describe the number of rare occurrences (e.g., counts), such as the number of arrivals in a constant time interval, or the number of individuals occurring in a fixed area. The "spaces" between discrete Poisson random variables, such as the time or distance between Poisson events, can be described as continuous exponential random variables. The probability distribution function for an exponential random variable X has only one parameter, usually written as β, and has the form $P(X) = \beta e^{-\beta X}$. The expected value of an exponential random variable = $1/\beta$, and its variance = $1/\beta^2$.[13] Table 2.4 summarizes the properties of these common continuous distributions.

[13] It is easy to simulate an exponential random variable on your computer by taking advantage of the fact that if U is a uniform random variable defined on the closed unit interval [0,1], then $-\ln(U)/\beta$ is an exponential random variable with parameter β.

Other continuous random variables and their probability distribution functions are used extensively in statistical analyses. These include the Student-*t*, chi-square, F, gamma, inverse gamma, and beta. We will discuss these later in the book when we use them with particular analytical techniques.

The Central Limit Theorem

The **Central Limit Theorem** is one of the cornerstones of probability and statistical analysis.[14] Here is a brief description of the theorem. Let S_n be the sum or the average of any set of n independent, identically distributed random variables X_i:

$$S_n = \sum_{i=1}^{n} X_i$$

each of which has the same expected value μ and all of which have the same variance σ^2. Then, S_n has the expected value of $n\mu$ and variance $n\sigma^2$. If we standardize S_n by subtracting the expected value from each observation and dividing by the square root of the variance,

$$S_{std} = \frac{S_n - n\mu}{\sigma\sqrt{n}} = \frac{\sum_{i=1}^{n} X_i - n\mu}{\sigma\sqrt{n}}$$

then the distribution of a set of S_{std} values approximates a standard normal variable.

Abraham De Moivre

Pierre Laplace

[14] The initial formulation of the Central Limit Theorem is due to Abraham De Moivre (biographical information in Footnote 11) and Pierre Laplace. In 1733, De Moivre proved his version of the Central Limit Theorem for a Bernoulli random variable. Pierre Laplace (1749–1827) extended De Moivre's result to any binary random variable. Laplace is better remembered for his *Mécanique céleste* (*Celestial Mechanics*), which translated Newton's system of geometrical studies of mechanics into a system based on calculus. According to Boyer's *History of Mathematics* (1968), after Napoleon had read *Mécanique céleste*, he asked Laplace why there was no mention of God in the work. Laplace is said to have responded that he had no need for that hypothesis. Napoleon later appointed Laplace to be Minister of the Interior, but eventually dismissed him with the comment that "he carried the spirit of the infinitely small into the management of affairs" (*fide* Boyer 1968). The Russian mathematician Pafnuty Chebyshev (1821–1884) proved the Central Limit Theorem for any random variable, but his complex proof is virtually unknown today. The modern, accessible proof of the Central Limit Theorem is due to Chebyshev's students Andrei Markov (1856–1922) and Alexander Lyapounov (1857–1918).

This is a powerful result. The Central Limit Theorem asserts that standardizing *any* random variable that itself is a sum or average of a set of independent random variables results in a new random variable that is "nearly the same as"[15] a standard normal one. We already used this technique when we generated a standard normal random variable (Z) from a normal random variable (X). The beauty of the Central Limit Theorem is that it allows us to use statistical tools that require our sample observations to be drawn from a sample space that is normally distributed, even though the underlying data themselves may not be normally distributed. The only caveats are that the sample size must be "large enough,"[16] and that the observations themselves must be independent and all drawn from a distribution with common expectation and variance. We will demonstrate the importance of the Central Limit Theorem when we discuss the different statistical techniques used by ecologists and environmental scientists in Chapters 9–12.

Summary

Random variables take on a variety of measurements, but their distributions can be characterized by their expectation and variance. Discrete distributions such as the Bernoulli, binomial, and Poisson apply to data that are discrete counts, whereas continuous distributions such as the uniform, normal, and exponen-

[15] More formally, the Central Limit Theorem states that for any standardized variable

$$Y_i = \frac{S_i - n\mu}{\sigma\sqrt{n}}$$

the area under the standard normal probability distribution over the open interval (a,b) equals

$$\lim_{i \to \infty} P(a < Y_i < b)$$

[16] An important question for practicing ecologists (and statisticians) is how quickly the probability $P(a < Y_i < b)$ converges to the area under the standard normal probability distribution. Most ecologists (and statisticians) would say that the sample size i should be at least 10, but recent studies suggest that i must exceed 10,000 before the two converge to even the first two decimal places! Fortunately, most statistical tests are fairly robust to the assumption of normality, so we can make use of the Central Limit Theorem even though our standardized data may not exhibit a perfectly normal distribution. Hoffmann-Jørgensen (1994; see also Chapter 5) provides a thorough and technical review of the Central Limit Theorem.

tial apply to data measured on a continuous scale. Regardless of the underlying distribution, the Central Limit Theorem asserts that the sums or averages of large, independent samples will follow a normal distribution if they are standardized. For a wide variety of data, including those collected most commonly by ecologists and environmental scientists, the Central Limit Theorem supports the use of statistical tests that assume normal distributions.

CHAPTER 3

Summary Statistics: Measures of Location and Spread

Data are the essence of scientific investigations, but rarely do we report all the data that we collect. Rather, we summarize our data using **summary statistics**. Biologists and statisticians distinguish between two kinds of summary statistics: measures of **location** and measures of **spread**. Measures of location illustrate where the majority of data are found; these measures include means, medians, and modes. In contrast, measures of spread describe how variable the data are; these measures include the sample standard deviation, variance, and standard errors. In this chapter, we introduce the most common summary statistics and illustrate how they arise directly from the **Law of Large Numbers**, one of the most important theorems of probability.

Henceforth, we will adopt standard statistical notation when describing random variables and statistical quantities or estimators. Random variables will be designated as Y, where each individual observation is indexed with a subscript, Y_i. The subscript i indicates the ith observation. The size of the sample will be denoted by n, and so i can take on any integer value between 1 and n. The arithmetic mean is written as $\overline{Y}$. Unknown **parameters** (or population statistics) of distributions, such as expected values and variances, will be written with Greek letters (such as μ for the expected value, σ^2 for the expected variance, σ for the expected standard deviation), whereas statistical estimators of those parameters (based on real data) will be written with italic letters (such as $\overline{Y}$ for the arithmetic mean, s^2 for the sample variance, and s for the sample standard deviation).

Throughout this chapter, we use as our example the data illustrated in Figure 2.6, the simulated measurement of tibial spines of 50 linyphiid spiders. These data, sorted in ascending order, are illustrated in Table 3.1.

TABLE 3.1 **Ordered measurements of tibial spines of 50 linyphiid spiders (millimeters)**

0.155	0.207	0.219	0.228	0.241	0.249	0.263	0.276	0.292	0.307
0.184	0.208	0.219	0.228	0.243	0.250	0.268	0.277	0.292	0.308
0.199	0.212	0.221	0.229	0.247	0.251	0.270	0.280	0.296	0.328
0.202	0.212	0.223	0.235	0.247	0.253	0.274	0.286	0.301	0.329
0.206	0.215	0.226	0.238	0.248	0.258	0.275	0.289	0.306	0.368

This simulated dataset is used throughout this chapter to illustrate measures of summary statistics and probability distributions. Although raw data of this sort form the basis for all of our calculations in statistics, the raw data are rarely published because they are too extensive and too difficult to comprehend. Summary statistics, if they are properly used, concisely communicate and summarize patterns in raw data without enumerating each individual observation.

Measures of Location

The Arithmetic Mean

There are many ways to summarize a set of data. The most familiar is the average, or **arithmetic mean** of the observations. The arithmetic mean is calculated as the sum of the observations (Y_i) divided by the number of observations (n) and is denoted by $\overline{Y}$:

$$\overline{Y} = \frac{\sum_{i=1}^{n} Y_i}{n} \quad (3.1)$$

For the data in Table 3.1, $\overline{Y} = 0.253$. Equation 3.1 looks similar to, but is not quite equivalent to, Equation 2.6, which was used in Chapter 2 to calculate the expected value of a discrete random variable:

$$E(Y) = \sum_{i=1}^{n} Y_i p_i$$

where the Y_i's are the values that the random variable can have, and the p_i's are their probabilities. For a continuous variable in which each Y_i occurs only once, with $p_i = 1/n$, Equations 3.1 and 2.6 give identical results.

For example, let *Spine length* be the set consisting of the 50 observations in Table 3.1: *Spine length* = {0.155, 0.184, ..., 0.329, 0.368}. If each element (or event) in *Spine length* is independent of all others, then the probability p_i of any of these 50 independent observations is 1/50. Using Equation 2.6, we can calculate the expected value of *Spine length* to be

$$E(Y) = \sum_{i=1}^{n} Y_i p_i$$

where Y_i is the ith element and $p_i = 1/50$. This sum,

$$E(Y) = \sum_{i=1}^{n} Y_i \times \frac{1}{50}$$

is now equivalent to Equation 3.1, used to calculate the arithmetic mean of n observations of a random variable Y:

$$\bar{Y} = \sum_{i=1}^{n} p_i Y_i = \sum_{i=1}^{n} Y_i \times \frac{1}{50} = \frac{1}{50}\sum_{i=1}^{n} Y_i$$

To calculate this expected value of *Spine length*, we used the formula for the expected value of a discrete random variable (Equation 2.6). However, the data given in Table 3.1 represent observations of a continuous, normal random variable. All we know about the expected value of a normal random variable is that it has some underlying true value, which we denote as μ. Does our calculated value of the mean of *Spine length* have any relationship to the unknown value of μ?

If three conditions are satisfied, the arithmetic mean of the observations in our sample is an **unbiased estimator** of μ. These three conditions are:

1. Observations are made on randomly selected individuals.
2. Observations in the sample are independent of each other.
3. Observations are drawn from a larger population that can be described by a normal random variable.

The fact that $\bar{Y}$ of a sample approximates μ of the population from which the sample was drawn is a special case of the second fundamental theorem of probability, the **Law of Large Numbers.**[1]

Here is a description of the Law of Large Numbers. Consider an infinite set of random samples of size n, drawn from a random variable Y. Thus, Y_1 is a sample from Y with 1 datum, $\{y_1\}$. Y_2 is a sample of size 2, $\{y_1, y_2\}$, etc. The Law of Large Numbers establishes that, as the sample size n increases, the arithmetic

Andrei Kolmogorov

[1] The modern (or "strong") version of the Law of Large Numbers was proven by the Russian mathematician Andrei Kolmogorov (1903–1987), who also studied **Markov processes** such as those used in modern computational Bayesian analysis (see Chapter 5) and fluid mechanics.

mean of Y_i (Equation 3.1) approaches the expected value of Y, $E(Y)$. In mathematical notation, we write

$$\lim_{n \to \infty} \left(\frac{\sum_{i=1}^{n} y_i}{n} = \bar{Y}_n \right) = E(Y) \tag{3.2}$$

In words, we say that as n gets very large, the average of the Y_i's equals $E(Y)$ (see Figure 3.1).

In our example, the tibial spine lengths of all individuals of linyphiid spiders in a population can be described as a normal random variable with expected value $= \mu$. We cannot measure all of these (infinitely many) spines, but we can measure a subset of them; Table 3.1 gives $n = 50$ of these measurements. If each spine measured is from a single individual spider, each spider chosen for measurement is chosen at random, and there is no bias in our measurements, then the expected value for each observation should be the same (because they come from the same infinitely large population of spiders). The Law of Large Numbers states that the average spine length of our 50 measurements approximates the expected value of the spine length in the entire population. Hence, we can estimate the unknown expected value μ with the average of our observations. As Figure 3.1 shows, the estimate of the true population mean is more reliable as we accumulate more data.

Other Means

The arithmetic average is not the only measure of location of a set of data. In some cases, the arithmetic average will generate unexpected answers. For example, suppose a population of mule deer (*Odocoileus hemionus*) increases in size by 10% in one year and 20% in the next year. What is the average population growth rate each year?[2] The answer is not 15%!

You can see this discrepancy by working through some numbers. Suppose the initial population size is 1000 deer. After one year, the population size (N_1) will be $(1.10) \times 1000 = 1100$. After the second year, the population size (N_2) will be $(1.20) \times 1100 = 1320$. However, if the average growth rate were 15% per

[2] In this analysis, we use the finite rate of increase, λ, as the parameter for population growth. λ is a multiplier that operates on the population size each year, such that $N_{t+1} = \lambda N_t$. Thus, if the population increases by 10% every year, $\lambda = 1.10$, and if the population decreases by 5% every year, $\lambda = 0.95$. A closely related measure of population growth rate is the instantaneous rate of increase, r, whose units are individuals/(individuals$\times$time). Mathematically, $\lambda = e^r$ and $r = \ln(\lambda)$. See Gotelli (2008) for more details.

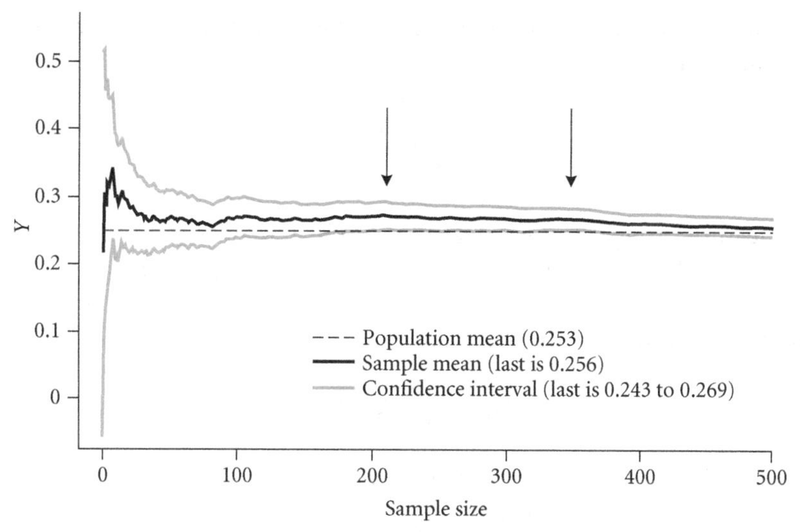

Figure 3.1 Illustration of the Law of Large Numbers and the construction of confidence intervals using the spider tibial spine data of Table 3.1. The population mean (0.253) is indicated by the dotted line. The sample mean for samples of increasing size n is indicated by the central solid line and illustrates the Law of Large Numbers: as the sample size increases, the sample mean approaches the true population mean. The upper and lower solid lines illustrate 95% confidence intervals about the sample mean. The width of the confidence interval decreases with increasing sample size. 95% of confidence intervals constructed in this way should contain the true population mean. Notice, however, that there are samples (between the arrows) for which the confidence interval does not include the true population mean. Curve constructed using algorithms and R code based on Blume and Royall (2003).

year, the population size would be $(1.15)\times 1000 = 1150$ after one year and then $(1.15)\times 1150 = 1322.50$ after 2 years. These numbers are close, but not identical; after several more years, the results diverge substantially.

THE GEOMETRIC MEAN In Chapter 2, we introduced the log-normal distribution: if Y is a random variable with a log-normal distribution, then the random variable $Z = \ln(Y)$ is a normal random variable. If we calculate the arithmetic mean of Z,

$$\bar{Z} = \frac{1}{n}\sum_{i=1}^{n} Z_i \tag{3.3}$$

what is this value expressed in units of Y? First, recognize that if $Z = \ln(Y)$, then $Y = e^Z$, where e is the base of the natural logarithm and equals ~2.71828… . Thus, the value of $\bar{Z}$ in units of Y is $e^{\bar{Z}}$. This so-called **back-transformed mean** is called the **geometric mean** and is written as GM_Y.

The simplest way to calculate the geometric mean is to take the antilog of the arithmetic mean:

$$GM_Y = e^{\left[\frac{1}{n}\sum_{i=1}^{n}\ln(Y_i)\right]} \tag{3.4}$$

A nice feature of logarithms is that the sum of the logarithms of a set of numbers equals the logarithm of their products: $\ln(Y_1) + \ln(Y_2) + ... = \ln(Y_1Y_2...Y_n)$. So another way of calculating the geometric mean is to take the nth root of the product of the observations:

$$GM_Y = \sqrt[n]{Y_1Y_2...Y_n} \tag{3.5}$$

Just as we have a special symbol for adding up a series of numbers:

$$\sum_{i=1}^{n} Y_i = Y_1 + Y_2 + ... + Y_n$$

we also have a special symbol for multiplying a series of numbers:

$$\prod_{i=1}^{n} Y_i = Y_1 \times Y_2 \times \ ... \ \times Y_n$$

Thus, we could also write our formula for the geometric mean as

$$GM_Y = \sqrt[n]{\prod_{i=1}^{n} Y_i}$$

Let's see if the geometric mean of the population growth rates does a better job of predicting average population growth rate than the arithmetic average does. First, if we express population growth rates as multipliers, the annual growth rates of 10% and 20% become 1.10 and 1.20, and the natural logarithms of these two values are $\ln(1.10) = 0.09531$ and $\ln(1.20) = 0.18232$. The arithmetic average of these two numbers is 0.138815. Back-calculating gives us a geometric mean of $GM_Y = e^{0.138815} = 1.14891$, which is slightly less than the arithmetic mean of 1.20.

Now we can calculate population growth rate over two years using this geometric mean growth rate. In the first year, the population would grow to $(1.14891)\times(1000) = 1148.91$, and in the second year to $(1.14891)\times(1148.91) = 1319.99$. This is the same answer we got with 10% growth in the first year and 20% growth in the second year $[(1.10)\times(1000)\times(1.20)] = 1320$. The values would match perfectly if we had not rounded the calculated growth rate. Notice also that although population size is always an integer variable (0.91 deer can be seen only in a theoretical forest), we treat it as a continuous variable to illustrate these calculations.

Why does GM_Y give us the correct answer? The reason is that population growth is a multiplicative process. Note that

$$\frac{N_2}{N_0} = \left(\frac{N_2}{N_1}\right) \times \left(\frac{N_1}{N_0}\right) \neq \left(\frac{N_2}{N_1}\right) + \left(\frac{N_1}{N_0}\right)$$

However, numbers that are multiplied together on an arithmetic scale can be added together on a logarithmic scale. Thus

$$\ln\left[\left(\frac{N_2}{N_1}\right) \times \left(\frac{N_1}{N_0}\right)\right] = \ln\left(\frac{N_2}{N_1}\right) + \ln\left(\frac{N_1}{N_0}\right)$$

THE HARMONIC MEAN A second kind of average can be calculated in a similar way, using the reciprocal transformation (1/Y). The reciprocal of the arithmetic mean of the reciprocals of a set of observations is called the **harmonic mean**:[3]

$$H_Y = \frac{1}{\frac{1}{n}\sum \frac{1}{Y_i}} \tag{3.6}$$

For the spine data in Table 3.1, $GM_Y = 0.249$ and $H_Y = 0.246$. Both of these means are smaller than the arithmetic mean (0.253); in general, these means are ordered as $\bar{Y} > GM_Y > H_Y$. However, if all the observations are equal ($Y_1 = Y_2 = Y_3 = \ldots Y_n$), all three of these means are identical as well ($\bar{Y} = GM_Y = H_Y$).

[3] The harmonic mean turns up in conservation biology and population genetics in the calculation of effective population size, which is the equivalent size of a population with completely random mating. If the effective population size is small (< 50), random changes in allele frequency due to genetic drift potentially are important. If population size changes from one year to the next, the harmonic mean gives the effective population size. For example, suppose a stable population of 100 sea otters passes through a severe bottleneck and is reduced to a population size of 12 for a single year. Thus, the population sizes are 100, 100, 12, 100, 100, 100, 100, 100, 100, and 100. The arithmetic mean of these numbers is 91.2, but the harmonic mean is only 57.6, an effective population size at which genetic drift could be important. Not only is the harmonic mean less than the arithmetic mean, the harmonic mean is especially sensitive to extreme values that are small. Incidentally, sea otters on the Pacific coast of North America did pass through a severe population bottleneck when they were overhunted in the eighteenth and nineteenth centuries. Although sea otter populations have recovered in size, they still exhibit low genetic diversity, a reflection of this past bottleneck (Larson et al. 2002). (Photograph by Warren Worthington, soundwaves.usgs.gov/2002/07/.)

Other Measures of Location: The Median and the Mode

Ecologists and environmental scientists commonly use two other measures of location, the median and the mode, to summarize datasets. The **median** is defined as the value of a set of ordered observations that has an equal number of observations above and below it. In other words, the median divides a dataset into two halves with equal number of observations in each half. For an odd number of observations, the median is simply the central observation. Thus, if we considered only the first 49 observations in our spine-length data, the median would be the 25th observation (0.248). But with an even number of observations, the median is defined as the midpoint between the $(n/2)$th and $[(n/2)+1]$th observation. If we consider all 50 observations in Table 3.1, the median would be the average of the 25th and 26th observations, or 0.2485.

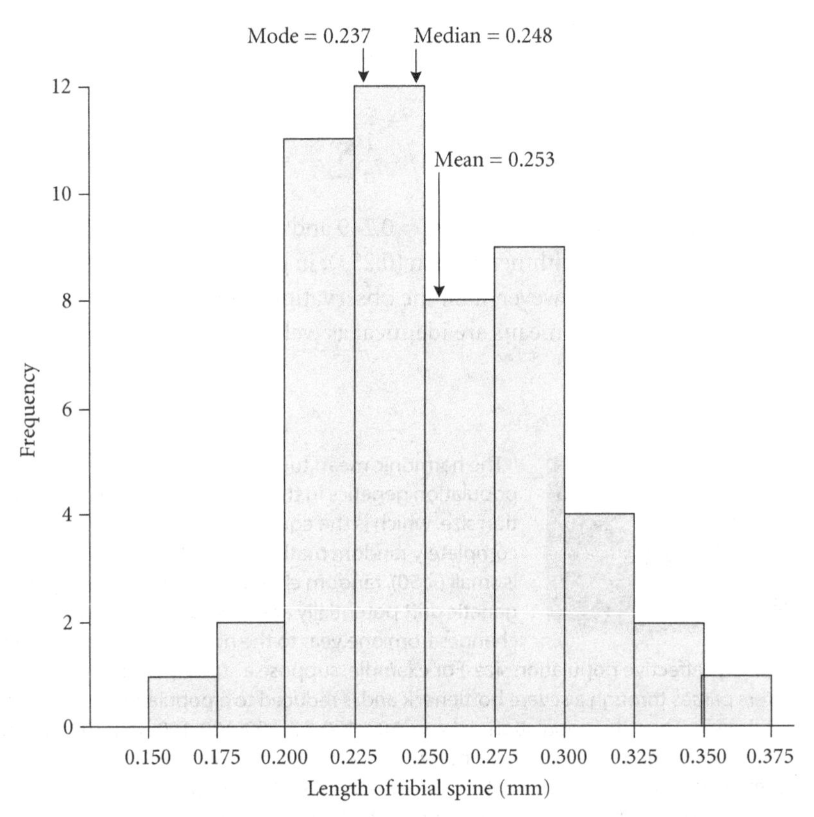

Figure 3.2 Histogram of the tibial spine data from Table 3.1 (n = 50) illustrating the arithmetic mean, median, and mode. The mean is the expectation of the data, calculated as the average of the continuous measurements. The median is the midpoint of the ordered set of observations. Half of all observations are larger than the median and half are smaller. The mode is the most frequent observation.

The **mode,** on the other hand, is the value of the observations that occurs most frequently in the sample. The mode can be read easily off of a histogram of the data, as it is the peak of the histogram. Figure 3.2 illustrates the arithmetic mean, the median, and the mode in a histogram of the tibial spine data.

When to Use Each Measure of Location

Why choose one measure of location over another? The arithmetic mean (Equation 3.1) is the most commonly used measure of location, in part because it is familiar. A more important justification is that the Central Limit Theorem (Chapter 2) shows that arithmetic means of large samples of random variables conform to a normal or Gaussian distribution, even if the underlying random variable does not. This property makes it easy to test hypotheses on arithmetic means.

The geometric mean (Equations 3.4 and 3.5) is more appropriate for describing multiplicative processes such as population growth rates or abundance classes of species (before they are logarithmically transformed; see the discussion of the log-normal distribution in Chapter 2 and the discussion of data transformations in Chapter 8). The harmonic mean (Equation 3.6) turns up in calculations used by population geneticists and conservation biologists.

The median or the mode better describe the location of the data when distributions of observations cannot be fit to a standard probability distribution, or when there are extreme observations. This is because the arithmetic, geometric, and harmonic means are very sensitive to extreme (large or small) observations, whereas the median and the mode tend to fall in the middle of the distribution regardless of its spread and shape. In symmetrical distributions such as the normal distribution, the arithmetic mean, median, and mode all are equal. But in asymmetrical distributions, such as that shown in Figure 3.2 (a relatively small random sample from an underlying normal distribution), the mean occurs toward the largest tail of the distribution, the mode occurs in the heaviest part of the distribution, and the median occurs between the two.[4]

[4] People also use different measures of location to support different points of view. For example, the average household income in the United States is considerably higher than the more typical (or median) income. This is because income has a log-normal distribution, so that averages are weighted by the long right-hand tail of the curve, representing the ultrarich. Pay attention to whether the mean, median, or mode of a data set is reported, and be suspicious if it is reported without any measure of spread or variation.

Measures of Spread

It is never sufficient simply to state the mean or other measure of location. Because there is variation in nature, and because there is a limit to the precision with which we can make measurements, we must also quantify and publish the spread, or variability, of our observations.

The Variance and the Standard Deviation

We introduced the concept of variance in Chapter 2. For a random variable Y, the variance $\sigma^2(Y)$ is a measurement of how far the observations of this random variable differ from the expected value. The variance is defined as $E[Y - E(Y)]^2$ where $E(Y)$ is the expected value of Y. As with the mean, the true variance of a population is an unknown quantity. Just as we calculated an estimate $\bar{Y}$ of the population mean μ using our data, we can calculate an estimate s^2 of the population variance σ^2 using our data:

$$s^2 = \frac{1}{n}\sum(Y_i - \bar{Y})^2 \tag{3.7}$$

This value is also referred to as the **mean square**. This term, along with its companion, the **sum of squares**,

$$SS_Y = \sum_{i=1}^{n}(Y_i - \bar{Y})^2 \tag{3.8}$$

will crop up again when we discuss regression and analysis of variance in Chapters 9 and 10. And just as we defined the standard deviation σ of a random variable as the (positive) square root of its variance, we can estimate it as $s = \sqrt{s^2}$. The square root transformation ensures that the units of standard deviation are the same as the units of the mean.

We noted earlier that the arithmetic mean $\bar{Y}$ provides an unbiased estimate of μ. By unbiased, we mean that if we sampled the population repeatedly (infinitely many times) and computed the arithmetic mean of each sample (regardless of sample size), the grand average of this set of arithmetic means should equal μ. However, our initial estimates of variance and standard deviation are not unbiased estimators of σ^2 and σ, respectively. In particular, Equation 3.7 consistently underestimates the actual variance of the population.

The bias in Equation 3.7 can be illustrated with a simple thought experiment. Suppose you draw a single observation Y_1 from a population and try to estimate μ and $\sigma^2(Y)$. Your estimate of μ is the average of your observations, which in this case is simply Y_1 itself. However, if you estimate $\sigma^2(Y)$ using Equation 3.7, your answer will always equal 0.0 because your lone observation is the same as your estimate of the mean! The problem is that, with a sample size of 1, we have already used our data to estimate μ, and we effectively have no additional information to estimate $\sigma^2(Y)$.

This leads directly to the concept of **degrees of freedom**. The degrees of freedom represent the number of independent pieces of information that we have in a dataset for estimating statistical parameters. In a dataset of sample size 1, we do not have enough independent observations that can be used to estimate the variance.

The unbiased estimate of the variance, referred to as the **sample variance**, is calculated by dividing the sums of squares by $(n - 1)$ instead of dividing by n. Hence, the unbiased estimate of the variance is

$$s^2 = \frac{1}{n-1}\sum (Y_i - \bar{Y})^2 \quad (3.9)$$

and the unbiased estimate of the standard deviation, referred to as the **sample standard deviation,**[5] is

$$s = \sqrt{\frac{1}{n-1}\sum (Y_i - \bar{Y})^2} \quad (3.10)$$

Equations 3.9 and 3.10 adjust for the degrees of freedom in the calculation of the sample variance and the standard deviation. These equations also illustrate that you need at least two observations to estimate the variance of a distribution.

For the tibial spine data given in Table 3.1, $s^2 = 0.0017$ and $s = 0.0417$.

The Standard Error of the Mean

Another measure of spread, used frequently by ecologists and environmental scientists, is the **standard error of the mean**. This measure of spread is abbreviated as $s_{\bar{Y}}$ and is calculated by dividing the sample standard deviation by the square root of the sample size:

$$s_{\bar{Y}} = \frac{s}{\sqrt{n}} \quad (3.11)$$

[5] This unbiased estimator of the standard deviation is itself unbiased only for relatively large sample sizes ($n > 30$). For smaller sample sizes, Equation 3.10 modestly tends to underestimate the population value of σ (Gurland and Tripathi 1971). Rohlf and Sokal (1995) provide a look-up table of correction factors by which s should be multiplied if $n < 30$. In practice, most biologists do not apply these corrections. As long as the sample sizes of the groups being compared are not greatly different, no serious harm is done by ignoring the correction to s.

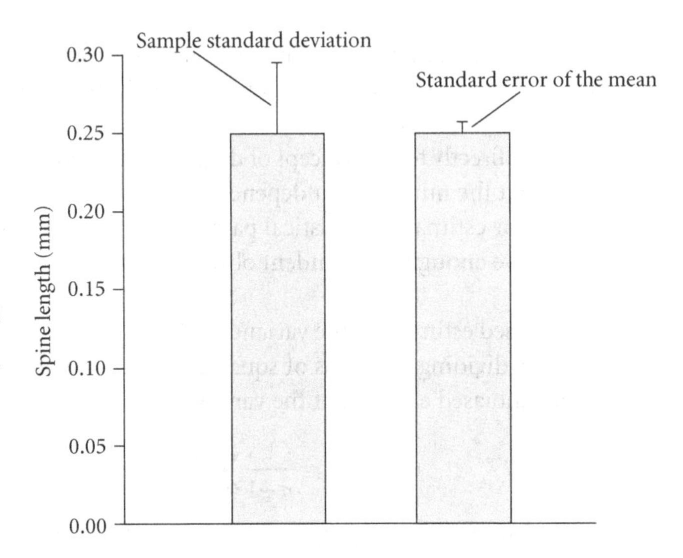

Figure 3.3 Bar chart showing the arithmetic mean for the spine data in Table 3.1 ($n = 50$), along with error bars indicating the sample standard deviation (left bar) and standard error of the mean (right bar). Whereas the standard deviation measures the variability of the individual measurements about the mean, the standard error measures the variability of the estimate of the mean itself. The standard error equals the standard deviation divided by $\sqrt{n}$, so it will always be smaller than the standard deviation, often considerably so. Figure legends and captions should always provide sample sizes and indicate clearly whether the standard deviation or the standard error has been used to construct error bars.

The Law of Large Numbers proves that for an infinitely large number of observations, $\Sigma Y_i/n$ approximates the population mean μ, where $Y_n = \{Y_i\}$ is a sample of size n from a random variable Y with expected value $E(Y)$. Similarly, the variance of $Y_n = \sigma^2/n$. Because the standard deviation is simply the square root of the variance, the standard deviation of Y_n is

$$\sqrt{\frac{\sigma^2}{n}} = \frac{\sigma}{\sqrt{n}}$$

which is the same as the standard error of the mean. Therefore, the standard error of the mean is an estimate of the standard deviation of the population mean μ.

Unfortunately, some scientists do not understand the distinction between the standard deviation (abbreviated in figure legends as SD) and the standard error of the mean (abbreviated as SE).[6] Because the standard error of the mean is always smaller than the sample standard deviation, means reported with standard errors appear less variable than those reported with standard deviations (Figure 3.3). However, the decision as to whether to present the sample standard deviation s or the standard error of the mean $s_{\bar{Y}}$ depends on what

[6] You may have noticed that we referred to the standard error of the mean, and not simply the standard error. The standard error of the mean is equal to the standard deviation of a set of means. Similarly, we could compute the standard deviation of a set of variances or other summary statistics. Although it is uncommon to see other standard errors in ecological and environmental publications, there may be times when you need to report, or at least consider, other standard errors. In Figure 3.1, the standard error of the median = $1.2533 \times s_{\bar{Y}}$, and the standard error of the standard deviation = $0.7071 \times s_{\bar{Y}}$. Sokal and Rohlf (1995) provide formulas for standard errors of other common statistics.

inference you want the reader to draw. If your conclusions based on a single sample are representative of the entire population, then report the standard error of the mean. On the other hand, if the conclusions are limited to the sample at hand, it is more honest to report the sample standard deviation. Broad observational surveys covering large spatial scales with large number of samples more likely are representative of the entire population of interest (hence, report $s_{\bar{Y}}$), whereas small, controlled experiments with few replicates more likely are based on a unique (and possibly unrepresentative) group of individuals (hence, report s).

We advocate the reporting of the sample standard deviation, s, which more accurately reflects the underlying variability of the actual data and makes fewer claims to generality. However, as long as you provide the sample size in your text, figure, or figure legend, readers can compute the standard error of the mean from the sample standard deviation and vice versa.

Skewness, Kurtosis, and Central Moments

The standard deviation and the variance are special cases of what statisticians (and physicists) call **central moments.** A central moment (*CM*) is the average of the deviations of all observations in a dataset from the mean of the observations, raised to a power r:

$$CM = \frac{1}{n}\sum_{i=1}^{n}(Y_i - \bar{Y})^r \tag{3.12}$$

In Equation 3.12, n is the number of observations, Y_i is the value of each individual observation, $\bar{Y}$ is the arithmetic mean of the n observations, and r is a positive integer. The first central moment ($r = 1$) is the sum of the differences of each observation from the sample average (arithmetic mean), which always equals 0. The second central moment ($r = 2$) is the variance (Equation 3.5).

The third central moment ($r = 3$) divided by the standard deviation cubed (s^3) is called the **skewness** (denoted as g_1):

$$g_1 = \frac{1}{ns^3}\sum_{i=1}^{n}(Y_i - \bar{Y})^3 \tag{3.13}$$

Skewness describes how the sample differs in shape from a symmetrical distribution. A normal distribution has $g_1 = 0$. A distribution for which $g_1 > 0$ is **right-skewed**: there is a long tail of observations greater than (i.e., to the right of) the mean. In contrast, $g_1 < 0$, is **left-skewed**: there is a long tail of observations less than (i.e., to the left of) the mean (Figure 3.4).

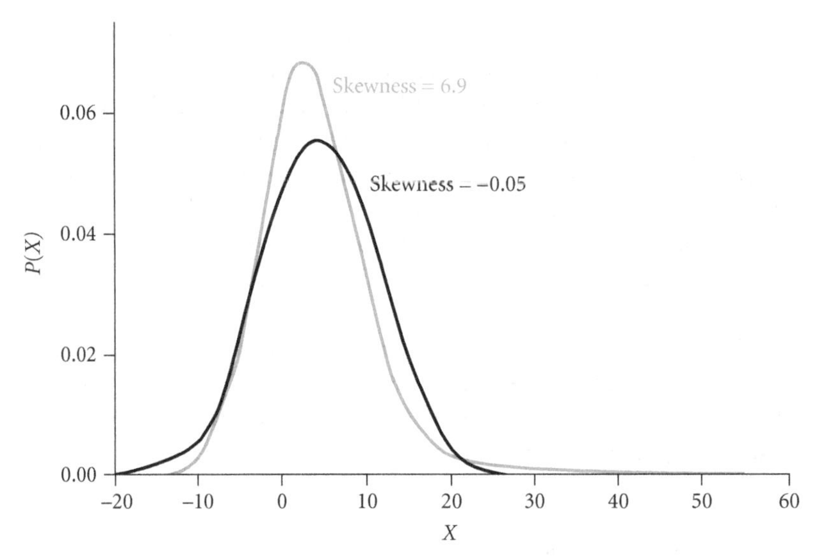

Figure 3.4 Continuous distributions illustrating skewness (g_1). Skewness measures the extent to which a distribution is asymmetric, with either a long right- or left-hand probability tail. The blue curve is the log-normal distribution illustrated in Figure 2.8; it has positive skewness, with many more observations to the right of the mean than to the left (a long right tail), and a skewness measure of 6.9. The black curve represents a sample of 1000 observations from a normal random variable with identical mean and standard deviation as the log-normal distribution. Because these data were drawn from a symmetric normal distribution, they have approximately the same number of observations on either side of the mean, and the measured skewness is approximately 0.

The **kurtosis** is based on the fourth central moment ($r = 4$):

$$g_2 = \left[\frac{1}{ns^4}\sum_{i=1}^{n}(Y_i - \overline{Y})^4\right] - 3 \quad \textbf{(3.14)}$$

Kurtosis measures the extent to which a probability density is distributed in the tails versus the center of the distribution. Clumped or **platykurtic** distributions have $g_2 < 0$; compared to a normal distribution, there is more probability mass in the center of the distribution, and less probability in the tails. In contrast, **leptokurtic** distributions have $g_2 > 0$. Leptokurtic distributions have less probability mass in the center, and relatively fat probability tails (Figure 3.5).

Although skewness and kurtosis were often reported in the ecological literature prior to the mid-1980s, it is uncommon to see these values reported now. Their statistical properties are not good: they are very sensitive to outliers, and

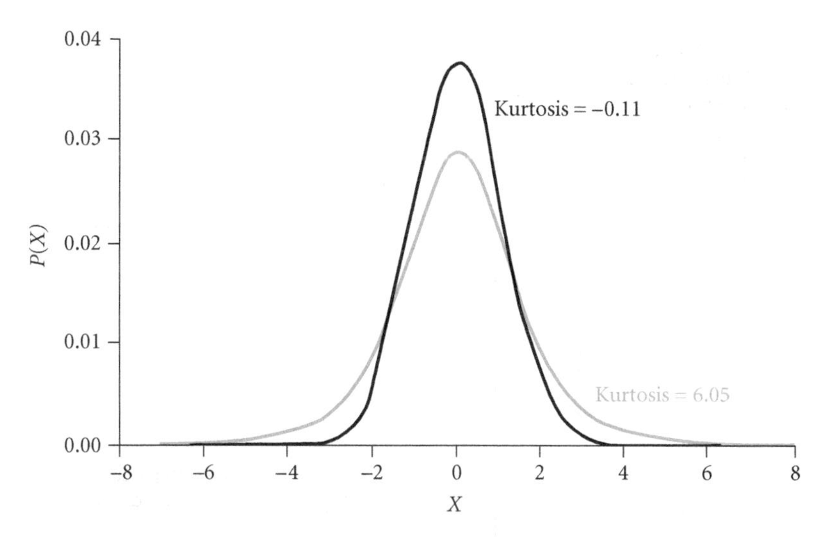

Figure 3.5 Distributions illustrating kurtosis (g_2). Kurtosis measures the extent to which the distribution is fat-tailed or thin-tailed compared to a standard normal distribution. Fat-tailed distributions are leptokurtic, and contain relative more area in the tails of the distribution and less in the center. Leptokurtic distributions have positive values for g_2. Thin-tailed distributions are platykurtic, and contain relatively less area in the tails of the distribution and more in the center. Platykurtic distributions have negative values for g_2. The black curve represents a sample of 1000 observations from a normal random variable with mean = 0 and standard deviation = 1 ($X \sim N(0,1)$); its kurtosis is nearly 0. The blue curve is a sample of 1000 observations from a *t* distribution with 3 degrees of freedom. The *t* distribution is leptokurtic and has a positive kurtosis ($g_2 = 6.05$ in this example).

to differences in the mean of the distribution. Weiner and Solbrig (1984) discuss the problem of using measures of skewness in ecological studies.

Quantiles

Another way to illustrate the spread of a distribution is to report its **quantiles.** We are all familiar with one kind of quantile, the **percentile**, because of its use in standardized testing. When a test score is reported as being in the 90th percentile, 90% of the scores are lower than the one being reported, and 10% are above it. Earlier in this chapter we saw another example of a percentile—the median, which is the value located at the 50th percentile of the data. In presentations of statistical data, we most commonly report upper and lower **quartiles**—the values for the 25th and 75th percentiles—and upper and lower **deciles**—the values for the 10th and 90th percentiles. These values for the spine data are illustrated concisely in a **box plot** (Figure 3.6). Unlike the variance and standard

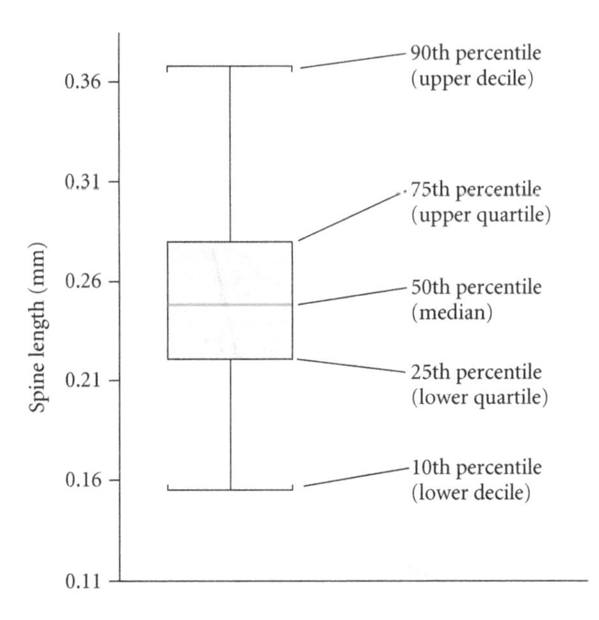

Figure 3.6 Box plot illustrating quantiles of data from Table 3.1 ($n = 50$). The line indicates the 50th percentile (median), and the box encompasses 50% of the data, from the 25th to the 75th percentile. The vertical lines extend from the 10th to the 90th percentile.

deviation, the values of the quantiles do not depend on the values of the arithmetic mean or median. When distributions are asymmetric or contain **outliers** (extreme data points that are not characteristic of the distribution they were sampled from; see Chapter 8), box plots of quantiles can portray the distribution of the data more accurately than conventional plots of means and standard deviations.

Using Measures of Spread

By themselves, measures of spread are not especially informative. Their primary utility is for comparing data from different populations or from different treatments within experiments. For example, analysis of variance (Chapter 10) uses the values of the sample variances to test hypotheses that experimental treatments differ from one another. The familiar *t*-test uses sample standard deviations to test hypotheses that the means of two populations differ from each other. It is not straightforward to compare variability itself across populations or treatment groups because the variance and standard deviation depend on the sample mean. However, by discounting the standard deviation by the mean, we can calculate an independent measure of variability, called the **coefficient of variation**, or *CV*.

The *CV* is simply the sample standard deviation divided by the mean, $s/\bar{Y}$, and is conventionally multiplied by 100 to give a percentage. The *CV* for our spine data = 16.5%. If another population of spiders had a *CV* of tibial spine length = 25%, we would say that our first population is somewhat less variable than the second population.

A related index is the **coefficient of dispersion**, which is calculated as the sample variance divided by the mean ($s^2/\bar{Y}$). The coefficient of dispersion can be used with discrete data to assess whether individuals are clumped or hyperdispersed in space, or whether they are spaced at random as predicted by a Poisson distribution. Biological forces that violate independence will cause observed distributions to differ from those predicted by the Poisson.

For example, some marine invertebrate larvae exhibit an aggregated settling response: once a juvenile occupies a patch, that patch becomes very attractive as a settlement surface for subsequent larvae (Crisp 1979). Compared to a Poisson distribution, these aggregated or clumped distributions will tend to have too many samples with high numbers of occurrence, *and* too many samples with 0 occurrences. In contrast, many ant colonies exhibit strong territoriality and kill or drive off other ants that try to establish colonies within the territory (Levings and Traniello 1981). This segregative behavior also will push the distribution away from the Poisson. In this case the colonies will be hyperdispersed: there will be too few samples in the 0 frequency class *and* too few samples with high numbers of occurrence.

Because the variance and mean of a Poisson random variable both equal λ, the coefficient of dispersion (CD) for a Poisson random variable = $\lambda/\lambda = 1$. On the other hand, if the data are clumped or aggregated, $CD > 1.0$, and if the data are hyperdispersed or segregated, $CD < 1.0$. However, the analysis of spatial pattern with Poisson distributions can become complicated because the results depend not only on the degree of clumping or segregation of the organisms, but also on the size, number, and placement of the sampling units. Hurlbert (1990) discusses some of the issues involved in fitting spatial data to a Poisson distribution.

Some Philosophical Issues Surrounding Summary Statistics

The fundamental summary statistics—the sample mean, standard deviation, and variance—are estimates of the actual population-level **parameters**, μ, σ, and σ^2, which we obtain directly from our data. Because we can never sample the entire population, we are forced to estimate these unknown parameters by $\bar{Y}$, s, and s^2. In doing so, we make a fundamental assumption: that there is a true fixed value for each of these parameters. The Law of Large Numbers proves that if we sampled our population infinitely many times, the average of the infinitely many $\bar{Y}$'s that we calculated from our infinitely many samples would equal μ.

The Law of Large Numbers forms the foundation for what has come to be known as **parametric, frequentist**, or **asymptotic statistics**. Parametric statistics are so called because the assumption is that the measured variable can be described by a random variable or probability distribution of known form with defined, fixed parameters. Frequentist or asymptotic statistics are so called because we assume that if the experiment were repeated infinitely many times, the most frequent estimates of the parameters would converge on (reach an asymptote at) their true values.

But what if this fundamental assumption—that the underlying parameters have true, fixed values—is false? For example, if our samples were taken over a long period of time, there might be changes in spine length of spider tibias because of phenotypic plasticity in growth, or even evolutionary change due to natural selection. Or, perhaps our samples were taken over a short period of time, but each spider came from a distinct microhabitat, for which there was a unique expectation and variance of spider tibia length. In such a case, is there any real meaning to an estimate of a single value for the average length of a tibial spine in the spider population? Bayesian statistics begin with the fundamental assumption that population-level parameters such as μ, σ, and σ^2 are themselves random variables. A Bayesian analysis produces estimates not only of the values of the parameters but also of the inherent variability in these parameters.

The distinction between the frequentist and Bayesian approaches is far from trivial, and has resulted in many years of acrimonious debate, first among statisticians, and more recently among ecologists. As we will see in Chapter 5, Bayesian estimates of parameters as random variables often require complex computer calculations. In contrast, frequentist estimates of parameters as fixed values use simple formulas that we have outlined in this chapter. Because of the computational complexity of the Bayesian estimates, it was initially unclear whether the results of frequentist and Bayesian analyses would be quantitatively different. However, with fast computers, we are now able to carry out complex Bayesian analyses. Under certain conditions, the results of the two types of analyses are quantitatively similar. The decision of which type of analysis to use, therefore, should be based more on a philosophical standpoint than on a quantitative outcome (Ellison 2004). However, the interpretation of statistical results may be quite different from the Bayesian and frequentist perspectives. An example of such a difference is the construction and interpretation of confidence intervals for parameter estimates.

Confidence Intervals

Scientists often use the sample standard deviation to construct a **confidence interval** around the mean (see Figure 3.1). For a normally distributed random variable, approximately 67% of the observations occur within ±1 standard deviation of the mean, and approximately 96% of the observations occur within ±2 standard deviations of the mean.[7] We use this observation to create a 95% confidence interval, which for large samples is the interval bounded by $(\overline{Y} - 1.96s_{\overline{Y}}, \overline{Y} + 1.96s_{\overline{Y}})$. What does this interval represent? It means that the probability that the true population mean μ falls within the confidence interval = 0.95:

$$P(\overline{Y} - 1.96s_{\overline{Y}} \leq \mu \leq \overline{Y} + 1.96s_{\overline{Y}}) = 0.95 \quad (3.15)$$

Because our sample mean and sample standard error of the mean are derived from a single sample, this confidence interval will change if we sample the population again (although if our sampling is random and unbiased, it should not change by very much). Thus, this expression is asserting that the probability that the true population mean μ falls within a single calculated confidence interval = 0.95. By extension, if we were to repeatedly sample the population (keeping the sample size constant), 5% of the time we would expect that the true population mean μ would lie outside of this confidence interval.

Interpreting a confidence interval is tricky. A common misinterpretation of the confidence interval is "There is a 95% chance that the true population mean μ occurs within this interval." Wrong. The confidence interval either does or does not contain μ; unlike Schrödinger's quantum cat (see Footnote 9 in Chapter 1), μ cannot be both in and out of the confidence interval simultaneously. What you can say is that 95% of the time, an interval calculated in this way will contain the fixed value of μ. Thus, if you carried out your sampling experiment 100 times, and created 100 such confidence intervals, approximately 95 of them would contain μ and 5 would not (see Figure 3.1 for an example of when a 95% confidence interval does not include the true population mean μ). Blume and Royall (2003) provide further examples and a more detailed pedagogical description.

[7] Use the "two standard deviation rule" when you read the scientific literature, and get into the habit of quickly estimating rough confidence intervals for sample data. For example, suppose you read in a paper that average nitrogen content of a sample of plant tissues was 3.4% ± 0.2, where 0.2 is the sample standard deviation. Two standard deviations = 0.4, which is then added to and subtracted from the mean. Therefore, approximately 95% of the observations were between 3.0% and 3.8%. You can use this same trick when you examine bar graphs in which the standard deviation is plotted as an error bar. This is an excellent way to use summary statistics to spot check reported statistical differences among groups.

This rather convoluted explanation is not satisfying, and it is not exactly what you would like to assert when you construct a confidence interval. Intuitively, you would like to be saying how confident you are that the mean is inside of your interval (i.e., you're 95% sure that the mean is in the interval). A frequentist statistician, however, can't assert that. If there is a fixed population mean μ, then it's either inside the interval or not, and the probability statement (Equation 3.15) asserts how probable it is that this particular confidence interval includes μ. On the other hand, a Bayesian statistician turns this around. Because the confidence interval is fixed (by your sample data), a Bayesian statistician can calculate the probability that the population mean (itself a random variable) occurs within the confidence interval. Bayesian statisticians refer to these intervals as **credibility intervals**, in order to distinguish them from frequentist confidence intervals. See Chapter 5, Ellison (1996), and Ellison and Dennis (2010) for further details.

Generalized Confidence Intervals

We can, of course, construct any percentile confidence interval, such as a 90% confidence interval or a 50% confidence interval. The general formula for an n% confidence interval is

$$P(\bar{Y} - t_{\alpha[n-1]}s_{\bar{Y}} \leq \mu \leq \bar{Y} + t_{\alpha[n-1]}s_{\bar{Y}}) = (1-\alpha) \quad \textbf{(3.16)}$$

where $t_{\alpha[n-1]}$ is the critical value of a ***t*-distribution** with probability $P = \alpha$, and sample size n. This probability expresses the percentage of the area under the two tails of the curve of a t-distribution (Figure 3.7). For a standard normal curve (a t-distribution with $n = \infty$), 95% of the area under the curve lies within ± 1.96 standard deviations of the mean. Thus 5% of the area ($P = 0.05$) under the curve remains in the two tails beyond the points ± 1.96.

So what is this t-distribution? Recall from Chapter 2 that an arithmetic transformation of a normal random variable is itself a normal random variable. Consider the set of sample means $\{\bar{Y}_k\}$ resulting from a set of replicate groups of measurements of a normal random variable with unknown mean μ. The deviation of the sample means from the population mean $\bar{Y}_k - \mu$ is also a normal random variable. If this latter random variable $(\bar{Y}_k - \mu)$ is divided by the unknown population standard deviation σ, the result is a standard normal random variable (mean = 0, standard deviation = 1). However, we don't know the population standard deviation σ, and must instead divide the deviations of each

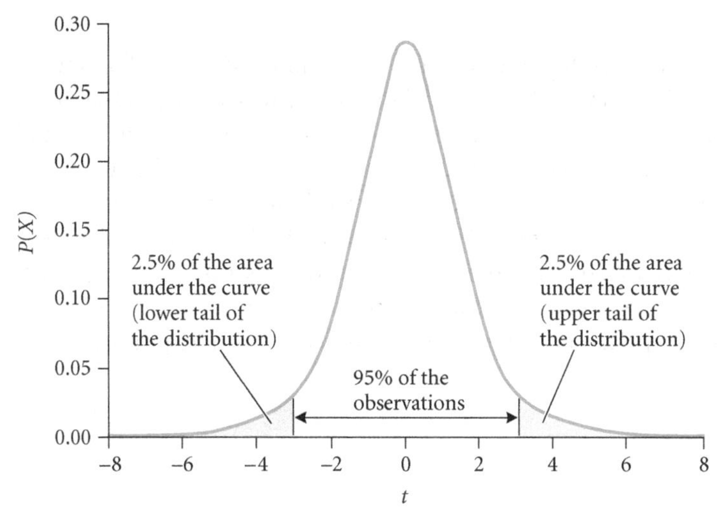

Figure 3.7 *t*-distribution illustrating that 95% of the observations, or probability mass lies within ±1.96 standard deviations of the mean (mean = 0) percentiles. The two tails of the distribution each contain 2.5% of the observations or probability mass of the distribution. Their sum is 5% of the observations, and the probability $P = 0.05$ that an observation falls within these tails. This distribution is identical to the *t*-distribution illustrated in Figure 3.5.

mean by the estimate of the standard error of the mean of each sample ($s_{\bar{Y}_k}$). The resulting *t*-distribution is similar, but not identical, to a standard normal distribution. This *t*-distribution is leptokurtic, with longer and heavier tails than a standard normal distribution.[8]

[8] This result was first demonstrated by the statistician W. S. Gossett, who published it using the pseudonym "Student." Gossett at the time was employed at the Guinness Brewery, which did not allow its employees to publish trade secrets; hence the need for a pseudonym. This modified standard normal distribution, which Gossett called a *t*-distribution, is also known as the Student's distribution, or the Student's *t*-distribution. As the number of samples increases, the *t*-distribution approaches the standard normal distribution in shape. The construction of the *t*-distribution requires the specification of the sample size n and is normally written as $t_{[n]}$. For $n = \infty$, $t_{[\infty]} \sim N(0,1)$.

Because a *t*-distribution for small n is leptokurtic (see Figure 3.5), the width of a confidence interval constructed from it will shrink as sample size increases. For example, for $n = 10$, 95% of the area under the curve falls between ±2.228. For $n = 100$, 95% of the area under the curve falls between ±1.990. The resulting confidence interval is 12% wider for $n = 10$ than for $n = 100$.

Summary

Summary statistics describe expected values and variability of a random sample of data. Measures of location include the median, mode, and several means. If the samples are random or independent, the arithmetic mean is an unbiased estimator of the expected value of a normal distribution. The geometric mean and harmonic mean are also used in special circumstances. Measures of spread include the variance, standard deviation, standard error, and quantiles. These measures describe the variation of the observations around the expected value. Measures of spread can also be used to construct confidence or credibility intervals. The interpretations of these intervals differ between frequentist and Bayesian statisticians.

CHAPTER 4

Framing and Testing Hypotheses

Hypotheses are potential explanations that can account for our observations of the external world. They usually describe cause-and-effect relationships between a proposed mechanism or process (the cause) and our observations (the effect). Observations are data—what we see or measure in the real world. Our goal in undertaking a scientific study is to understand the cause(s) of observable phenomena. Collecting observations is a means to that end: we accumulate different kinds of observations and use them to distinguish among different possible causes. Some scientists and statisticians distinguish between observations made during manipulative experiments and those made during **observational** or **correlative studies**. However, in most cases, the statistical treatment of such data is identical. The distinction lies in the confidence we can place in **inferences**[1] drawn from those studies. Well-designed manipulative experiments allow us to be confident in our inferences; we have less confidence in data from poorly designed experiments, or from studies in which we were unable to directly manipulate variables.

If observations are the "what" of science, hypotheses are the "how." Whereas observations are taken from the real world, hypotheses need not be. Although our observations may suggest hypotheses, hypotheses can also come from the existing body of scientific literature, from the predictions of theoretical models, and from our own intuition and reasoning. However, not all descriptions of cause-and-effect relationships constitute valid scientific hypotheses. A scientific hypothesis must be testable: in other words, there should be some set of additional observations or experimental results that we could collect that would cause

[1] In logic, an *inference* is a conclusion that is derived from premises. Scientists make inferences (draw conclusions) about causes based on their data. These conclusions may be suggested, or implied, by the data. But remember that it is the scientist who infers, and the data that imply.

us to modify, reject, or discard our working hypothesis.[2] Metaphysical hypotheses, including the activities of omnipotent gods, do not qualify as scientific hypotheses because these explanations are taken on faith, and there are no observations that would cause a believer to reject these hypotheses.[3]

In addition to being testable, a good scientific hypothesis should generate novel predictions. These predictions can then be tested by collecting additional observations. However, the same set of observations may be predicted by more than one hypothesis. Although hypotheses are chosen to account for our initial observations, a good scientific hypothesis also should provide a unique set of predictions that do not emerge from other explanations. By focusing on these unique predictions, we can collect more quickly the critical data that will discriminate among the alternatives.

Scientific Methods

The "scientific method" is the technique used to decide among hypotheses on the basis of observations and predictions. Most textbooks present only a single scientific method, but practicing scientists actually use several methods in their work.

[2] A scientific *hypothesis* refers to a particular mechanism or cause-and-effect relationship; a scientific *theory* is much broader and more synthetic. In its early stages, not all elements of a scientific theory may be fully articulated, so that explicit hypotheses initially may not be possible. For example, Darwin's theory of natural selection required a mechanism of inheritance that conserved traits from one generation to the next while still preserving variation among individuals. Darwin did not have an explanation for inheritance, and he discussed this weakness of his theory in *The Origin of Species* (1859). Darwin did not know that precisely such a mechanism had in fact been discovered by Gregor Mendel in his experimental studies (ca. 1856) of inheritance in pea plants. Ironically, Darwin's grandfather, Erasmus Darwin, had published work on inheritance two generations earlier, in his *Zoonomia, or the Laws of Organic Life* (1794–1796). However, Erasmus Darwin used snapdragons as his experimental organism, whereas Mendel used pea plants. Inheritance of flower color is simpler in pea plants than in snapdragons, and Mendel was able to recognize the particulate nature of genes, whereas Erasmus Darwin could not.

[3] Although many philosophies have attempted to bridge the gap between science and religion, the contradiction between reason and faith is a critical fault line separating the two. The early Christian philosopher Tertullian (~155–222 AD) seized upon this contradiction and asserted "*Credo quai absurdum est*" ("I believe because it is absurd"). In Tertullian's view, that the son of God died is to be believed because it is contradictory; and that he rose from the grave has certitude because it is impossible (Reese 1980).

Deduction and Induction

Deduction and induction are two important modes of scientific reasoning, and both involve drawing inferences from data or models. **Deduction** proceeds from the general case to the specific case. The following set of statements provides an example of classic deduction:

1. All of the ants in the Harvard Forest belong to the genus *Myrmica.*
2. I sampled this particular ant in the Harvard Forest.
3. This particular ant is in the genus *Myrmica.*

Statements 1 and 2 are usually referred to as the *major and minor premises*, and statement 3 is the *conclusion*. The set of three statements is called a **syllogism**, an important logical structure developed by Aristotle. Notice that the sequence of the syllogism proceeds from the general case (all of the ants in the Harvard Forest) to the specific case (the particular ant that was sampled).

In contrast, **induction** proceeds from the specific case to the general case:[4]

1. All 25 of these ants are in the genus *Myrmica.*
2. All 25 of these ants were collected in the Harvard Forest.
3. All of the ants in the Harvard Forest are in the genus *Myrmica.*

Sir Francis Bacon

[4] The champion of the inductive method was Sir Francis Bacon (1561–1626), a major legal, philosophical, and political figure in Elizabethan England. He was a prominent member of parliament, and was knighted in 1603. Among scholars who question the authorship of Shakespeare's works (the so-called anti-Stratfordians), some believe Bacon was the true author of Shakespeare's plays, but the evidence isn't very compelling. Bacon's most important scientific writing is the *Novum organum* (1620), in which he urged the use of induction and empiricism as a way of knowing the world. This was an important philosophical break with the past, in which explorations of "natural philosophy" involved excessive reliance on deduction and on published authority (particularly the works of Aristotle). Bacon's inductive method paved the way for the great scientific breakthroughs by Galileo and Newton in the Age of Reason. Near the end of his life, Bacon's political fortunes took a turn for the worse; in 1621 he was convicted of accepting bribes and was removed from office. Bacon's devotion to empiricism eventually did him in. Attempting to test the hypothesis that freezing slows the putrefaction of flesh, Bacon ventured out in the cold during the winter of 1626 to stuff a chicken with snow. He became badly chilled and died a few days later at the age of 65.

Some philosophers define deduction as certain inference and induction as probable inference. These definitions certainly fit our example of ants collected in the Harvard Forest. In the first set of statements (deduction), the conclusion must be logically true if the first two premises are true. But in the second case (induction), although the conclusion is *likely* to be true, it may be false; our confidence will increase with the size of our sample, as is always the case in statistical inference. Statistics, by its very nature, is an inductive process: we are always trying to draw general conclusions based on a specific, limited sample.

Both induction and deduction are used in all models of scientific reasoning, but they receive different emphases. Even using the inductive method, we probably will use deduction to derive specific predictions from the general hypothesis in each turn of the cycle.

Any scientific inquiry begins with an observation that we are trying to explain. The inductive method takes this observation and develops a single hypothesis to explain it. Bacon himself emphasized the importance of using the data to suggest the hypothesis, rather than relying on conventional wisdom, accepted authority, or abstract philosophical theory. Once the hypothesis is formulated, it generates—through deduction—further predictions. These predictions are then tested by collecting additional observations. If the new observations match the predictions, the hypothesis is supported. If not, the hypothesis must be modified to take into account both the original observation and the new observations. This cycle of hypothesis–prediction–observation is repeatedly traversed. After each cycle, the modified hypothesis should come closer to the truth[5] (Figure 4.1).

Two advantages of the inductive method are (1) it emphasizes the close link between data and theory; and (2) it explicitly builds and modifies hypotheses based on previous knowledge. The inductive method is *confirmatory* in that we

[5] Ecologists and environmental scientists rely on induction when they use statistical software to fit non-linear (curvy) functions to data (see Chapter 9). The software requires that you specify not only the equation to be fit, but also a set of initial values for the unknown parameters. These initial values need to be "close to" the actual values because the algorithms are *local* estimators (i.e., they solve for local minima or maxima in the function). Thus, if the initial estimates are "far away" from the actual values of the function, the curve-fitting routines may either fail to converge on a solution, or will converge on a non-sensical one. Plotting the fitted curve along with the data is a good safeguard to confirm that the curve derived from the estimated parameters actually fits the original data.

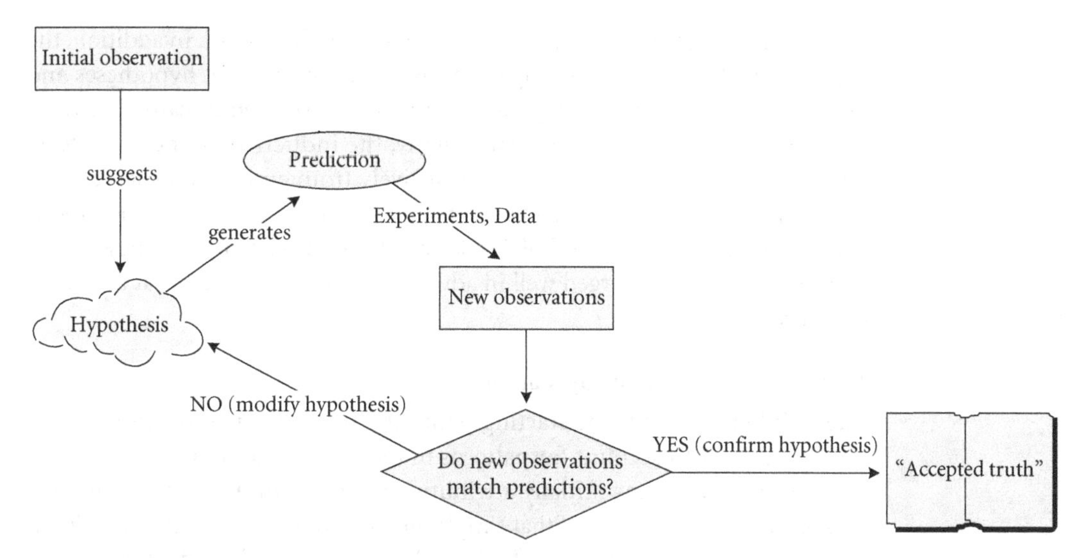

Figure 4.1 The inductive method. The cycle of hypothesis, prediction, and observation is repeatedly traversed. Hypothesis confirmation represents the theoretical endpoint of the process. Compare the inductive method to the hypothetico-deductive method (see Figure 4.4), in which multiple working hypotheses are proposed and emphasis is placed on falsification rather than verification.

seek data that support the hypothesis, and then we modify the hypothesis to conform with the accumulating data.[6]

There are also several disadvantages of the inductive method. Perhaps the most serious is that the inductive method considers only a single starting hypothesis; other hypotheses are only considered later, in response to additional data and observations. If we "get off on the wrong foot" and begin exploring an incorrect hypothesis, it may take a long time to arrive at the correct answer

Robert H. MacArthur

[6] The community ecologist Robert H. MacArthur (1930–1972) once wrote that the group of researchers interested in making ecology a science "arranges ecological data as examples testing the proposed theories and spends most of its time patching up the theories to account for as many of the data as possible" (MacArthur 1962). This quote characterizes much of the early theoretical work in community ecology. Later, theoretical ecology developed as a discipline in its own right, and some interesting lines of research blossomed without any reference to data or the real world. Ecologists disagree about whether such a large body of purely theoretical work has been good or bad for our science (Pielou 1981; Caswell 1988).

through induction. In some cases we may never get there at all. In addition, the inductive method may encourage scientists to champion pet hypotheses and perhaps hang on to them long after they should have been discarded or radically modified (Loehle 1987). And finally, the inductive method—at least Bacon's view of it—derives theory exclusively from empirical observations. However, many important theoretical insights have come from theoretical modeling, abstract reasoning, and plain old intuition. Important hypotheses in all sciences have often emerged well in advance of the critical data that are needed to test them.[7]

Modern-Day Induction: Bayesian Inference

The **null hypothesis** is the starting point of a scientific investigation. A null hypothesis tries to account for patterns in the data in the simplest way possible, which often means initially attributing variation in the data to randomness or measurement error. If that simple null hypothesis can be rejected, we can move on to entertain more complex hypotheses.[8] Because the inductive method begins with an observation that suggests an hypothesis, how do we generate an appropriate null hypothesis? Bayesian inference represents a modern, updated version of the inductive method. The principals of Bayesian inference can be illustrated with a simple example.

The photosynthetic response of leaves to increases in light intensity is a well-studied problem. Imagine an experiment in which we grow 15 mangrove seedlings, each under a different light intensity (expressed as photosynthetic photon flux density, or PPFD, in μmol photons per m^2 of leaf tissue exposed to

[7] For example, in 1931 the Austrian physicist Wolfgang Pauli (1900–1958) hypothesized the existence of the neutrino, an electrically neutral particle with negligible mass, to account for apparent inconsistencies in the conservation of energy during radioactive decay. Empirical confirmation of the existence of neutrino did not come until 1956.

Sir William of Ockham

[8] The preference for simple hypotheses over complex ones has a long history in science. Sir William of Ockham's (1290–1349) Principle of Parsimony states that "[Entities] are not to be multiplied beyond necessity." Ockham believed that unnecessarily complex hypotheses were vain and insulting to God. The Principle of Parsimony is sometimes known as Ockham's Razor, the razor shearing away unnecessary complexity. Ockham lived an interesting life. He was educated at Oxford and was a member of the Franciscan order. He was charged with heresy for some of the writing in his Master's thesis. The charge was eventually dropped, but when Pope John XXII challenged the Franciscan doctrine of apostolic poverty, Ockham was excommunicated and fled to Bavaria. Ockham died in 1349, probably a victim of the bubonic plague epidemic.

light each second) and measure the photosynthetic response of each plant (expressed as μmol CO_2 fixed per m^2 of leaf tissue exposed to light each second). We then plot the data with light intensity on the x-axis (the *predictor variable*) and photosynthetic rate on the y-axis (the *response variable*). Each point represents a different leaf for which we have recorded these two numbers.

In the absence of any information about the relationship between light and photosynthetic rates, the simplest null hypothesis is that there is no relationship between these two variables (Figure 4.2). If we fit a line to this null hypothesis, the slope of the line would equal 0. If we collected data and found some other relationship between light availability and photosynthetic rate, we would then use those data to modify our hypothesis, following the inductive method.

But is it really necessary to frame the null hypothesis as if you had no information at all? Using just a bit of knowledge about plant physiology, we can formulate a more realistic initial hypothesis. Specifically, we expect there to be some

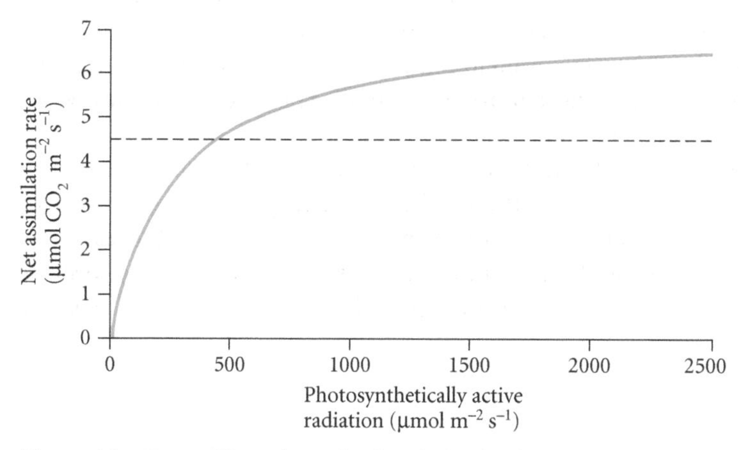

Figure 4.2 Two null hypotheses for the relationship between light intensity (measured as photosynthetically active radiation) and photosynthetic rate (measured as net assimilation rate) in plants. The simplest null hypothesis is that there is no association between the two variables (dashed line). This null hypothesis is the starting point for a hypothetico-deductive approach that assumes no prior knowledge about the relationship between the variables and is the basis for a standard linear regression model (see Chapter 9). In contrast, the blue curve represents a Bayesian approach of bringing prior knowledge to create an informed null hypothesis. In this case, the "prior knowledge" is of plant physiology and photosynthesis. We expect that the assimilation rate will rise rapidly at first as light intensity increases, but then reach an asymptote or saturation level. Such a relationship can be described by a Michaelis-Menten equation [$Y = kX/(D + X)$], which includes parameters for an asymptotic assimilation rate (k) and a half saturation constant (D) that controls the steepness of the curve. Bayesian methods can incorporate this type of prior information into the analysis.

maximum photosynthetic rate that the plant can achieve. Beyond this point, increases in light intensity will not yield additional photosynthate, because some other factor, such as water or nutrients, becomes limiting. Even if these factors were supplied and the plant were grown under optimal conditions, photosynthetic rates will still level out because there are inherent limitations in the rates of biochemical processes and electron transfers that occur during photosynthesis. (In fact, if we keep increasing light intensity, excessive light energy can damage plant tissues and reduce photosynthesis. But in our example, we limit the upper range of light intensities to those that the plant can tolerate.)

Thus, our informed null hypothesis is that the relationship between photosynthetic rate and light intensity should be non-linear, with an asymptote at high light intensities (see Figure 4.2). Real data could then be used to test the degree of support for this more realistic null hypothesis (Figure 4.3). To determine which null hypothesis to use, we also must ask what, precisely, is the point of the study? The simple null hypothesis (linear equation) is appropriate if we merely want to establish that a non-random relationship exists between light intensity and photosynthetic rate. The informed null hypothesis (Michaelis-Menten equation) is appropriate if we want to compare saturation curves among species or to test theoretical models that make quantitative predictions for the asymptote or half-saturation constant.

Figures 4.2 and 4.3 illustrate how a modern-day inductivist, or Bayesian statistician, generates an hypothesis. The Bayesian approach is to use prior knowledge or information to generate and test hypotheses. In this example, the prior knowledge was derived from plant physiology and the expected shape of the light saturation curve. However, prior knowledge might also be based

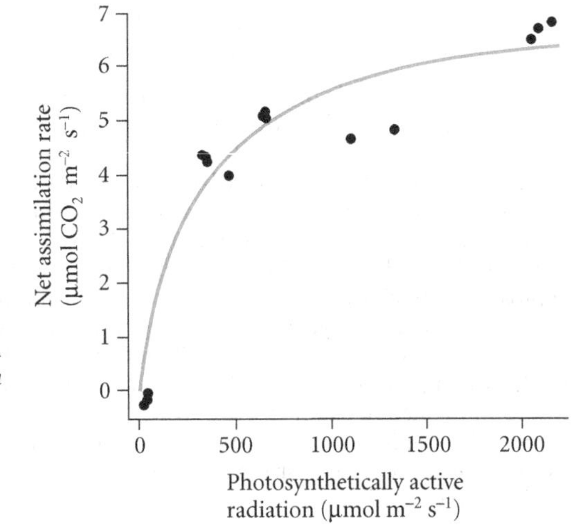

Figure 4.3 Relationship between light intensity and photosynthetic rate. The data are measurements of net assimilation rate and photosynthetically active radiation for $n = 15$ young sun leaves of the mangrove *Rhizophora mangle* in Belize (Farnsworth and Ellison 1996b). A Michaelis-Menten equation of the form $Y = kX/(D + X)$ was fit to the data. The parameter estimates $\pm$ 1 standard error are $k = 7.3 \pm 0.58$ and $D = 313 \pm 86.6$.

on the extensive base of published literature on light saturation curves (Björkman 1981; Lambers et al. 1998). If we had empirical parameter estimates from other studies, we could quantify our prior estimates of the threshold and asymptote values for light saturation. These estimates could then be used to further specify the initial hypothesis for fitting the asymptote value to our experimental data.

Use of prior knowledge in this way is different from Bacon's view of induction, which was based entirely on an individual's own experience. In a Baconian universe, if you had never studied plants before, you would have no direct evidence on the relationship between light and photosynthetic rate, and you would begin with a null hypothesis such as the flat line shown in Figure 4.2. This is actually the starting point for the hypothetico-deductive method presented in the next section.

The strict Baconian interpretation of induction is the basis of the fundamental critique of the Bayesian approach: that the prior knowledge used to develop the initial model is arbitrary and subjective, and may be biased by preconceived notions of the investigator. Thus, the hypothetico-deductive method is viewed by some as more "objective" and hence more "scientific." Bayesians counter this argument by asserting that the statistical null hypotheses and curve-fitting techniques used by hypothetico-deductivists are just as subjective; these methods only seem to be more objective because they are familiar and uncritically accepted. For a further discussion of these philosophical issues, see Ellison (1996, 2004), Dennis (1996), and Taper and Lele (2004).

The Hypothetico-Deductive Method

The **hypothetico-deductive method** (Figure 4.4) developed from the works of Sir Isaac Newton and other seventeenth-century scientists and was championed by the philosopher of science Karl Popper.[9] Like the inductive method, the hypothetico-deductive method begins with an initial observation that we are trying to explain. However, rather than positing a single hypothesis and working

Karl Popper

[9] The Austrian philosopher of science Karl Popper (1902–1994) was the most articulate champion of the hypothetico-deductive method and falsifiability as the cornerstone of science. In *The Logic of Scientific Discovery* (1935), Popper argued that falsifiability is a more reliable criterion of truth than verifiability. In *The Open Society and Its Enemies* (1945), Popper defended democracy and criticized the totalitarian implications of induction and the political theories of Plato and Karl Marx.

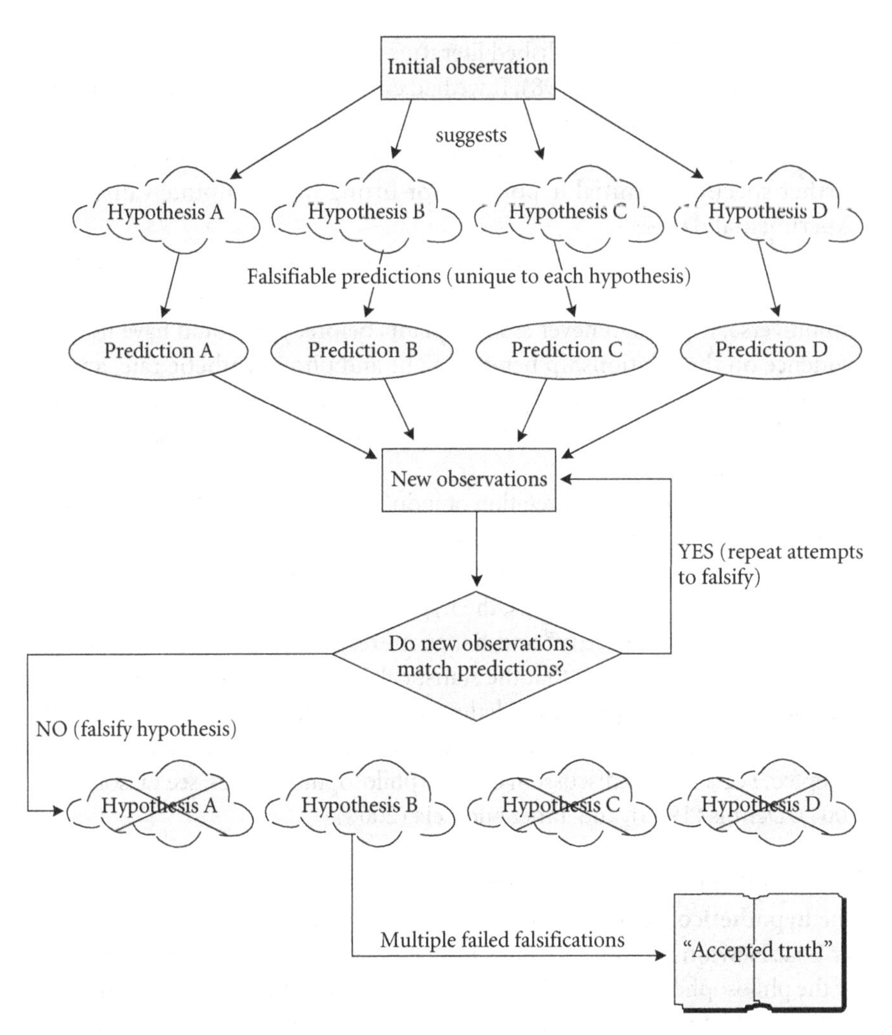

Figure 4.4 The hypothetico-deductive method. Multiple working hypotheses are proposed and their predictions tested with the goal of falsifying the incorrect hypotheses. The correct explanation is the one that stands up to repeated testing but fails to be falsified.

forward, the hypothetico-deductive method asks us to propose multiple, working hypotheses. All of these hypotheses account for the initial observation, but they each make additional unique predictions that can be tested by further experiments or observations. The goal of these tests is not to confirm, but to falsify, the hypotheses. Falsification eliminates some of the explanations, and the list is winnowed down to a smaller number of contenders. The cycle of predictions and new observations is repeated. However, the hypothetico-deductive method never confirms an hypothesis; the accepted scientific explanation is the hypothesis that successfully withstands repeated attempts to falsify it.

The two advantages of the hypothetico-deductive method are: (1) it forces a consideration of multiple working hypotheses right from the start; and (2) it highlights the key predictive differences between them. In contrast to the inductive method, hypotheses do not have to be built up from the data, but can be developed independently or in parallel with data collection. The emphasis on falsification tends to produce simple, testable hypotheses, so that parsimonious explanations are considered first, and more complicated mechanisms only later.[10]

The disadvantages of the hypothetico-deductive method are that multiple working hypotheses may not always be available, particularly in the early stages of investigation. Even if multiple hypotheses are available, the method does not really work unless the "correct" hypothesis is among the alternatives. In contrast, the inductive method may begin with an incorrect hypothesis, but can reach the correct explanation through repeated modification of the original hypothesis, as informed by data collection. Another useful distinction is that the inductive method gains strength by comparing many datasets to a single hypothesis, whereas the hypothetico-deductive method is best for comparing a single dataset to multiple hypotheses. Finally, both the inductive method and hypothetico-deductive method place emphasis on a single correct hypothesis, making it difficult to evaluate cases in which multiple factors are at work. This is less of a problem with the inductive approach, because multiple explanations can be incorporated into more complex hypotheses.

[10] The **logic tree** is a well-known variant of the hypothetico-deductive method that you may be familiar with from chemistry courses. The logic tree is a dichotomous decision tree in which different branches are followed depending on the results of experiments at each fork in the tree. The terminal branch tips of the tree represent the different hypotheses that are being tested. The logic tree also can be found in the familiar dichotomous taxonomic key for identifying to species unknown plants or animals: "If the animal has 3 pairs of walking legs, go to couplet *x*; if it has 4 or more pairs, go to couplet *y*." The logic tree is not always practical for complex ecological hypotheses; there may be too many branch points, and they may not all be dichotomous. However, it is always an excellent exercise to try and place your ideas and experiments in such a comprehensive framework. Platt (1964) champions this method and points to its spectacular success in molecular biology; the discovery of the helical structure of DNA is a classic example of the hypothetico-deductive method (Watson and Crick 1953).

Neither scientific method is the correct one, and some philosophers of science deny that either scenario really describes how science operates.[11] However, the hypothetico-deductive and inductive methods do characterize much science in the real world (as opposed to the abstract world of the philosophy of science). The reason for spending time on these models is to understand their relationship to statistical tests of an hypothesis.

Testing Statistical Hypotheses

Statistical Hypotheses versus Scientific Hypotheses

Using statistics to test hypotheses is only a small facet of the scientific method, but it consumes a disproportionate amount of our time and journal space. We use statistics to describe patterns in our data, and then we use statistical tests to decide whether the predictions of an hypothesis are supported or not. Establishing hypotheses, articulating their predictions, designing and executing valid experiments, and collecting, organizing, and summarizing the data all occur before we use statistical tests. We emphasize that accepting or rejecting a statistical hypothesis is quite distinct from accepting or rejecting a scientific hypothesis. The statistical null hypothesis is usually one of "no pattern," such as no difference between groups, or no relationship between two continuous variables. In contrast, the alternative hypothesis is that pattern exists. In other words, there

Thomas Kuhn

[11] No discussion of Popper and the hypothetico-deductive method would be complete without mention of Popper's philosophical nemesis, Thomas Kuhn (1922–1996). In *The Structure of Scientific Revolutions* (1962), Kuhn called into question the entire framework of hypothesis testing, and argued that it did not represent the way that science was done. Kuhn believed that science was done within the context of major **paradigms**, or research frameworks, and that the domain of these paradigms was implicitly adopted by each generation of scientists. The "puzzle-solving" activities of scientists constitute "ordinary science," in which empirical anomalies are reconciled with the existing paradigm. However, no paradigm can encompass all observations, and as anomalies accumulate, the paradigm becomes unwieldy. Eventually it collapses, and there is a scientific revolution in which an entirely new paradigm replaces the existing framework. Taking somewhat of an intermediate stance between Popper and Kuhn, the philosopher Imre Lakatos (1922–1974) thought that scientific research programs (SRPs) consisted of a core of central principles that generated a belt of surrounding hypotheses that make more specific predictions. The predictions of the hypotheses can be tested by the scientific method, but the core is not directly accessible (Lakatos 1978). Exchanges between Kuhn, Popper, Lakatos, and other philosophers of science can be read in Lakatos and Musgrave (1970). See also Horn (1986) for further discussion of these ideas.

are distinct differences in measured values between groups, or a clear relationship exists between two continuous variables. You must ask how such patterns relate to the scientific hypothesis you are testing.

For example, suppose you are evaluating the scientific hypothesis that waves scouring a rocky coast create empty space by removing competitively dominant invertebrate species. The open space can be colonized by competitively subordinate species that would otherwise be excluded. This hypothesis predicts that species diversity of marine invertebrates will change as a function of level of disturbance (Sousa 1979). You collect data on the number of species on disturbed and undisturbed rock surfaces. Using an appropriate statistical test, you find no difference in species richness in these two groups. In this case, you have *failed to reject* the statistical null hypothesis, and the pattern of the data *fail to support* one of the predictions of the disturbance hypothesis. Note, however, that absence of evidence is not evidence of absence; failure to reject a null hypothesis is not equivalent to accepting a null hypothesis (although it is often treated that way).

Here is a second example in which the statistical pattern is the same, but the scientific conclusion is different. The ideal free distribution is an hypothesis that predicts that organisms move between habitats and adjust their density so that they have the same mean fitness in different habitats (Fretwell and Lucas 1970). One testable prediction of this hypothesis is that the fitness of organisms in different habitats is similar, even though population density may differ. Suppose you measure population growth rate of birds (an important component of avian fitness) in forest and field habitats as a test of this prediction (Gill et al. 2001). As in the first example, you fail to reject the statistical null hypothesis, so that there is no evidence that growth rates differ among habitats. But in this case, *failing to reject* the statistical null hypothesis actually *supports* a prediction of the ideal free distribution.

Naturally, there are many additional observations and tests we would want to make to evaluate the disturbance hypothesis or the ideal free distribution. The point here is that the scientific and statistical hypotheses are distinct entities. In any study, you must determine whether supporting or refuting the statistical null hypothesis provides positive or negative evidence for the scientific hypothesis. Such a determination also influences profoundly how you set up your experimental study or observational sampling protocols. The distinction between the statistical null hypothesis and the scientific hypothesis is so important that we will return to it later in this chapter.

Statistical Significance and *P*-Values

It is nearly universal to report the results of a statistical test in order to assert the importance of patterns we observe in the data we collect. A typical assertion is: "The control and treatment groups differed significantly from one another

($P = 0.01$)." What, precisely, does "$P = 0.01$" mean, and how does it relate to the concepts of probability that we introduced in Chapters 1 and 2?

AN HYPOTHETICAL EXAMPLE: COMPARING MEANS A common assessment problem in environmental science is to determine whether or not human activities result in increased stress in animals. In vertebrates, stress can be measured as levels of the glucocorticoid hormones (GC) in the bloodstream or feces. For example, wolves that are not exposed to snowmobiles have 872.0 ng GC/g, whereas wolves exposed to snowmobiles have 1468.0 ng GC/g (Creel et al. 2002). Now, how do you decide whether this difference is large enough to be attributed to the presence of snowmobiles?[12]

Here is where you could conduct a conventional statistical test. Such tests can be very simple (such as the familiar *t*-test), or rather complex (such as tests for interaction terms in an analysis of variance). But all such statistical tests produce as their output a **test statistic**, which is just the numerical result of the test, and a **probability value** (or ***P*-value**) that is associated with the test statistic.

THE STATISTICAL NULL HYPOTHESIS Before we can define the probability of a statistical test, we must first define the statistical null hypothesis, or H_0. We noted above that scientists favor parsimonious or simple explanations over more complex ones. What is the simplest explanation to account for the difference in the means of the two groups? In our example, the simplest explanation is that the differences represent random variation between the groups and do not reflect any systematic effect of snowmobiles. In other words, if we were to divide the wolves into two groups but not expose individuals in either

[12] Many people try to answer this question by simply comparing the means. However, we cannot evaluate a difference between means unless we also have some feeling for how much individuals within a treatment group differ. For example, if several of the individuals in the no-snowmobile group have GC levels as low as 200 ng/g and others have GC levels as high as 1544 ng/g (the average, remember, was 872), then a difference of 600 ng/g between the two exposure groups may not mean much. On the other hand, if most individuals in the no-snowmobile group have GC levels between 850 and 950 ng/g, then a 600 ng/g difference is substantial. As we discussed in Chapter 3, we need to know not only the difference in the means, but the variance about those means—the amount that a typical individual differs from its group mean. Without knowing something about the variance, we cannot say anything about whether differences between the means of two groups are meaningful.

group to snowmobiles, we might still find that the means differ from each other. Remember that it is extremely unlikely that the means of two samples of numbers will be the same, even if they were sampled from a larger population using an identical process.

Glucocorticoid levels will differ from one individual to another for many reasons that cannot be studied or controlled in this experiment, and all of this variation—including variation due to measurement error—is what we label random variation. We want to know if there is any evidence that the observed difference in the mean GC levels of the two groups is larger than we would expect given the random variation among individuals. Thus, a typical statistical null hypothesis is that "differences between groups are no greater than we would expect due to random variation." We call this a **statistical null hypothesis** because the hypothesis is that a specific mechanism or force—some force *other* than random variation—does *not* operate.

THE ALTERNATIVE HYPOTHESIS Once we state the statistical null hypothesis, we then define one or more *alternatives* to the null hypothesis. In our example, the natural **alternative hypothesis** is that the observed difference in the average GC levels of the two groups is too large to be accounted for by random variation among individuals. Notice that the alternative hypothesis is *not* that snowmobile exposure is responsible for an increase in GC! Instead, the alternative hypothesis is focused simply on the *pattern* that is present in the data. The investigator can *infer* mechanism from the pattern, but that inference is a separate step. The statistical test merely reveals whether the pattern is likely or unlikely, given that the null hypothesis is true. Our ability to assign causal mechanisms to those statistical patterns depends on the quality of our experimental design and our measurements.

For example, suppose the group of wolves exposed to snowmobiles had also been hunted and chased by humans and their hunting dogs within the last day, whereas the unexposed group included wolves from a remote area uninhabited by humans. The statistical analysis would probably reveal significant differences in GC levels between the two groups regardless of exposure to snowmobiles. However, it would be dangerous to conclude that the difference between the means of the two groups was caused by snowmobiles, even though we can reject the statistical null hypothesis that the pattern is accounted for by random variation among individuals. In this case, the treatment effect is **confounded** with other differences between the control and treatment groups (exposure to hunting dogs) that are potentially related to stress levels. As we will discuss in

Chapters 6 and 7, an important goal of good experimental design is to avoid such confounded designs.

If our experiment was designed and executed correctly, it may be safe to infer that the difference between the means is caused by the presence of snowmobiles. But even here, we cannot pin down the precise physiological mechanism if all we did was measure the GC levels of exposed and unexposed individuals. We would need much more detailed information on hormone physiology, blood chemistry, and the like if we want to get at the underlying mechanisms.[13] Statistics help us establish convincing patterns, and from those patterns we can begin to draw inferences or conclusions about cause-and-effect relationships.

In most tests, the alternative hypothesis is not explicitly stated because there is usually more than one alternative hypothesis that could account for the patterns in the data. Rather, we consider the set of alternatives to be "not H_0." In a Venn diagram, all outcomes of data can then be classified into either H_0 or not H_0.

THE *P*-VALUE In many statistical analyses, we ask whether the null hypothesis of random variation among individuals can be rejected. The *P*-value is a guide to making that decision. A statistical *P*-value measures the probability that observed or more extreme differences would be found *if the null hypothesis were true*. Using the notation of conditional probability introduced in Chapter 1, $P\text{-value} = P(\text{data} \mid H_0)$.

Suppose the *P*-value is relatively small (close to 0.0). Then it is unlikely (the probability is small) that the observed differences could have been obtained if the null hypothesis were true. In our example of wolves and snowmobiles, a low *P*-value would mean that it is unlikely that a difference of 600 ng/g in GC levels would have been observed between the exposed and unexposed groups if there was only random variation among individuals and no consistent effect of snowmobiles (i.e., if the null hypothesis is true). Therefore, with a small *P*-value, the results would be improbable given the null hypothesis, so we reject it. Because we had only one alternative hypothesis in our study, our conclusion is that snow-

[13] Even if the physiological mechanisms were elucidated, there would still be questions about ultimate mechanisms at the molecular or genetic level. Whenever we propose a mechanism, there will always be lower-level processes that are not completely described by our explanation and have to be treated as a "black box." However, not all higher-level processes can be explained successfully by reductionism to lower-level mechanisms.

mobiles (or something associated with them) could be responsible for the difference between the treatment groups.[14]

On the other hand, suppose that the calculated P-value is relatively large (close to 1.0). Then it is likely that the observed difference could have occurred given

[14] Accepting an alternative hypothesis based on this mechanism of testing a null hypothesis is an example of the fallacy of "affirming the consequent" (Barker 1989). Formally, the P-value = $P(\text{data} \mid H_0)$. If the null hypothesis is true, it would result in (or in the terms of logic, *imply*) a particular set of observations (here, the data). We can write this formally as $H_0 \Rightarrow$ null data, where the arrow is read as "implies." If your observations are different from those expected under H_0, then a low P-value suggests that $H_0 \nRightarrow$ your data, where the crossed arrow is read as "does not imply." Because you have set up only one alternative hypothesis, H_a, then you are further asserting that $H_a = \neg H_0$ (where the symbol $\neg$ means "not"), and the only possibilities for data are those data possible under H_0 ("null data") and those not possible under H_0 ("$\neg$null data" = "your data"). Thus, you are asserting the following logical progression:

1. Given: $H_0 \Rightarrow$ null data
2. Observe: $\neg$null data
3. Conclude: $\neg$null data $\Rightarrow \neg H_0$
4. Thus: $\neg H_0$ (= H_a) $\Rightarrow \neg$null data

But really, all you can conclude is point 3: $\neg$null data $\Rightarrow \neg H_0$ (the so-called *contrapositive* of 1). In 3, the alternative hypothesis (H_a) is the "consequent," and you cannot assert its truth simply by observing its "predicate" ($\neg$null data in 3); many other possible causes could have yielded your results ($\neg$null data). You can affirm the consequent (assert H_a is true) if and only if there is *only one* possible alternative to your null hypothesis. In the simplest case, where H_0 asserts "no effect" and H_a asserts "some effect," proceeding from 3 to 4 makes sense. But biologically, it is usually of more interest to know what is the actual effect (as opposed to simply showing there is "some effect").

Consider the ant example earlier in this chapter. Let H_0 = all 25 ants in the Harvard Forest are *Myrmica*, and H_a = 10 ants in the forest are not *Myrmica*. If you collect a specimen of *Camponotus* in the forest, you can conclude that the data imply that the null hypothesis is false (observation of a *Camponotus* in the forest $\Rightarrow \neg H_0$). But you cannot draw any conclusion about the alternative hypothesis. You could support a less stringent alternative hypothesis, H_a = not all ants in the forest are *Myrmica*, but affirming this alternative hypothesis does not tell you anything about the actual distribution of ants in the forest, or the identity of the species and genera that are present.

This is more than splitting logical hairs. Many scientists appear to believe that when they report a P-value that they are giving the probability of observing the null hypothesis given the data [$P(H_0 \mid \text{data})$] or the probability that the alternative hypothesis is false, given the data [$1 - P(H_a \mid \text{data})$]. But, in fact, they are reporting something completely different—the probability of observing the data given the null hypothesis: $P(\text{data} \mid H_0)$. Unfortunately, as we saw in Chapter 1, $P(\text{data} \mid H_0) \neq P(H_0 \mid \text{data}) \neq 1 - P(H_a \mid \text{data})$; in the words of the immortal anonymous philosopher from Maine, *you can't get there from here*. However, it is possible to compute directly $P(H_0 \mid \text{data})$ or $P(H_a \mid \text{data})$ using Bayes' Theorem (see Chapter 1) and the Bayesian methods outlined in Chapter 5.

that the null hypothesis is true. In this example, a large *P*-value would mean that a 600-ng/g difference in GC levels likely would have been observed between the exposed and unexposed groups even if snowmobiles had no effect and there was only random variation among individuals. That is, with a large *P*-value, the observed results would be likely under the null hypothesis, so we do not have sufficient evidence to reject it. Our conclusion is that differences in GC levels between the two groups can be most parsimoniously attributed to random variation among individuals.

Keep in mind that when we calculate a statistical *P*-value, we are viewing the data through the lens of the null hypothesis. If the patterns in our data are likely under the null hypothesis (large *P*-value), we have no reason to reject the null hypothesis in favor of more complex explanations. On the other hand, if the patterns are unlikely under the null hypothesis (small *P*-value), it is more parsimonious to reject the null hypothesis and conclude that something more than random variation among subjects contributes to the results.

WHAT DETERMINES THE *P*-VALUE? The calculated *P*-value depends on three things: the number of observations in the samples (n), the difference between the means of the samples ($\bar{Y}_i - \bar{Y}_j$), and the level of variation among individuals (s^2). The more observations in a sample, the lower the *P*-value, because the more data we have, the more likely it is we are estimating the true population means and can detect a real difference between them, if it exists (see the Law of Large Numbers in Chapter 3). The *P*-value also will be lower the more different the two groups are in the variable we are measuring. Thus, a 10-ng/g difference in mean GC levels between control and treatment groups will generate a lower *P*-value than a 2-ng/g difference, all other things being equal. Finally, the *P*-value will be lower if the variance among individuals within a treatment group is small. The less variation there is from one individual to the next, the easier it will be to detect differences among groups. In the extreme case, if the GC levels for all individuals within the group of wolves exposed to snowmobiles were identical, and the GC levels for all individuals within the unexposed group were identical, then any difference in the means of the two groups, no matter how small, would generate a low *P*-value.

WHEN IS A *P*-VALUE SMALL ENOUGH? In our example, we obtained a *P*-value = 0.01 for the probability of obtaining the observed difference in GC levels between wolves exposed to and not exposed to snowmobiles. Thus, if the null hypothesis were true and there was only random variation among individuals in the data, the chance of finding a 600-ng/g difference in GC between exposed and unexposed groups is only 1 in 100. Stated another way, if the null hypothesis were true, and we conducted this experiment 100 times, using different subjects each time, in only *one* of the experiments would we expect to see a difference as large or larger than what we actually observed. Therefore, it seems unlikely the null hypothesis is true, and we reject it. If our experiment was properly designed, we can safely conclude that snowmobiles cause increases in GC levels, although we cannot specify what it is about snowmobiles that causes this response. On the other hand, if the calculated statistical probability were $P = 0.88$, then we would expect a result similar to what we found in 88 out of 100 experiments due to random variation among individuals; our observed result would not be at all unusual under the null hypothesis, and there would be no reason to reject it.

But what is the precise cutoff point that we should use in making the decision to reject or not reject the null hypothesis? This is a judgment call, as there is no natural critical value below which we should always reject the null hypothesis and above which we should never reject it. However, after many decades of custom, tradition, and vigilant enforcement by editors and journal reviewers, the operational critical value for making these decisions equals 0.05. In other words, if the statistical probability $P \leq 0.05$, the convention is to reject the null hypothesis, and if the statistical probability $P > 0.05$, the null hypothesis is not rejected. When scientists report that a particular result is "significant," they mean that they rejected the null hypothesis with a P-value ≤ 0.05.[15]

A little reflection should convince you that a critical value of 0.05 is relatively low. If you used this rule in your everyday life, you would never take an umbrella with you unless the forecast for rain was at least 95%. You would get wet a lot more often than your friends and neighbors. On the other hand, if your friends and neighbors saw you carrying your umbrella, they could be pretty confident of rain.

In other words, setting a critical value = 0.05 as the standard for rejecting a null hypothesis is very conservative. We require the evidence to be very strong in order to reject the statistical null hypothesis. Some investigators are unhappy about using an arbitrary critical value, and about setting it as low as 0.05. After all, most of us would take an umbrella with a 90% forecast of rain, so why shouldn't we be a bit less rigid in our standard for rejecting the null hypothesis? Perhaps we should set the critical critical value = 0.10, or perhaps we should use different critical values for different kinds of data and questions.

[15] When scientists discuss "significant" results in their work, they are really speaking about how confident they are that a statistical null hypothesis has been correctly rejected. But the public equates "significant" with "important." This distinction causes no end of confusion, and it is one of the reasons that scientists have such a hard time communicating their ideas clearly in the popular press.

A defense of the 0.05-cutoff is the observation that scientific standards need to be high so that investigators can build confidently on the work of others. If the null hypothesis is rejected with more liberal standards, there is a greater risk of falsely rejecting a true null hypothesis (a Type I error, described in more detail below). If we are trying to build hypotheses and scientific theories based on the data and results of others, such mistakes slow down scientific progress. By using a low critical value, we can be confident that the patterns in the data are quite strong. However, even a low critical value is not a safeguard against a poorly designed experiment or study. In such cases, the null hypothesis may be rejected, but the patterns in the data reflect flaws in the sampling or manipulations, not underlying biological differences that we are seeking to understand.

Perhaps the strongest argument in favor of requiring a low critical value is that we humans are psychologically predisposed to recognizing and seeing patterns in our data, even when they don't exist. Our vertebrate sensory system is adapted for organizing data and observations into "useful" patterns, generating a built-in bias toward rejecting null hypotheses and seeing patterns where there is really randomness (Sale 1984).[16] A low critical value is a safeguard against such activity. A low critical value also helps act as a gatekeeper on the rate of scientific publications because non-significant results are much less likely to be reported or published.[17] We emphasize, however, that no law *requires* a critical value to be ≤ 0.05 in order for the results to be declared significant. In many cases, it may be more useful to report the exact *P*-value and let the readers decide for themselves how important the results are. However, the practical reality is that reviewers and editors will usually not allow you to discuss mechanisms that are not supported by a $P \leq 0.05$ result.

[16] A fascinating illustration of this is to ask a friend to draw a set of 25 randomly located points on a piece of paper. If you compare the distribution of those points to a set of truly random points generated by a computer, you will often find that the drawings are distinctly non-random. People have a tendency to space the points too evenly across the paper, whereas a truly random pattern generates apparent "clumps" and "holes." Given this tendency to see patterns everywhere, we should use a low critical value to ensure we are not deceiving ourselves.

[17] The well-known tendency for journals to reject papers with non-significant results (Murtaugh 2002a) and authors to therefore not bother trying to publish them is not a good thing. In the hypothetico-deductive method, science progresses through the elimination of alternative hypotheses, and this can often be done when we fail to reject a null hypothesis. However, this approach requires authors to specify and test the unique predictions that are made by competing alternative hypotheses. Statistical tests based on H_0 versus not H_0 do not often allow for this kind of specificity.

STATISTICAL HYPOTHESES VERSUS SCIENTIFIC HYPOTHESES REDUX The biggest difficulty in using *P*-values results from the failure to distinguish statistical null hypotheses from scientific hypotheses. Remember that a *scientific hypothesis* poses a formal mechanism to account for patterns in the data. In this case, our scientific hypothesis is that snowmobiles cause stress in wolves, which we propose to test by measuring GC levels. Higher levels of GC might come about by complex changes in physiology that lead to changes in GC production when an animal is under stress. In contrast, the *statistical null hypothesis* is a statement about patterns in the data and the likelihood that these patterns could arise by chance or random processes that are not related to the factors we are explicitly studying.

We use the methods of probability when deciding whether or not to reject the statistical null hypothesis; think of this process as a method for establishing pattern in the data. Next, we draw a conclusion about the validity of our scientific hypothesis based on the statistical pattern in this data. The strength of this inference depends very much on the details of the experiment and sampling design. In a well-designed and replicated experiment that includes appropriate controls and in which individuals have been assigned randomly to clear-cut treatments, we can be fairly confident about our inferences and our ability to evaluate the scientific hypothesis we are considering. However, in a sampling study in which we have not manipulated any variables but have simply measured differences among groups, it is difficult to make solid inferences about the underlying scientific hypotheses, even if we have rejected the statistical null hypothesis.[18]

We think the more general issue is not the particular critical value that is chosen, but whether we always should be using an hypothesis-testing framework. Certainly, for many questions statistical hypothesis tests are a powerful way to establish what patterns do or do not exist in the data. But in many studies, the real issue may not be hypothesis testing, but **parameter estimation**. For example, in the stress study, it may be more important to determine the range of GC levels expected for wolves exposed to snowmobiles rather than merely to establish that snowmobiles significantly increases GC levels. We also should establish the level of confidence or certainty in our parameter estimates.

[18] In contrast to the example of the snowmobiles and wolves, suppose we measured the GC levels of 10 randomly chosen old wolves and 10 randomly chosen young ones. Could we be as confident about our inferences as in the snowmobile experiment? Why or why not? What are the differences, if any, between experiments in which we manipulate individuals in different groups (exposed wolves versus unexposed wolves) and sampling surveys in which we measure variation among groups but do not directly manipulate or change conditions for those groups (old wolves versus young wolves)?

Errors in Hypothesis Testing

Although statistics involves many precise calculations, it is important not to lose sight of the fact that statistics is a discipline steeped in uncertainty. We are trying to use limited and incomplete data to make inferences about underlying mechanisms that we may understand only partially. In reality, the statistical null hypothesis is either true or false; if we had complete and perfect information, we would know whether or not it were true and we would not need statistics to tell us. Instead, we have only our data and methods of statistical inference to decide whether or not to reject the statistical null hypothesis. This leads to an interesting 2 × 2 table of possible outcomes whenever we test a statistical null hypothesis (Table 4.1)

Ideally, we would like to end up in either the upper left or lower right cells of Table 4.1. In other words, when there is only random variation in our data, we would hope to not reject the statistical null hypothesis (upper left cell), and when there is something more, we would hope to reject it (lower right cell). However, we may find ourselves in one of the other two cells, which correspond to the two kinds of errors that can be made in a statistical decision.

TYPE I ERROR If we falsely reject a null hypothesis that is true (upper right cell in Table 4.1), we have made a false claim that some factor above and beyond random variation is causing patterns in our data. This is a Type I error, and by convention, the probability of committing a Type I error is denoted by α. When you calculate a statistical *P*-value, you are actually estimating α. So, a more precise

TABLE 4.1 **The quadripartite world of statistical testing**

	Retain H_0	**Reject H_0**
H_0 true	Correct decision	Type I error (α)
H_0 false	Type II error (β)	Correct decision

Underlying null hypotheses are either true or false, but in the real world we must use sampling and limited data to make a decision to accept or reject the null hypothesis. Whenever a statistical decision is made, one of four outcomes will result. A correct decision results when we retain a null hypothesis that is true (upper left-hand corner) or reject a null hypothesis that is false (lower right-hand corner). The other two possibilities represent errors in the decision process. If we reject a null hypothesis that is true, we have committed a Type I error (upper right-hand corner). Standard parametric tests seek to control α, the probability of a Type I error. If we retain a null hypothesis that is false, we have committed a Type II error (lower left-hand corner). The probability of a Type II error is β.

definition of a *P*-value is that it is the chance we will make a Type I error by falsely rejecting a true null hypothesis.[19] This definition lends further support for asserting statistical significance only when the *P*-value is very small. The smaller the *P*-value, the more confident we can be that we will not commit a Type I error if we reject H_0. In the glucocorticoid example, the risk of making a Type I error by rejecting the null hypothesis is 1%. As we noted before, scientific publications use a standard of a maximum of a 5% risk of Type I error for rejecting a null hypothesis. In environmental impact assessment, a Type I error would be a "false positive" in which, for example, an effect of a pollutant on human health is reported but does not, in fact, exist.

TYPE II ERROR AND STATISTICAL POWER The lower left cell in Table 4.1 represents a Type II error. In this case, the investigator has incorrectly failed to reject a null hypothesis that is false. In other words, there are systematic differences between the groups being compared, but the investigator has failed to reject the null hypothesis and has concluded incorrectly that only random variation among observations is present. By convention, the probability of committing a Type II error is denoted by β. In environmental assessment, a Type II error would be a "false negative" in which, for example, there is an effect of a pollutant on human health, but it is not detected.[20]

[19] We have followed standard statistical treatments that equate the calculated *P*-value with the estimate of Type I error rate α. However, Fisher's evidential *P*-value may not be strictly equivalent to Neyman and Pearson's α. Statisticians disagree whether the distinction is an important philosophical issue or simply a semantic difference. Hubbard and Bayarri (2003) argue that the incompatibility is important, and their paper is followed by discussion, comments, and rebuttals from other statisticians. Stay tuned!

[20] The relationship between Type I and Type II errors informs discussions of the precautionary principle of environmental decision making. Historically, for example, regulatory agencies have assumed that new chemical products were benign until proven harmful. Very strong evidence was required to reject the null hypothesis of no effect on health and well-being. Manufacturers of chemicals and other potential pollutants are keen to minimize the probability of committing a Type I error. In contrast, environmental groups that serve the general public are interested in minimizing the probability that the manufacturer committed a Type II error. Such groups assume that a chemical is harmful until proven benign, and are willing to accept a larger probability of committing a Type I error if this means they can be more confident that the manufacturer has not falsely accepted the null hypothesis. Following such reasoning, in assessing quality control of industrial production, Type I and Type II errors are often known as producer and consumer errors, respectively (Sokal and Rohlf 1995).

A concept related to the probability of committing a Type II error is the **power** of a statistical test. Power is calculated as $1 - \beta$, and equals the probability of correctly rejecting the null hypothesis when it is false. We want our statistical tests to have good power so that we have a good chance of detecting significant patterns in our data when they are present.

WHAT IS THE RELATIONSHIP BETWEEN TYPE I AND TYPE II ERROR? Ideally, we would like to minimize both Type I and Type II errors in our statistical inference. However, strategies designed to reduce Type I error inevitably increase the risk of Type II error, and vice versa. For example, suppose you decide to reject the null hypothesis only if $P < 0.01$—a fivefold more stringent standard than the conventional criterion of $P < 0.05$. Although your risk of committing a Type I error is now much lower, there is a much greater chance that when you fail to reject the null hypothesis, you may be doing so incorrectly (i.e., you will be committing a Type II error). Although Type I and Type II errors are inversely related to one another, there is no simple mathematical relationship between them, because the probability of a Type II error depends in part on what the alternative hypothesis is, how large an effect we hope to detect (Figure 4.5), the sample size, and the wisdom of our experimental design or sampling protocol.

WHY ARE STATISTICAL DECISIONS BASED ON TYPE I ERROR? In contrast to the probability of committing a Type I error, which we determine with standard statistical tests, the probability of committing a Type II error is not often calculated or reported, and in many scientific papers, the probability of committing a Type II error is not even discussed. Why not? To begin with, we often cannot calculate the probability of a Type II error unless the alternative hypotheses are completely specified. In other words, if we want to determine the risk of falsely accepting the null hypothesis, the alternatives have to be fleshed out more than just "not H_0." In contrast, calculating the probability of a Type I error does not require this specification; instead we are required only to meet some assumptions of normality and independence (see Chapters 9 and 10).

On a philosophical basis, some authors have argued that a Type I error is a more serious mistake in science than a Type II error (Shrader-Frechette and McCoy 1992). A Type I error is an error of falsity, in which we have incorrectly rejected a null hypothesis and made a claim about a more complex mechanism. Others may follow our work and try to build their own studies based on that false claim. In contrast, a Type II error is an error of ignorance. Although we have not rejected the null hypothesis, someone else with a better experiment or more data may be able to do so in the future, and the science will progress

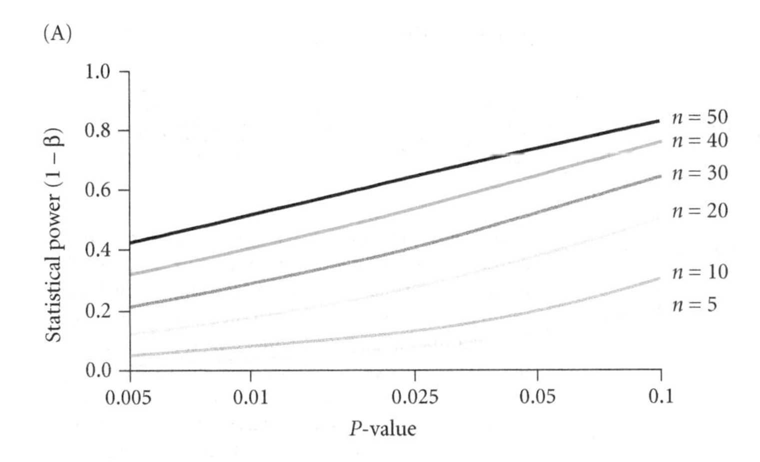

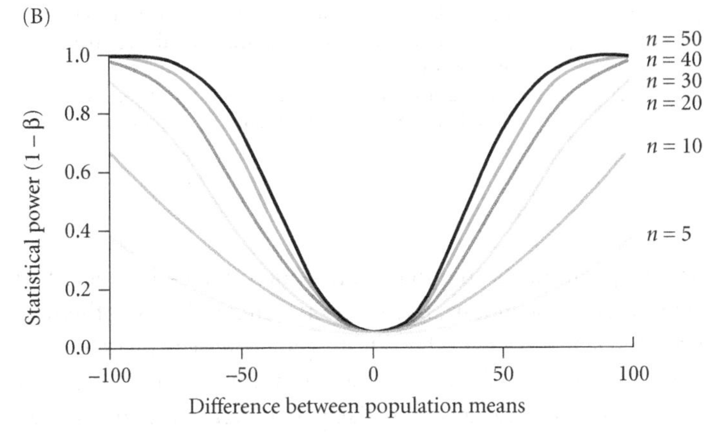

Figure 4.5 The relationship between statistical power, *P*-values, and observable effect sizes as a function of sample size. (A) The *P*-value is the probability of incorrectly rejecting a true null hypothesis, whereas statistical power is the probability of correctly rejecting a false null hypothesis. The general result is that the lower the *P*-value used for rejection of the null hypothesis, the lower the statistical power of correctly detecting a treatment effect. At a given *P*-value, statistical power is greater when the sample size is larger. (B) The smaller the observable effect of the treatment (i.e., the smaller the difference between the treatment group and the control group), the larger the sample size necessary for good statistical power to detect a treatment effect.[21]

from that point. However, in many applied problems, such as environmental monitoring or disease diagnosis, Type II errors may have more serious consequences because diseases or adverse environmental effects would not be correctly detected.

[21] We can apply these graphs to our example comparing glucocorticoid hormone levels for populations of wolves exposed to snowmobiles (treatment group) versus wolves that were not exposed to snowmobiles (control group). In the original data (Creel et al. 2002), the standard deviation of the control population of wolves unexposed to snowmobiles was 73.1, and that of wolves exposed to snowmobiles was 114.2. Panel (A) suggests that if there were a 50-ng/g difference between the experimental populations, and if sample size was $n = 50$ in each group, the experimenters would have correctly accepted the alternative hypothesis only 51% of the time for $P = 0.01$. Panel (B) shows that power increases steeply as the populations become more different. In the actual well-designed study (Creel et al. 2002), sample size was 193 in the unexposed group and 178 in the exposed group, the difference between population means was 598 ng/g, and the actual power of the statistical test was close to 1.0 for $P = 0.01$.

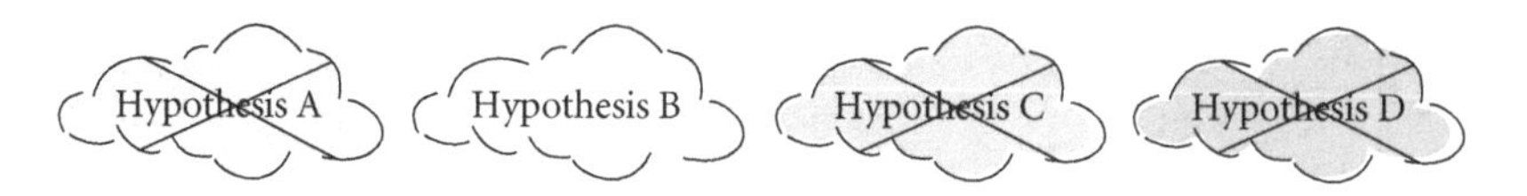

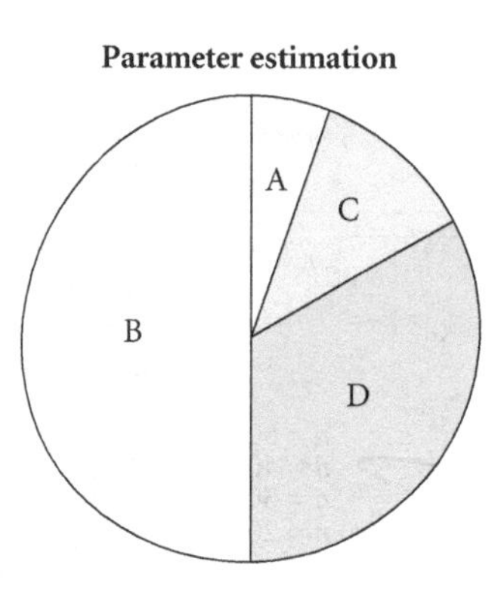

Figure 4.6 Hypothesis testing versus parameter estimation. Parameter estimation more easily accommodates multiple mechanisms and may allow for an estimate of the relative importance of different factors. Parameter estimation may involve the construction of confidence or credibility intervals (see Chapter 3) to estimate the strength of an effect. A related technique in the analysis of variance is to decompose the total variation in the data into proportions that are explained by different factors in the model (see Chapter 10). Both methods quantify the relative importance of different factors, whereas hypothesis testing emphasizes a binary yes/no decision as to whether a factor has a measurable effect or not.

Parameter Estimation and Prediction

All the methods for hypothesis testing that we have described—the inductive method (and its modern descendant, Bayesian inference), the hypothetico-deductive method, and statistical hypothesis testing are concerned with choosing a single explanatory "answer" from an initial set of multiple hypotheses. In ecology and environmental science, it is more likely that many mechanisms may be operating simultaneously to produce observed patterns; an hypothesis-testing framework that emphasizes single explanations may not be appropriate. Rather than try to test multiple hypotheses, it may be more worthwhile to estimate the relative contributions of each to a particular pattern. This approach is sketched in Figure 4.6, in which we partition the effects of each hypothesis on the observed patterns by estimating how much each cause contributes to the observed effect.

In such cases, rather than ask whether a particular cause has some effect versus no effect (i.e., is it significantly different from 0.0?), we ask what is the best estimate of the **parameter** that expresses the magnitude of the effect.[22] For example, in Figure 4.3, measured photosynthetic rates for young sun leaves of

[22] Chapters 9, 13, and 14 will introduce some of the strategies used for fitting curves and estimating parameters from data. See Hilborn and Mangel (1997) for a detailed discussion. Clark et al. (2003) describe Bayesian strategies to curve fitting.

Rhizophora mangle were fit to a Michaelis-Menten equation. This equation is a simple model that describes a variable rising smoothly to an asymptote. The Michaelis-Menten equation shows up frequently in biology, being used to describe everything from enzyme kinematics (Real 1977) to invertebrate foraging rates (Holling 1959).

The Michaelis-Menten equation takes the form

$$Y = \frac{kX}{X + D}$$

where k and D are the two fitted parameters of the model, and X and Y are the independent and dependent variables. In this example, k represents the asymptote of the curve, which in this case is the maximum assimilation rate; the independent variable X is light intensity; and the dependent variable Y is the net assimilation rate. For the data in Figure 4.3, the parameter estimate for k is a maximum assimilation rate of 7.1 μmol CO_2 m^{-2} s^{-1}. This accords well with an "eyeball estimate" of where the asymptote would be on this graph.

The second parameter in the Michaelis-Menten equation is D, the half-saturation constant. This parameter gives the value of the X variable that yields a Y variable that is half of the asymptote. The smaller D is, the more quickly the curve rises to the asymptote. For the data in Figure 4.3, the parameter estimate for D is photosynthetically active radiation (PAR) of 250 μmol CO_2 m^{-2} s^{-1}.

We also can measure the uncertainty in these parameter estimates by using estimates of standard error to construct confidence or credibility intervals (see Chapter 3). The estimated standard error for $k = 0.49$, and for $D = 71.3$. Statistical hypothesis testing and parameter estimation are related, because if the confidence interval of uncertainty includes 0.0, we usually are not able to reject the null hypothesis of no effect for one of the mechanisms. For the parameters k and D in Figure 4.3, the P-values for the test of the null hypothesis that the parameter does not differ from 0.0 are 0.0001 and 0.004, respectively. Thus, we can be fairly confident in our statement that these parameters are greater than 0.0. But, for the purposes of evaluating and fitting models, the numerical values of the parameters are more informative than just asking whether they differ or not from 0.0. In later chapters, we will give other examples of studies in which model parameters are estimated from data.

Summary

Science is done using inductive and hypothetico-deductive methods. In both methods, observed data are compared with data predicted by the hypotheses. Through the inductive method, which includes modern Bayesian analyses, a single hypothesis is repeatedly tested and modified; the goal is to confirm or assert the probability of a particular hypothesis. In contrast, the hypothetico-deductive method requires the simultaneous statement of multiple hypotheses. These are tested against observations with the goal of falsifying or eliminating all but one of the alternative hypotheses. Statistics are used to test hypotheses objectively, and can be used in both inductive and hypothetico-deductive approaches.

Probabilities are calculated and reported with virtually all statistical tests. The probability values associated with statistical tests may allow us to infer causes of the phenomena that we are studying. Tests of statistical hypotheses using the hypothetico-deductive method yield estimates of the chance of obtaining a result equal to or more extreme than the one observed, given that the null hypothesis is true. This *P*-value is also the probability of incorrectly rejecting a true null hypothesis (or committing a Type I statistical error). By convention and tradition, 0.05 is the cutoff value in the sciences for claiming that a result is statistically significant. The calculated *P*-value depends on the number of observations, the difference between the means of the groups being compared, and the amount of variation among individuals within each group. Type II statistical errors occur when a false null hypothesis is incorrectly accepted. This kind of error may be just as serious as a Type I error, but the probability of Type II errors is reported rarely in scientific publications. Tests of statistical hypotheses using inductive or Bayesian methods yield estimates of the probability of the hypothesis or hypotheses of interest given the observed data. Because these are confirmatory methods, they do not give probabilities of Type I or Type II errors. Rather, the results are expressed as the odds or likelihood that a particular hypothesis is correct.

Regardless of the method used, all science proceeds by articulating testable hypotheses, collecting data that can be used to test the predictions of the hypotheses, and relating the results to underlying cause-and-effect relationships.

CHAPTER 5

Three Frameworks for Statistical Analysis

In this chapter, we introduce three major frameworks for statistical analysis: **Monte Carlo analysis**, **parametric analysis**, and **Bayesian analysis**. In a nutshell, Monte Carlo analysis makes minimal assumptions about the underlying distribution of the data. It uses randomizations of observed data as a basis for inference. Parametric analysis assumes the data were sampled from an underlying distribution of known form, such as those described in Chapter 2, and estimates the parameters of the distribution from the data. Parametric analysis estimates probabilities from observed frequencies of events and uses these probabilities as a basis for inference. Hence, it is a type of frequentist inference. Bayesian analysis also assumes the data were sampled from an underlying distribution of known form. It estimates parameters not only from the data, but also from prior knowledge, and assigns probabilities to these parameters. These probabilities are the basis for Bayesian inference. Most standard statistics texts teach students parametric analysis, but the other two are equally important, and Monte Carlo analysis is actually easier to understand initially. To introduce these methods, we will use each of them to analyze the same sample problem.

Sample Problem

Imagine you are trying to compare the nest density of ground-foraging ants in two habitats—field and forest. In this sample problem, we won't concern ourselves with the scientific hypothesis that you are testing (perhaps you don't even have one at this point); we will simply follow through the process of gathering and analyzing the data to determine whether there are consistent differences in the density of ant nests in the two habitats.

You visit a forest and an adjacent field and estimate the average density of ant nests in each, using replicated sampling. In each habitat, you choose a random location, place a square quadrat of 1-m^2 area, and carefully count all of the ant nests that occur within the quadrat. You repeat this procedure several times in

TABLE 5.1 Sample dataset used to illustrate Monte Carlo, parametric, and Bayesian analyses

ID number	Habitat	Number of ant nests per quadrat
1	Forest	9
2	Forest	6
3	Forest	4
4	Forest	6
5	Forest	7
6	Forest	10
7	Field	12
8	Field	9
9	Field	12
10	Field	10

Each row is an independent observation. The first column identifies the replicate with a unique ID number, the second column indicates the habitat sampled, and the third column gives the number of ant nests recorded in the replicate.

each habitat. The issue of choosing random locations is very important for any type of statistical analysis. Without more complicated methods, such as stratified sampling, randomization is the only safeguard to ensure that we have a representative sample from a population (see Chapter 6).

The spatial scale of the sampling determines the scope of inference. Strictly speaking, this sampling design will allow you to discuss differences between forest and field ant densities *only at this particular site*. A better design would be to visit several different forests and fields and sample one quadrat within each of them. Then the conclusions could be more readily generalized.

Table 5.1 illustrates the data in a **spreadsheet**. The data are arranged in a table, with labeled rows and columns. Each row of the table contains all of the information on a particular observation. Each column of the table indicates a different variable that has been measured or recorded for each observation. In this case, your original intent was to sample 6 field and 6 forest quadrats, but the field quadrats were more time-consuming than you expected and you only managed to collect 4 field samples. Thus the data table has 10 rows (in addition to the first row, which displays labels) because you collected 10 different samples (6 from the forest and 4 from the field). The table has 3 columns. The first column contains the unique ID number assigned to each replicate. The other columns contain the two pieces of information recorded for each replicate: the habitat in which the replicate was sampled (field or forest); and the number of ant nests recorded in the quadrat.[1]

[1] Many statistics texts would show these data as two columns of numbers, one for forest and one for field. However, the layout we have shown here is the one that is recognized most commonly by statistical software for data analysis.

TABLE 5.2 **Summary statistics for the sample data in Table 5.1**

Habitat	*N*	Mean	Standard deviation
Forest	6	7.00	2.19
Field	4	10.75	1.50

Following the procedures in Chapter 3, calculate the mean and standard deviation for each sample (Table 5.2). Plot the data, using a conventional bar chart of means and standard deviations (Figure 5.1), or the more informative box plot described in Chapter 3 (Figure 5.2).

Although the numbers collected in the forest and field show some overlap, they appear to form two distinct groups: the forest samples with a mean of 7.0 nests per quadrat, and the field samples with a mean of 10.75 nests per quadrat. On the other hand, our sample size is very small ($n = 6$ forest and $n = 4$ field samples). Perhaps these differences could have arisen by chance or random sampling. We need to conduct a statistical test before deciding whether these differences are significant or not.

Monte Carlo Analysis

Monte Carlo analysis involves a number of methods in which data are randomized or reshuffled so that observations are randomly reassigned to differ-

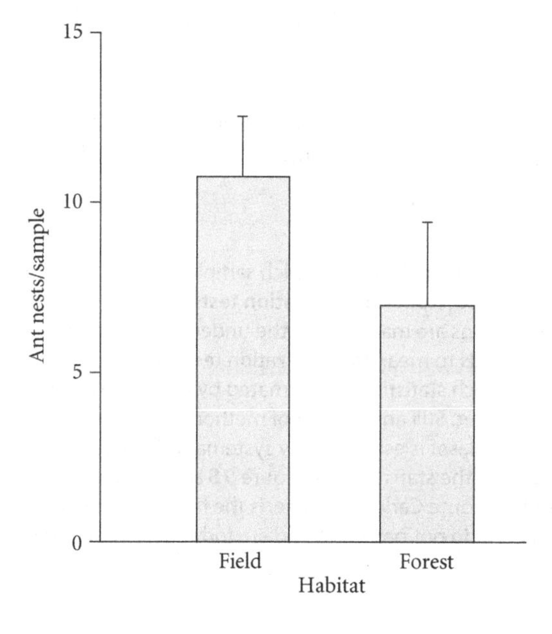

Figure 5.1 Standard bar chart for sample data in Table 5.1. The height of the bar is the mean of the sample and the vertical line indicates one standard deviation above the mean. Monte Carlo, parametric, and Bayesian analyses can all be used to evaluate the difference in means between the groups.

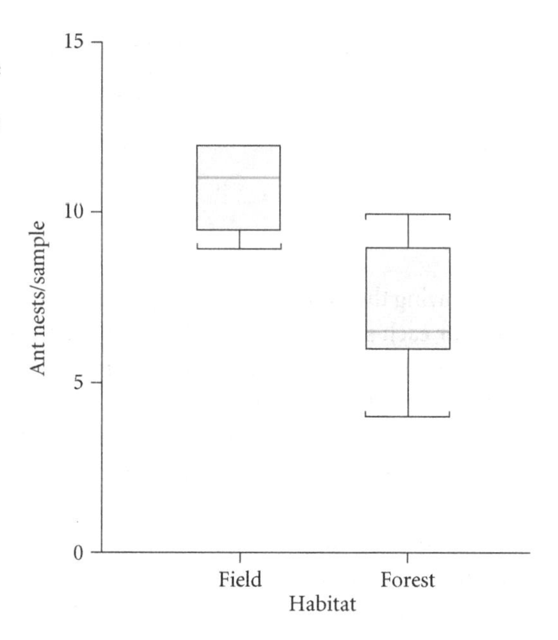

Figure 5.2 Box plot of data in Table 5.1. In a box plot, the central line within the box is the median of the data. The box includes 50% of the data. The top of the box indicates the 75th percentile (upper quartile) of the data, and the bottom of the box indicates the 25th percentile (lower quartile) of the data. The vertical lines extend to the upper and lower deciles (90th and 10th percentiles). For the field sample, there are so few data that the 75th and 90th percentiles do not differ. When data have asymmetric distributions or outliers, box plots may be more informative than standard bar graphs such as Figure 5.1.

ent treatments or groups. This randomization[2] specifies the null hypothesis under consideration: that the pattern in the data is no different from that which we would expect if the observations were assigned randomly to the different groups. There are four steps in Monte Carlo analysis:

1. Specify a test statistic or index to describe the pattern in the data.
2. Create a distribution of the test statistic that would be expected under the null hypothesis.
3. Decide on a one- or two-tailed test.
4. Compare the observed test statistic to a distribution of simulated values and estimate the appropriate P-value as a tail probability (as described in Chapter 3).

[2] Some statisticians distinguish Monte Carlo methods, in which samples are drawn from a known or specified statistical distribution, from **randomization tests**, in which existing data are reshuffled but no assumptions are made about the underlying distribution. In this book, we use Monte Carlo methods to mean randomization tests. Another set of methods includes **bootstrapping**, in which statistics are estimated by repeatedly subsampling with replacement from a dataset. Still another set of methods includes **jackknifing**, in which the variability in the dataset is estimated by systematically deleting each observation and then recalculating the statistics (see Figure 9.8 and the section "Discriminant Analysis" in Chapter 12). Monte Carlo, of course, is the famous gambling resort city on the Riviera, whose citizens do not pay taxes and are forbidden from entering the gaming rooms.

Step 1: Specifying the Test Statistic

For this analysis, we will use as a measure of pattern the absolute difference in the means of the forest and field samples, or DIF:

$$DIF_{obs} = |10.75 - 7.00| = 3.75$$

The subscript "obs" indicates that this DIF value is calculated for the observed data. The null hypothesis is that a DIF_{obs} equal to 3.75 is about what would be expected by random sampling. The alternative hypothesis would be that a DIF_{obs} equal to 3.75 is larger than would be expected by chance.

Step 2: Creating the Null Distribution

Next, estimate what DIF would be if the null hypothesis were true. To do this, use the computer (or a deck of playing cards) to randomly reassign the forest and field labels to the dataset. In the randomized dataset, there will still be 4 field and 6 forest samples, but those labels (Field and Forest) will be randomly reassigned (Table 5.3). Notice that in the randomly reshuffled dataset, many of the observations were placed in the same group as the original data. This will happen by chance fairly often in small datasets. Next, calculate the sample statistics for this randomized dataset (Table 5.4). For this dataset, $DIF_{sim} = |7.75 - 9.00| = 1.25$. The subscript "sim" indicates that this value of DIF is calculated for the randomized, or simulated, data.

In this first simulated dataset, the difference between the means of the two groups ($DIF_{sim} = 1.25$) is smaller than the difference observed in the real data ($DIF_{obs} = 3.75$). This result suggests that means of the forest and field samples may differ more than expected under the null hypothesis of random assignment.

TABLE 5.3 **Monte Carlo randomization of the habitat labels in Table 5.1**

Habitat	Number of ant nests per quadrat
Field	9
Field	6
Forest	4
Forest	6
Field	7
Forest	10
Forest	12
Field	9
Forest	12
Forest	10

In the Monte Carlo analysis, the sample labels in Table 5.1 are reshuffled randomly among the different samples. After reshuffling, the difference in the means of the two groups, DIF, is recorded. This procedure is repeated many times, generating a distribution of DIF values (see Figure 5.3).

TABLE 5.4 **Summary statistics for the randomized data in Table 5.3**

Habitat	N	Mean	Standard deviation
Forest	6	9.00	3.286
Field	4	7.75	1.500

In the Monte Carlo analysis, these values represent the mean and standard deviation in each group after a single reshuffling of the labels. The difference between the two means (DIF = |7.75 – 9.00| = 1.25) is the test statistic.

However, there are many different random combinations that can be produced by reshuffling the sample labels.[3] Some of these have relatively large values of DIF_{sim} and some have relatively small values of DIF_{sim}. Repeat this reshuffling exercise many times (usually 1000 or more), then illustrate the distribution of the simulated DIF values with a histogram (Figure 5.3) and summary statistics (Table 5.5).

The mean DIF of the simulated datasets was only 1.46, compared to the observed value of 3.75 for the original data. The standard deviation of 2.07 could be used to construct a confidence interval (see Chapter 3), but it should not be, because the distribution of simulated values (see Figure 5.3) has a long right-hand tail and does not follow a normal distribution. An approximate 95% confidence interval can be derived from this Monte Carlo analysis by identifying the upper and lower 2.5 percentiles of the data from the set of simulated DIF values and using these values as the upper and lower bounds of the confidence interval (Efron 1982).

Step 3: Deciding on a One- or Two-Tailed Test

Next, decide whether to use a **one-tailed test** or a **two-tailed test**. The "tail" refers to the extreme left- or right-hand areas under the probability density function (see Figure 3.7). It is these tails of the distribution that are used to determine the cutoff points for a statistical test at $P = 0.05$ (or any other probability level). A one-tailed test uses only one tail of the distribution to estimate the P-value. A two-tailed test uses both tails of the distribution to estimate the P-value. In a

[3] With 10 samples split into a group of 6 and a group of 4, there are

$$\binom{10}{4} = 210$$

combinations that can be created by reshuffling the labels (see Chapter 2). However, the number of unique values of DIF that are possible is somewhat less than this because some of the samples had identical nest counts.

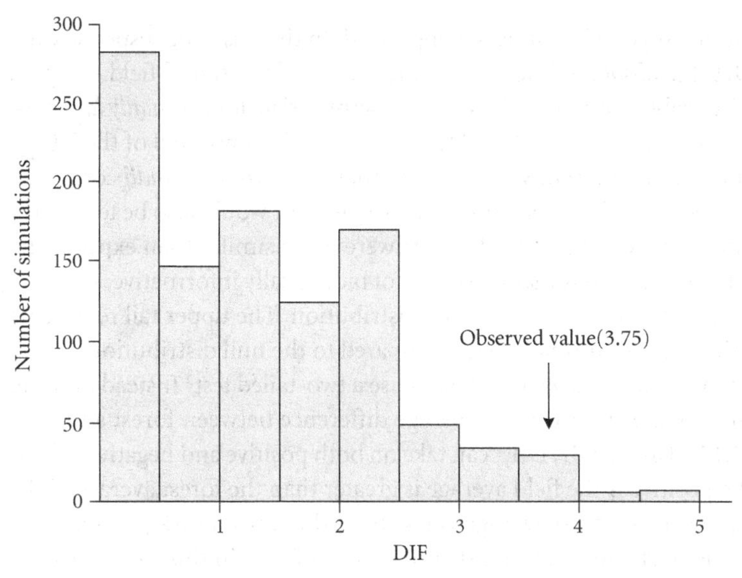

Figure 5.3 Monte Carlo analysis of the data in Table 5.1. For each randomization, the data labels (Forest, Field) were randomly reshuffled among the replicates. Next, the difference (DIF) between the means of the two groups was calculated. This histogram illustrates the distribution of DIF values from 1000 such randomizations. The arrow indicates the single value of DIF observed in the real dataset (3.75). The observed DIF sits well in the right-hand tail of the distribution. The observed DIF of 3.75 was larger than or equal to all but 36 of the simulated DIF values. Therefore, under the null hypothesis of random assignment of samples to groups, the tail probability of finding this observation (or one more extreme) is 36/1000 = 0.036.

one-tailed test, all 5% of the area under the curve is located in one tail of the distribution. In a two-tailed test, each tail of the distribution would encompass 2.5% of the area. Thus, a two-tailed test requires more extreme values to achieve statistical significance than does a one-tailed test.

However, the cutoff value is not the most important issue in deciding whether to use a one- or a two-tailed test. Most important are the nature of the response

TABLE 5.5 **Summary statistics for 1000 simulated values of DIF in Figure 5.3**

Variable	*N*	Mean	Standard deviation
DIF_{sim}	1000	1.46	2.07

In the Monte Carlo analysis, each of these values was created by reshuffling the data labels in Table 5.1, and calculating the difference between the means of the two groups (DIF). Calculation of the *P*-value does not require that DIF follow a normal distribution, because the *P*-value is determined directly by the location of the observed DIF statistic in the histogram (see Table 5.6).

variable and the precise hypothesis being tested. In this case, the response variable was DIF, the *absolute* difference in the means of forest and field samples. For the DIF variable, a one-tailed test is most appropriate for *unusually large* values of DIF. Why not use a two-tailed test with DIF? The lower tail of the DIF_{sim} distribution would represent values of DIF that are *unusually small* compared to the null hypothesis. In other words, a two-tailed test would also be testing for the possibility that forest and field means were more similar than expected by chance. This test for extreme similarity is not biologically informative, so attention is restricted to the upper tail of the distribution. The upper tail represents cases in which DIF is unusually large compared to the null distribution.

How could the analysis be modified to use a two-tailed test? Instead of using the absolute value of DIF, use the average difference between forest and field samples (DIF*). Unlike DIF, DIF* can take on both positive and negative values. DIF* will be positive if the field average is greater than the forest average. DIF* will be negative if the field average is less than the forest average. As before, randomize the data and create a distribution of DIF^*_{sim}. In this case, however, a two-tailed test would detect cases in which the field mean was unusually large compared to the forest mean (DIF* positive) and cases in which the field mean was unusually small compared to the forest mean (DIF* negative).

Whether you are using Monte Carlo, parametric, or Bayesian analysis, you should study carefully the response variable you are using. How would you interpret an extremely large or an extremely small value of the response variable relative to the null hypothesis? The answer to this question will help you decide whether a one- or a two-tailed test is most appropriate.

Step 4: Calculating the Tail Probability

The final step is to estimate the probability of obtaining DIF_{obs} or a value more extreme, given that the null hypothesis is true [$P(\text{data}|H_0)$]. To do this, examine the set of simulated DIF values (plotted as the histogram in Figure 5.3), and tally up the number of times that the DIF_{obs} was greater than, equal to, or less than each of the 1000 values of DIF_{sim}.

In 29 of 1000 randomizations, $DIF_{sim} = DIF_{obs}$, so the probability of obtaining DIF_{obs} under the null hypothesis is $29/1000 = 0.029$ (Table 5.6). However, when we calculate a statistical test, we usually are not interested in this **exact probability** as much as we are the **tail probability**. That is, we want to know the chances of obtaining an observation *as large or larger than* the real data, given that the null hypothesis is true. In 7 of 1000 randomizations, $DIF_{sim} > DIF_{obs}$. Thus, the probability that $DIF_{obs} \geq 3.75$ is $(7 + 29)/1000 = 0.036$. This tail probability is the frequency of obtaining the observed value (29/1000) plus the frequency of obtaining a more extreme result (7/1000).

TABLE 5.6 **Calculation of tail probabilities in Monte Carlo analysis**

Inequality	N
$DIF_{sim} > DIF_{obs}$	7
$DIF_{sim} = DIF_{obs}$	29
$DIF_{sim} < DIF_{obs}$	964

Comparisons of DIF_{obs} (absolute difference in the mean of the two groups in the original data) with DIF_{sim} (absolute difference in the mean of the two groups after randomizing the group assignments). N is the number of simulations (out of 1000) for which the inequality was obtained. Because $DIF_{sim} \geq DIF_{obs}$ in $7 + 29 = 36$ trials out of 1000, the tail probability under the null hypothesis of finding DIF_{obs} this extreme is $36/1000 = 0.036$.

Follow the procedures and interpretations of *P*-values that we discussed in Chapter 4. With a tail probability of 0.036, it is unlikely that these data would have occurred given the null hypothesis is true.

Assumptions of Monte Carlo Methods

Monte Carlo methods rest on three assumptions:

1. The data collected represent random, independent samples.
2. The test statistic describes the pattern of interest.
3. The randomization creates an appropriate null distribution for the question.

Assumptions 1 and 2 are common to all statistical analyses. Assumption 1 is the most critical, but it is also the most difficult to confirm, as we will discuss in Chapter 6. Assumption 3 is easy to meet in this case. The sampling structure and null hypothesis being tested are very simple. For more complex questions, however, the appropriate randomization method may not be obvious, and there may be more than one way to construct the null distribution (Gotelli and Graves 1996).

Advantages and Disadvantages of Monte Carlo Methods

The chief conceptual advantage of Monte Carlo methods is that it makes clear and explicit the underlying assumptions and the structure of the null hypothesis. In contrast, conventional parametric analyses often gloss over these features, perhaps because the methods are so familiar. Another advantage of Monte Carlo methods over parametric analysis is that it does not require the assumption that the data are sampled from a specified probability distribution, such as the normal. Finally, Monte Carlo simulations allow you to tailor your statistical test to particular questions and datasets, rather than having to shoehorn them into a conventional test that may not be the most powerful method for your question, or whose assumptions may not match the sampling design of your data.

The chief disadvantage of Monte Carlo methods is that it is computer-intensive and is not included in most traditional statistical packages (but see Gotelli and Entsminger 2003). As computers get faster and faster, older limitations on computational methods are disappearing, and there is no reason that even very complex statistical analyses cannot be run as a Monte Carlo simulation. However, until such routines become widely available, Monte Carlo methods are available only to those who know a programming language and can write their own programs.[4]

A second disadvantage of Monte Carlo analysis is psychological. Some scientists are uneasy about Monte Carlo methods because different analyses of the same dataset can yield slightly different answers. For example, we re-ran the analysis on the ant data in Table 5.1 ten times and got *P*-values that ranged from 0.030 to 0.046. Most researchers are more comfortable with parametric analyses, which have more of an air of objectivity because the same *P*-value results each time the analysis is repeated.

A final weakness is that the domain of inference for a Monte Carlo analysis is subtly more restrictive than that for a parametric analysis. A parametric analysis assumes a specified distribution and allows for inferences about the underlying parent population from which the data were sampled. Strictly speaking, inferences from Monte Carlo analyses (at least those based on simple randomization tests) are limited to the specific data that have been collected. However, if a sample is representative of the parent population, the results can be generalized cautiously.

[4] One of the most important things you can do is to take the time to learn a real programming language. Although some individuals are proficient at programming macros in spreadsheets, macros are practical only for the most elementary calculations; for anything more complex, it is actually simpler to write a few lines of computer code than to deal with the convoluted (and error-prone) steps necessary to write macros. There are now many computer languages available for you to choose from. We both prefer to use R, an open-source package that includes many built-in mathematical and statistical functions (R-project.org). Unfortunately, learning to program is like learning to speak a foreign language—it takes time and practice, and there is no immediate payoff. Sadly, our academic culture doesn't encourage the learning of programming skills (or languages other than English). But if you can overcome the steep learning curve, the scientific payoff is tremendous. The best way to begin is not to take a class, but to obtain software, work through examples in the manual, and try to code a problem that interests you. Hilborn and Mangel's *The Ecological Detective* (1997) contains an excellent series of ecological exercises to build your programming skills in any language. Bolker (2008) provides a wealth of worked examples using R for likelihood analyses and stochastic simulation models. Not only will programming free you from the chains of canned software packages, it will sharpen your analytical skills and give you new insights into ecological and statistical models. You will have a deeper understanding of a model or a statistic once you have successfully programmed it!

Parametric Analysis

Parametric analysis refers to the large body of statistical tests and theory built on the assumption that the data being analyzed were sampled from a specified distribution. Most statistical tests familiar to ecologists and environmental scientists specify the normal distribution. The parameters of the distribution (e.g., the population mean μ and variance σ^2) are then estimated and used to calculate tail probabilities for a true null hypothesis. A large statistical framework has been built around the simplifying assumption of normality of data. As much as 80% to 90% of what is taught in standard statistics texts falls under this umbrella. Here we use a common parametric method, the **analysis of variance (ANOVA)**, to test for differences in the group means of the sample data.[5] There are three steps in parametric analysis:

1. Specify the test statistic.
2. Specify the null distribution.
3. Calculate the tail probability.

Step 1: Specifying the Test Statistic

Parametric analysis of variance assumes that the data are drawn from a normal, or Gaussian, distribution. The mean and variance of these curves can be estimated from the sample data (see Table 5.1) using Equations 3.1 and 3.9. Figure 5.4 shows the distributions used in parametric analysis of variance. The original data are arrayed on the x-axis, and each color represents a different habitat (black circles for the forest samples, blue circles for the field samples).

Sir Ronald Fisher

[5] The framework for modern parametric statistical theory was largely developed by the remarkable Sir Ronald Fisher (1890–1962), and the F-ratio is named in his honor (although Fisher himself felt that the ratio needed further study and refinement). Fisher held the Balfour Chair in Genetics at Cambridge University from 1943 to 1957 and made fundamental contributions to the theory of population genetics and evolution. In statistics, he developed the analysis of variance (ANOVA) to analyze crop yields in agricultural systems, in which it may be difficult or impossible to replicate treatments. Many of the same constraints face ecologists in the design of their experiments today, which is why Fisher's methods continue to be so useful. His classic book, *The Design of Experiments* (1935) is still enjoyable and worthwhile reading. It is ironic that Fisher became uneasy about his own methods when dealing with experiments he could design well, whereas today many ecologists apply his methods to ad hoc observations in poorly designed natural experiments (see Chapters 4 and 6). (Photograph courtesy of the Ronald Fisher Memorial Trust.)

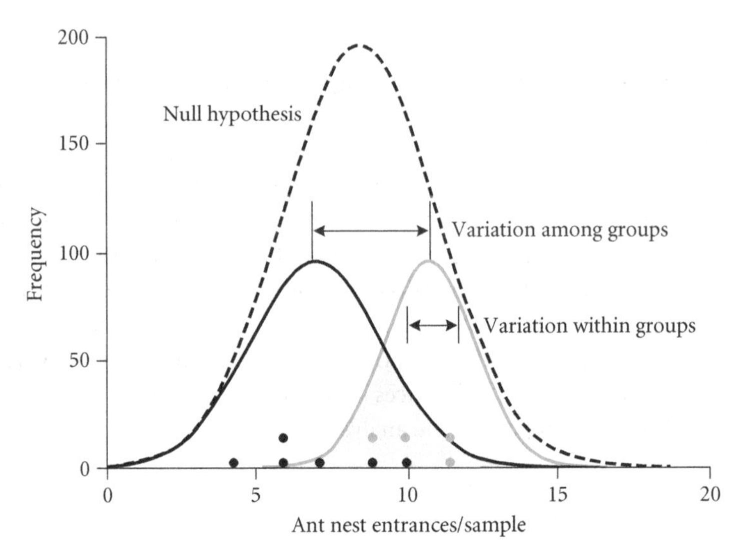

Figure 5.4 Normal distributions based on sample data in Table 5.1 The data in Table 5.1 are shown as symbols (black circles and curve, forest samples; blue circles and curve, field samples) that indicate the number of ant nests counted in each quadrat. The null hypothesis is that all the data were drawn from the same population whose normal distribution is indicated by the dashed line. The alternative hypothesis is that each habitat has its own distinct mean (and variance), indicated by the two smaller normal distributions. The smaller the shaded overlap of the two distributions, the less likely it is that the null hypothesis is true. Measures of variation among and within groups are used to calculate the F-ratio and test the null hypothesis.

First consider the null hypothesis: that both sets of data were drawn from a single underlying normal distribution, which is estimated from the mean and variance of all the data (dashed curve; mean = 8.5, standard deviation = 2.54). The alternative hypothesis is that the samples were drawn from two different populations, each of which can be characterized by a different normal distributions, one for the forest and one for the field. Each distribution has its own mean, although we assume the variance is the same (or similar) for each of the two groups. These two curves are also illustrated in Figure 5.4, using the summary statistics calculated in Table 5.2.

How is the null hypothesis tested? The closer the two curves are for the forest and field data, the more likely the data would be collected given the null hypothesis is true, and the single dashed curve best represents the data. Conversely, the more separate the two curves are, the less likely it is that the data represent a single population with a common mean and variance. The area of overlap between these two distributions (shaded in Figure 5.4) should be a measure of how close or how far apart the distributions are.

Fisher's contribution was to quantify that overlap as a ratio of two variables. The first is the amount of variation among the groups, which we can think of as the variance (or standard deviation) of the means of the two groups. The second is the amount of variation within each group, which we can think of as the variance of the observations around their respective means. **Fisher's F-ratio** can be interpreted as a ratio of these two sources of variation:

$$F = (\text{variance among groups} + \text{variance within groups}) / \text{variance within groups} \quad (5.1)$$

In Chapter 10, we will explain in detail how to calculate the numerator and denominator of the F-ratio. For now, we simply emphasize that the ratio measures the relative size of two sources of variation in the data: variation among groups and within groups. For these data, the F-ratio is calculated as 33.75/ 3.84 = 8.78.

In an ANOVA, the F-ratio is the test statistic that describes (as a single number) the pattern of differences among the means of the different groups being compared.

Step 2: Specifying the Null Distribution

The null hypothesis is that all the data were drawn from the same population, so that any differences between the means of the groups are no larger than would be expected by chance. If this null hypothesis is true, then the variation among groups will be small, and we expect to find an F-ratio of 1.0. The F-ratio will be correspondingly larger than 1.0 if the means of the groups are widely separated (large among-group variation) relative to the variation within groups.[6] In this example, the observed F-ratio of 8.78 is almost 10 times larger than the expected value of 1.0, a result that seems unlikely if the null hypothesis were true.

Step 3: Calculating the Tail Probability

The *P*-value is an estimate of the probability of obtaining an F-ratio ≥ 8.78, given that the null hypothesis is true. Figure 5.5 shows the theoretical distribution of the F-ratio and the observed F-ratio of 8.78, which lies in the extreme right hand tail of the distribution. What is its tail probability (or *P*-value)?

[6] As in Monte Carlo analysis, there is the theoretical possibility of obtaining an F-ratio that is smaller than expected by chance. In such a case, the means of the groups are unusually similar, and differ less than expected if the null hypothesis were true. Unusually small F-ratios are rarely seen in the ecological literature, although Schluter (1990) used them as an index of species-for-species matching of body sizes and community convergence.

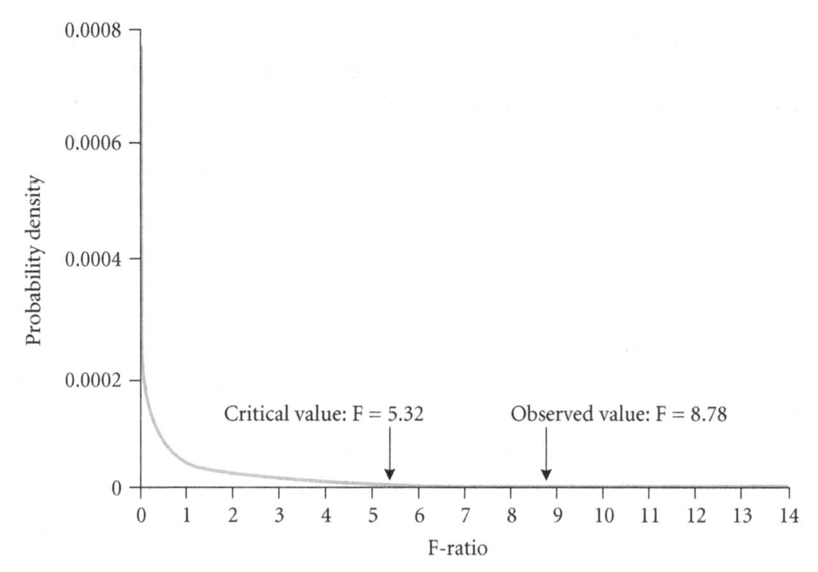

Figure 5.5 Theoretical F-distribution. The larger the observed F-ratio, the more unlikely it would be if the null hypothesis were true. The critical value for this distribution equals 5.32; the area under the curve beyond this point is equal to 5% of the area under the entire curve. The observed F-ratio of 8.78 lies beyond the critical value, so $P(\text{ant data} \mid \text{null hypothesis}) \leq 0.05$. In fact, the actual probability of $P(\text{ant data} \mid \text{null hypothesis}) = 0.018$, because the area under the curve to the right of the observed F-ratio represents 1.8% of the total area under the curve. Compare this result to the *P*-value of 0.036 from the Monte Carlo analysis (see Figure 5.3).

The *P*-value of this F-ratio is calculated as the probability mass of the **F-ratio distribution** (or area under the curve) equal to or greater than the observed F-ratio. For these data, the probability of obtaining an F-ratio as large or larger than 8.78 (given two groups and a total *N* of 10) equals 0.018. As the *P*-value is smaller than 0.05, we consider it unlikely that such a large F-ratio would occur by chance alone, and we reject the null hypothesis that our data were sampled from a single population.

Assumptions of the Parametric Method

There are two assumptions for all parametric analyses:

1. The data collected represent random, independent samples.
2. The data were sampled from a specified distribution.

As we noted for Monte Carlo analysis, the first assumption of random, independent sampling is always the most important in any analysis. The second assumption is usually satisfied because normal (bell-shaped) distributions are ubiquitous and turn up frequently in the real world.

Specific parametric tests usually include additional assumptions. For example, ANOVA further assumes that variances within each group being compared are equal (see Chapter 10). If sample sizes are large, this assumption can be modestly violated and the results will still be robust. However, if sample sizes are small (as in this example), this assumption is more important.

Advantages and Disadvantages of the Parametric Method

The advantage of the parametric method is that it uses a powerful framework based on known probability distributions. The analysis we presented was very simple, but there are many parametric tests appropriate for complex experimental and sampling designs (see Chapter 7).

Although parametric analysis is intimately associated with the testing of statistical null hypotheses, it may not be as powerful as sophisticated Monte Carlo models that are tailored to particular questions or data. In contrast to Bayesian analysis, parametric analysis rarely incorporates a priori information or results from other experiments. Bayesian analysis will be our next topic—after a brief detour into non-parametric statistics.

Non-Parametric Analysis: A Special Case of Monte Carlo Analysis

Non-parametric statistics are based on the analysis of ranked data. In the ant example, we would rank the observations from largest to smallest and then calculate statistics based on the sums, distributions, or other synthetic measures of those ranks. Non-parametric analyses do not assume a specified parametric distribution (hence the name), but they still require independent, random sampling, (as do all statistical analyses). A non-parametric test is in essence a Monte Carlo analysis of ranked data, and non-parametric statistical tables give *P*-values that would be obtained by a randomization test on ranks. Thus, we have already described the general rationale and procedures for such tests.

Although they are used commonly by some ecologists and environmental scientists, we do not favor non-parametric analyses for three reasons. First, using ranked data wastes information that is present in the original observations. A Monte Carlo analysis of the raw data is much more informative, and often more powerful. One justification that is offered for a non-parametric analysis is that the ranked data may be more robust to measurement error. However, if the original observations are so error-prone that only the ranks are reliable, it is probably a good idea to re-do the experiment using measurement methods that offer greater accuracy. Second, relaxing the assumption of a parametric distribution (e.g., normality) is not such a great advantage, because parametric analyses often are robust to violations of this assumption (thanks to the Central Limit Theorem). Third, non-parametric methods are available only for extremely simple experimental designs, and cannot easily incorporate covariates or blocking structures. We have found that virtually all needs for statistical analysis of ecological and environmental data can be met by parametric, Monte Carlo, or Bayesian approaches.

Bayesian Analysis

Bayesian analysis is the third major framework for data analysis. Scientists often believe that their methods are "objective" because they treat each experiment as a *tabula rasa* (blank slate): the simple statistical null hypothesis of random variation reflects ignorance about cause and effect. In our example of ant nest densities in forests and fields, our null hypothesis is that the two are equal, or that being in forests and fields has no consistent effect on ant nest density. Although it is possible that no one has ever investigated ant nests in forests and fields before, it is extremely unlikely; our reading of the literature on ant biology prompted us to conduct this particular study. So why not use data that already exist to frame our hypotheses? If our only goal is the hypothetico-deductive one of falsification of a null hypothesis, and if previous data all suggest that forests and fields differ in ant nest densities, it is very likely that we, too, will falsify the null hypothesis. Thus, we needn't waste time or energy doing the study yet again.

Bayesians argue that we could make more progress by specifying the observed difference (e.g., expressed as the DIF or the F-ratio described in the previous sections), and then using our data to extend earlier results of other investigators. Bayesian analysis allows us to do just this, as well as to quantify the probability of the observed difference. This is the most important difference between Bayesian and frequentist methods.

There are six steps in Bayesian inference:

1. Specify the hypothesis.
2. Specify parameters as random variables.
3. Specify the prior probability distribution.
4. Calculate the likelihood.
5. Calculate the posterior probability distribution.
6. Interpret the results.

Step 1: Specifying the Hypothesis

The primary goal of a Bayesian analysis is to determine the probability of the hypothesis *given* the data that have been collected: $P(\mathrm{H} \mid \mathrm{data})$. The hypothesis needs to be quite specific, and it needs to be quantitative. In our parametric analysis of ant nest density, the hypothesis of interest (i.e., the alternative hypothesis) was that the samples were drawn from two populations with different means and equal variances, one for the forest and one for the field. We did not test this hypothesis directly. Instead, we tested the null hypothesis: the observed value of F was no larger than that expected if the samples were drawn from a single population. We found that the observed F-ratio was improbably large ($P = 0.018$), and we rejected the null hypothesis.

We could specify more precisely the null hypothesis and the alternative hypothesis as hypotheses about the value of the F-ratio. Before we can specify these hypotheses, we need to know the critical value for the F-distribution in Figure 5.5. In other words, how large does an F-ratio have to be in order to have a P-value ≤ 0.05?

For 10 observations of ants in two groups (field and forest), the critical value of the F-distribution (i.e., that value for which the area under the curve equals 5% of the total area) equals 5.32 (see Figure 5.5). Thus, any observed F-ratio greater than or equal to 5.32 would be grounds for rejecting the null hypothesis. Remember that the general hypothetico-deductive statement of the probability of the null hypothesis is $P(\text{data} \mid H_0)$. In the ant nest example, the data result in an F-ratio equal to 8.78. If the null hypothesis is true, the observed F-ratio should be a random sample from the F-distribution shown in Figure 5.5. Therefore we ask what is $P(F_{obs} = 8.78 \mid F_{theoretical})$?

In contrast, Bayesian analysis proceeds by inverting this probability statement: what is the probability of the hypothesis *given* the data we collected [$P(H \mid \text{data})$]? The ant nest data can be expressed as $F = 8.78$. How are the hypotheses expressed in terms of the F-distribution? The null hypothesis is that the ants were sampled from a single population. In this case, the expected value of the F-ratio is small ($F < 5.32$, the critical value). The alternative hypothesis is that the ants were sampled from two populations, in which case the F-ratio would be large ($F = 5.32$). Therefore, the Bayesian analysis of the alternative hypothesis calculates $P(F \geq 5.32 \mid F_{obs} = 8.78)$. By the First Axiom of Probability, $P(F < 5.32 \mid F_{obs}) = 1 - P(F \geq 5.32 \mid F_{obs})$.

A modification of Bayes' Theorem (introduced in Chapter 1) allows us to directly calculate $P(\text{hypothesis} \mid \text{data})$:

$$P(\text{hypothesis} \mid \text{data}) = \frac{P(\text{hypothesis})P(\text{data} \mid \text{hypothesis})}{P(\text{data})} \quad (5.2)$$

In Equation 5.2, $P(\text{hypothesis} \mid \text{data})$ on the left-hand side of the equation is called the **posterior probability distribution** (or simply the posterior), and is the quantity of interest. The right-hand side of the equation consists of a fraction. In the numerator, the term $P(\text{hypothesis})$ is referred to as the **prior**

probability distribution (or simply the prior), and is the probability of the hypothesis of interest *before* you conducted the experiment. The next term in the numerator, P(data | hypothesis), is referred to as the **likelihood** of the data; it reflects the probability of observing the data given the hypothesis.[7] The denominator, P(data) is a normalizing constant that reflects the probability of the data given all possible hypotheses.[8] Because it is simply a normalizing constant (and so scales our posterior probability to the range [0,1]),

$$P(\text{hypothesis} \mid \text{data}) \propto P(\text{hypothesis})P(\text{data} \mid \text{hypothesis})$$

(where $\propto$ means "is proportional to") and we can focus our attention on the numerator.

Returning to the ants and their F-ratios, we focus on $P(F \geq 5.32 \mid F_{obs} = 8.78)$. We have now specified our hypothesis quantitatively in terms of the relationship between the F-ratio we observe (the data) and the critical value of $F \geq 5.32$ (the hypothesis). This is a more precise hypothesis than the hypothesis dis-

[7] Fisher developed the concept of likelihood as a response to his discomfort with Bayesian methods of inverse probability:

> *What has now appeared, is that the mathematical concept of probability is inadequate to express our mental confidence or diffidence in making such inferences, and that the mathematical quantity which appears to be appropriate for measuring our order of preference among different possible populations does not in fact obey the laws of probability. To distinguish it from probability, I have used the term 'Likelihood' to designate this quantity. (Fisher 1925, p. 10).*

The likelihood is written as L(hypothesis | data) and is directly proportional (but not equal to) the probability of the *observed* data given the hypothesis of interest: $L(\text{hypothesis} \mid \text{data}) = cP(\text{data}_{obs} \mid \text{hypothesis})$. In this way, the likelihood differs from a frequentist P-value, because the P-value expresses the probability of the infinitely many possible samples of the data given the statistical null hypothesis (Edwards 1992). Likelihood is used extensively in information-theoretic approaches to statistical inference (e.g., Hilborn and Mangel 1997; Burnham and Anderson 2010), and it is a central part of Bayesian inference. However, likelihood does not follow the axioms of probability. Because the language of probability is a more consistent way of expressing our confidence in a particular outcome, we feel that statements of the probabilities of different hypotheses (which are scaled between 0 and 1) are more easily interpreted than likelihoods (which are not).

[8] The denominator is calculated as

$$\int_i P(H_i)P(data \mid H_i)dH$$

cussed in the previous two sections, that there is a difference between the density of ants in fields and in forests.

Step 2: Specifying Parameters as Random Variables

A second fundamental difference between frequentist analysis and Bayesian analysis is that, in a frequentist analysis, parameters (such as the true population means μ_{forest} and μ_{field}, their standard deviations σ^2, or the F-ratio) are *fixed*. In other words, we assume there is a *true value* for the density of ant nests in fields and forests (or at least in the field and forest that we sampled), and we estimate those parameters from our data. In contrast, Bayesian analysis considers these parameters to be *random variables*, with their own associated parameters (e.g., means, variances). Thus, for example, instead of the population mean of ant colonies in the field being a fixed value μ_{field}, the mean could be expressed as a normal random variable with its own mean and variance: $\mu_{field} \sim N(\lambda_{field}, \sigma^2)$. Note that the random variable representing ant colonies does not have to be normal. The type of random variable used for each population parameter should reflect biological reality, not statistical or mathematical convenience. In this example, however, it is reasonable to describe ant colony densities as normal random variables: $\mu_{field} \sim N(\lambda_{field}, \sigma^2)$, $\mu_{forest} \sim N(\lambda_{forest}, \sigma^2)$.

Step 3: Specifying the Prior Probability Distribution

Because our parameters are random variables, they have associated probability distributions. Our unknown population means (the μ_{field} and μ_{forest} terms) themselves have normal distributions, with associated unknown means and variances. To do the calculations required by Bayes' Theorem, we have to specify the prior probability distributions for these parameters—that is, what are probability distributions for these random variables *before* we do the experiment?[9]

[9] The specification of priors is a fundamental division between frequentists and Bayesians. To a frequentist, specifying a prior reflects subjectivity on the part of the investigator, and thus the use of a prior is considered unscientific. Bayesians argue that specifying a prior makes explicit all the hidden assumptions of an investigation, and so it is a more honest and objective approach to doing science. This argument has lasted for centuries (see reviews in Effron 1986 and in Berger and Berry 1988), and was one reason for the marginalization of Bayesians within the statistical community. However, the advent of modern computational techniques allowed Bayesians to work with uninformative priors, such as the ones we use here. It turns out that, with uninformative priors, Bayesian and frequentist results are very similar, although their final interpretations remain different. These findings have led to a recent renaissance in Bayesian statistics and relatively easy-to-use software for Bayesian calculations is now widely available (Kéry 2010).

We have two basic choices for specifying the prior. First, we can comb and reanalyze data in the literature, talk to experts, and come up with reasonable estimates for the density of ant nests in fields and forests. Alternatively, we can use an uninformative prior, for which we initially estimate the density of ant nests to be equal to zero and the variances to be very large. (In this example, we set the population variances to be equal to 100,000.) Using an uninformative prior is equivalent to saying that we have no prior information, and that the mean could take on virtually any value with roughly equal probability.[10] Of course, if you have more information, you can be more specific with your prior. Figure 5.6 illustrates the uninformative prior and a (hypothetical) more informative one.

Similarly, the standard deviation σ for μ_{field} and μ_{forest} also has a prior probability distribution. Bayesian inference usually specifies an inverse gamma distribution[11] for the variances; as with the priors on the means, we use an uninformative prior for the variance. We write this symbolically as $\sigma^2 \sim IG(1{,}000, 1{,}000)$ (read "the variance is distributed as an inverse gamma distribution with parameters 1,000 and 1,000). We also calculate the *precision* of our estimate of variance, which we symbolize as τ, where $\tau = 1/\sigma^2$. Here τ is a gamma random variable (the inverse of an inverse gamma is a gamma), and we write this symbolically as $\tau \sim \Gamma(0.001, 0.001)$ (read "tau is distributed as a gamma distribution with parameters 0.001, 0.001"). The form of this distribution is illustrated in Figure 5.7.

[10] You might ask why we don't use a uniform distribution, in which all values have equal probability. The reason is that the uniform distribution is an improper prior. Because the integral of a uniform distribution is undefined, we cannot use it to calculate a posterior distribution using Bayes' Theorem. The uninformative prior N(0,100,000) is nearly uniform over a huge range, but it can be integrated. See Carlin and Louis (2000) for further discussion of improper and uninformative priors.

[11] The gamma distribution for precision and the inverse gamma distribution for variance are used for two reasons. First, the precision (or variance) needs to take on only positive values. Thus, any probability density function that has only positive values could be used for priors for precision or variance. For continuous variables, such distributions include a uniform distribution that is restricted to positive numbers and the gamma distribution. The gamma distribution is somewhat more flexible than the uniform distribution and allows for better incorporation of prior knowledge.

Second, before the use of high-speed computation, most Bayesian analyses were done using conjugate analysis. In a conjugate analysis, a prior probability distribution is sought that has the same form as the posterior probability distribution. This convenient mathematical property allows for closed-form (analytical) solutions to the complex integration involved in Bayes' Theorem. For data that are normally distributed, the conjugate prior for the parameter that specifies the mean is a normal distribution, and the conjugate prior for the parameter that specifies the precision (= 1/variance) is a gamma distribution. For data that are drawn from a Poisson distribution, the gamma distribu-

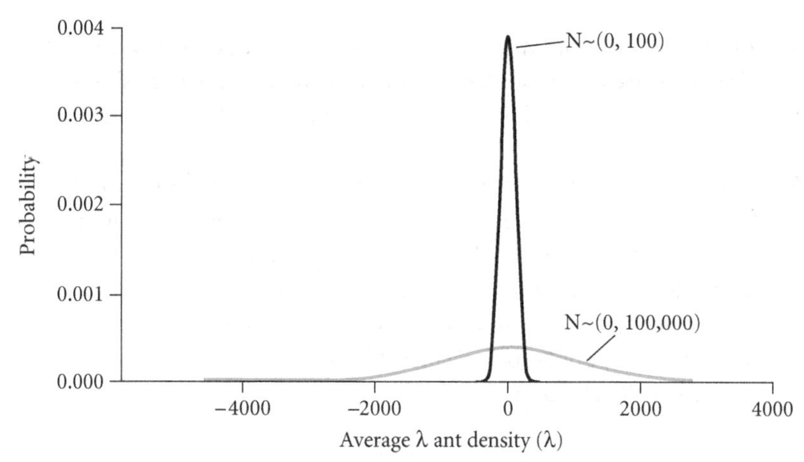

Figure 5.6 Prior probability distributions for Bayesian analysis. Bayesian analysis requires specification of prior probability distributions for the statistical parameters of interest. In this analysis of the data in Table 5.1, the parameter is average ant density, λ. We begin with a simple uninformative prior probability distribution that average ant density is described by a normal distribution with mean 0 and standard deviation of 100,000 (blue curve). Because the standard deviation is so large, the distribution is nearly uniform over a large range of values: between –1500 and +1500, the probability is essentially constant (~0.0002), which is appropriate for an uninformative prior. The black curve represents a more precise prior probability distribution. Because the standard deviation is much smaller (100), the probability is no longer constant over a larger range of values, but instead decreases more sharply at extreme values.

tion is the conjugate prior for the parameter that defines the mean (or rate) of the Poisson distribution. For further discussion, see Gelman et al. (1995).

The gamma distribution is a two-parameter distribution, written as $\Gamma(a,b)$, where a is referred to the shape parameter and b is the scale parameter. The probability density function of the gamma distribution is

$$P(X) = \frac{b^a}{\Gamma(a)} X^{(a-1)} e^{-bX}, \text{ for } X > 0$$

where $\Gamma(a)$ is the gamma function $\Gamma(n) = (n-1)!$ for integers $n > 0$. More generally, for real numbers z, the gamma function is defined as

$$\Gamma(z) = \int_0^1 \left[\ln \frac{1}{t} \right]^{z-1} dt$$

The gamma distribution has expected value $E(X) = a/b$ and variance $= a/b^2$. Two distributions used commonly by statisticians are special cases of the gamma distribution. The χ^2 distribution with ν degrees of freedom is equal to $\Gamma(\nu/2, 0.5)$. The exponential distribution with parameter β that was discussed in Chapter 2 is equal to $\Gamma(1,\beta)$.

Finally, if the random variable $1/X \sim \Gamma(a,b)$, then X is said to have an inverse gamma (IG) distribution. To obtain an uninformative prior for the variance of a normal random variable, we take the limit of the IG distribution as a and b both approach 0. This is the reason we use $a = b = 0.001$ as the prior parameters for the gamma distribution describing the precision of the estimate.

This use of precision should make some intuitive sense; because our estimate of variability decreases when we have more information, the precision of our estimate increases. Thus, a high value of precision equals a low variance of our estimated parameter.

We now have our prior probability distributions. The unknown population means of ants in fields and forests, μ_{field} and μ_{forest}, are both normal random variables with expected values equal to λ_{field} and λ_{forest} and unknown (but equal) variances, σ^2. These expected means themselves are normal random variables with expected values of 0 and variances of 100,000. The population precision τ is the reciprocal of the population variance σ^2, and is a gamma random variable with parameters (0.001, 0.001):

$$\mu_i \sim N(\lambda_i, \sigma^2)$$

$$\lambda_i \sim N(0, 100{,}000)$$

$$\tau = 1/\sigma^2 \sim \Gamma(0.001, 0.001)$$

These equations completely specify P(hypothesis) in the numerator of Equation 5.2. If we had real prior information, such as the density of ant nests in other fields and forests, we could use those values to more accurately specify the expected means and variances of the λ's.

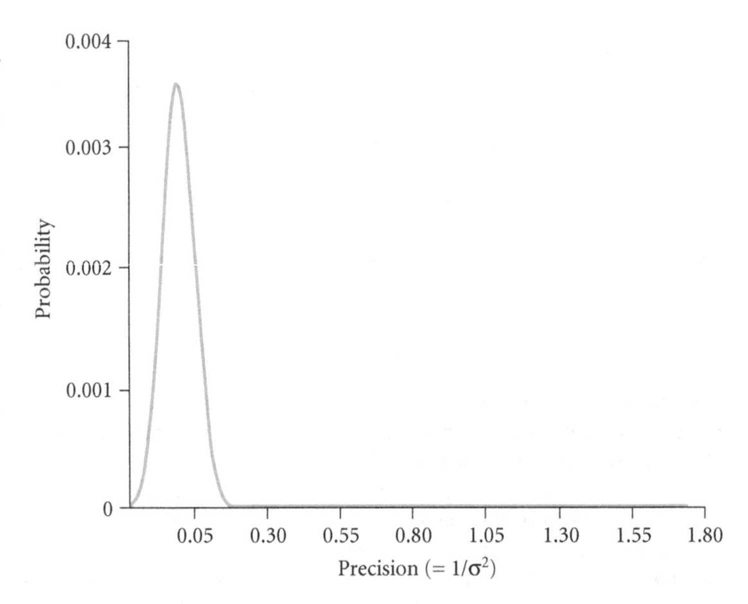

Figure 5.7 Uninformative prior probability distribution for the precision (= 1/variance). Bayesian inference requires not only a specification of a prior distribution for the mean of the variable (see Figure 5.6), but also a specification for the precision (= 1/variance). Bayesian inference usually specifies an inverse gamma distribution for the variances. As with the distribution of the means, an uninformative prior is used for the variance. In this case, the variance is distributed as an inverse gamma distribution with parameters 1000 and 1000: $\sigma^2 \sim IG(1{,}000, 1{,}000)$. Because the variance is very large, the precision is small.

Step 4: Calculating the Likelihood

The other quantity in the numerator of Bayes' Theorem (see Equation 5.2) is the likelihood, $P(\text{data}_{obs} \mid \text{hypothesis})$. The likelihood is a distribution that is proportional to the probability of the observed data given the hypothesis.[12] Each parameter λ_i and τ of our prior probability distribution has its own likelihood function. In other words, the different values of λ_i have likelihood functions that are normal random variables with means equal to the observed means (here, 7 ant nests per quadrat in the forest and 10.75 ant nests per quadrat in the field; see Table 5.2). The variances are equal to the sample variances (4.79 in the forest and 2.25 in the field). The parameter τ has a likelihood function that is a gamma random variable. Finally, the F-ratio is an F-random variable with expected value (or **maximum likelihood** estimate)[13] equal to 8.78 (calculated from the data using Equation 5.1).

Step 5: Calculating the Posterior Probability Distribution

To calculate the posterior probability distribution, $P(\text{H} \mid \text{data})$, we apply Equation 5.2, multiply the prior by the likelihood, and divide by the normalizing constant (or marginal likelihood). Although this multiplication is straightforward for well-behaved distributions like the normal, computational methods are used

[12] There is a key difference between the likelihood function and a probability distribution. The probability of data given a hypothesis, $P(\text{data} \mid \text{H})$, is the probability of any set of random data given a specific hypothesis, usually the statistical null hypothesis. The associated probability density function (see Chapter 2) conforms to the First Axiom of Probability—that the sum of all probabilities = 1. In contrast, the likelihood is based on only one dataset (the observed sample) and may be calculated for many different hypotheses or parameters. Although it is a function, and results in a distribution of values, the distribution is not a probability distribution, and the sum of all likelihoods does not necessarily sum to 1.

[13] The maximum likelihood is the value for our parameter that maximizes the likelihood function. To obtain this value, take the derivative of the likelihood, set it to 0, and solve for the parameter values. Frequentist parameter estimates are usually equal to maximum-likelihood estimates for the parameters of the specified probability density functions. Fisher claimed, in his system of fiducial inference, that the maximum-likelihood estimate gave a realistic probability of the alternative (or null) hypothesis. Fisher based this claim on a statistical axiom that he defined by saying that given observed data Y, the likelihood function $L(\text{H} \mid Y)$ contains all the relevant information about the hypothesis H. See Chapter 14 for further application of maximum likelihood and Bayesian methods. Berger and Wolpert (1984), Edwards (1992), and Bolker (2008) provide additional discussion of likelihood methods.

to iteratively estimate the posterior distribution for any prior distribution (Carlin and Louis 2000; Kéry 2010).

In contrast to the results of a parametric or Monte Carlo analysis, the result of a Bayesian analysis is a probability *distribution*, not a single *P*-value. Thus, in this example, we express $P(F \geq 5.32 \mid F_{obs})$ as a *random variable* with expected mean and variance. For the data in Table 5.1, we calculated posterior estimates for all the parameters: λ_{field}, λ_{forest}, σ^2 $(= 1/\tau)$ (Table 5.7). Because we used uninformative priors, the parameter estimates for the Bayesian and parametric analyses are similar, though not identical.

The hypothesized F-distribution with expected value equal to 5.32 is shown in Figure 5.8. To compute $P(F \geq 5.32 \mid F_{obs})$, we simulated 20,000 F-ratios using a Monte Carlo algorithm. The average, or expected value, of all of these F-ratios is 9.77; this number is somewhat larger than the frequentist (maximum likelihood) estimate of 8.78 because our sample size is very small ($N = 10$). The spread about the mean is large: SD = 7.495; hence the precision of our estimate is relatively low (0.017).

Step 6: Interpreting the Results

We now return to the motivating question: What is the probability of obtaining an F-ratio ≥ 5.32, given the data on ant nest density in Table 5.1? In other words, how probable is it that the mean ant nest densities in the two habitats really differ? We can answer this question directly by asking what percentage of values in Figure 5.8 are greater than or equal to 5.32. The answer is 67.3%. This doesn't look quite as convincing as the *P*-value of 0.018 (1.8%) obtained in the parametric analysis in the previous section. In fact, the percentage of values in Figure 5.8 that are ≥8.78, the observed value for which we found $P = 0.018$ in the parametric analysis section, is 46.5. In other words, the Bayesian analysis (see Figure 5.8)

TABLE 5.7 **Parametric and Bayesian estimators for the means and standard deviations of the data in Table 5.1**

	Estimator			
Analysis	λ_{Forest}	λ_{Field}	σ_{Forest}	σ_{Field}
Parametric	7.00	10.75	2.19	1.50
Bayesian (uninformed prior)	6.97	10.74	0.91	1.13
Bayesian (informed prior)	7.00	10.74	1.01	1.02

The standard deviation estimators from the Bayesian analysis are slightly smaller because the Bayesian analysis incorporates information from the prior probability distribution. Bayesian analysis may give different results, depending on the shape of the prior distribution and the sample size.

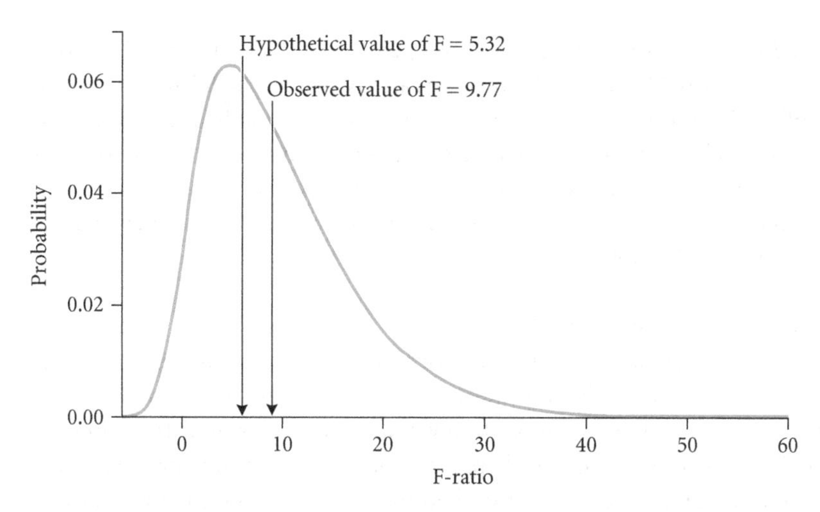

Figure 5.8 Hypothesized F-distribution with an expected value of 5.32. We are interested in determining the probability of $F \geq 5.32$ (the critical value for $P < 0.05$ in a standard F-ratio test), given the data on ant nest densities in Table 5.1. This is the inverse of the traditional null hypothesis, which asks: what is the probability of obtaining the data, given the null hypothesis? In the Bayesian analysis, the posterior probability of $F \geq 5.32$, given the data in Table 5.1, is the proportion of the area under the curve to the right of F = 5.32, which is 0.673. In other words, *P*(hypothesis that the fields and forests differ in average density of ant nests | observed data in Table 5.1) = 0.673. The most likely posterior value of the F-ratio is 9.77. The proportion of area under the curve to the right of this value is 0.413. The parametric analysis says that the observed data are unlikely given a null distribution specified by the F-ratio [$P(\text{data} \mid H_0) = 0.018$], whereas the Bayesian analysis says that the probability of observing an F-ratio of 9.77 or larger is not unlikely given the data [$P(F \geq 5.32 \mid \text{data}) = 0.673$].

indicates $P = 0.67$ that ant nest densities in the two habitats are truly different, given the Bayesian estimate of F = 9.77 [$P(F \geq 5.32 \mid F_{obs}) = 0.67$]. In contrast, the parametric analysis (see Figure 5.5) indicates $P = 0.018$ that the parametric estimate of F = 8.78 (or a greater F-ratio) would be found given the null hypothesis that the ant densities in the two habitats are the same [$P(F_{obs} \mid H_0) = 0.018$].

Using Bayesian analysis, a different answer would result if we used a different prior distribution rather than the uninformative prior of means of 0 with large variances. For example, if we used prior means of 15 for the forest and 7 for the field, an among-group variance of 10, and a within-group variance of 0.001, then $P(F \geq 5.32 \mid \text{data}) = 0.57$. Nevertheless, you can see that the posterior probability does depend on the priors that are used in the analysis (see Table 5.7).

Finally, we can estimate a 95% Bayesian **credibility interval** around our estimate of the observed F-ratio. As with Monte Carlo methods, we estimate the

95% credibility interval as the 2.5 and 97.5 percentiles of the simulated F-ratios. These values are 0.28 and 28.39. Thus, we can say that we are 95% sure that the value of the F-ratio for this experiment lies in the interval [0.28, 28.39]. Note that the spread is large because the precision of our estimate of the F-ratio is low, reflecting the small sample size in our analysis. You should compare this interpretation of a credibility interval with the interpretation of a confidence interval presented in Chapter 3.

Assumptions of Bayesian Analysis

In addition to the standard assumptions of all statistics methods (random, independent observations), the key assumption of Bayesian analysis is that the parameters to be estimated are random variables with known distributions. In our analysis, we also assumed little prior information (uninformative priors), and therefore the likelihood function had more influence on the final calculation of the posterior probability distribution than did the prior. This should make intuitive sense. On the other hand, if we had a lot of prior information, our prior probability distribution (e.g., the black curve in Figure 5.6) would have low variance and the likelihood function would not substantially change the variance of the posterior probability distribution. If we had a lot of prior information, and we were confident in it, we would not have learned much from the experiment. A well-designed experiment should decrease the posterior estimate of the variance relative to the prior estimate of the variance.

The relative contributions of prior and likelihood to the posterior estimate of the probability of the mean density of nests in the forest are illustrated in Figure 5.9. In this figure, the prior is flat over the range of the data (i.e., it is an unin-

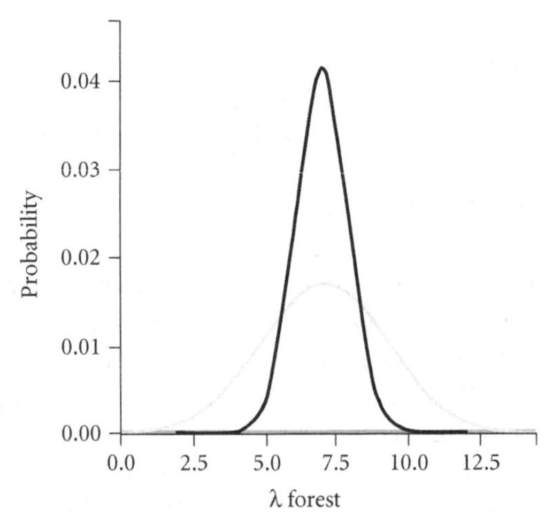

Figure 5.9 Probability densities for the prior, likelihood, and posterior for the mean number of ant nests in the forest plots. In the Bayesian analysis of the data in Table 5.1, we used an uninformative prior distribution with a mean of 0 and a variance of 100,000. This normal distribution generated an essentially uniform range of prior probability values (dark blue) over the range of values for λ_{forest}, the density of forest ants. The likelihood (light blue) represents the probability based on the observed data (see Table 5.1), and the posterior probability (black) is the product of the two. Notice that the posterior distribution is more precise than the likelihood because it takes into account the (modest) information contained in the prior.

formative prior), the likelihood is the distribution based on the observed data (see Table 5.1), and the posterior is the product of the two. Note that the variance of the posterior is smaller than the variance of the likelihood because we had some prior information. However, the expected values of the likelihood (7.0) and the posterior (6.97) are very close, because all values of the prior were approximately equally likely over the range of the data.

After doing this experiment, we have new information on each of the parameters that could be used in analysis of subsequent experiments. For example, if we were to repeat the experiment, we could use the posterior in Figure 5.9 as our prior for the average density of ant nests in other forests. To do this, we would use the values for λ_i and σ_i in Table 5.7 as the estimates of λ_i and σ_i in setting up the new prior probability distributions.

Advantages and Disadvantages of Bayesian Analysis

Bayesian analysis has a number of advantages relative to parametric and Monte Carlo approaches conducted in a frequentist framework. Bayesian analysis allows for the explicit incorporation of prior information, and the results from one experiment (the posterior) can be used to inform (as a prior) subsequent experiments. The results of Bayesian analysis are interpreted in an intuitively straightforward way, and the inferences obtained are conditional on both the observed data and the prior information.

Disadvantages to Bayesian analysis are its computational challenges (Albert 2007; Clark 2007; Kéry 2007) and the requirement to condition the hypothesis on the data [i.e., $P(\text{hypothesis} \mid \text{data})$]. The most serious disadvantage of Bayesian analysis is its potential lack of objectivity, because different results will be obtained using different priors. Consequently, different investigators may obtain different results from the same dataset if they start with different preconceptions or prior information. The use of uninformative priors addresses this criticism, but increases the computational complexity.

Summary

Three major frameworks for statistical analysis are Monte Carlo, parametric, and Bayesian. All three assume that the data were sampled randomly and independently. In Monte Carlo analysis, the data are randomized or reshuffled, so that individuals are randomly re-assigned to groups. Test statistics are calculated for these randomized datasets, and the reshuffling is repeated many times to generate a distribution of simulated values. The tail probability of the observed test statistic is then estimated from this distribution. The advantage of Monte Carlo analysis is that it makes no assumptions about the distribution of the data, it makes the null hypothesis clear and explicit, and it can be tailored to individual datasets and hypotheses. The disadvantage is that it is not a general solution and usually requires computer programming to be implemented.

In parametric analysis, the data are assumed to have been drawn from an underlying known distribution. An observed test statistic is compared to a theoretical distribution based on a null hypothesis of random variation. The advantage of parametric analysis is that it provides a unifying framework for statistical tests of classical null hypotheses. Parametric analysis is also familiar to most ecologists and environmental scientists and is widely implemented in statistical software. The disadvantage of parametric analysis is that the tests do not specify the probability of alternative hypotheses, which often is of greater interest than the null hypothesis.

Bayesian analysis considers parameters to be random variables as opposed to having fixed values. It can take explicit advantage of prior information, although modern Bayesian methods rely on uninformative priors. The results of Bayesian analysis are expressed as probability distributions, and their interpretation conforms to our intuition. However, Bayesian analysis requires complex computation and often requires the investigators to write their own programs.

PART II

Designing Experiments

CHAPTER 6

Designing Successful Field Studies

The proper analysis of data goes hand in hand with an appropriate sampling design and experimental layout. If there are serious errors or problems in the design of the study or in the collection of the data, rarely is it possible to repair these problems after the fact. In contrast, if the study is properly designed and executed, the data can often be analyzed in several different ways to answer different questions. In this chapter, we discuss the broad issues that you need to consider when designing an ecological study. We can't overemphasize the importance of thinking about these issues *before* you begin to collect data.

What Is the Point of the Study?

Although it may seem facetious and the answer self-evident, many studies are initiated without a clear answer to this central question. Most answers will take the form of a more focused question.

Are There Spatial or Temporal Differences in Variable *Y*?

This is the most common question that is addressed with survey data, and it represents the starting point of many ecological studies. Standard statistical methods such as analysis of variance (ANOVA) and regression are well-suited to answer this question. Moreover, the conventional testing and rejection of a simple null hypothesis (see Chapter 4) yields a dichotomous yes/no answer to this question. It is difficult to even discuss mechanisms without some sense of the spatial or temporal pattern in your data. Understanding the forces controlling biological diversity, for example, requires at a minimum a spatial map of species richness. The design and implementation of a successful ecological survey requires a great deal of effort and care, just as much as is needed for a successful experimental study. In some cases, the survey study will address all of your research goals; in other cases, a survey study will be the first step in a research project. Once you have documented spatial and temporal patterns in your data, you will conduct experiments or collect additional data to address the mechanisms responsible for those patterns.

What Is the Effect of Factor *X* on Variable *Y*?

This is the question directly answered by a manipulative experiment. In a field or laboratory experiment, the investigator actively establishes different levels of Factor *X* and measures the response of Variable *Y*. If the experimental design and statistical analysis are appropriate, the resulting *P*-value can be used to test the null hypothesis of no effect of Factor *X*. Statistically significant results suggest that Factor *X* influences Variable *Y*, and that the "signal" of Factor *X* is strong enough to be detected above the "noise" caused by other sources of natural variation.[1] Certain natural experiments can be analyzed in the same way, taking advantage of natural variation that exists in Factor *X*. However, the resulting inferences are usually weaker because there is less control over confounding variables. We discuss natural experiments in more detail later in this chapter.

Are the Measurements of Variable *Y* Consistent with the Predictions of Hypothesis H?

This question represents the classic confrontation between theory and data (Hilborn and Mangel 1997). In Chapter 4, we discussed two strategies we use for this confrontation: the inductive approach, in which a single hypothesis is recursively modified to conform to accumulating data, and the hypothetico-deductive approach, in which hypotheses are falsified and discarded if they do not predict the data. Data from either experimental or observational studies can be used to ask whether observations are consistent with the predictions of a mechanistic hypothesis. Unfortunately, ecologists do not always state this question so plainly. Two limitations are (1) many ecological hypotheses do not generate simple, falsifiable predictions; and (2) even when an hypothesis does generate predictions, they are rarely unique. Therefore, it may not be possible to definitively test Hypothesis H using only data collected on Variable *Y*.

[1] Although manipulative experiments allow for strong inferences, they may not reveal explicit mechanisms. Many ecological experiments are simple "black box" experiments that measure the *response* of the Variable *Y* to changes in Factor *X*, but do not elucidate lower-level mechanisms causing that response. Such a mechanistic understanding may require additional observations or experiments addressing a more focused question about process.

Using the Measurements of Variable Y, What Is the Best Estimate of Parameter θ in Model Z?

Statistical and mathematical models are powerful tools in ecology and environmental science. They allow us to forecast how populations and communities will change through time or respond to altered environmental conditions (e.g., Sjögren-Gulve and Ebenhard 2000). Models can also help us to understand how different ecological mechanisms interact simultaneously to control the structure of communities and populations (Caswell 1988). Parameter estimation is required for building predictive models and is an especially important feature of Bayesian analysis (see Chapter 5). Rarely is there a simple one-to-one correspondence between the value of Variable Y measured in the field and the value of Parameter θ in our model. Instead, those parameters have to be extracted and estimated indirectly from our data. Unfortunately, some of the most common and traditional designs used in ecological experiments and field surveys, such as the analysis of variance (see Chapter 10), are not very useful for estimating model parameters. Chapter 7 discusses some alternative designs that are more useful for parameter estimation.

Manipulative Experiments

In a **manipulative experiment**, the investigator first alters levels of the predictor variable (or factor), and then measures how one or more variables of interest respond to these alterations. These results are then used to test hypotheses of cause and effect. For example, if we are interested in testing the hypothesis that lizard predation controls spider density on small Caribbean islands, we could alter the density of lizards in a series of enclosures and measure the resulting density of spiders (e.g., Spiller and Schoener 1998). We could then plot these data in a graph in which the x-axis (= independent variable) is lizard density, and the y-axis (= dependent variable) is spider density (Figure 6.1A,B).

Our null hypothesis is that there is no relationship between these two variables (Figure 6.1A). That is, spider density might be high or low in a particular enclosure, but it is not related to the density of lizards that were established in the enclosure. Alternatively, we might observe a negative relationship between spider and lizard density: enclosures with the highest lizard density have the fewest spiders, and vice-versa (Figure 6.1B). This pattern then would have to be subject to a statistical analysis such as regression (see Chapter 9) to determine whether or not the evidence was sufficient to reject the null hypothesis of no relationship between lizard and spider densities. From these data we could also estimate regression model parameters that quantify the strength of the relationship.

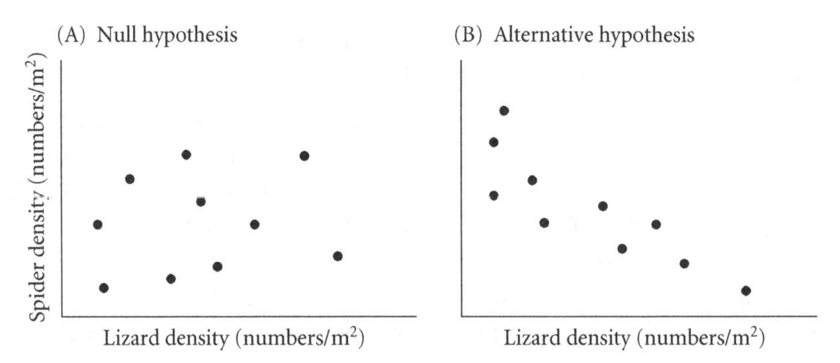

Figure 6.1 Relationship between lizard density and spider density in manipulative and natural field experiments. Each point represents a plot or quadrat in which both spider density and lizard density have been measured. (A) The null hypothesis is that lizard density has no effect on spider density. (B) The alternative hypothesis is that lizard predation controls spider density, leading to a negative relationship between these two variables.

Although field experiments are popular and powerful, they have several important limitations. First, it is challenging to conduct experiments on large spatial scales; over 80% of field experiments have been conducted in plots of less than 1 m^2 (Kareiva and Anderson 1988; Wiens 1989). When experiments are conducted on large spatial scales, replication is inevitably sacrificed (Carpenter 1989). Even when they are properly replicated, experiments conducted on small spatial scales may not yield results that are representative of patterns and processes occurring at larger spatial scales (Englund and Cooper 2003).

Second, field experiments are often restricted to relatively small-bodied and short-lived organisms that are easy to manipulate. Although we always want to generalize the results of our experiments to other systems, it is unlikely that the interaction between lizards and spiders will tell us much about the interaction between lions and wildebeest. Third, it is difficult to change one and only one variable at a time in a manipulative experiment. For example, cages can exclude other kinds of predators and prey, and introduce shading. If we carelessly compare spider densities in caged plots versus uncaged "controls," the effects of lizard predation are **confounded** with other physical differences among the treatments. We discuss solutions to confounding variables later in this chapter.

Finally, many standard experimental designs are simply unwieldy for realistic field experiments. For example, suppose we are interested in investigating competitive interactions in a group of eight spider species. Each treatment in such an experiment would consist of a unique combination of species. Although the number of species in each treatment ranges from only 1 to 8, the number of unique combinations is $2^8 - 1 = 255$. If we want to establish even 10 replicates of each treatment (see "The Rule of 10," discussed later in this chapter), we need 2550 plots. That may not be possible because of constraints on space, time, or labor. Because of all these potential limitations, many important questions in community ecology cannot be addressed with field experiments.

Natural Experiments

A **natural experiment** (Cody 1974) is not really an experiment at all. Instead, it is an observational study in which we take advantage of natural variation that is present in the variable of interest. For example, rather than manipulate lizard densities directly (a difficult, expensive, and time-consuming endeavor), we could census a set of plots (or islands) that vary naturally in their density of lizards (Schoener 1991). Ideally, these plots would vary *only* in the density of lizards and would be identical in all other ways. We could then analyze the relationship between spider density and lizard density as illustrated in Figure 6.1.

Natural experiments and manipulative experiments superficially generate the same kinds of data and are often analyzed with the same kinds of statistics. However, there are often important differences in the interpretation of natural and manipulative experiments. In a manipulative experiment, if we have established valid controls and maintained the same environmental conditions among the replicates, any consistent differences in the response variable (e.g., spider density) can be attributed confidently to differences in the manipulated factor (e.g., lizard density).

We don't have this same confidence in interpreting results of natural experiments. In a natural experiment, we do not know the direction of cause and effect, and we have not controlled for other variables that surely will differ among the replicates. For the lizard–spider example, there are at least four hypotheses that could account for a negative association between lizard and spider densities:

1. Lizards may control spider density. This was the alternative hypothesis of interest in the original field experiment.
2. Spiders may directly or indirectly control lizard density. Suppose, for example, that large hunting spiders consume small lizards, or that spiders are also preyed upon by birds that feed on lizards. In both cases, increasing spider density may decrease lizard density, even though lizards do feed on spiders.

3. Both spider and lizard densities are controlled by an unmeasured environmental factor. For example, suppose that spider densities are highest in wet plots and lizard densities are highest in dry plots. Even if lizards have little effect on spiders, the pattern in Figure 6.1B will emerge: wet plots will have many spiders and few lizards, and dry plots will have many lizards and few spiders.
4. Environmental factors may control the strength of the interaction between lizards and spiders. For example, lizards might be efficient predators on spiders in dry plots, but inefficient predators in wet plots. In such cases, the density of spiders will depend on both the density of lizards and the level of moisture in the plot (Spiller and Schoener 1995).

These four scenarios are only the simplest ones that might lead to a negative relationship between lizard density and spider density (Figure 6.2). If we add double-headed arrows to these diagrams (lizards and spiders reciprocally affect one another's densities), there is an even larger suite of hypotheses that could account for the observed relationships between spider density and lizard density (see Figure 6.1).

All of this does not mean that natural experiments are hopeless, however. In many cases we can collect additional data to distinguish among these hypotheses. For example, if we suspect that environmental variables such as moisture

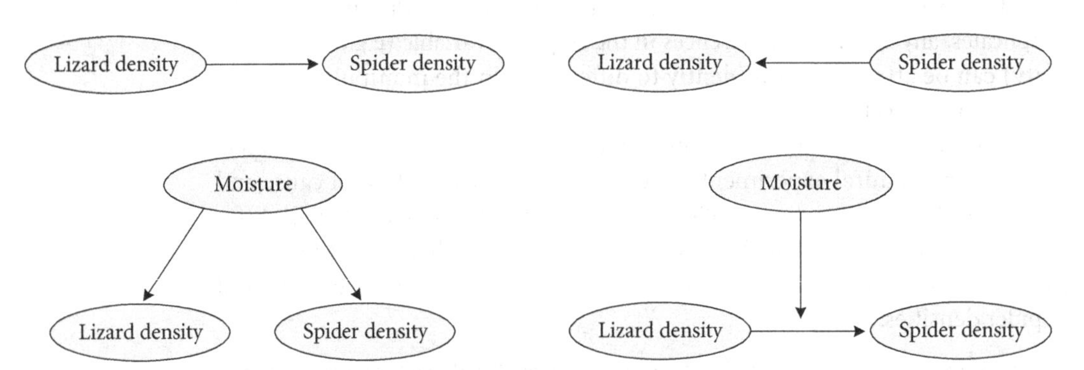

Figure 6.2 Mechanistic hypotheses to account for correlations between lizard density and spider density (see Figure 6.1). The cause-and-effect relationship might be from predator to prey (upper left) or prey to predator (upper right). More complicated models include the effects of other biotic or abiotic variables. For example, there might be no interaction between spiders and lizards, but densities of both are controlled by a third variable, such as moisture (lower left). Alternatively, moisture might have an indirect effect by altering the interaction of lizards and spiders (lower right).

are important, we either can restrict the survey to a set of plots with comparable moisture levels, or (better still) measure lizard density, spider density, and moisture levels in a series of plots censused over a moisture gradient. Confounding variables and alternative mechanisms also can be problematic in field experiments. However, their impacts will be reduced if the investigator conducts the experiment at an appropriate spatial and temporal scale, establishes proper controls, replicates adequately, and uses randomization to locate replicates and assign treatments.

Overall, manipulative experiments allow for greater confidence in our inferences about cause and effect, but they are limited to relatively small spatial scales and short time frames. Natural experiments can be conducted at virtually any spatial scale (small quadrats to entire continents) and over any time interval (weekly field measurements, to annual censuses, to fossil strata). However, it is more challenging to tease apart cause-and-effect relationships in natural experiments.[2]

Snapshot versus Trajectory Experiments

Two variants of the natural experiment are the **snapshot experiment** and the **trajectory experiment** (Diamond 1986). Snapshot experiments are replicated in space, and trajectory experiments are replicated in time. For the data in Figure 6.1, suppose we censused 10 different plots in a single day. This is a snapshot experiment in which the replication is spatial; each observation represents a different plot censused at the same time. On the other hand, suppose we visited a single plot in 10 different years. This is a trajectory experiment in which the replication is temporal; each observation represents a different year in the study.

The advantages of a snapshot experiment are that it is rapid, and the spatial replicates arguably are more statistically independent of one another than are

[2] In some cases, the distinction between manipulative and natural field experiments is not clear-cut. Human activity has generated many unintended large-scale experiments including eutrophication, habitat alteration, global climate change, and species introductions and removals. Imaginative ecologists can take advantage of these alterations to design studies in which the confidence in the conclusions is very high. For example, Knapp et al. (2001) studied the impacts of trout introductions to lakes in the Sierra Nevada by comparing invertebrate communities in naturally fishless lakes, stocked lakes, and lakes that formerly were stocked with fish. Many comparisons of this kind are possible to document consequences of human activity. However, as human impacts become more widespread and pervasive, it may be harder and harder to find sites that can be considered unmanipulated controls.

the temporal replicates of a trajectory experiment. The majority of ecological data sets are snapshot experiments, reflecting the 3- to 5-year time frame of most research grants and dissertation studies.[3] In fact, many studies of temporal change are actually snapshot studies, because variation in space is treated as a proxy variable for variation in time. For example, successional change in plant communities can be studied by sampling from a chronosequence—a set of observations, sites, or habitats along a spatial gradient that differ in the time of origin (e.g., Law et al. 2003).

The advantage of a trajectory experiment is that it reveals how ecological systems change through time. Many ecological and environmental models describe precisely this kind of change, and trajectory experiments allow for stronger comparisons between model predictions and field data. Moreover, many models for conservation and environmental forecasting are designed to predict future conditions, and data for these models are derived most reliably from trajectory experiments. Many of the most valuable data sets in ecology are long time-series data for which populations and communities at a site are sampled year after year with consistent, standardized methods. However, trajectory experiments that are restricted to a single site are unreplicated in space. We don't know if the temporal trajectories described from that site are typical for what we might find at other sites. Each trajectory is essentially a sample size of *one* at a given site.[4]

The Problem of Temporal Dependence

A more difficult problem with trajectory experiments is the potential non-independence of data collected in a temporal sequence. For example, suppose you measure tree diameters each month for one year in a plot of redwood trees. Red-

[3] A notable exception to short-term ecological experiments is the coordinated set of studies developed at Long Term Ecological Research (LTER) sites. The National Science Foundation (NSF) funded the establishment of these sites throughout the 1980s and 1990s specifically to address the need for ecological research studies that span decades to centuries. See www.lternet.edu/.

[4] Snapshot and trajectory designs show up in manipulative experiments as well. In particular, some designs include a series of measurements taken before and after a manipulation. The "before" measurements serve as a type of "control" that can be compared to the measurements taken after the manipulation or intervention. This sort of BACI design (Before-After, Control-Impact) is especially important in environmental impact analysis and in studies where spatial replication may be limited. For more on BACI, see the section "Large Scale Studies and Environmental Impacts" later in this chapter, and see Chapter 7.

woods grow very slowly, so the measurements from one month to the next will be virtually identical. Most foresters would say that you don't have 12 independent data points, you have only one (the average diameter for that year). On the other hand, monthly measurements of a rapidly developing freshwater plankton community reasonably could be viewed as statistically independent of one another. Naturally, the further apart in time the samples are separated from one another, the more they function as independent replicates.

But even if the correct census interval is used, there is still a subtle problem in how temporal change should be modeled. For example, suppose you are trying to model changes in population size of a desert annual plant for which you have access to a nice trajectory study, with 100 years of consecutive annual censuses. You could fit a standard linear regression model (see Chapter 9) to the time series

$$N_t = \beta_0 + \beta_1 t + \varepsilon \quad (6.1)$$

In this equation, population size (N_t) is a linear function of the amount of time (t) that has passed. The coefficients β_0 and β_1 are the intercept and slope of this straight line. If β_1 is less than 0.0, the population is shrinking with time, and if $\beta_1 > 0$, N is increasing. Here ε is a normally distributed **white noise**[5] error term that incorporates both measurement error and random variation in population size. Chapter 9 will explain this model in much greater detail, but we introduce it now as a simple way to think about how population size might change in a linear fashion with the passage of time.

However, this model does not take into account that population size changes through births and deaths affecting *current* population size. A **time-series model** would describe population growth as

$$N_{t+1} = \beta_0 + \beta_1 N_t + \varepsilon \quad (6.2)$$

[5] White noise is a type of error distribution in which the errors are independent and uncorrelated with one another. It is called white noise as an analogy to white light, which is an equal mixture of short and long wavelengths. In contrast, red noise is dominated by low-frequency perturbations, just as red light is dominated by low-frequency light waves. Most time series of population sizes exhibit a reddened noise spectrum (Pimm and Redfearn 1988), so that variances in population size increase when they are analyzed at larger temporal scales. Parametric regression models require normally distributed error terms, so white noise distributions form the basis for most stochastic ecological models. However, an entire spectrum of colored noise distributions (1/*f* noise) may provide a better fit to many ecological and evolutionary datasets (Halley 1996).

In this model, the population size in the next time step (N_{t+1}) depends not simply on the amount of time t that has passed, but rather on the population size at the last time step (N_t). In this model, the constant β_1 is a multiplier term that determines whether the population is exponentially increasing ($\beta_1 > 1.0$) or decreasing ($\beta_1 < 1.0$). As before, ε is a white noise error term.

The linear model (Equation 6.1) describes a simple *additive* increase of N with time, whereas the time-series, or **autoregressive** model (Equation 6.2) describes an *exponential* increase, because the factor β_1 is a multiplier that, on average, gives a constant percentage increase in population size at each time step. The more important difference between the two models, however, is that the differences between the observed and predicted population sizes (i.e., the **deviations**) in the time-series model are correlated with one another. As a consequence, there tend to be runs, or periods of consecutive increases followed by periods of consecutive decreases. This is because the growth trajectory has a "memory"—each consecutive observation (N_{t+1}) depends directly on the one that came before it (the N_t term in Equation 6.2). In contrast, the linear model has no memory, and the increases are a function only of time (and ε), and not of N_t. Hence, the positive and negative deviations follow one another in a purely random fashion (Figure 6.3). Correlated deviations, which are typical of data collected in trajectory studies, violate the assumptions of most conventional statistical analyses.[6] Analytical and computer-intensive methods have been developed for analyzing both sample data and experimental data collected through time (Ives et al. 2003; Turchin 2003).

This does not mean we cannot incorporate time-series data into conventional statistical analyses. In Chapters 7 and 10, we will discuss additional ways to analyze time-series data. These methods require that you pay careful attention to both the sampling design and the treatment of the data after you have collected them. In this respect, time-series or trajectory data are just like any other data.

Press versus Pulse Experiments

In manipulative studies, we also distinguish between **press experiments** and **pulse experiments** (Bender et al. 1984). In a press experiment, the altered conditions in the treatment are maintained through time and are re-applied as necessary to ensure that the strength of the manipulation remains constant. Thus,

[6] Actually, spatial autocorrelation generates the same problems (Legendre and Legendre 1998; Lichstein et al. 2003). However, tools for spatial autocorrelation analysis have developed more or less independently of time-series analyses, perhaps because we perceive time as a strictly one-dimensional variable and space as a two- or three-dimensional variable.

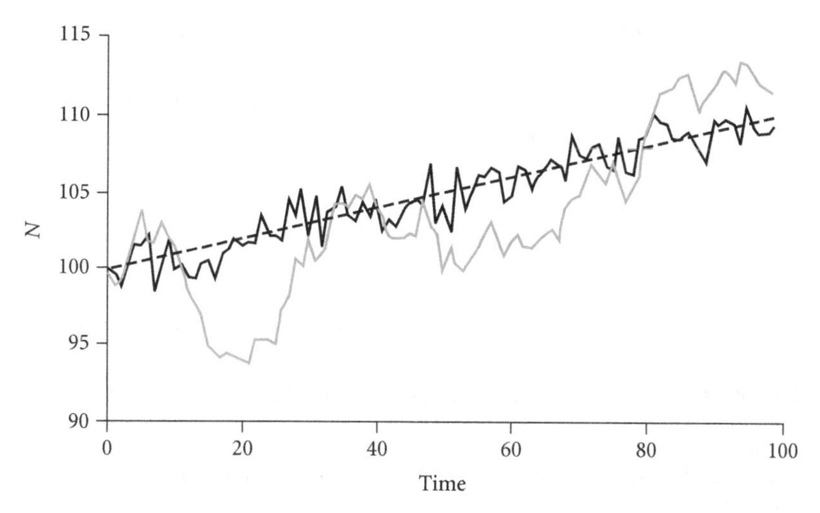

Figure 6.3 Examples of deterministic and stochastic time series, with and without autocorrelation. Each population begins with 100 individuals. A linear model without error (dashed line) illustrates a constant upward trend in population data. A linear model with a stochastic white noise error term (black line) adds temporally uncorrelated variability. Finally, an autocorrelated model (blue line) describes population size in the next time step ($t + 1$) as a function of the population size in the current time step (t) plus random noise. Although the error term in this model is still a simple random variable, the resulting time series shows autocorrelation—there are runs of population increases followed by runs of population decreases. For the linear model and the stochastic white noise model, the equation is $N_t = a + bt + \varepsilon$, with $a = 100$ and $b = 0.10$. For the autocorrelated model, $N_{t+1} = a + bN_t + \varepsilon$, with $a = 0.0$ and $b = 1.0015$. For both models with error, ε is a normal random variable: $\varepsilon \sim N(0,1)$.

fertilizer may have to be re-applied to plants, and animals that have died or disappeared from a plot may have to be replaced. In contrast, in a pulse experiment, experimental treatments are applied only once, at the start of the study. The treatment is not re-applied, and the replicate is allowed to "recover" from the manipulation (Figure 6.4A).

Press and pulse experiments measure two different responses to the treatment. The press experiment (Figure 6.4B) measures the resistance of the system to the experimental treatment: the extent to which it resists change in the constant environment created by the press experiment. A system with low resistance will exhibit a large response in a press experiment, whereas a system with high resistance will exhibit little difference between control and manipulated treatments.

The pulse experiment measures the resilience of the system to the experimental treatment: the extent to which the system recovers from a single perturbation. A system with high resilience will show a rapid return to control con-

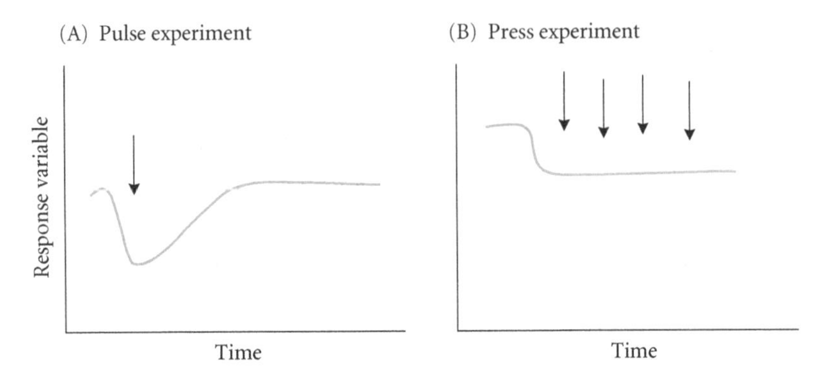

Figure 6.4 Ecological pulse and press experiments. The arrow indicates a treatment application, and the line indicates the temporal trajectory of the response variable. The pulse experiment (A) measures the response to a single treatment application (resilience), whereas the press experiment (B) measures the response under constant conditions (resistance).

ditions, whereas a system with low resilience will take a long time to recover; control and manipulated plots will continue to differ for a long time after the single treatment application.

The distinction between press and pulse experiments is not in the number of treatment applications used, but in whether the altered conditions are maintained through time in the treatment. If environmental conditions remain constant following a single perturbation for the duration of the experiment, the design is effectively a press experiment. Another distinction between press and pulse experiments is that the press experiment measures the response of the system under equilibrium conditions, whereas the pulse experiment records transient responses in a changing environment.

Replication

How Much Replication?

This is one of the most common questions that ecologists and environmental scientists ask of statisticians. The correct response is that the answer depends on the variance in the data and the **effect size**—the difference that you wish to detect between the averages of the groups being compared. Unfortunately, these two quantities may be difficult to estimate, although you always should consider what effect size would be reasonable to observe.

To estimate variances, many statisticians will recommend that you conduct a pilot study. Unfortunately, pilot studies usually are not feasible—you rarely have the freedom to set up and run a costly or lengthy study more than once.

Field seasons and grant proposals are too short for this sort of luxury. However, you may be able to estimate reasonable ranges of variances and effect sizes from previously published studies and from discussions with colleagues. You can then use these values to determine the statistical power (see Chapter 4) that will result from different combinations of replicates, variances, and effect sizes (see Figure 4.5 for an example). At a minimum, however, you need to first answer the following question:

How Many Total Replicates Are Affordable?

It takes time, labor, and money to collect either experimental or survey data, and you need to determine precisely the total sample size that you can afford. If you are conducting expensive tissue or sample analyses, the dollar cost may be the limiting factor. However, in many studies, time and labor are more limiting than money. This is especially true for geographical surveys conducted over large spatial scales, for which you (and your field crew if you are lucky enough to have one) may spend as much time traveling to study sites as you do collecting field data. Ideally, all of the replicates should be measured simultaneously, giving you a perfect snapshot experiment. The more time it takes to collect all the data, the more conditions will have changed from the first sample to the last. For experimental studies, if the data are not collected all at once, then the amount of time that has passed since treatment application is no longer identical for all replicates.

Obviously, the larger the spatial scale of the study, the harder it is to collect all of the data within a reasonable time frame. Nevertheless, the payoff may be greater because the scope of inference is tied to the spatial scale of analysis: conclusions based on samples taken only at one site may not be valid at other sites. However, there is no point in developing an unrealistic sampling design. Carefully map out your project from start to finish to ensure it will be feasible.[7] Only once you know the total number of replicates or observations that you can collect can you begin to design your experiment by applying the rule of 10.

[7] It can be very informative to use a stopwatch to time carefully how long it takes to complete a single replicate measurement of your study. Like the efficiency expert father in *Cheaper By The Dozen* (Gilbreth and Carey 1949), we put great stock in such numbers. With these data, we can accurately estimate how many replicates we can take in an hour, and how much total field time we will need to complete the census. The same principle applies to sample processing, measurements that we make back in the laboratory, the entry of data into the computer, and the long-term storage and curation of data (see Chapter 8). All of these activities take time that needs to be accounted for when planning an ecological study.

The Rule of 10

The **Rule of 10** is that you should collect at least 10 replicate observations for each category or treatment level. For example, suppose you have determined that you can collect 50 total observations in a experiment examining photosynthetic rates among different plant species. A good design for a one-way ANOVA would be to compare photosynthetic rates among not more than five species. For each species, you would choose randomly 10 plants and take one measurement from each plant.

The Rule of 10 is not based on any theoretical principle of experimental design or statistical analysis, but is a reflection of our hard-won field experience with designs that have been successful and those that have not. It is certainly possible to analyze data sets with less than 10 observations per treatment, and we ourselves often break the rule. Balanced designs with many treatment combinations but only four or five replicates may be quite powerful. And certain one-way designs with only a few treatment levels may require more than 10 replicates per treatment if variances are large.

Nevertheless, the Rule of 10 is a solid starting point. Even if you set up the design with 10 observations per treatment level, it is unlikely that you will end up with that number. In spite of your best efforts, data may be lost for a variety of reasons, including equipment failures, weather disasters, plot losses, human disturbances or errors, improper data transcription, and environmental alterations. The Rule of 10 at least gives you a fighting chance to collect data with reasonable statistical power for revealing patterns.[8] In Chapter 7, we will discuss efficient sample designs and strategies for maximizing the amount of information you can squeeze out of your data.

Large-Scale Studies and Environmental Impacts

The Rule of 10 is useful for small-scale manipulative studies in which the study units (plots, leaves, etc.) are of manageable size. But it doesn't apply to large-scale ecosystem experiments, such as whole-lake manipulations, because replicates may be unavailable or too expensive. The Rule of 10 also does not apply to many environmental impact studies, where the assessment of an impact is required at a single site. In such cases, the best strategy is to use a **BACI design** (Before-After, Control-Impact). In some BACI designs, the replication is achieved through time:

[8] Another useful rule is the Rule of 5. If you want to estimate the curvature or non-linearity of a response, you need to use at least five levels of the predictor variable. As we will discuss in Chapter 7, a better solution is to use a regression design, in which the predictor variable is continuous, rather than categorical with a fixed number of levels.

the control and impact sites are censused repeatedly both before and after the impact. The lack of spatial replication restricts the inferences to the impact site itself (which may be the point of the study), and requires that the impact is not confounded with other factors that may be co-varying with the impact. The lack of spatial replication in simple BACI designs is controversial (Underwood 1994; Murtaugh 2002b), but in many cases they are the best design option (Stewart-Oaten and Bence 2001), especially if they are used with explicit time-series modeling (Carpenter et al. 1989). We will return to BACI and its alternatives in Chapters 7 and 10.

Ensuring Independence

Most statistical analyses assume that replicates are independent of one another. By **independence**, we mean that the observations collected in one replicate do not have an influence on the observations collected in another replicate. Non-independence is most easily understood in an experimental context. Suppose you are studying the response of hummingbird pollinators to the amount of nectar produced by flowers. You set up two adjacent 5 m × 5 m plots. One plot is a control plot; the adjacent plot is a nectar removal plot in which you drain all of the nectar from the flowers. You measure hummingbird visits to flowers in the two plots. In the control plot, you measure an average of 10 visits/hour, compared to only 5 visits/hour in the removal plot.

However, while collecting the data, you notice that once birds arrive at the removal plot, they immediately leave, and *the same birds* then visit the adjacent control plot (Figure 6.5A). Clearly, the two sets of observations are not independent of one another. If the control and treatment plots had been more widely separated in space, the numbers might have come out differently, and the average in the control plots might have been only 7 visits/hour instead of 10 visits/hour (Figure 6.5B). When the two plots are adjacent to one another, non-independence inflates the difference between them, perhaps leading to a spuriously low *P*-value, and a Type I error (incorrect rejection of a true null hypothesis; see Chapter 4). In other cases, non-independence may decrease the apparent differences between treatments, contributing to a Type II error (incorrect acceptance of a false null hypothesis). Unfortunately, non-independence inflates or deflates both *P*-values and power to unknown degrees.

The best safeguard against non-independence is to ensure that replicates within and among treatments are separated from one another by enough space or time to ensure that the they do not affect one another. Unfortunately, we rarely know what that distance or spacing should be, and this is true for both experimental and observational studies. We should use common sense and as much

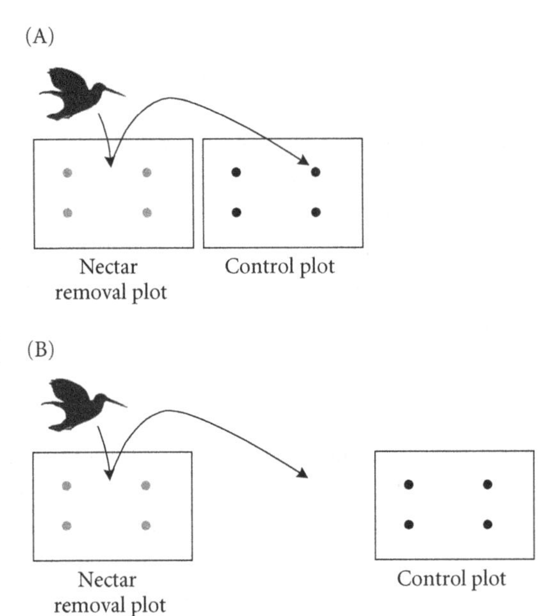

Figure 6.5 The problem of non-independence in ecological studies is illustrated by an experimental design in which hummingbirds forage for nectar in control plots and in plots from which nectar has been removed from all of the flowers. (A) In a non-independent layout, the nectar removal and control plots are adjacent to one another, and hummingbirds that enter the nectar removal plot immediately leave and begin foraging in the adjacent control plot. As a consequence, the data collected in the control plot are not independent of the data collected in the nectar removal plot: the responses in one treatment influence the responses in the other. (B) If the layout is modified so that the two plots are widely separated, hummingbirds that leave the nectar removal plot do not necessarily enter the control plot. The two plots are independent, and the data collected in one plot are not influenced by the presence of the other plot. Although it is easy to illustrate the potential problem of non-independence, in practice it is can be very difficult to know ahead of time the spatial and temporal scales that will ensure statistical independence.

biological knowledge as possible. Try to look at the world from the organism's perspective to think about how far to separate samples. Pilot studies, if feasible, also can suggest appropriate spacing to ensure independence.

So why not just maximize the distance or time between samples? First, as we described earlier, it becomes more expensive to collect data as the distance between samples increases. Second, moving the samples very far apart can introduce new sources of variation because of differences (heterogeneity) within or among habitats. We want our replicates close enough together to ensure we are sampling relatively homogenous or consistent conditions, but far enough apart to ensure that the responses we measure are independent of one another.

In spite of its central importance, the independence problem is almost never discussed explicitly in scientific papers. In the Methods section of a paper, you are likely to read a sentence such as, "We measured 100 randomly selected seedlings growing in full sunlight. Each measured seedling was at least 50 cm from its nearest neighbor." What the authors mean is, "We don't know how far apart the observations would have to have been in order to ensure independence. However, 50 cm seemed like a fair distance for the tiny seedlings we studied. If we had chosen distances greater than 50 cm, we could not have collected all of our data in full sunlight, and some of the seedlings would have been collected in the shade, which obviously would have influenced our results."

Avoiding Confounding Factors

When factors are confounded with one another, their effects cannot be easily disentangled. Let's return to the hummingbird example. Suppose we prudently separated the control and nectar removal plots, but inadvertently placed the removal plot on a sunny hillside and the control plot in a cool valley (Figure 6.6). Hummingbirds forage less frequently in the removal plot (7 visits/hour), and the two plots are now far enough apart that there is no problem of independence. However, hummingbirds naturally tend to avoid foraging in the cool valley, so the foraging rate also is low in these plots (6 visits/hour). Because the treatments are confounded with temperature differences, we cannot tease apart the effects of foraging preferences from those of thermal preferences. In this case, the two forces largely cancel one another, leading to comparable foraging rates in the two plots, although for very different reasons.

This example may seem a bit contrived. Knowing the thermal preferences of hummingbirds, we would not have set up such an experiment. The problem is that there are likely to be unmeasured or unknown variables—even in an apparently homogenous environment—that can have equally strong effects on our experiment. And, if we are conducting a natural experiment, we are stuck with whatever confounding factors are present in the environment. In an observational study of hummingbird foraging, we may not be able to find plots that differ only in their levels of nectar rewards but do not also differ in temperature and other factors known to affect foraging behavior.

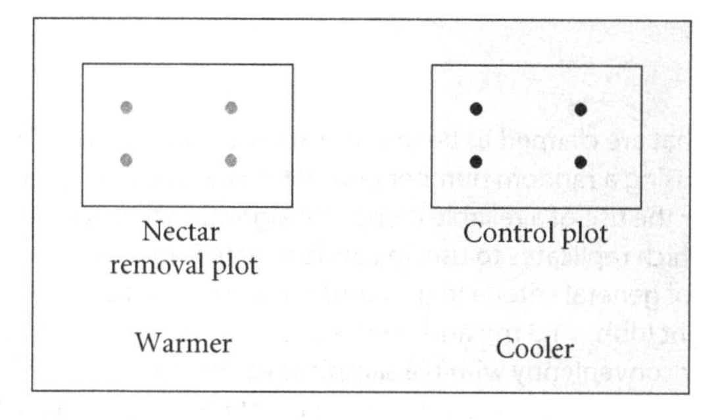

Figure 6.6 A confounded experimental design. As in Figure 6.5, the study establishes control and experimental nectar removal plots to evaluate the responses of foraging hummingbirds. In this design, although the plots are far enough apart to ensure independence, they have been placed at different points along a thermal gradient. Consequently, the treatment effects are confounded with differences in the thermal environment. The net result is that the experiment compares data from a warm nectar removal plot with data from a cool control plot.

Replication and Randomization

The dual threats of confounding factors and non-independence would seem to threaten all of our statistical conclusions and render even our experimental studies suspect. Incorporating **replication** and **randomization** into experimental designs can largely offset the problems introduced by confounding factors and non-independence. By replication, we mean the establishment of multiple plots or observations within the same treatment or comparison group. By randomization, we mean the random assignment of treatments or selection of samples.[9]

Let's return one more time to the hummingbird example. If we follow the principles of randomization and replication, we will set up many replicate control and removal plots (ideally, a minimum of 10 of each). The location of each of these plots in the study area will be random, and the assignment of the treatment (control or removal) to each plot also will be random (Figure 6.7).[10]

How will randomization and replication reduce the problem of confounding factors? Both the warm hillside, the cool valley, and several intermediate sites each will have multiple plots from both control and nectar removal treatments. Thus, the temperature factor is no longer confounded with the treatment, as all treatments occur within each level of temperature. As an additional benefit, this design will also allow you to test the effects of temperature as a covariate on hummingbird foraging behavior, independent of the levels of nectar (see Chapters 7 and 10). It is true that hummingbird visits will still be more frequent on the warm hillside than in the cool valley, but that will be true for replicates of both the control and nectar removal. The temperature will add more variation to the data, but it will not bias the results because the warm and cool plots will

[9] Many samples that are claimed to be random are really **haphazard**. Truly random sampling means using a random number generator (such as the flip of a fair coin, the roll of a fair die, or the use of a reliable computer algorithm for producing random numbers) to decide which replicates to use. In contrast, with haphazard sampling, an ecologist follows a set of general criteria [e.g., mature trees have a diameter of more than 3 cm at breast height (dbh = 1.3 m)] and selects sites or organisms that are spaced homogenously or conveniently within a sample area. Haphazard sampling is often necessary at some level because random sampling is not efficient for many kinds of organisms, especially if their distribution is spatially patchy. However, once a set of organisms or sites is identified, randomization should be used to sample or to assign replicates to different treatment groups.

[10] Randomization takes some time, and you should do as much of it as possible in advance, before you get into the field. It is easy to generate random numbers and simulate random sampling with computer spreadsheets. But it is often the case that you will need to generate random numbers in the field. Coins and dice (especially 10-sided

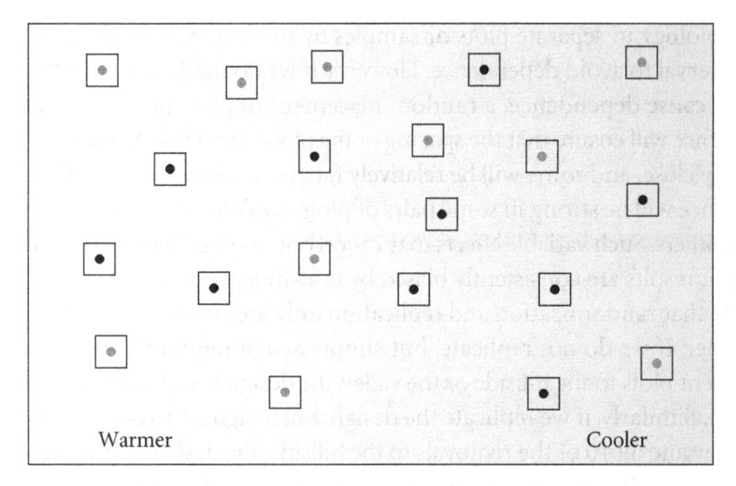

Figure 6.7 A properly replicated and randomized experimental design. The study establishes plots as in Figures 6.6. Each square represents a replicate control plot (black dots) or nectar removal plot (gray dots). The plots are separated by enough distance to ensure independence, but their location within the temperature gradient has been randomized. There are 10 replicates for each of the two treatments. The spatial scale of the drawing is larger than in Figure 6.6.

be distributed approximately equally between the control and removal treatments. Of course, if we knew ahead of time that temperature was an important determinant of foraging behavior, we might not have used this design for the experiment. Randomization minimizes the confounding of treatments with unknown or unmeasured variables in the study area.

It is less obvious how randomization and replication reduce the problem of non-independence among samples. After all, if the plots are too close together, the foraging visits will not be independent, regardless of the amount of replication or randomization. Whenever possible, we should use common sense and

gaming dice) are useful for this purpose. A clever trick is to use a set of coins as a binary random number generator. For example, suppose you have to assign each of your replicates to one of 8 different treatments, and you want to do so randomly. Toss 3 coins, and convert the pattern of heads and tails to a binary number (i.e., a number in base 2). Thus, the first coin indicates the 1s, the second coin indicates the 2s, the third coin indicates the 4s, and so on. Tossing 3 coins will give you a random integer between 0 and 7. If your three tosses are heads, tails, heads (HTH), you have a 1 in the one's place, a 0 in the two's place, and a 1 in the four's place. The number is $1 + 0 + 4 = 5$. A toss of (THT) is $0 + 2 + 0 = 2$. Three tails gives you a 0 $(0 + 0 + 0)$ and three heads give you a 7 $(1 + 2 + 4)$. Tossing 4 coins will give you 16 integers, and 5 coins will give you 32.

An even easier method is to take a digital stopwatch into the field. Let the watch run for a few seconds and then stop it without looking at it. The final digit that measures time in $1/100^{th}$ of a second can be used as a random uniform digit from 0 to 9. A statistical analysis of 100 such random digits passed all of the standard diagnostic tests for randomness and uniformity (B. Inouye, personal communication).

knowledge of biology to separate plots or samples by some minimum distance or sampling interval to avoid dependence. However if we do not know all of the forces that can cause dependence, a random placement of plots beyond some minimum distance will ensure that the spacing of the plots is variable. Some plots will be relatively close, and some will be relatively far apart. Therefore, the effect of the dependence will be strong in some pairs of plots, weak in others, and non-existent in still others. Such variable effects may cancel one another and can reduce the chances that results are consistently biased by non-independence.

Finally, note that randomization and replication only are effective when they are used together. If we do not replicate, but simply assign randomly the control and treatment plots to the hillside or the valley, the design is still confounded (see Figure 6.6). Similarly, if we replicate the design, but assign all 10 of the controls to the valley and all 10 of the removals to the hillside, the design is also confounded (Figure 6.8). It is only when we use multiple plots and assign the treatments randomly that the confounding effect of temperature is removed from the design (see Figure 6.7). Indeed, it is fair to say that any unreplicated design is always going to be confounded with one or more environmental factors.[11]

Although the concept of randomization is straightforward, it must be applied at several stages in the design. First, randomization applies only to a well-defined, initially non-random sample space. The sample space doesn't simply mean the physical area from which replicates are sampled (although this is an important aspect of the sample space). Rather, the sample space refers to a set of elements that have experienced similar, though not identical, conditions.

Examples of a sample space might include individual cutthroat trout that are reproductively mature, lightfall gaps created by fires, old-fields abandoned 10–20

[11] Although confounding is easy to recognize in a field experiment of this sort, it may not be apparent that the same problem exists in laboratory and greenhouse experiments. If we rear insect larvae at high and low temperatures in two environmental chambers, this is a confounded design because all of the high temperature larvae are in one chamber and all of the low temperature larvae are in the other. If environmental factors other than temperature also differ between the chambers, their effects are confounded with temperature. The correct solution would be to rear each larva in its own separate chamber, thereby ensuring that each replicate is truly independent and that temperature is not confounded with other factors. But this sort of design simply is too expensive and wasteful of space ever to be used. Perhaps the argument can be made that environmental chambers and greenhouses really do differ only in temperature and no other factors, but that is only an assumption that should be tested explicitly. In many cases, the environment in environmental chambers is surprisingly heterogeneous, both within and between chambers. Potvin (2001) discusses how this variability can be measured and then used to design better laboratory experiments.

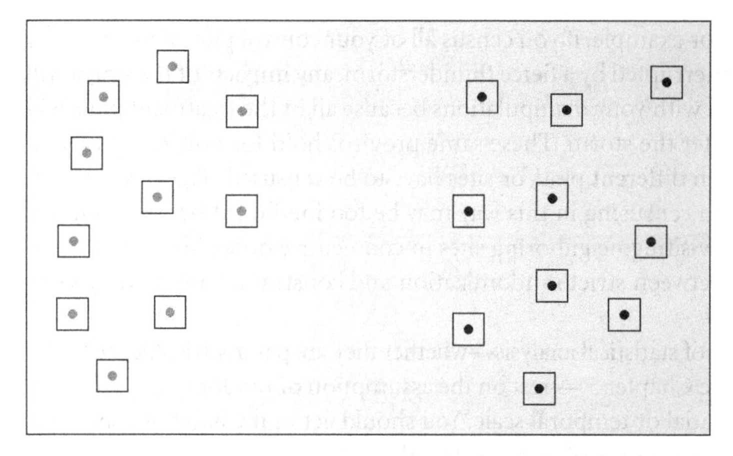

Figure 6.8 A replicated, but confounded, design. As in Figures 6.5, 6.6, and 6.7, the study establishes control and experimental nectar removal plots to evaluate the responses of foraging hummingbirds. Each square represents a replicate control plot (black dots) or nectar removal plot (gray dots). If treatments are replicated but not assigned randomly, the design still confounds treatments with underlying environmental gradients. Replication combined with randomization and sufficient spacing of replicates (see Figure 6.7) is the only safeguard against non-independence (see Figure 6.5) and confounding (see Figures 6.6 and 6.8).

years ago, or large bleached coral heads at 5–10 meters depth. Once this sample space has been defined clearly, sites, individuals, or replicates that meet the criteria should be chosen at random. As we noted in Chapter 1, the spatial and temporal boundaries of the study will dictate not only the sampling effort involved, but also the domain of inference for the conclusions of the study.

Once sites or samples are randomly selected, treatments should be assigned to them randomly, which ensures that different treatments are not clumped in space or confounded with environmental variables.[12] Samples should also be collected and treatments applied in a random sequence. That way, if environmental conditions change during the experiment, the results will not be

[12] If the sample size is too small, even a random assignment can lead to spatial clumping of treatments. One solution would be to set out the treatments in a repeated order (...123123...), which ensures that there is no clumping. However, if there is any non-independence among treatments, this design may exaggerate its effects, because Treatment 2 will always occur spatially between Treatments 1 and 3. A better solution would be to repeat the randomization and then statistically test the layout to ensure there is no clumping. See Hurlbert (1984) for a thorough discussion of the numerous hazards that can arise by failing to properly replicate and randomize ecological experiments.

confounded. For example, if you census all of your control plots first, and your field work is interrupted by a fierce thunderstorm, any impacts of the storm will be confounded with your manipulations because all of the treatment plots will be censused after the storm. These same provisos hold for non-experimental studies in which different plots or sites have to be censused. The caveat is that strictly random censusing in this way may be too inefficient because you will usually not be visiting neighboring sites in consecutive order. You may have to compromise between strict randomization and constraints imposed by sampling efficiency.

All methods of statistical analysis—whether they are parametric, Monte Carlo, or Bayesian (see Chapter 5)—rest on the assumption of random sampling at an appropriate spatial or temporal scale. You should get in the habit of using randomization whenever possible in your work.

Designing Effective Field Experiments and Sampling Studies

Here are some questions to ask when designing field experiments and sampling studies. Although some of these questions appear to be specific to manipulative experiments, they are also relevant to certain natural experiments, where "controls" might consist of plots lacking a particular species or set of abiotic conditions.

Are the Plots or Enclosures Large Enough to Ensure Realistic Results?

Field experiments that seek to control animal density must necessarily constrain the movement of animals. If the enclosures are too small, the movement, foraging, and mating behaviors of the animals may be so unrealistic that the results obtained will be uninterpretable or meaningless (MacNally 2000a). Try to use the largest plots or cages that are feasible and that are appropriate for the organism you are studying. The same considerations apply to sampling studies: the plots need to be large enough and sampled at an appropriate spatial scale to answer your question.

What Is the Grain and Extent of the Study?

Although much importance has been placed on the spatial scale of an experiment or a sampling study, there are actually two components of spatial scale that need to be addressed: grain and extent. **Grain** is the size of the smallest unit of study, which will usually be the size of an individual replicate or plot. **Extent** is the total area encompassed by all of the sampling units in the study. Grain and extent can be either large or small (Figure 6.9). There is no single combination of grain and extent that is necessarily correct. However, ecological studies with

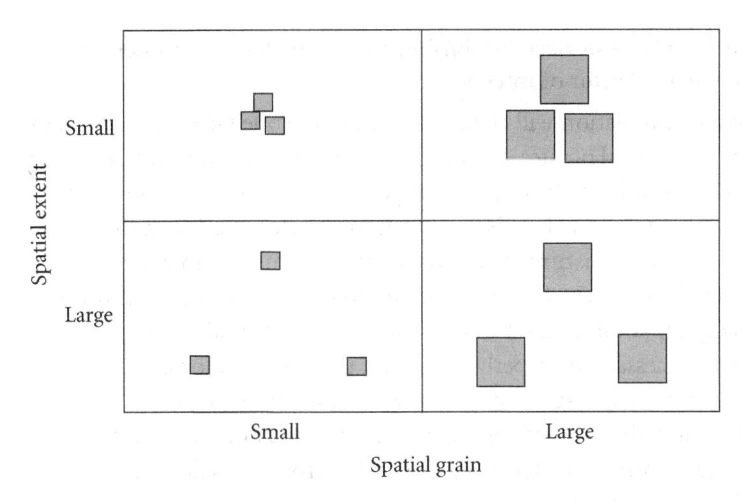

Figure 6.9 Spatial grain and spatial extent in ecological studies. Each square represents a single plot. Spatial grain measures the size of the sampling units, represented by small or large squares. Spatial extent measures the area encompassing all of the replicates of the study, represented by closely grouped or widely spaced squares.

both a small grain and a small extent, such as pitfall catches of beetles in a single forest plot, may sometimes be too limited in scope to allow for broad conclusions. On the other hand, studies with a large grain but a small extent, such as whole-lake manipulations in a single valley, may be very informative. Our own preference is for studies with a small grain, but a medium or large extent, such as ant and plant censuses in small plots (5 m × 5 m) across New England (Gotelli and Ellison 2002a,b) or eastern North America (Gotelli and Arnett 2000), or on small mangrove islands in the Caribbean (Farnsworth and Ellison 1996a). The small grain allows for experimental manipulations and observations taken at scales that are relevant to the organism, but the large extent expands the domain of inference for the results. In determining grain and extent, you should consider both the question you are trying to ask and the constraints on your sampling.

Does the Range of Treatments or Census Categories Bracket or Span the Range of Possible Environmental Conditions?

Many field experiments describe their manipulations as "bracketing or spanning the range of conditions encountered in the field." However, if you are trying to model climate change or altered environments, it may be necessary to also include conditions that are outside the range of those normally encountered in the field.

Have Appropriate Controls Been Established to Ensure that Results Reflect Variation Only in the Factor of Interest?

It is rare that a manipulation will change one and only one factor at a time. For example, if you surround plants with a cage to exclude herbivores, you have also altered the shading and moisture regime. If you simply compare these plants to unmanipulated controls, the herbivore effects are confounded with the differences in shading and moisture. The most common mistake in experimental designs is to establish a set of unmanipulated plots and then treat those as a control. Usually, an additional set of control plots that contain some minimal alteration will be necessary to properly control for the manipulations. In the example described above, an open-sided cage roof will allow herbivores access to plants, but will still include the shading effects of the cage. With this simple design of three treatments (Unmanipulated, Cage control, Herbivore exclusion), you can make the following contrasts:

1. *Unmanipulated versus Cage control.* This comparison reveals the extent to which shading and physical changes due to the cage per se are affecting plant growth and responses.
2. *Cage control versus Herbivore exclusion.* This comparison reveals the extent to which herbivory alters plant growth. Both the Control and Herbivore exclusion plots experience the shading effects of the cage, so any difference between them can be attributed to the effect of herbivores.
3. *Unmanipulated versus Herbivore exclusion.* This comparison measures the *combined effect* of both the herbivores and the shading on plant growth. Because the experiment is designed to measure only the herbivore effect, this particular comparison confounds treatment and caging effects.

In Chapter 10, we will explain how you use can use contrasts after analysis of variance to quantify these comparisons.

Have All Replicates Been Manipulated in the Same Way Except for the Intended Treatment Application?

Again, appropriate controls usually require more than lack of manipulation. If you have to push back plants to apply treatments, you should push back plants in the control plots as well (Salisbury 1963; Jaffe 1980). In a reciprocal transplant experiment with insect larvae, live animals may be sent via overnight courier to distant sites and established in new field populations. The appropriate control is a set of animals that are re-established in the populations from which they were collected. These animals will also have to receive the "UPS treatment" and be sent through the mail system to ensure they receive the same stress as the animals that were transplanted to distant sites. If you are not careful to ensure that all organisms are treated identically in your experiments, your treatments will be confounded with differences in handling effects (Cahill et al. 2000).

Have Appropriate Covariates Been Measured in Each Replicate?

Covariates are continuous variables (see Chapter 7) that potentially affect the response variable, but are not necessarily controlled or manipulated by the investigator. Examples include variation among plots in temperature, shade, pH, or herbivore density. Different statistical methods, such as analysis of covariance (see Chapter 10), can be used to quantify the effect of covariates.

However, you should avoid the temptation to measure every conceivable covariate in a plot just because you have the instrumentation (and the time) to do so. You will quickly end up with a dataset in which you have more variables measured than you have replicates, which causes additional problems in the analysis (Burnham and Anderson 2010). It is better to choose ahead of time the most biologically relevant covariates, measure only those covariates, and use sufficient replication. Remember also that the measurement of covariates is useful, but it is not a substitute for proper randomization and replication.

Summary

The sound design of an ecological experiment first requires a clear statement of the question being asked. Both manipulative and observational experiments can answer ecological questions, and each type of experiment has its own strengths and weaknesses. Investigators should consider the appropriateness of using a press versus a pulse experiment, and whether the replication will be in space (snapshot experiment), time (trajectory experiment), or both. Non-independence and confounding factors can compromise the statistical analysis of data from both manipulative and observational studies. Randomization, replication, and knowledge of the ecology and natural history of the organisms are the best safeguards against non-independence and confounding factors. Whenever possible, try to use at least 10 observations per treatment group. Field experiments usually require carefully designed controls to account for handling effects and other unintended alterations. Measurement of appropriate environmental covariates can be used to account for uncontrolled variation, although it is not a substitute for randomization and replication.

CHAPTER 7

A Bestiary of Experimental and Sampling Designs

In an experimental study, we have to decide on a set of biologically realistic manipulations that include appropriate controls. In an observational study, we have to decide which variables to measure that will best answer the question we have asked. These decisions are very important, and were the subject of Chapter 6. In this chapter we discuss specific designs for experimental and sampling studies in ecology and environmental science. The design of an experiment or observational study refers to how the replicates are physically arranged in space, and how those replicates are sampled through time. The design of the experiment is intimately linked to the details of replication, randomization, and independence (see Chapter 6). Certain kinds of designs have proven very powerful for the interpretation and analysis of field data. Other designs are more difficult to analyze and interpret. However, you cannot draw blood from a stone, and even the most sophisticated statistical analysis cannot rescue a poor design.

We first present a simple framework for classifying designs according to the types of independent and dependent variables. Next, we describe a small number of useful designs in each category. We discuss each design and the kinds of questions it can be used to address, illustrate it with a simple dataset, and describe the advantages and disadvantages of the design. The details of how to analyze data from these designs are postponed until Chapters 9–12.

The literature on experimental and sampling designs is vast (e.g., Cochran and Cox 1957; Winer 1991; Underwood 1997; Quinn and Keough 2002), and we present only a selective coverage in this chapter. We restrict ourselves to those designs that are practical and useful for ecologists and environmental scientists, and that have proven to be most successful in field studies.

Categorical versus Continuous Variables

We first distinguish between **categorical variables** and **continuous variables**. Categorical variables are classified into one of two or more unique categories. Ecological examples include sex (male, female), trophic status (producer, herbivore, carnivore), and habitat type (shade, sun). Continuous variables are measured on a continuous numerical scale; they can take on a range of real number or integer values. Examples include measurements of individual size, species richness, habitat coverage, and population density.

Many statistics texts make a further distinction between purely categorical variables, in which the categories are not ordered, and ranked (or ordinal) variables, in which the categories are ordered based on a numerical scale. An example of an ordinal variable would be a numeric score (0, 1, 2, 3, or 4) assigned to the amount of sunlight reaching the forest floor: 0 for 0–5% light; 1 for 6–25% light; 2 for 26–50% light; 3 for 51–75% light; and 4 for 76–100% light. In many cases, methods used for analyzing continuous data also can be applied to ordinal data. In a few cases, however, ordinal data are better analyzed with Monte Carlo methods, which were discussed in Chapter 5. In this book, we use the term *categorical variable* to refer to both ordered and unordered categorical variables.

The distinction between categorical and continuous variables is not always clear-cut; in many cases, the designation depends simply on how the investigator chooses to measure the variable. For example, a categorical habitat variable such as sun/shade could be measured on a continuous scale by using a light meter and recording light intensity in different places. Conversely, a continuous variable such as salinity could be classified into three levels (low, medium, and high) and treated as a categorical variable. Recognizing the kind of variable you are measuring is important because different designs are based on categorical and continuous variables.

In Chapter 2, we distinguished two kinds of random variables: discrete and continuous. What's the difference between discrete and continuous random variables on the one hand, and categorical and continuous variables on the other? Discrete and continuous random variables are mathematical functions for generating values associated with probability distributions. In contrast, categorical and continuous variables describe the kinds of data that we actually measure in the field or laboratory. Continuous variables usually can be modeled as continuous random variables, whereas both categorical and ordinal variables usually can be modeled as discrete random variables. For example, the categorical variable *sex* can be modeled as a binomial random variable; the numerical variable *height* can be modeled as normal random variable; and the ordinal variable *light reaching the forest floor* can be modeled as a binomial, Poisson, or uniform random variable.

Dependent and Independent Variables

After identifying the types of variables with which you are working, the next step is to designate **dependent** and **independent variables.** The assignment of dependent and independent variables implies an hypothesis of cause and effect that you are trying to test. The dependent variable is the **response variable** that you are measuring and for which you are trying to determine a cause or causes. In a scatterplot of two variables, the dependent or response variable is called the *Y* variable, and it usually is plotted on the **ordinate** (vertical or *y*-axis). The independent variable is the **predictor variable** that you hypothesize is responsible for the variation in the response variable. In the same scatterplot of two variables, the independent or predictor variable is called the *X* variable, and it usually is plotted on the **abscissa** (horizontal or *x*-axis).[1]

In an experimental study, you typically manipulate or directly control the levels of the independent variable and measure the response in the dependent variable. In an observational study, you depend on natural variation in the independent variable from one replicate to the next. In both natural and experimental studies, you don't know ahead of time the strength of the predictor variable. In fact, you are often testing the statistical null hypothesis that variation in the response variable is unrelated to variation in the predictor variable, and is no greater than that expected by chance or sampling error. The alternative hypothesis is that chance cannot entirely account for this variation, and that at least some of the variation can be attributed to the predictor variable. You also may be interested in estimating the size of the effect of the predictor or causal variable on the response variable.

Four Classes of Experimental Design

By combining variable types—categorical versus continuous, dependent versus independent—we obtain four different design classes (Table 7.1). When independent variables are continuous, the classes are either regression (continuous dependent variables) or logistic regression (categorical dependent variables). When independent variables are categorical, the classes are either ANOVA (continuous dependent variable) or tabular (categorical dependent variable). Not all designs fit nicely into these four categories. The analysis of covariance

[1] Of course, merely plotting a variable on the *x*-axis is does not guarantee that it is actually the predictor variable. Particularly in natural experiments, the direction of cause and effect is not always clear, even though the measured variables may be highly correlated (see Chapter 6).

TABLE 7.1 **Four classes of experimental and sampling designs**

Dependent variable	Independent variable	
	Continuous	**Categorical**
Continuous	Regression	ANOVA
Categorical	Logistic regression	Tabular

Different kinds of designs are used depending on whether the independent and dependent variables are continuous or categorical. When both the dependent and the independent variables are continuous, a regression design is used. If the dependent variable is categorical and the independent variable is continuous, a logistic regression design is used. The analysis of regression designs is covered in Chapter 9. If the independent variable is categorical and the dependent variable is continuous, an analysis of variance (ANOVA) design is used. The analysis of ANOVA designs is described in Chapter 10. Finally, if both the dependent and independent variables are categorical, a tabular design is used. Analysis of tabular data is described in Chapter 11.

(ANCOVA) is used when there are two independent variables, one of which is categorical and one of which is continuous (the covariate). ANCOVA is discussed in Chapter 10. Table 7.1 categorizes univariate data, in which there is a single dependent variable. If, instead, we have a vector of correlated dependent variables, we rely on a multivariate analysis of variance (MANOVA) or other multivariate methods that are described in Chapter 12.

Regression Designs

When independent variables are measured on continuous numerical scales (see Figure 6.1 for an example), the sampling layout is a **regression design**. If the dependent variable is also measured on a continuous scale, we use linear or nonlinear regression models to analyze the data. If the dependent variable is measured on an ordinal scale (an ordered response), we use logistic regression to analyze the data. These three types of regression models are discussed in detail in Chapter 9.

SINGLE-FACTOR REGRESSION A regression design is simple and intuitive. Collect data on a set of independent replicates. For each replicate, measure both the predictor and the response variables. In an observational study, neither of the two variables is manipulated, and your sampling is dictated by the levels of natural variation in the independent variable. For example, suppose your hypothesis is that the density of desert rodents is controlled by the availability of seeds (Brown and Leiberman 1973). You could sample 20 independent plots, each chosen to represent a different abundance level of seeds. In each plot, you measure the density of seeds and the density of desert rodents (Figure 7.1). The data are organized in a spreadsheet in which each row is a different plot, and each column is a different response or predictor variable. The entries in each row represent the measurements taken in a single plot.

In an experimental study, the levels of the predictor variable are controlled and manipulated directly, and you measure the response variable. Because your hypothesis is that seed density is responsible for desert rodent density (and not the other way around), you would manipulate seed density in an experimental study, either adding or removing seeds to alter their availability to rodents. In both the experimental study and the observational study, your assumption is that the predictor variable is a causal variable: changes in the value of the predictor (seed density) would *cause* a change in the value of the response (rodent density). This is very different from a study in which you would examine the correlation (statistical covariation) between the two variables. Correlation does not specify a cause-and-effect relationship between the two variables.[2]

In addition to the usual caveats about adequate replication and independence of the data (see Chapter 6), two principles should be followed in designing a regression study:

1. *Ensure that the range of values sampled for the predictor variable is large enough to capture the full range of responses by the response variable.* If the predictor variable is sampled from too limited a range, there may appear to be a weak or nonexistent statistical relationship between the predictor

[2] The sampling scheme needs to reflect the goals of the study. If the study is designed simply to document the relationship between seeds and rodent density, then a series of random plots can be selected, and **correlation** is used to explore the relationship between the two variables. However, if the hypothesis is that seed density is responsible for rodent density, then a series of plots that encompass a uniform range of seed densities should be sampled, and **regression** is used to explore the functional dependence of rodent abundance on seed density. Ideally, the sampled plots should differ from one another only in the density of seeds present. Another important distinction is that a true regression analysis assumes that the value of the independent variable is known exactly and is not subject to measurement error. Finally, standard linear regression (also referred to as Model I regression) minimizes residual deviations in the vertical (y) direction only, whereas correlation minimizes the perpendicular (x and y) distance of each point from the regression line (also referred to as Model II regression). The distinction between correlation and regression is subtle, and is often confusing because some statistics (such as the correlation coefficient) are identical for both kinds of analyses. See Chapter 9 for more details.

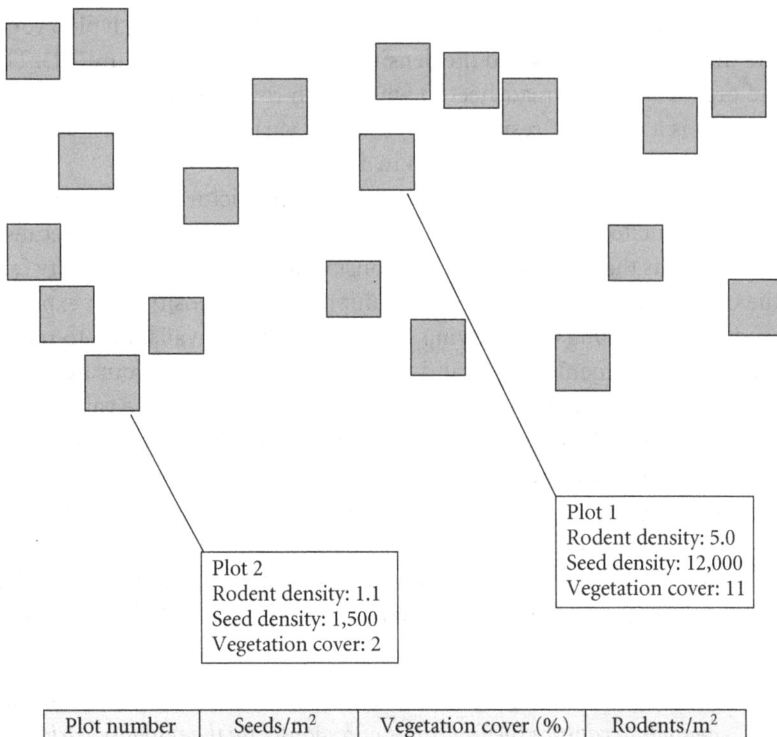

Plot number	Seeds/m^2	Vegetation cover (%)	Rodents/m^2
1	12,000	11	5.0
2	1,500	2	1.1
.	.	.	.
.	.	.	.
20	11,500	52	3.7

Figure 7.1 Spatial arrangement of replicates for a regression study. Each square represents a different 25-m^2 plot. Plots were sampled to ensure a uniform coverage of seed density (see Figures 7.2 and 7.3). Within each plot, the investigator measures rodent density (the response variable), and seed density and vegetation cover (the two predictor variables). The data are organized in a spreadsheet in which each row is a plot, and the columns are the measured variables within the plot.

and response variables even though they are related (Figure 7.2). A limited sampling range makes the study susceptible to a Type II statistical error (failure to reject a false null hypothesis; see Chapter 4).

2. *Ensure that the distribution of predictor values is approximately uniform within the sampled range.* Beware of datasets in which one or two of the values of the predictor variable are very different in size from the others. These influential points can dominate the slope of the regression and

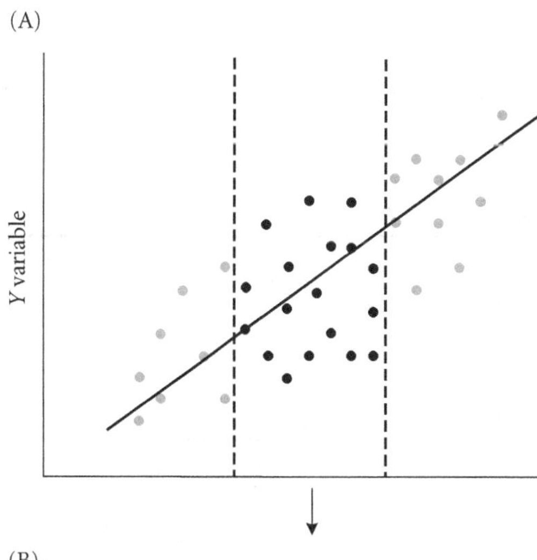

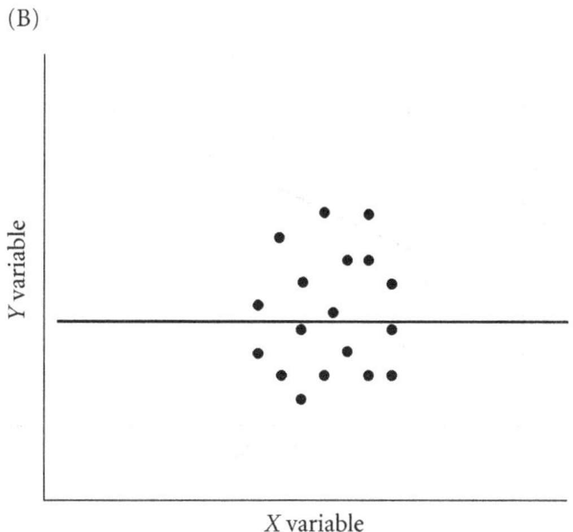

Figure 7.2 Inadequate sampling over a narrow range (within the dashed lines) of the *X* variable can create a spuriously non-significant regression slope, even though *X* and *Y* are strongly correlated with one another. Each point represents a single replicate for which a value has been measured for both the *X* and the *Y* variables. Blue circles represent possible data that were not collected for the analysis. Black circles represent the sample of replicates that were measured. (A) The full range of data. The solid line indicates the true linear relationship between the variables. (B) The regression line is fitted to the sample data. Because the *X* variable was sampled over a narrow range of values, there is limited variation in the resulting *Y* variable, and the slope of the fitted regression appears to be close to zero. Sampling over the entire range of the *X* variable will prevent this type of error.

generate a significant relationship where one does not really exist (Figure 7.3; see Chapter 8 for further discussion of such outliers). Sometimes influential data points can be corrected with a transformation of the predictor variable (see Chapter 8), but we re-emphasize that analysis cannot rescue a poor sampling design.

MULTIPLE REGRESSION The extension to multiple regression is straightforward. Two or more continuous predictor variables are measured for each replicate, along with the single response variable. Returning to the desert rodent example, you suspect that, in addition to seed availability, rodent density is also controlled by vegetation structure—in plots with sparse vegetation, desert rodents are vul-

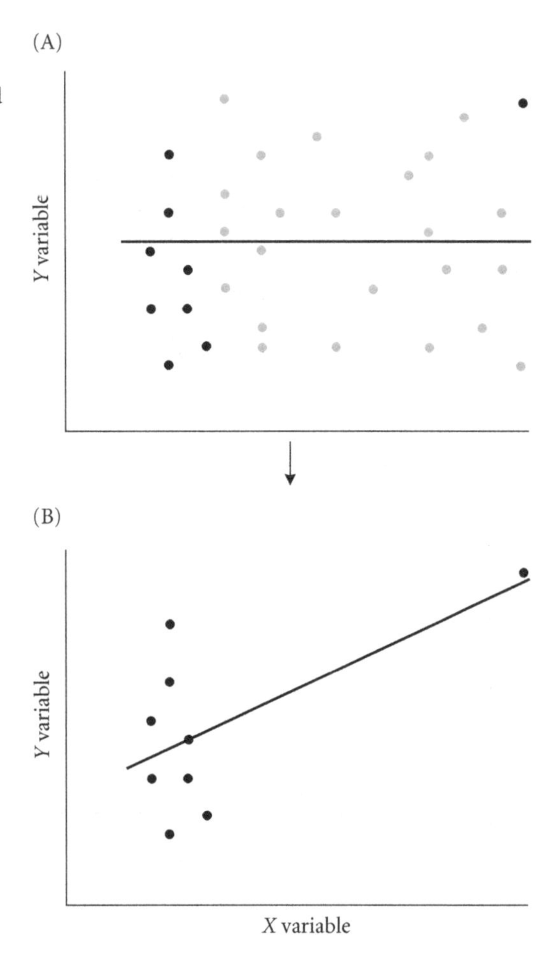

Figure 7.3 Failure to sample uniformly the entire range of a variable can lead to spurious results. As in Figure 7.2, each black point represents a single recorded observation; the blue points represent unobserved *X*, *Y* pairs. (A) The solid line indicates the true linear relationship between the variables. This relationship would have been revealed if the *X* variable had been sampled uniformly. (B) The regression line is fitted to the sample data alone (i.e., just the black points). Because only a single datum with a large value of *X* was measured, this point has an inordinate influence on the fitted regression line. In this case, the fitted regression line inaccurately suggests a positive relationship between the two variables.

nerable to avian predators (Abramsky et al. 1997). In this case, you would take three measurements in each plot: rodent density, seed density, and vegetation cover. Rodent density is still the response variable, and seed density and vegetation cover are the two predictor variables (see Figure 7.1). Ideally, the different predictor variables should be independent of one another. As in simple regression designs, the different values of the predictor variables should be established evenly across the full range of possible values. This is straightforward in an experimental study, but rarely is achievable in an observational study. In an observational study, it is often the case that the predictor variables themselves will be correlated with each other. For example, plots with high vegetation density are likely to have high seed density. There may be few or no plots in which vegetation density is high and seed density is low (or vice versa). This **collinearity** makes it difficult to estimate accurately regression parameters[3] and to tease apart how much variation in the response variable is actually associated with each of the predictor variables.

As always, replication becomes important as we add more predictor variables to the analysis. Following the Rule of 10 (see Chapter 6), you should try to obtain at least 10 replicates for each predictor variable in your study. But in many studies, it is a lot easier to measure additional predictor variables than it is to obtain additional independent replicates. However, you should avoid the temptation to measure everything that you can just because it is possible. Try to select variables that are biologically important and relevant to the hypothesis or question you are asking. It is a mistake to think that a model selection algorithm, such as stepwise multiple regression, can identify reliably the "correct" set of predictor variables from a large dataset (Burnham and Anderson 2010). Moreover, large datasets often suffer from **multicollinearity**: many of the predictor variables are correlated with one another (Graham 2003).

ANOVA Designs

If your predictor variables are categorical (ordered or unordered) and your response variables are continuous, your design is called an **ANOVA** (for *an*alysis *o*f *va*riance). ANOVA also refers to the statistical analysis of these types of designs (see Chapter 10).

TERMINOLOGY ANOVA is rife with terminology. **Treatments** refer to the different categories of the predictor variables that are used. In an experimental study, the treatments represent the different manipulations that have been performed. In an observational study, the treatments represent the different groups that are being compared. The number of treatments in a study equals the number of categories being compared. Within each treatment, multiple observations will be made, and each of these observations is a **replicate**. In standard ANOVA designs, each replicate should be independent, both statistically and biologically, of the other replicates within and among treatments. Later in this

[3] In fact, if one of the predictor variables can be described as a perfect linear function of the other one, it is not even algebraically possible to solve for the regression coefficients. Even when the problem is not this severe, correlations among predictor variables make it difficult to test and compare models. See MacNally (2000b) for a discussion of correlated variables and model-building in conservation biology.

chapter, we will discuss certain ANOVA designs that relax the assumption of independence among replicates.

We also distinguish between **single-factor designs** and **multifactor designs**. In a single-factor design, each of the treatments represents variation in a single predictor variable or **factor**. Each value of the factor that represents a particular treatment is called a **treatment level**. For example, a single-factor ANOVA design could be used to compare growth responses of plants raised at 4 different levels of nitrogen, or the growth responses of 5 different plant species to a single level of nitrogen. The treatment groups may be ordered (e.g., 4 nitrogen levels) or unordered (e.g., 5 plant species).

In a multifactor design, the treatments cover two (or more) different factors, and each factor is applied in combination in different treatments. In a multifactor design, there are different levels of the treatment for each factor. As in the single-factor design, the treatments within each factor may be either ordered or unordered. For example, a two-factor ANOVA design would be necessary if you wanted to compare the responses of plants to 4 levels of nitrogen (Factor 1) *and* 4 levels of phosphorus (Factor 2). In this design, each of the $4 \times 4 = 16$ treatment levels represents a different combination of nitrogen level and phosphorus level. Each combination of nutrients is applied to all of the replicates within the treatment (Figure 7.4).

Although we will return to this topic later, it is worth asking at this point what the advantage is of using a two-factor design. Why not just run two separate experiments? For example, you could test the effects of phosphorus in a one-way ANOVA design with 4 treatment levels, and you could test the effects of nitrogen in a separate one-way ANOVA design, also with 4 treatment levels. What is the advantage of using a two-way design with 16 phosphorus–nitrogen treatment combinations in a single experiment?

One advantage of the two-way design is efficiency. It is likely to be more cost-effective to run a single experiment—even one with 16 treatments—than to run two separate experiments with 4 treatments each. A more important advantage is that the two-way design allows you to test both for main effects (e.g., the effects of nitrogen and phosphorus on plant growth) and for interaction effects (e.g., interactions between nitrogen and phosphorus).

The **main effects** are the additive effects of each level of one treatment averaged over all of the levels of the other treatment. For example, the additive effect of nitrogen would represent the response of plants at each nitrogen level, averaged over the responses to the phosphorus levels. Conversely, the additive effect of phosphorus would be measured as the response of plants at each phosphorus level, averaged over the responses to the different nitrogen levels.

Nitrogen treatment (*one-way layout*)			
0.00 mg	0.10 mg	0.50 mg	1.00 mg
10	10	10	10

Phosphorous treatment (*one-way layout*)			
0.00 mg	0.05 mg	0.10 mg	0.25 mg
10	10	10	10

(*Simultaneous N and P treatments in a two-way layout*)		Nitrogen treatment			
		0.00 mg	0.10 mg	0.50 mg	1.00 mg
Phosphorus treatment	0.00 mg	10	10	10	10
	0.05 mg	10	10	10	10
	0.10 mg	10	10	10	10
	0.25 mg	10	10	10	10

Figure 7.4 Treatment combinations in single-factor designs (upper two panels) and in a two-factor design (lower panel). In all designs, the number in each cell indicates the number of independent replicate plots to be established. In the two single-factor designs (one-way layouts), the four treatment levels represent four different nitrogen or phosphorous concentrations (mg/L). The total sample size is 40 plots in each single-factor experiment. In the two-factor design, the $4 \times 4 = 16$ treatments represent different combinations of nitrogen *and* phosphorous concentrations that are applied simultaneously to a replicate plot. This fully crossed two-factor ANOVA design with 10 replicates per treatment combination would require a total sample size of 160 plots. See Figures 7.9 and 7.10 for other examples of a crossed two-factor design.

Interaction effects represent unique responses to particular treatment combinations that cannot be predicted simply from knowing the main effects. For example, the growth of plants in the high nitrogen–high phosphorus treatment might be synergistically greater than you would predict from knowing the simple additive effects of nitrogen and phosphorus at high levels. Interaction effects are frequently the most important reason for using a factorial design. Strong interactions are the driving force behind much ecological and evolutionary change, and often are more important than the main effects. Chapter 10 will discuss analytical methods and interaction terms in more detail.

SINGLE-FACTOR ANOVA The single-factor ANOVA is one of the simplest, but most powerful, experimental designs. After describing the basic one-way layout, we also explain the randomized block and nested ANOVA designs. Strictly speaking, the randomized block and nested ANOVA are two-factor designs, but the second factor (blocks, or subsamples) is included only to control for sampling variation and is not of primary interest.

The **one-way layout** is used to compare means among two or more treatments or groups. For example, suppose you want to determine whether the recruitment of barnacles in the intertidal zone of a shoreline is affected by different kinds of rock substrates (Caffey 1982). You could start by obtaining a set of slate, granite, and concrete tiles. The tiles should be identical in size and shape, and differ only in material. Following the Rule of 10 (see Chapter 6), set out 10 replicates of each substrate type ($N = 30$ total). Each replicate is placed in the mid-intertidal zone at a set of spatial coordinates that were chosen with a random number generator (Figure 7.5).

After setting up the experiment, you return 10 days later and count the number of new barnacle recruits inside a 10 cm×10 cm square centered in the middle of each tile. The data are organized in a spreadsheet in which each row is a

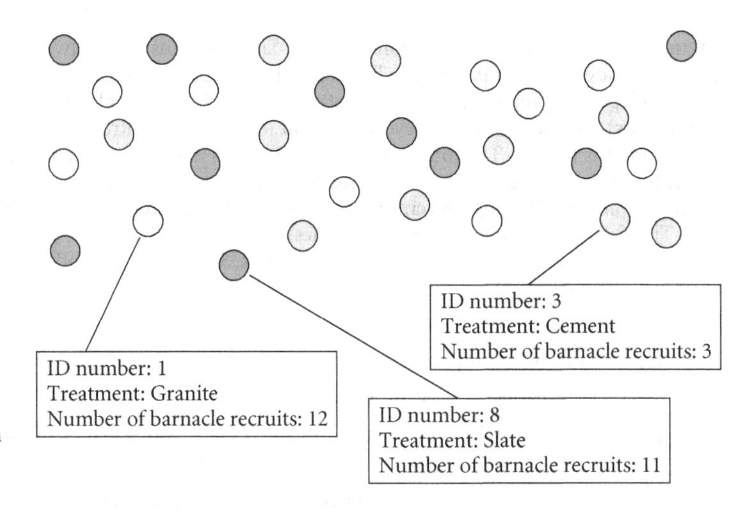

ID number	Treatment	Replicate	Number of barnacle recruits
1	Granite	1	12
2	Slate	1	10
3	Cement	1	3
4	Granite	2	14
5	Slate	2	10
6	Cement	2	8
7	Granite	3	11
8	Slate	3	11
9	Cement	3	7
.	.	.	.
.	.	.	.
30	Cement	10	8

Figure 7.5 Example of a one-way layout. This experiment is designed to test for the effect of substrate type on barnacle recruitment in the rocky intertidal (Caffey 1982). Each circle represents an independent rock substrate. There are 10 randomly placed replicates of each of three treatments, represented by the three shades of blue. The number of barnacle recruits is sampled from a 10-cm square in the center of each rock surface. The data are organized in a spreadsheet in which each row is an independent replicate. The columns indicate the ID number of each replicate (1–30), the treatment group (Cement, Slate, or Granite), the replicate number within each treatment (1–10), and the number of barnacle recruits (the response variable).

replicate. The first few columns contain identifying information associated with the replicate, and the last column of the spreadsheet gives the number of barnacles that recruited into the square. Although the details are different, this is the same layout used in the study of ant density described in Chapter 5: multiple, independent replicate observations are obtained for each treatment or sampling group.

The one-way layout is one of the simplest but most powerful experimental designs, and it can readily accommodate studies in which the number of replicates per treatment is not identical (unequal sample sizes). The one-way layout allows you to test for differences among treatments, as well as to test more specific hypotheses about which particular treatment group means are different and which are similar (see "Comparing Means" in Chapter 10).

The major disadvantage of the one-way layout is that it does not explicitly accommodate environmental heterogeneity. Complete randomization of the replicates within each treatment implies that they will sample the entire array of background conditions, all of which may affect the response variable. On the one hand, this is a good thing because it means that the results of the experiment can be generalized across all of these environments. On the other hand, if the environmental "noise" is much stronger than the "signal" of the treatment, the experiment will have low power; the analysis may not reveal treatment differences unless there are many replicates. Other designs, including the randomized block and the two-way layout, can be used to accommodate environmental variability.

A second, more subtle, disadvantage of the one-way layout is that it organizes the treatment groups along a single factor. If the treatments represent distinctly different kinds of factors, then a two-way layout should be used to tease apart main effects and interaction terms. Interaction terms are especially important because the effect of one factor often depends on the levels of another. For example, the pattern of recruitment onto different substrates may depend on the levels of a second factor (such as predator density).

RANDOMIZED BLOCK DESIGNS One effective way to incorporate environmental heterogeneity is to modify the one-way ANOVA and use a **randomized block design**. A **block** is a delineated area or time period within which the environmental conditions are relatively homogeneous. Blocks may be placed randomly or systematically in the study area, but they should be arranged so that environmental conditions are more similar within blocks than between them.

Once the blocks are established, replicates will still be assigned randomly to treatments, but there is a restriction on the randomization: a single replicate from each of the treatments is assigned to each block. Thus, in a simple randomized block design, each block contains exactly one replicate of all the treatments in the experiment. Within each block, the placement of the treatment replicates should be randomized. Figure 7.6 illustrates the barnacle experiment laid out as a randomized block design. Because there are 10 replicates, there are

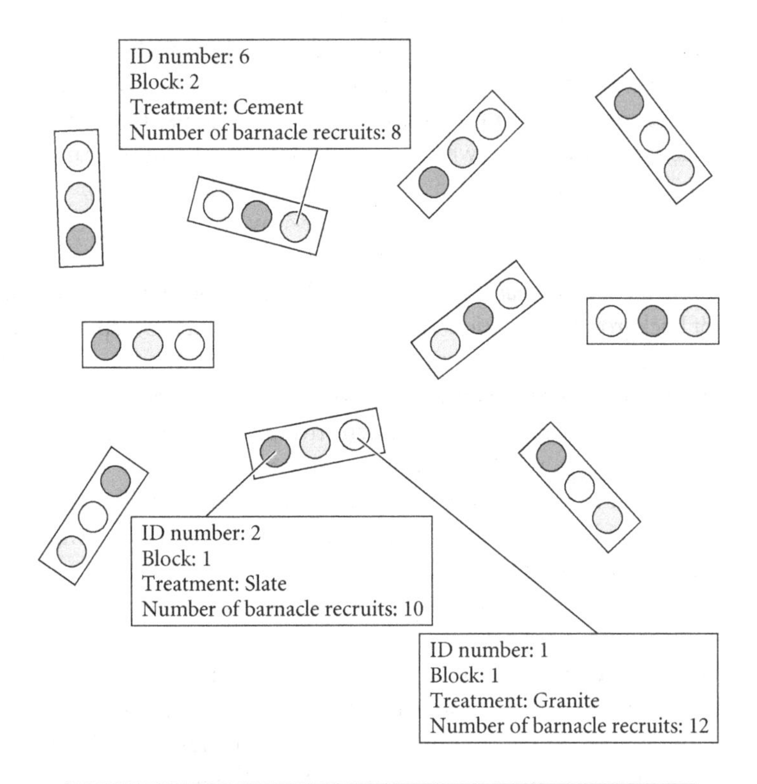

ID number	Treatment	Block	Number of barnacle recruits
1	Granite	1	12
2	Slate	1	10
3	Cement	1	3
4	Granite	2	14
5	Slate	2	10
6	Cement	2	8
7	Granite	3	11
8	Slate	3	11
9	Cement	3	7
.	.	.	.
.	.	.	.
30	Cement	10	8

Figure 7.6 Example of a randomized block design. The 10 replicates of each of the three treatments are grouped in blocks—physical groupings of one replicate of each of the three treatments. Both the placement of blocks and placement of treatments within blocks are randomized. Data organization in the spreadsheet is identical to the one-way layout (Figure 7.5), but the replicate column is replaced by a column indicating the block with which each replicate is associated.

10 blocks (fewer, if you replicate within each block), and each block will contain one replicate of each of the three treatments. The spreadsheet layout for these data is the same as for the one-way layout, except the replicate column is now replaced by a column indicating the block.

Each block should be small enough to encompass a relatively homogenous set of conditions. However, each block must also be large enough to accommodate a single replicate of each of the treatments. Moreover, there must be room within the block to allow sufficient spacing between replicates to ensure their independence (see Figure 6.5). The blocks themselves also have to be far enough apart from one another to ensure independence of replicates among blocks.

If there are geographic gradients in environmental conditions, then each block should encompass a small interval of the gradient. For example, there are strong environmental gradients along a mountainside, so we might set up an experiment with three blocks, one each at high, medium, and low elevation (Figure 7.7A). But it would not be appropriate to create three blocks that run "across the grain" from high to low elevation (Figure 7.7B); each block encompasses conditions that are too heterogeneous. In other cases, the environmental variation may be patchy, and the blocks should be arranged to reflect that patchiness. For example, if an experiment is being conducted in a wetland complex, each semi-isolated fen could be treated as a block. Finally, if the spatial organization of environmental heterogeneity is unknown, the blocks can be arranged randomly within the study area.[4]

The randomized block design is an efficient and very flexible design that provides a simple control for environmental heterogeneity. It can be used to control for environmental gradients and patchy habitats. As we will see in Chapter 10, when environmental heterogeneity is present, the randomized block design is more efficient than a completely randomized one-way layout, which may require a great deal more replication to achieve the same statistical power.

[4] The randomized block design allows you to set up your blocks to encompass environmental gradients in a single spatial dimension. But what if the variation occurs in two dimensions? For example, suppose there is a north-to-south moisture gradient in a field, but also an east-to-west gradient in predator density. In such cases, more complex randomized block designs can be used. For example, the **Latin square** is a block design in which the n treatments are placed in the field in an $n \times n$ square; each treatment appears exactly once in every row and once in every column of the layout. Sir Ronald Fisher (see Footnote 5 in Chapter 5) pioneered these kinds of designs for agricultural studies in which a single field is partitioned and treatments applied to the contiguous subplots. These designs have not been used often by ecologists because the restrictions on randomization and layout are difficult to achieve in field experiments.

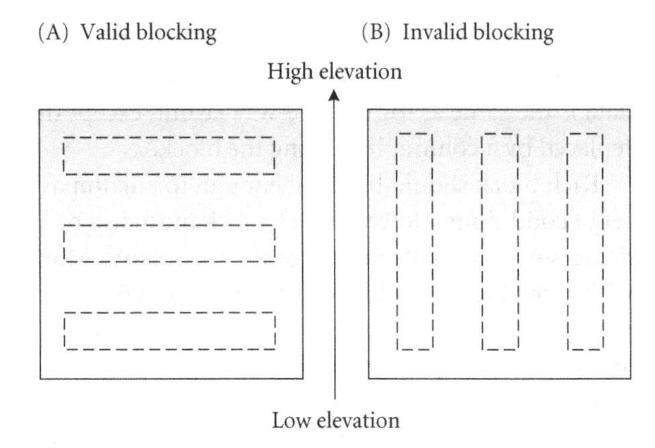

Figure 7.7 Valid and invalid blocking designs. (A) Three properly oriented blocks, each encompassing a single elevation on a mountainside or other environmental gradient. Environmental conditions are more similar within than among blocks. (B) These blocks are oriented improperly, going "across the grain" of the elevational gradient. Conditions are as heterogeneous within the blocks as between them, so no advantage is gained by blocking.

The randomized block design is also useful when your replication is constrained by space or time. For example, suppose you are running a laboratory experiment on algal growth with 8 treatments and you want to complete 10 replicates per treatment. However, you have enough space in your laboratory to run only 12 replicates at a time. What can you do? You should run the experiment in blocks, in which you set up a single replicate of each of the 8 treatments. After the result is recorded, you set up the experiment again (including another set of randomizations for treatment establishment and placement) and continue until you have accumulated 10 blocks. This design controls for inevitable changes in environmental conditions that occur in your laboratory through time, but still allows for appropriate comparison of treatments. In other cases, the limitation may not be space, but organisms. For example, in a study of mating behavior of fish, you may have to wait until you have a certain number of sexually mature fish before you can set up and run a single block of the experiment. In both examples, the randomized block design is the best safeguard against variation in background conditions during the course of your experiment.

Finally, the randomized block design can be adapted for a **matched pairs layout**. Each block consists of a group of individual organisms or plots that have been deliberately chosen to be most similar in background characteristics. Each replicate in the group receives one of the assigned treatments. For example, in a simple experimental study of the effects of abrasion on coral growth, a pair of coral heads of similar size would be considered a single block. One of the coral heads would be randomly assigned to the control group, and the other would be assigned to the abrasion group. Other matched pairs would be chosen in the same way and the treatments applied. Even though the individuals in each pair are not part of a spatial or a temporal block, they are probably going to be more similar than individuals in other such blocks because they have been matched on the basis of colony size or other characteristics. For this reason, the analysis will use a randomized block design. The matched pairs approach is a very effective method when the responses of the replicates potentially are very heterogeneous. Matching the individuals controls for that heterogeneity, making it easier to detect treatment effects.

There are four disadvantages to the randomized block design. The first is that there is a statistical cost to running the experiment with blocks. If the sample size is small and the block effect is weak, the randomized block design is less powerful than a simple one-way layout (see Chapter 10). The second disadvantage is that if the blocks are too small you may introduce non-independence by physically crowding the treatments together. As we discussed in Chapter 6, randomizing the placement of the treatments within the block will help with this problem, but won't eliminate it entirely. The third disadvantage of the randomized block design is that if any of the replicates are lost, the data from that block cannot be used unless the missing values can be estimated indirectly.

The fourth—and most serious—disadvantage of the randomized block design is that it assumes there is no interaction between the blocks and the treatments. The blocking design accounts for additive differences in the response variable and assumes that the rank order of the responses to the treatment does not change from one block to the next. Returning to the barnacle example, the randomized block model assumes that if recruitment in one of the blocks is high, all of the observations in that block will have elevated recruitment. However, the treatment effects are assumed to be consistent from one block to the next, so that the rank order of barnacle recruitment among treatments (Granite > Slate > Cement) is the same, regardless of any differences in the overall recruitment levels among blocks. But suppose that in some blocks recruitment is highest on the cement substrate and in other blocks it is highest on the granite substrate. In this case, the randomized block design may fail to properly characterize the main treatment effects. For this reason, some authors (Mead 1988; Underwood 1997) have argued that the simple randomized block design should not be used unless there is replication within blocks. With replication, the design becomes a two-factor analysis of variance, which we discuss below.

Replication within blocks will indeed tease apart main effects, block effects, and the interaction between blocks and treatments. Replication will also address the problem of missing or lost data from within a block. However, ecologists often do not have the luxury of replication within blocks, particularly when the blocking factor is not of primary interest. The simple randomized block design (without replication) will at least capture the additive component (often the most important component) of environmental variation that would otherwise be lumped with pure error in a simple one-way layout.

NESTED DESIGNS A **nested design** refers to any design in which there is subsampling within each of the replicates. We will illustrate it with the barnacle example. Suppose that, instead of measuring recruitment for a replicate in a single 10 cm×10 cm square, you decided to take three such measurements for each of the 30 tiles in the study (Figure 7.8). Although the number of replicates has

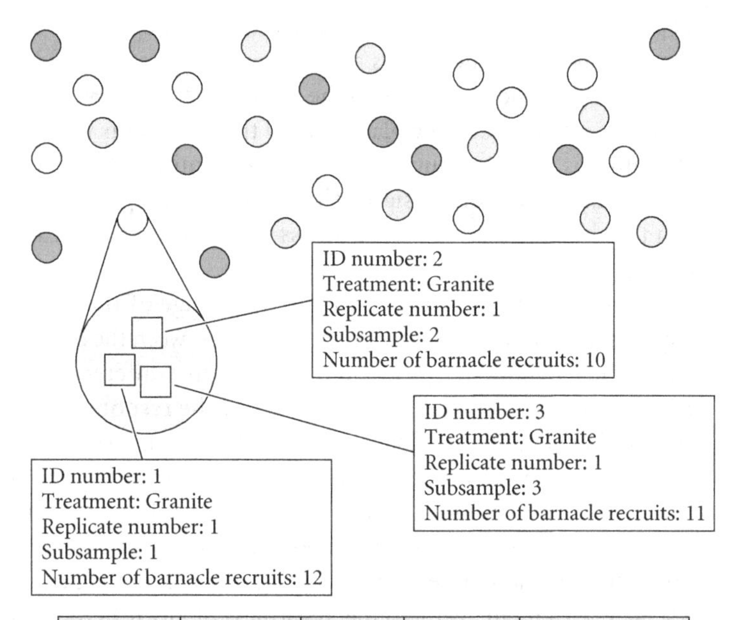

ID number	Treatment	Replicate number	Subsample	Number of barnacle recruits
1	Granite	1	1	12
2	Granite	1	2	10
3	Granite	1	3	11
4	Slate	2	1	14
5	Slate	2	2	10
6	Slate	2	3	7
7	Cement	3	1	5
8	Cement	3	2	6
9	Cement	3	3	10
.	.	.	.	.
.	.	.	.	.
90	Cement	30	3	6

Figure 7.8 Example of a nested design. The study is the same as that shown in Figures 7.5 and 7.6. The layout is identical to the one-way layout in Figure 7.5, but here three subsamples are taken for each independent replicate. In the spreadsheet, an additional column is added to indicate the subsample number, and the total number of observations is increased from 30 to 90.

not increased, the number of observations has increased from 30 to 90. In the spreadsheet for these data, each row now represents a different subsample, and the columns indicate from which replicate and from which treatment the subsample was taken.

This is the first design in which we have included subsamples that are clearly not independent of one another. What is the rationale for such a sampling scheme? The main reason is to increase the precision with which we estimate the response for each replicate. Because of the Law of Large Numbers (see Chapter 3), the more subsamples we use, the more precisely we will estimate the mean for each replicate. The increase in precision should increase the power of the test.

There are three advantages to using a nested design. The first advantage, as we noted, is that subsampling increases the precision of the estimate for each replicate in the design. Second, the nested design allows you to test two hypotheses: first, is there variation among treatments? And, second, is there variation among the replicates *within* a treatment? The first hypothesis is equivalent to a one-way design that uses the *subsample averages* as the observation for each replicate. The second hypothesis is equivalent to a one-way design that uses the subsamples to test for differences among replicates *within* treatments.[5]

Finally, the nested design can be extended to a hierarchical sampling design. For example, you could, in a single study, census subsamples nested within replicates, replicates nested within intertidal zones, intertidal zones nested within shores, shores nested within regions, and even regions nested within continents (Caffey 1985). The reason for carrying out this kind of sampling is that the variation in the data can be partitioned into components that represent each of the hierarchical levels of the study (see Chapter 10). For example, you might be able to show that 80% of the variation in the data occurs at the level of intertidal zones within shores, but only 2% can be attributed to variation among shores within a region. This would mean that barnacle density varies strongly from the high to the low intertidal, but doesn't vary much from one shoreline to the next. Such statements are useful for assessing the relative importance of different mechanisms in producing pattern (Petraitis 1998; see also Figure 4.6).

Nested designs potentially are dangerous in that they are often analyzed incorrectly. One of the most serious and common mistakes in ANOVA is for

[5] You can think of this second hypothesis as a one-way design at a lower hierarchical level. For example, suppose you used the data only from the four replicates of the granite treatment. Consider each replicate as a different "treatment" and each subsample as a different "replicate" for that treatment. The design is now a one-way design that compares the replicates of the granite treatment.

investigators to treat each subsample as an independent replicate and analyze the nested design as a one-way design (Hurlbert 1984). The non-independence of the subsamples artificially boosts the sample size (by threefold in our example, in which we took three subsamples from each tile) and badly inflates the probability of a Type I statistical error (i.e., falsely rejecting a true null hypothesis). A second, less serious problem, is that the nested design can be difficult or even impossible to analyze properly if the sample sizes are not equal in each group. Even with equal numbers of samples and subsamples, nested sampling in more complex layouts, such as the two-way layout or the split-plot design, can be tricky to analyze; the simple default settings for statistical software usually are not appropriate.

But the most serious disadvantage of the nested design is that it often represents a case of misplaced sampling effort. As we will see in Chapter 10, the power of ANOVA designs depends much more on the number of independent replicates than on the precision with which each replicate is measured. It is a much better strategy to invest your sampling effort in obtaining more independent replicates than subsampling within each replicate. By carefully specifying your sampling protocol (e.g., "only undamaged fruits from uncrowded plants growing in full shade"), you may be able to increase the precision of your estimates more effectively than by repeated subsampling.

That being said, you should certainly go ahead and subsample if it is quick and cheap to do so. However, our advice is that you then average (or pool) those subsamples so that you have a single observation for each replicate and then treat the experiment as a one-way design. As long as the numbers aren't too unbalanced, averaging also can alleviate problems of unequal sample size among subsamples and improve the fit of the errors to a normal distribution. It is possible, however, that after averaging among subsamples within replicates you no longer have sufficient replicates for a full analysis. In that case, you need a design with more replicates that are truly independent. Subsampling is no solution to the problem of inadequate replication!

MULTIPLE-FACTOR DESIGNS: TWO-WAY LAYOUT Multifactor designs extend the principles of the one-way layout to two or more treatment factors. Issues of randomization, layout, and sampling are identical to those discussed for the one-way, randomized block, and nested designs. Indeed, the only real difference in the design is in the assignment of the treatments to two or more factors instead of to a single factor. As before, the factors can represent either ordered or unordered treatments.

Returning again to the barnacle example, suppose that, in addition to substrate effects, you wanted to test the effects of predatory snails on barnacle recruitment. You could set up a second one-way experiment in which you established four treatments: unmanipulated, cage control,[6] predator exclusion, and predator inclusion. Instead of running two separate experiments, however, you decide to examine both factors in a single experiment. Not only is this a more efficient use of your field time, but also the effect of predators on barnacle recruitment might differ depending on the substrate type. Therefore, you establish treatments in which you simultaneously apply a different substrate and a different predation treatment.

This is an example of a **factorial design** in which two or more factors are tested simultaneously in one experiment. The key element of a proper factorial design is that the treatments are **fully crossed** or **orthogonal**: every treatment level of the first factor (substrate) must be represented with every treatment level of the second factor (predation; Figure 7.9). Thus, the two-factor experiment has $3 \times 4 = 12$ distinct treatment combinations, as opposed to only 3 treatments for the single-factor substrate experiment or 4 treatments for the single-factor predation experiment. Notice that each of these single-factor experiments would be restricted to only one of the treatment combinations of the other factor. In other words, the substrate experiment that we described above was conducted with the unmanipulated predation treatment, and the predation treatment would be conducted on only a single substrate type. Once we have determined the treatment combinations, the physical set up of the experiment would be the same as for a one-way layout with 12 treatment combinations (Figure 7.10).

In the two-factor experiment, it is critical that all of the crossed treatment combinations be represented in the design. If some of the treatment combinations are missing, we end up with a confounded design. As an extreme example, suppose we set up only the granite substrate–predator exclusion treatment and the slate substrate–predator inclusion treatment. Now the predator effect is confounded with the substrate effect. Whether the results are statistically significant or not, we cannot tease apart whether the pattern is due to the effect of the predator, the effect of the substrate, or the interaction between them.

This example highlights an important difference between manipulative experiments and observational studies. In the observational study, we would gather data on variation in predator and prey abundance from a range of samples.

Cage and cage control

[6] In a cage control, investigators attempt to mimic the physical conditions generated by the cage, but still allow organisms to move freely in and out of the plot. For example, a cage control might consist of a mesh roof (placed over a plot) that allows predatory snails to enter from the sides. In an exclusion treatment, all predators are removed from a mesh cage, and in the inclusion treatment, predators are placed inside each mesh cage. The accompanying figure illustrates a cage (upper panel) and cage control (lower panel) in a fish exclusion experiment in a Venezuelan stream (Flecker 1996).

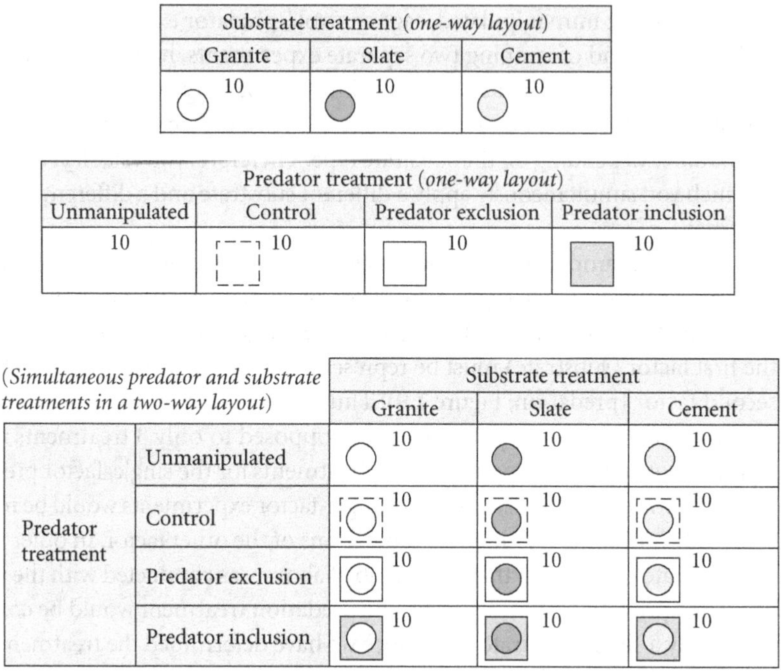

Substrate treatment (*one-way layout*)		
Granite	Slate	Cement
10	10	10

Predator treatment (*one-way layout*)			
Unmanipulated	Control	Predator exclusion	Predator inclusion
10	10	10	10

(*Simultaneous predator and substrate treatments in a two-way layout*)

		Substrate treatment		
		Granite	Slate	Cement
Predator treatment	Unmanipulated	10	10	10
	Control	10	10	10
	Predator exclusion	10	10	10
	Predator inclusion	10	10	10

Figure 7.9 Treatment combinations in two single-factor designs and in a fully crossed two-factor design. This experiment is designed to test for the effect of substrate type (Granite, Slate, or Cement) and predation (Unmanipulated, Control, Predator exclusion, Predator inclusion) on barnacle recruitment in the rocky intertidal. The number 10 indicates the total number of replicates in each treatment. The three shaded colors of the circles represent the three substrate treatments, and the patterns of the squares represent the four predation treatments. The two upper panels illustrate two one-way designs, in which only one of the two factors is systematically varied. In the two-factor design (lower panel), the $4 \times 3 = 12$ treatments represent different combinations of substrate *and* predation. The symbol in each cell indicates the combination of predation and substrate treatment that is applied.

But predators often are restricted to only certain microhabitats or substrate types, so that the presence or absence of the predator is indeed naturally confounded with differences in substrate type. This makes it difficult to tease apart cause and effect (see Chapter 6). The strength of multifactor field experiments is that they break apart this natural covariation and reveal the effects of multiple factors separately *and* in concert. The fact that some of these treatment combinations may be artificial and rarely, if ever, found in nature actually is a strength of the experiment: it reveals the independent contribution of each factor to the observed patterns.

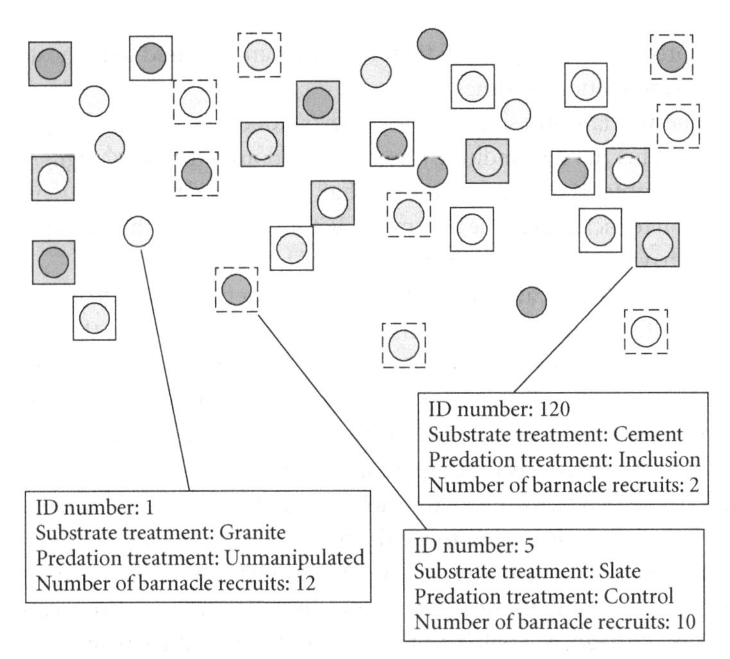

ID number	Substrate treatment	Predation treatment	Number of barnacle recruits
1	Granite	Unmanipulated	12
2	Slate	Unmanipulated	10
3	Cement	Unmanipulated	8
4	Granite	Control	14
5	Slate	Control	10
6	Cement	Control	8
7	Granite	Predator exclusion	50
8	Slate	Predator exclusion	68
9	Cement	Predator exclusion	39
.	.	.	.
.	.	.	.
120	Cement	Predator inclusion	2

Figure 7.10 Example of a two-way design. Treatment symbols are given in Figure 7.9. The spreadsheet contains columns to indicate which substrate treatment and which predation treatment were applied to each replicate. The entire design includes $4 \times 3 \times 10 = 120$ replicates total, but only 36 replicates (three per treatment combination) are illustrated.

The key advantage of two-way designs is the ability to tease apart main effects and interactions between two factors. As we will discuss in Chapter 10, the interaction term represents the non-additive component of the response. The interaction measures the extent to which different treatment combinations act additively, synergistically, or antagonistically.

Perhaps the main disadvantage of the two-way design is that the number of treatment combinations can quickly become too large for adequate replication.

In the barnacle predation example, 120 total replicates are required to replicate each treatment combination 10 times.

As with the one-way layout, a simple two-way layout does not account for spatial heterogeneity. This can be handled by a simple randomized block design, in which every block contains one replicate each of all 12 treatment combinations. Alternatively, if you replicate all of the treatments within each block, this becomes a three-way design, with the blocks forming the third factor in the analysis.

A final limitation of two-way designs is that it may not be possible to establish all orthogonal treatment combinations. It is somewhat surprising that for many common ecological experiments, the full set of treatment combinations may not be feasible or logical. For example, suppose you are studying the effects of competition between two species of salamanders on salamander survival rate. You decide to use a simple two-way design in which each species represents one of the factors. Within each factor, the two treatments are the presence or absence of the species. This fully crossed design yields four treatments (Table 7.2). But what are you going to measure in the treatment combination that has neither Species A nor Species B? By definition, there is nothing to measure in this treatment combination. Instead, you will have to establish the other three treatments ([Species A Present, Species B Absent], [Species A Absent, Species B Present], [Species A Present, Species B Present]) and analyze the design as a one-way ANOVA. The two-way design is possible only if we change the response variable. If the response variable is the abundance of salamander prey remaining

TABLE 7.2 **Treatment combinations in a two-way layout for simple species addition and removal experiments**

Species B	Species A	
	Absent	**Present**
Absent	10	10
Present	10	10

The entry in each cell is the number of replicates of each treatment combination. If the response variable is some property of the species themselves (e.g., survivorship, growth rate), then the treatment combination Species A Absent–Species B Absent (boxed) is not logically possible, and the analysis will have to use a one-way layout with three treatment groups (Species A Present–Species B Present, Species A Present–Species B Absent, and Species A Absent–Species B Present). If the response variable is some property of the environment that is potentially affected by the species (e.g., prey abundance, pH), then all four treatment combinations can be used and analyzed as a two-way ANOVA with two orthogonal factors (Species A and Species B), each with two treatment levels (Absent, Present).

in each plot at the end of the experiment, rather than salamander survivorship, we can then establish the treatment with no salamanders of either species and measure prey levels in the fully crossed two-way layout. Of course, this experiment now asks an entirely different question.

Two-species competition experiments like our salamander example have a long history in ecological and environmental research (Goldberg and Scheiner 2001). A number of subtle problems arise in the design and analysis of two-species competition experiments. These experiments attempt to distinguish between a focal species, for which the response variable is measured, an associative species, whose density is manipulated, and background species, which may be present, but are not experimentally manipulated.

The first issue is what kind of design to use: **additive**, **substitutive**, or **response surface** (Figure 7.11; Silvertown 1987). In an additive design, the density of the focal species is kept constant while the density of the experimental species is varied. However, this design confounds both density and frequency effects. For example, if we compare a control plot (5 individuals of Species A, 0 individuals of Species B) to an addition plot (5 individuals of Species A, 5 individuals of Species B), we have confounded total density (10 individuals) with the presence of the competitor (Underwood 1986; Bernardo et al. 1995). On the other hand, some

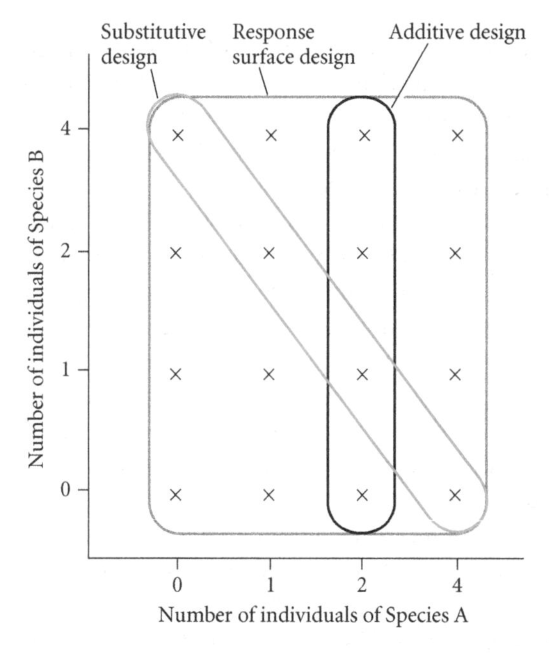

Figure 7.11 Experimental designs for competition experiments. The abundance of Species A and B are each set at 0, 1, 2, or 4 individuals. Each × indicates a different treatment combination. In an additive design, the abundance of one species is fixed (2 individuals of Species A) and the abundance of the competitor is varied (0, 1, 2, or 4 individuals of Species B). In a substitutive design, the total abundance of both competitors is held constant at 4 individuals, but the species composition in the different treatments is altered (0,4; 1,3; 2,2; 3,1; 4,0). In a response surface design, all abundance combinations of the two competitors are established in different treatments (4×4 = 16 treatments). The response surface design is preferred because it follows the principle of a good two-way ANOVA: the treatment levels are fully orthogonal (all abundance levels of Species A are represented with all abundance levels of Species B). (See Inouye 2001 for more details. Figure modified from Goldberg and Scheiner 2001.)

authors have argued that such changes in density are indeed observed when a new species enters a community and establishes a population, so that adjusting for total density is not necessarily appropriate (Schluter 1995).

In a substitutive design, total density of organisms is kept constant, but the relative proportions of the two competitors are varied. These designs measure the relative intensity of inter- and intraspecific competition, but they do not measure the absolute strength of competition, and they assume responses are comparable at different density levels.

The response-surface design is a fully-crossed, two-way design that varies both the relative proportion and density of competitors. This design can be used to measure both relative intensity and absolute strength of inter- and intraspecific competitive interactions. However, as with all two-factor experiments with many treatment levels, adequate replication may be a problem. Inouye (2001) thoroughly reviews response-surface designs and other alternatives for competition studies.

Other issues that need to be addressed in competition experiments include: how many density levels to incorporate in order to estimate accurately competitive effects; how to deal with the non-independence of individuals within a treatment replicate; whether to manipulate or control for background species; and how to deal with residual carry-over effects and spatial heterogeneity that are generated by removal experiments, in which plots are established based on the presence of a species (Goldberg and Scheiner 2001).

SPLIT-PLOT DESIGNS The split-plot design is an extension of the randomized block design to two experimental treatments. The terminology comes from agricultural studies in which a single plot is split into subplots, each of which receives a different treatment. For our purposes, such a split-plot is equivalent to a block that contains within it different treatment replicates.

What distinguishes a split-plot design from a randomized block design is that a second treatment factor is also applied, this time at the level of the entire plot. Let's return one last time to the barnacle example. Once again, you are going to set up a two-way design, testing for predation and substrate effects. However, suppose that the cages are expensive and time-consuming to construct, and that you suspect there is a lot of microhabitat variation in the environment that is affecting your results. In a split-plot design, you would group the three substrates together, just as you did in the randomized block design. However, you would then place a *single cage* over all three of the substrate replicates within a single block. In this design, the predation treatment is referred to as the **whole-plot factor** because a single predation treatment is applied to an entire block. The substrate treatment is referred to as the **subplot factor** because all substrate treatments are applied within a single block. The split-plot design is illustrated in Figure 7.12.

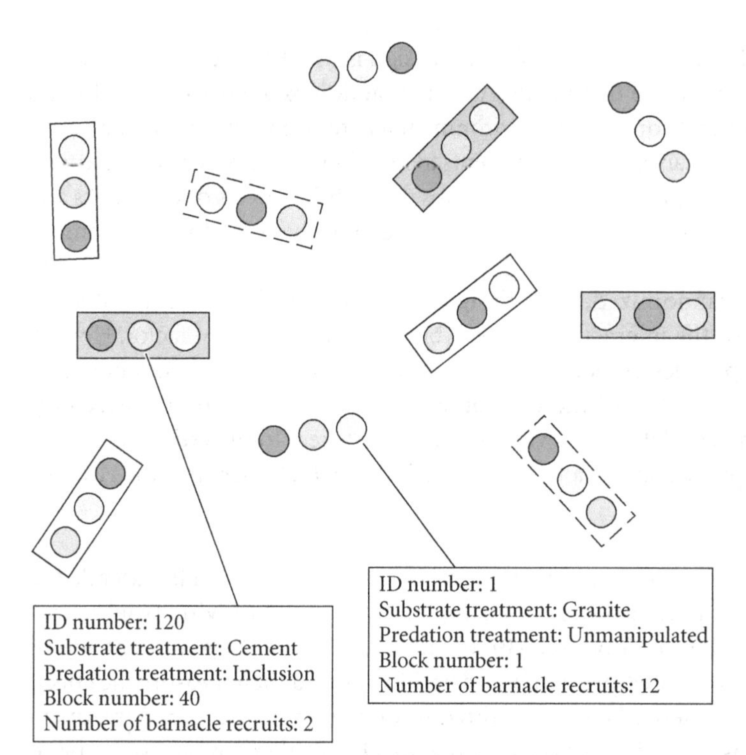

Figure 7.12 Example of a split-plot design. Treatment symbols are given in Figure 7.9. The three substrate treatments (subplot factor) are grouped in blocks. The predation treatment (whole plot factor) is applied to an entire block. The spreadsheet contains columns to indicate the substrate treatment, predation treatment, and block identity for each replicate. Only a subset of the blocks in each predation treatment are illustrated. The split-plot design is similar to a randomized block design (see Figure 7.6), but in this case a second treatment factor is applied to the entire block (= plot).

ID number	Substrate treatment	Predation treatment	Block number	Number of barnacle recruits
1	Granite	Unmanipulated	1	12
2	Slate	Unmanipulated	1	10
3	Cement	Unmanipulated	1	8
4	Granite	Control	2	14
5	Slate	Control	2	10
6	Cement	Control	2	8
7	Granite	Predator exclusion	3	50
8	Slate	Predator exclusion	3	68
9	Cement	Predator exclusion	3	39
.	.	.	.	.
.	.	.	.	.
120	Cement	Predator inclusion	40	2

You should compare carefully the two-way layout (Figure 7.10), and the split-plot layout (Figure 7.12) and appreciate the subtle difference between them. The distinction is that, in the two-way layout, each replicate receives the treatment applications independently and separately. In the split-plot layout, one of the treatments is applied to entire blocks or plots, and the other treatment is applied to replicates within blocks.

The chief advantage of the split-plot design is the efficient use of blocks for the application of two treatments. As in the randomized block design, this is a simple layout that controls for environmental heterogeneity. It also may be less labor-intensive than applying treatments to individual replicates in a simple two-way design. The split-plot design removes the additive effects of the blocks and allows for tests of the main effects and interactions between the two manipulated factors.[7]

As in the randomized block design, the split-plot design does not allow you to test for the interaction between blocks and the subplot factor. However, the split-plot design does let you test for the main effect of the whole-plot factor, the main effect of the subplot factor, and the interaction between the two. As with nested designs, a very common mistake is for investigators to analyze a split-plot design as a two-factor ANOVA, which increases the risk of a Type I error.

DESIGNS FOR THREE OR MORE FACTORS The two-way design can be extended to three or even more factors. For example, if you were studying trophic cascades in a freshwater food web (Brett and Goldman 1997), you might add or remove top carnivores, predators, and herbivores, and then measure the effects on the producer level. This simple three-way design generates $2^3 = 8$ treatment combinations, including one combination that has neither top carnivores, predators, nor herbivores (Table 7.3). As we noted above, if you set up a randomized block design with a two-way layout and then replicate within blocks, the blocks then become a third factor in the analysis. However, three-factor (and higher) designs are used rarely in ecological studies. There are simply too many treatment combinations to make these designs logistically feasible. If you find

[7] Although the example we presented used two experimentally manipulated factors, the split-plot design is also effective when one of the two factors represents a source of natural variation. For example, in our research, we have studied the organization of aquatic food webs that develop in the rain-filled leaves of the pitcher plant *Sarracenia purpurea*. A new pitcher opens about once every 20 days, fills with rainwater, and quickly develops an associated food web of invertebrates and microorganisms.

In one of our experiments, we manipulated the disturbance regime by adding or removing water from the leaves of each plant (Gotelli and Ellison 2006; see also Figure 9.15). These water manipulations were applied to all of the pitchers of a plant. Next, we recorded food web structure in the first, second, and third pitchers. These data are analyzed as a split-plot design. The whole-plot factor is water treatment (5 levels) and the subplot factor is pitcher age (3 levels). The plant served as a natural block, and it was efficient and realistic to apply the water treatments to all the leaves of a plant.

TABLE 7.3 **Treatment combinations in a three-way layout for a food web addition and removal experiment**

	Carnivore absent		Carnivore present	
	Herbivore absent	Herbivore present	Herbivore absent	Herbivore present
Producer absent	10	10	10	10
Producer present	10	10	10	10

In this experiment, the three trophic groups represent the three experimental factors (Carnivore, Herbivore, Producer), each of which has two levels (Absent, Present). The entry in each cell is the number of replicates of each treatment combination. If the response variable is some property of the food web itself, then the treatment combination in which all three trophic levels are absent (boxed) is not logically possible.

your design becoming too large and complex, you should consider breaking it down into a number of smaller experiments that address the key hypotheses you want to test.

INCORPORATING TEMPORAL VARIABILITY: REPEATED MEASURES DESIGNS In all of the designs we have described so far, the response variable is measured for each replicate at a single point in time at the end of the experiment. A **repeated measures design** is used whenever multiple observations on the same replicate are collected at different times. The repeated measures design can be thought of as a split-plot design in which a single replicate serves as a block, and the subplot factor is time. Repeated measures designs were first used in medical and psychological studies in which repeated observations were taken on an individual subject. Thus, in repeated measures terminology, the **between-subjects factor** corresponds to the whole-plot factor, and the **within-subjects factor** corresponds to the different times. In a repeated measures design, however, the multiple observations on a single individual are not independent of one another, and the analysis must proceed cautiously.

For example, suppose we used the simple one-way design for the barnacle study shown in Figure 7.5. But rather than censusing each replicate once, we measured the number of new barnacle recruits on each replicate for 4 consecutive weeks. Now, instead of 3 treatments $\times$ 10 replicates = 30 observations, we have 3 treatments $\times$ 10 replicates $\times$ 4 weeks = 120 observations (Table 7.4). If we only used data from one of the four censuses, the analysis would be identical to the one-way layout.

There are three advantages to a repeated measures design; the first is efficiency. Data are recorded at different times, but it is not necessary to have unique replicates for each time $\times$ treatment combination. Second, the repeated measures

TABLE 7.4 **Spreadsheet for a simple repeated measures analysis**

			Barnacle recruits			
ID number	**Treatment**	**Replicate**	**Week 1**	**Week 2**	**Week 3**	**Week 4**
1	Granite	1	12	15	17	17
2	Slate	1	10	6	19	32
3	Cement	1	3	2	0	2
4	Granite	2	14	14	5	11
5	Slate	2	10	11	13	15
6	Cement	2	8	9	4	4
7	Granite	3	11	13	22	29
8	Slate	3	11	17	28	15
9	Cement	3	7	7	7	6
⋮	⋮	⋮	⋮	⋮	⋮	⋮
30	Cement	10	8	0	0	3

This experiment is designed to test for the effect of substrate type on barnacle recruitment in the rocky intertidal (see Figure 7.5). The data are organized in a spreadsheet in which each row is an independent replicate. The columns indicate the ID number (1–30), the treatment group (Cement, Slate, or Granite), and the replicate number (1–10 within each treatment). The next four columns give the number of barnacle recruits recorded on a particular substrate in each of four consecutive weeks. The measurements at different times are not independent of one another because they are taken from the same replicate each week.

design allows each replicate to serve as its own block or control. When the replicates represent individuals (plants, animals, or humans), this effectively controls for variation in size, age, and individual history, which often have strong influences on the response variable. Finally, the repeated measures design allows us to test for interactions of time with treatment. For many reasons, we expect that differences among treatments may change with time. In a press experiment (see Chapter 6), there may be cumulative effects of the treatment that are not expressed until some time after the start of the experiment. In contrast, in a pulse experiment, we expect to see differences among treatments diminish as more time passes following the single, pulsed treatment application. Such complex effects are best seen in the interaction between time and treatment, and they may not be detected if the response variable is measured at only a single point in time.

Both the randomized block and the repeated measures designs make a special assumption of **circularity** for the within-subjects factor. Circularity (in the context of ANOVA) means that the variance of the difference between any two treatment levels in the subplot is the same. For the randomized block design, this means that the variance of the difference between any pair of treatments in the block is the same. If the treatment plots are large enough and adequately spaced, this is often a reasonable assumption. For the repeated measures design, the assumption of circularity means that the variance of the difference of observations between any pair of times is the same. This assumption of circularity is unlikely to be met for repeated measures; in most cases, the variance of the difference between two consecutive observations is likely to be much smaller than the variance of the difference between two observations that are widely separated in time. This is because time series measured on the same subject are likely to have a temporal "memory" such that current values are a function of values observed in the recent past. This premise of correlated observations is the basis for time-series analysis (see Chapter 6).

The chief disadvantage with repeated measures analysis is failure to meet the assumption of circularity. If the repeated measures are serially correlated, Type I error rates for F-tests will be inflated, and the null hypothesis may be incorrectly rejected when it is true. The best way to meet the circularity assumption is to use evenly spaced sampling times along with knowledge of the natural history of your organisms to select an appropriate sampling interval.

What are some alternatives to repeated measures analysis that do not rely on the assumption of circularity? One approach is to set out enough replicates so that a different set is censused at each time period. With this design, time can be treated as a simple factor in a two-way analysis of variance. If the sampling methods are destructive (e.g., collecting stomach contents of fish, killing and preserving an invertebrate sample, or harvesting plants), this is the only method for incorporating time into the design.

A second strategy is to use the repeated measures layout, but to be more creative in the design of the response variable. Collapse the correlated repeated measures into a single response variable for each individual, and then use a simple one-way analysis of variance. For example, if you want to test whether temporal trends differ among the treatments (the between-subjects factor), you could fit a regression line (with either a linear or time-series model) to the repeated measures data, and use the slope of the line as the response variable. A separate slope value would be calculated for each of the individuals in the study. The slopes would then be compared using a simple one-way analysis, treating each individual as an independent observation (which it is). Significant treatment effects would indicate different temporal trajectories for individuals in the different treatments; this test is very similar to the test for an interaction of time and treatment in a standard repeated measures analysis.

Although such composite variables are created from correlated observations collected from one individual, they are independent among individuals. Moreover, the Central Limit Theorem (see Chapter 2) tells us that averages of these values will follow an approximately normal distribution even if the underlying variables themselves do not. Because most repeated measures data do not meet the assumption of circularity, our advice is to be careful with these analyses. We prefer to collapse the temporal data to a single variable that is truly independent among observations and then use a simpler one-way design for the analysis.

ENVIRONMENTAL IMPACTS OVER TIME: BACI DESIGNS A special type of repeated measures design is one in which measurements are taken both before and after the application of a treatment. For example, suppose you are interested in measuring the effect of atrazine (a hormone-mimicking compound) on the body size of frogs (Allran and Karasov 2001). In a simple one-way layout, you could assign frogs randomly to control and treatment groups, apply the atrazine, and measure body size at the end of the experiment. A more sensitive design might be to establish the control and treatment groups, and take measurements of body size for one or more time periods before application of the treatment. After the treatment is applied, you again measure body size at several times in both the control and treatment groups.

These designs also are used for observational studies assessing environmental impacts. In impact assessment, the measurements are taken before and after the impact occurs. A typical assessment might be the potential responses of a marine invertebrate community to the operation of a nuclear power plant, which discharges considerable hot water effluent (Schroeter et al. 1993). Before the power plant begins operation, you take one or more samples in the area that will be affected by the plant and estimate the abundance of species of interest (e.g., snails, sea stars, sea urchins). Replication in this study could be spatial, temporal, or both. Spatial replication would require sampling several different plots, both within and beyond the projected plume of hot water discharge.[8] Temporal replication would require sampling a single site in the discharge area at several times before the plant came on-line. Ideally, multiple sites would be sampled several times in the pre-discharge period.

Once the discharge begins, the sampling protocol is repeated. In this assessment design, it is imperative that there be at least one control or reference site that is sampled at the same time, both before and after the discharge. Then, if you observe a

[8] A key assumption in this layout is that the investigator knows ahead of time the spatial extent of the impact. Without this information, some of the "control" plots may end up within the "treatment" region, and the effect of the hot water plume would be underestimated.

decline in abundance at the impacted site but not at the control site, you can test whether the decline is significant. Alternatively, invertebrate abundance might be declining for reasons that have nothing to do with hot water discharge. In this case, you would find lower abundance at both the control and the impacted sites.

This kind of repeated measures design is referred to as a **BACI design** (Before-After, Control-Impact). Not only is there replication of control and treatment plots, there is temporal replication with measurements before and after treatment application (Figure 7.13).

In its ideal form, the BACI design is a powerful layout for assessing environmental perturbations and monitoring trajectories before and after the impact. Replication in space ensures that the results will be applicable to other sites that may be perturbed in the same way. Replication through time ensures that the temporal trajectory of response and recovery can be monitored. The design is appropriate for both pulse and press experiments or disturbances.

Unfortunately, this idealized BACI design is rarely achieved in environmental impact studies. Often times, there is only a single site that will be impacted, and that site usually is not chosen randomly (Stewart-Oaten and Bence 2001; Murtaugh 2002b). Spatial replicates within the impacted area are not independent replicates because there is only a single impact studied at a single site (Underwood 1994). If the impact represents an environmental accident, such as an oil spill, no pre-impact data may be available, either from reference or impact sites.

The potential control of randomization and treatment assignments is much better in large-scale experimental manipulations, but even in these cases there may be little, if any, spatial replication. These studies rely on more intensive temporal replication, both before and after the manipulation. For example, since 1983, Brezonik et al. (1986) have conducted a long-term acidification experiment on Little Rock Lake, a small oligotrophic seepage lake in northern Wisconsin. The lake was divided into a treatment and reference basin with an impermeable vinyl curtain. Baseline (pre-manipulation) data were collected in both basins from August 1983 through April 1985. The treatment basin was then acidified with sulfuric acid in a stepwise fashion to three target pH levels (5.6, 5.1, 4.7). These pH levels were maintained in a press experiment at two-year intervals.

Because there is only one treatment basin and one control basin, conventional ANOVA methods cannot be used to analyze these data.[9] There are two general

[9] Of course, if one assumes the samples are independent replicates, conventional ANOVA could be used. But there is no wisdom in forcing data into a model structure they do not fit. One of the key themes of this chapter is to choose simple designs for your experiments and surveys whose assumptions best meet the constraints of your data.

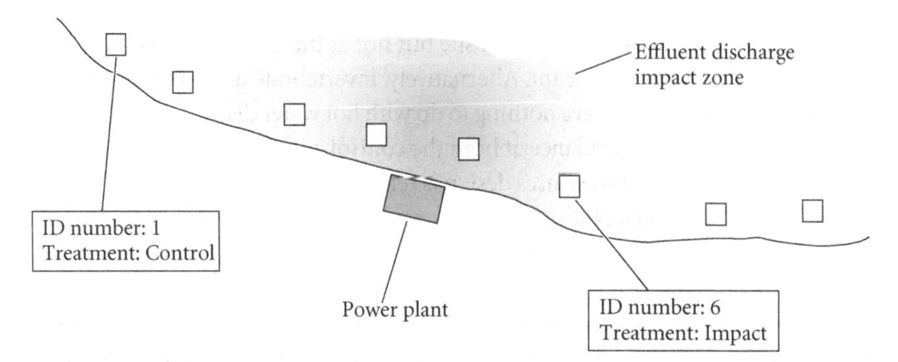

ID		Pre-impact sampling				Post-impact sampling			
number	Treatment	Week 1	Week 2	Week 3	Week 4	Week 5	Week 6	Week 7	Week 8
1	Control	106	108	108	120	122	123	130	190
2	Control	104	88	84	104	106	119	135	120
3	Impact	99	97	102	192	150	140	145	150
4	Impact	120	122	98	120	137	135	155	165
5	Impact	88	90	92	94	0	7	75	77
6	Impact	100	120	129	82	2	3	66	130
7	Control	66	70	70	99	45	55	55	109
8	Control	130	209	220	250	100	90	88	140

Figure 7.13 Example of a spatial arrangement of replicates for a BACI design. Each square represents a sample plot on a shoreline that will potentially be affected by hot water discharged from a nuclear power plant. Permanent plots are established within the hot water effluent zone (shaded area), and in adjacent control zones (unshaded areas). All plots are sampled weekly for 4 weeks before the plant begins discharging hot water and for 4 weeks afterward. Each row of the spreadsheet represents a different replicate. Two columns indicate the replicate ID number and the treatment (Control or Impact). The remaining columns give the invertebrate abundance data collected at each of the 8 sampling dates (4 sample dates pre-discharge, 4 sample dates post-discharge).

strategies for analysis. **Randomized intervention analysis** (RIA) is a Monte Carlo procedure (see Chapter 5) in which a test statistic calculated from the time series is compared to a distribution of values created by randomizing or reshuffling the time-series data among the treatment intervals (Carpenter et al. 1989). RIA relaxes the assumption of normality, but it is still susceptible to temporal correlations in the data (Stewart-Oaten et al. 1992).

A second strategy is to use time-series analysis to fit simple models to the data. The **autoregressive integrated moving average** (ARIMA) model describes the correlation structure in temporal data with a few parameters (see Chapter 6). Additional model parameters estimate the stepwise changes that occur with the experimental interventions, and these parameters can then be tested against the null hypothesis that they do not differ from 0. ARIMA models can be fit individually to the control and manipulation time series data, or to a derived data series created by taking the ratio of the treatment/control data at each time step (Rasmussen et al. 1993). Bayesian methods can also be used to analyze data from BACI designs (Carpenter et al. 1996; Rao and Tirtotjondro 1996; Reckhow 1996; Varis and Kuikka 1997; Fox 2001).

RIA, ARIMA, and Bayesian methods are powerful tools for detecting treatment effects in time-series data. However, without replication, it is still problematic to generalize the results of the analysis. What might happen in other lakes? Other years? Additional information, including the results of small-scale experiments (Frost et al. 1988), or snapshot comparisons with a large number of unmanipulated control sites (Schindler et al. 1985; Underwood 1994) can help to expand the domain of inference from BACI study.

Alternatives to ANOVA: Experimental Regression

The literature on experimental design is dominated by ANOVA layouts, and modern ecological science has been referred to sarcastically as little more than the care and curation of ANOVA tables. Although ANOVA designs are convenient and powerful for many purposes, they are not always the best choice. ANOVA has become so popular that it may act as an intellectual straightjacket (Werner 1998), and cause scientists to neglect other useful experimental designs.

We suggest that ANOVA designs often are employed when a regression design would be more appropriate. In many ANOVA designs, a continuous predictor variable is tested at only a few values so that it can be treated as a categorical predictor variable, and shoehorned into an ANOVA design. Examples include treatment levels that represent different nutrient concentrations, temperatures, or resource levels.

In contrast, an experimental regression design (Figure 7.14) uses many different levels of the continuous independent variable, and then uses regression to fit a line, curve, or surface to the data. One tricky issue in such a design (the same problem is present in an ANOVA) is to choose appropriate levels for the predictor variable. A uniform selection of predictor values within the desired range should ensure high statistical power and a reasonable fit to the regression line. However, if the response is expected to be multiplicative rather than linear (e.g., a 10% decrease in growth with every doubling of concentration), it might be better to set the predictor values on an evenly spaced logarithmic scale. In this design, you will have more data collected at low concentrations, where changes in the response variable might be expected to be steepest.

One of the chief advantages of regression designs is efficiency. Suppose you are studying responses of terrestrial plant and insect communities to nitrogen

(Simultaneous N and P treatments in a two-way ANOVA design)		Nitrogen treatment	
		0.00 mg	0.50 mg
Phosphorus treatment	0.00 mg	12	12
	0.05 mg	12	12

(Two-way experimental regression design)		Nitrogen treatment						
		0.00 mg	0.05 mg	0.10 mg	0.20 mg	0.40 mg	0.80 mg	1.00 mg
Phosphorus treatment	0.00 mg	1	1	1	1	1	1	1
	0.01 mg	1	1	1	1	1	1	1
	0.05 mg	1	1	1	1	1	1	1
	0.10 mg	1	1	1	1	1	1	1
	0.20 mg	1	1	1	1	1	1	1
	0.40 mg	1	1	1	1	1	1	1
	0.50 mg	1	1	1	1	1	1	1

Figure 7.14 Treatment combinations in a two-way ANOVA design (upper panel) and an experimental regression design (lower panel). These experiments test for additive and interactive effects of nitrogen (N) and phosphorus (P) on plant growth or some other response variable. Each cell in the table indicates the number of replicate plots used. If the maximum number of replicates is 50 and a minimum of 10 replicates per treatment in required, only 2 treatment levels each for N and P are possible (each with 12 replicates) in the two-way ANOVA. In contrast, the experimental regression allows for 7 treatment levels each of N and P. Each of the $7 \times 7 = 49$ plots in the design receives a unique combination of N and P concentrations.

(N), and your total sample size is limited by available space or labor to 50 plots. If you try to follow the Rule of 10, an ANOVA design would force you to select only 5 different fertilization levels and replicate each one 10 times. Although this design is adequate for some purposes, it may not help to pinpoint critical threshold levels at which community structure changes dramatically in response to N. In contrast, a regression design would allow you to set up 50 different N levels, one in each of the plots. With this design, you could very accurately characterize changes that occur in community structure with increasing N levels; graphical displays may help to reveal threshold points and non-linear effects (see Chapter 9). Of course, even minimal replication of each treatment level is very desirable, but if the total sample size is limited, this may not be possible.

For a two-factor ANOVA, the experimental regression is even more efficient and powerful. If you want to manipulate nitrogen and phosphorus (P) as independent factors, and still maintain 10 replicates per treatment, you could have no more than 2 levels of N and 2 levels of P. Because one of those levels must be a control plot (i.e., no fertilization), the experiment isn't going to give you very much information about the role of changing levels of N and P on the system. If the result is statistically significant, you can say only that the community responds to those particular levels of N and P, which is something that you may have already known from the literature before you started. If the result is not statistically significant, the obvious criticism would be that the concentrations of N and P were too low to generate a response.

In contrast, an experimental regression design would be a fully crossed design with 7 levels of nitrogen and 7 levels of phosphorus, with one level of each corresponding to the control (no N or P). The $7 \times 7 = 49$ replicates each receive a unique concentration of N and P, with one of the 49 plots receiving neither N nor P (see Figure 7.14). This is a response surface design (Inouye 2001), in which the response variable will be modeled by multiple regression. With seven levels of each nutrient, this design provides a much more powerful test for additive and interactive effects of nutrients, and could also reveal non-linear responses. If the effects of N and P are weak or subtle, the regression model will be more likely to reveal significant effects than the two-way ANOVA.[10]

Efficiency is not the only advantage of an experimental regression design. By representing the predictor variable naturally on a continuous scale, it is much easier to detect non-linear, threshold, or asymptotic responses. These cannot be inferred reliably from an ANOVA design, which usually will not have enough treatment levels to be informative. If the relationship is something other than a straight line, there are a number of statistical methods for fitting non-linear responses (see Chapter 9).

A final advantage to using an experimental regression design is the potential benefits for integrating your results with theoretical predictions and ecological models. ANOVA provides estimates of means and variances for groups or particular levels of categorical variables. These estimates are rarely of interest or use in ecological models. In contrast, a regression analysis provides esti-

[10] There is a further hidden penalty in using the two-way ANOVA design for this experiment that often is not appreciated. If the treatment levels represent a small subset of many possible other levels that could have been used, then the design is referred to as a **random effects ANOVA** model. Unless there is something special about the particular treatment levels that were used, a random effects model is always the most appropriate choice when a continuous variable has been shoehorned into a categorical variable for ANOVA. In the random effects model, the denominator of the F-ratio test for treatment effects is the interaction mean square, not the error mean square that is used in a standard **fixed effects ANOVA** model. If there are not many treatment levels, there will not be very many degrees of freedom associated with the interaction term, regardless of the amount of replication within treatments. As a consequence, the test will be much less powerful than a typical fixed effects ANOVA. See Chapter 10 for more details on fixed and random effects ANOVA models and the construction of F-ratios.

mates of slope and intercept parameters that measure the change in the response Y relative to a change in predictor X (dY/dX). These derivatives are precisely what are needed for testing the many ecological models that are written as simple differential equations.

An experimental regression approach might not be feasible if it is very expensive or time-consuming to establish unique levels of the predictor variable. In that case, an ANOVA design may be preferred because only a few levels of the predictor variable can be established. An apparent disadvantage of the experimental regression is that it appears to have no replication! In Figure 7.14, each unique treatment level is applied to only a single replicate, and that seems to fly in the face of the principle of replication (see Chapter 6). If each unique treatment is unreplicated, the least-squares solution to the regression line still provides an estimate of the regression parameters and their variances (see Chapter 9). The regression line provides an unbiased estimate of the expected value of the response Y for a given value of the predictor X, and the variance estimates can be used to construct a confidence interval about that expectation. This is actually more informative than the results of an ANOVA model, which allow you to estimate means and confidence intervals for only the handful of treatment levels that were used.

A potentially more serious issue is that the regression design may not include any replication of controls. In our two-factor example, there is only one plot that contains no nitrogen and no phosphorus addition. Whether this is a serious problem or not depends on the details of the experimental design. As long as all of the replicates are treated the same and differ only in the treatment application, the experiment is still a valid one, and the results will estimate accurately the relative effects of different levels of the predictor variable on the response variable. If it is desirable to estimate the absolute treatment effect, then additional replicated control plots may be needed to account for any handling effects or other responses to the general experimental conditions. These issues are no different than those that are encountered in ANOVA designs.

Historically, regression has been used predominantly in the analysis of non-experimental data, even though its assumptions are unlikely to be met in most sampling studies. An experimental study based on a regression design not only meets the assumptions of the analysis, but is often more powerful and appropriate than an ANOVA design. We encourage you to "think outside the ANOVA box" and consider a regression design when you are manipulating a continuous predictor variable.

Tabular Designs

The last class of experimental designs is used when both predictor and response variables are categorical. The measurements in these designs are counts. The simplest such variable is a dichotomous (or binomial, see Chapter 2) response in a series of independent trials. For example, in a test of cockroach behavior, you could place an individual cockroach in an arena with a black and a white side, and then record on which side the animal spent the majority of its time. To ensure independence, each replicate cockroach would be tested individually.

More typically, a dichotomous response will be recorded for two or more categories of the predictor variable. In the cockroach study, half of the cockroaches might be infected experimentally with a parasite that is known to alter host behavior (Moore 1984). Now we want to ask whether the response of the cockroach differs between parasitized and unparasitized individuals (Moore 2001; Poulin 2000). This approach could be extended to a three-way design by adding an additional treatment and asking whether the difference between parasitized and unparasitized individuals changes in the presence or absence of a vertebrate predator.

We might predict that uninfected individuals are more likely to use the black substrate, which will make them less conspicuous to a visual predator. In the presence of a predator, uninfected individuals might shift even more toward the dark surfaces, whereas infected individuals might shift more toward white surfaces. Alternatively, the parasite might alter host behavior, but those alterations might be independent of the presence or absence of the predator. Still another possibility is that host behavior might be very sensitive to the presence of the predator, but not necessarily affected by parasite infection. A **contingency table analysis** (see Chapter 11) is used to test all these hypotheses with the same dataset.

In some tabular designs, the investigator determines the total number of individuals in each category of predictor variable, and these individuals will be classified according to their responses. The total for each category is referred to as the **marginal total** because it represents the column or row sum in the margin of the data table. In an observational study, the investigator might determine one or both of the marginal totals, or perhaps only the grand total of independent observations. In a tabular design, the grand total equals the sum of either the column or row marginal totals.

For example, suppose you are trying to determine the associations of four species of *Anolis* lizard with three microhabitat types (ground, tree trunks, tree branches; see Butler and Losos 2002). Table 7.5 shows the two-way layout of the data from such a study. Each row in the table represents a different lizard species, and each column represents a different habitat category. The entries in each cell represent the counts of a particular lizard species recorded in a particular habitat. The marginal row totals represent the total number of observations for each lizard species, summed across the three habitat types. The marginal column totals

TABLE 7.5 **Tabulated counts of the occurrence of four lizard species censused in three different microhabitats**

		Habitat			
		Ground	Tree trunk	Tree branch	Species totals
Lizard species	**Species A**	9	0	15	*24*
	Species B	9	0	12	*21*
	Species C	9	5	0	*14*
	Species D	9	10	3	*22*
	Habitat totals	*36*	*15*	*30*	*81*

Italicized values are the marginal totals for the two-way table. The total sample size is 81 observations. In these data, both the response variable (species identity) and the predictor variable (microhabitat category) are categorical.

represent the total number of observations in each habitat type, summed across the three habitats. The grand total in the table ($N = 81$) represents the total count of all lizard species observed in all habitats.

There are several ways that these data could have been collected, depending on whether the sampling was based on the marginal totals for the microhabitats, the marginal totals for the lizards, or the grand total for the entire sample.

In a sampling scheme built around the microhabitats, the investigator might have spent 10 hours sampling each microhabitat, and recording the number of different lizard species encountered in each habitat. Thus, in the census of tree trunks, the investigator found a total of 15 lizards: 5 of Species C and 10 of Species D; Species A and B were never encountered. In the ground census, the investigator found a total of 36 lizards, with all four species equally represented.

Alternatively, the sampling could have been based on the lizards themselves. In such a design, the investigator would put in an equal sampling effort for each species by searching all habitats randomly for individuals of a particular species of lizard and then recording in which microhabitat they occurred. Thus, in a search for Species B, 21 individuals were found, 9 on the ground, and 12 on tree branches. Another sampling variant is one in which the row and column totals are simultaneously fixed. Although this design is not very common in ecological studies, it can be used for an exact statistical test of the distribution of sampled values (Fisher's Exact Test; see Chapter 11). Finally, the sampling might have been based simply on the grand total of observations. Thus, the investigator might have taken a random sample of 81 lizards, and for each lizard encountered, recorded the species identity and the microhabitat.

Ideally, the marginal totals on which the sampling is based should be the same for each category, just as we try to achieve equal sample sizes in setting up an

ANOVA design. However, identical sample sizes are not necessary for analysis of tabular data. Nevertheless, the tests do require (as always) that the observations be randomly sampled and that the replicates be truly independent of one another. This may be very difficult to achieve in some cases. For example, if lizards tend to aggregate or move in groups, we cannot simply count individuals as they are encountered because an entire group is likely to be found in a single microhabitat. In this example, we also assume that all of the lizards are equally conspicuous to the observer in all of the habitats. If some species are more obvious in certain habitats than others, then the relative frequencies will reflect sampling biases rather than species' microhabitat associations.

SAMPLING DESIGNS FOR CATEGORICAL DATA In contrast to the large literature for regression and ANOVA designs, relatively little has been written, in an ecological context, about sampling designs for categorical data. If the observations are expensive or time-consuming, every effort should be made to ensure that each observation is independent, so that a simple two- or multiway layout can be used. Unfortunately, many published analyses of categorical data are based on nonindependent observations, some of it collected in different times or in different places. Many behavioral studies analyze multiple observations of the same individual. Such data clearly should not be treated as independent (Kramer and Schmidhammer 1992). If the tabular data are not independent, random samples from the same sampling space, you should explicitly incorporate the temporal or spatial categories as factors in your analysis.

Alternatives to Tabular Designs: Proportional Designs

If the individual observations are inexpensive and can be gathered in large numbers, there is an alternative to tabular designs. One of the categorical variables can be collapsed to a measure of proportion (number of desired outcomes/number of observations), which is a continuous variable. The continuous variable can then be analyzed using any of the methods described above for regression or ANOVA.

There are two advantages of using proportional designs in lieu of tabular ones. The first is that the standard set of ANOVA and regression designs can be used, including blocking. The second advantage is that the analysis of proportions can be used to accommodate frequency data that are not strictly independent. For example, suppose that, to save time, the cockroach experiment were set up with 10 individuals placed in the behavioral arena at one time. It would not be legitimate to treat the 10 individuals as independent replicates, for the same reason that subsamples from a single cage are not independent replicates for an ANOVA. However, the data from this run can be treated as a single replicate, for which we could calculate the proportion of individuals present on the black side of the arena. With multiple runs of the experiment, we can now test hypotheses about differences in the proportion among groups (e.g., parasitized versus unparasitized). The design is still problematic, because it is possible that substrate selection by solitary cockroaches may be different from substrate selection by groups of cockroaches. Nevertheless, the design at least avoids treating individuals within an arena as independent replicates, which they are not.

Although proportions, like probabilities, are continuous variables, they are bounded between 0.0 and 1.0. An arcsin square root transformation of proportional data may be necessary to meet the assumption of normality (see Chapter 8). A second consideration in the analysis of proportions is that it is very important to use at least 10 trials per replicate, and to make sure that sample sizes are as closely balanced as possible. With 10 individuals per replicate, the possible measures for the response variable are in the set {0.0, 0.1, 0.2, 0.3, 0.4, 0.5, 0.6, 0.7, 0.8, 0.9, 1.0}. But suppose the same treatment is applied to a replicate in which only three individuals were used. In this case, the only possible values are in the set {0.0, 0.33, 0.66, 1.0}. These small sample sizes will greatly inflate the measured variance, and this problem is not alleviated by any data transformation.

A final problem with the analysis of proportions arises if there are three or more categorical variables. With a dichotomous response, the proportion completely characterizes the data. However, if there are more than two categories, the proportion will have to be carefully defined in terms of only one of the categories. For example, if the arena includes vertical and horizontal black and white surfaces, there are now four categories from which the proportion can be measured. Thus, the analysis might be based on the proportion of individuals using the horizontal black surface. Alternatively, the proportion could be defined in terms of two or more summed categories, such as the proportion of individuals using any vertical surface (white or black).

Summary

Independent and dependent variables are either categorical or continuous, and most designs fit into one of four possible categories based on this classification. Analysis of variance (ANOVA) designs are used for experiments in which the independent variable is categorical and the dependent variable is continuous. Useful ANOVA designs include one- and two-way ANOVAs, randomized block, and split-plot designs. We do not favor the use of nested ANOVAs, in which non-independent subsamples are taken from within a replicate. Repeated measures designs can be used when repeated observations are collected on a single replicate through time. However, these data are often autocorrelated, so that the assumptions of the analysis may not be met. In such cases, the temporal data should be collapsed to a single independent measurement, or time-series analysis should be employed.

If the independent variable is continuous, a regression design is used. Regression designs are appropriate for both experimental and sampling studies, although they are used predominantly in the latter. We advocate increased use of experimental regression in lieu of ANOVA with only a few levels of the independent variable represented. Adequate sampling of the range of predictor values is important in designing a sound regression experiment. Multiple regression designs include two or more predictor variables, although the analysis becomes problematic if there are strong correlations (collinearity) among the predictor variables.

If both the independent and the dependent variables are categorical, a tabular design is employed. Tabular designs require true independence of the replicate counts. If the counts are not independent, they should be collapsed so that the response variable is a single proportion. The experimental design is then similar to a regression or ANOVA.

We favor simple experimental and sampling designs and emphasize the importance of collecting data from replicates that are independent of one another. Good replication and balanced sample sizes will improve the power and reliability of the analyses. Even the most sophisticated analysis cannot salvage results from a poorly designed study.

CHAPTER 8

Managing and Curating Data

Data are the raw material of scientific studies. However, we rarely devote the same amount of time and energy to data organization, management, and curation that we devote to data collection, analysis, and publication. This is an unfortunate state of affairs; it is far easier to use the original (raw) data themselves to test new hypotheses than it is to reconstruct the original data from summary tables and figures presented in the published literature. A basic requirement of any scientific study is that it be repeatable by others. Analyses cannot be reconstructed unless the original data are properly organized and documented, safely stored, and made available. Moreover, it is a legal requirement that any data collected as part of a project that is supported by public funds (e.g., federal or state grants and contracts) be made available to other scientists and to the public. Data sharing is legally mandated because the granting agency, not the investigator, technically owns the data. Most granting agencies allow ample time (usually one or more years) for investigators to publish their results before the data have to be made available publicly.[1] However, the Freedom of Information Act (5 U.S.C. § 552) allows anyone in the United States to request publicly funded data from scientists at any time. For all of these reasons, we encourage you to manage your data as carefully as you collect, analyze, and publish them.

[1] This policy is part of an unwritten gentleman's agreement between granting agencies and the scientific community. This agreement works because the scientific community so far has demonstrated its willingness to make data available freely and in a timely fashion. Ecologists and environmental scientists generally are slower to make data available than are scientists working in more market-driven fields (such as biotechnology and molecular genetics) where there is greater public scrutiny. Similarly, provisions for data access and standards for metadata production by the ecological and environmental community have lagged behind those of other fields. It could be argued that the lack of a GenBank for ecologists has slowed the progress of ecological and environmental sciences. However, ecological data are much more heterogeneous than gene sequences, so it has proven difficult to design a single data structure to facilitate rapid data sharing.

The First Step: Managing Raw Data

In the laboratory-based bench sciences, the bound lab notebook is the traditional medium for recording observations. In contrast, ecological and environmental data are recorded in many forms. Examples include sets of observations written in notebooks, scrawled on plastic diving slates or waterproof paper, or dictated into voice recorders; digital outputs of instruments streamed directly into data loggers, computers, or palm-pilots; and silver-oxide negatives or digital satellite images. Our first responsibility with our data, therefore, is to transfer them rapidly from their heterogeneous origins into a common format that can be organized, checked, analyzed, and shared.

Spreadsheets

In Chapter 5, we illustrated how to organize data in a **spreadsheet**. A spreadsheet is an electronic page[2] in which each row represents a single observation (e.g., the unit of study, such as an individual leaf, feather, organism, or island), and each column represents a single measured or observed variable. The entry in each **cell** of the spreadsheet, or row-by-column pair, is the value for the unit of study (the row) of the measurement or observation of the variable (the column).[3] All types of data used by ecologists and environmental scientists can be stored in spreadsheets, and the majority of experimental designs (see Chapter 7) can be represented in spreadsheet formats.

Field data should be entered into spreadsheets as soon as possible after collection. In our research, we try to transfer the data from our field notes or instruments into spreadsheets on the same day that we collect it. There are several important reasons to transcribe your data promptly. First, you can determine quickly if necessary observations or experimental units were missed due to researcher oversight or instrument failure. If it is necessary to go back and col-

[2] Although commercial spreadsheet software can facilitate data organization and management, it is not appropriate to use commercial software for the archival electronic copy of your data. Rather, archived data should be stored as ASCII (American Standard Code for Information Exchange) text files, which can be read by any machine without recourse to commercial packages that may not exist in 50 years.

[3] We illustrate and work only with two-dimensional (row, column) spreadsheets (which are a type of **flat files**). Regardless of whether data are stored in two- or three-dimensional formats, pages within electronic workbooks must be exported as flat files for analysis with statistical software. Although spreadsheet software often contains statistical routines and random number generators, we do not recommend the use of these programs for analysis of scientific data. Spreadsheets are fine for generating simple summary statistics, but their random number generators and statistical algorithms may not be reliable for more complex analyses (Knüsel 1998; McCullough and Wilson 1999).

lect additional observations, this must be done very quickly or the new data won't be comparable. Second, your data sheets usually will include marginal notes and observations, in addition to the numbers you record. These marginal notes make sense when you take them, but they have a short half life and rapidly become incomprehensible (and sometimes illegible). Third, once data are entered into a spreadsheet, you then have two copies of them; it is less likely you will misplace or lose both copies of your valuable results. Finally, rapid organization of data may promote timely analysis and publication; unorganized raw data certainly will not.

Data should be proofread as soon as possible after entry into a spreadsheet. However, proofreading usually does not catch all errors in the data, and once the data have been organized and documented, further checking of the data is necessary (discussed later in the chapter).

Metadata

At the same time that you enter the data into a spreadsheet, you should begin to construct the **metadata** that must accompany the data. Metadata are "data about data" and describe key attributes of the dataset. Minimal metadata[4] that should accompany a dataset include:

- Name and contact information for the person who collected the data
- Geographic information about where the data were collected
- The name of the study for which the data were collected
- The source of support that enabled the data to be collected
- A description of the organization of the data file, which should include:
  - A brief description of the methods used to collect the data
  - The types of experimental units
  - The units of measurement or observation of each of the variables
  - A description of any abbreviations used in the data file
  - An explicit description of what data are in columns, what data are in rows, and what character is used to separate elements within rows

[4] Three "levels" of metadata have been described by Michener et al. (1997) and Michener (2000). Level I metadata include a basic description of the dataset and the structure of the data. Level II metadata include not only descriptions of the dataset and its structure, but also information on the origin and location of the research, information on the version of the data, and instructions on how to access the data. Finally, Level III metadata are considered auditable and publishable. Level III metadata include all Level II metadata plus descriptions of how data were acquired, documentation of quality assurance and quality control (QA/QC), a full description of any software used to process the data, a clear description of how the data have been archived, a set of publications associated with the data, and a history of how the dataset has been used.

(e.g., tabs, spaces, or commas) and between columns (e.g., a "hard return").[5]

An alternative approach—more common in the laboratory sciences—is to create the metadata *before* the data are ever collected. A great deal of clear thinking can accompany the formal laying out and writing up of the methods, including plot design, intended number of observations, sample space, and organization of the data file. Such a priori metadata construction also facilitates the writing of the Methods section of the final report or publication of the study. We also encourage you to draw blank plots and tables that illustrate the relationships among variables that you propose to examine.

Regardless of whether you choose to write up your metadata before or after conducting your study, we cannot overemphasize their importance. Even though it is time-consuming to assemble metadata, and they would seem to contain information that is self-evident, we guarantee that you will find them invaluable. You may think that you will never forget that Nemo Swamp was where you collected the data for your opus on endangered orchids. And of course you can still recall all the back roads you drove to get there. But time will erode those memories as the months and years pass, and you move on to new projects and studies.

We all have experiences looking through lists of computer file names (NEMO, NEMO03, NEMO03NEW), trying to find the one we need. Even if we can find the needed data file, we may still be out of luck. Without metadata, there is no way to recollect the meaning of the alphabet soup of abbreviations and acronyms running across the top row of the file. Reconstructing datasets is frustrating, inefficient, and time-consuming, and impedes further data analysis. Undocumented datasets are nearly useless to you and to anyone else. Data without metadata cannot be stored in publicly-accessible data archives, and have no life beyond the summary statistics that describe them in a publication or report.

The Second Step: Storing and Curating the Data

Storage: Temporary and Archival

Once the data are organized and documented, the dataset—a combination of data and metadata—should be stored on permanent media. The most perma-

[5] Ecologists and environmental scientists continue to develop standards for metadata. These include the Content Standards for Digital Geospatial Metadata (FGDC 1998) for spatially-organized data (e.g., data from Geographic Information Systems, or GIS); standard descriptors for ecological and environmental data on soils (Boone et al. 1999); and Classification and Information Standards for data describing types of vegetation (FGDC 1997). Many of these have been incorporated within the Ecological Metadata Language, or EML, developed in late 2002 (knb.ecoinformatics.org/software/eml/) and which is now the de facto standard for documenting ecological data.

nent medium, and the only medium acceptable as truly archival, is acid-free paper. It is good practice to print out a copy of the raw data spreadsheet and its accompanying metadata onto acid-free paper using a laser printer.[6] This copy should be stored somewhere that is safe from damage. Electronic storage of the original dataset also is a good idea, but you should not expect electronic media to last more than 5–10 years.[7] Thus, electronic storage should be used primarily for working copies. If you intend the electronic copies to last more than a few years, you will have to copy the datasets onto newer electronic media on a regular basis. You must recognize that data storage has real costs, in space (where the data are stored), time (to maintain and transfer datasets among media), and money (because time and space are money).

Curating the Data

Most ecological and environmental data are collected by researchers using funds obtained through grants and contracts. Usually, these data technically are owned by the granting agency, and they must be made available to others within a reasonably short period of time. Thus, regardless of how they are stored, your datasets must be maintained in a state that makes them usable not only by you, but also by the broader scientific community and other interested individuals and parties. Like museum and herbarium collections of animals and plants, datasets must be curated so they are accessible to others. Multiple datasets gen-

[6] Inks from antique dot-matrix printers (and typewriters) last longer than those of ink-jet printers, but the electrostatic bonding of ink to paper by laser printers lasts longer than either. True Luddites prefer Higgins Aeterna India Ink (with a tip o' the hat to Henry Horn).

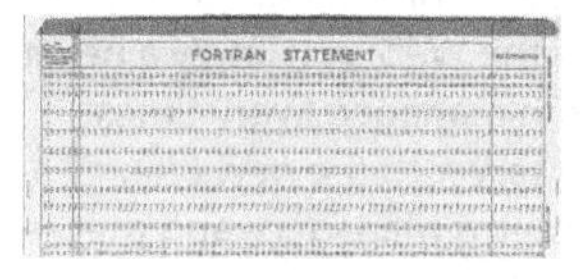

Computer punch card

[7] Few readers will remember floppy disks of any size (8", 5-1/4", 3-1/2"), much less paper tapes, punch cards (illustrated), or magnetic tape reels. When we wrote the first edition, CD-ROMs were being replaced rapidly by DVD-ROMs; in the intervening years, we have moved on to terabyte bricks, RAID arrays, and the seemingly ubiquitous "cloud." Although the actual lifespan of most magnetic media is on the order of years to decades, and that of most optical media is on the order of decades-to-centuries (only years in the humid tropics where fungi eat the glue between the laminated layers of material and etch the disk), the capitalist marketplace replaces them all on an approximately 5-year cycle. Thus, although we still have on our shelves readable 5-1/4" floppy disks and magnetic tapes with data from the 1980s and 1990s, it is nearly impossible now to find, much less buy, a disk drive with which to read them. And even if we could find one, the current operating systems of our computers wouldn't recognize it as usable hardware. A good resource for obsolete equipment and software is the Computer History Museum in Mountain View, California: www.computerhistory.org/.

erated by a project or research group can be organized into electronic data catalogs. These catalogs often can be accessed and searched from remote locations using the internet.

Nearly a decade after we wrote the first edition of this *Primer*, standards for curation of ecological and environmental data are still being developed. Most data catalogs are maintained on servers connected to the World Wide Web,[8] and can be found using search engines that are available widely. Like data storage, data curation is costly; it has been estimated that data management and curation should account for approximately 10% of the total cost of a research project (Michener and Haddad 1992). Unfortunately, although granting agencies require that data be made available, they have been reluctant to fund the full costs of data management and curation. Researchers themselves rarely budget these costs in their grant proposals. When budgets inevitably are cut, data management and curation costs often are the first items to be dropped.

The Third Step: Checking the Data

Before beginning to analyze a dataset, you must carefully examine it for outliers and errors.

The Importance of Outliers

Outliers are recorded values of measurements or observations that are outside the range of the bulk of the data (Figure 8.1). Although there is no standard level of "outside the range" that defines an outlier, it is common practice to consider

[8] Many scientists consider posting data on the World Wide Web or in the "cloud" to be a means of permanently archiving data. This is illusory. First, it is simply a transfer of responsibility from you to a computer system manager (or other information technology professional). By placing your electronic archival copy on the Web, you imply a belief that regular backups are made and maintained by the system manager. Every time a server is upgraded, the data have to be copied from the old server to the new one. Most laboratories or departments do not have their own systems managers, and the interests of computing centers in archiving and maintaining Web pages and data files do not necessarily parallel those of individual investigators. Second, server hard disks fail regularly (and often spectacularly). Last, the Web is neither permanent nor stable. GOPHER and LYNX have disappeared, FTP has all but been replaced by HTTP, and HTML, the original language of the Web, is steadily being replaced by (the not entirely compatible) XML. All of these changes to the functionality of the World Wide Web and the accessibility of files stored occurred within 10 years. It often is easier to recover data from notebooks that were hand-written in the nineteenth century than it is to recover data from Web sites that were digitally "archived" in the 1990s! In fact, it cost us more than $2000 in 2006 to recover from a 1997 magnetic tape the data used to illustrate cluster analysis and redundancy analysis in Chapter 12!

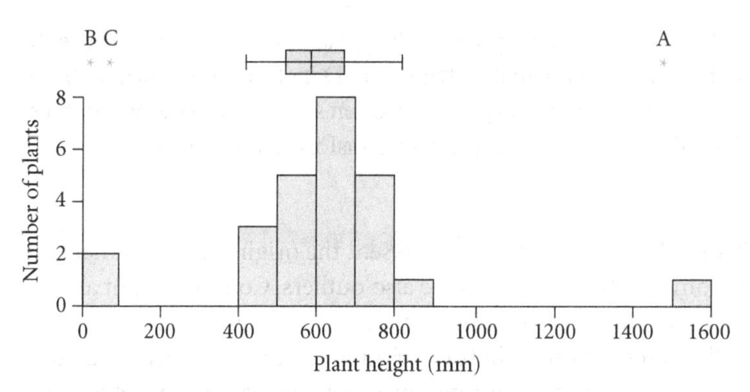

Figure 8.1 Histogram and box plot of measurements of plant height for samples of the carnivorous pitcher plant *Darlingtonia californica* (sample size $N = 25$; data from Table 8.1). The histogram shows the frequency of plants in size class intervals of 100 mm. Above the histogram is the box plot of the same data. The vertical line in the box indicates the median of the distribution. The box encompasses 50% of the data, and the whiskers encompass 90% of the data. A, B, and C indicate the three extreme data points and correspond to the labeled rows in Table 8.1 Basic histograms and box plots allow you to rapidly identify extreme values in your data.

values beyond the upper or lower deciles (90th percentile, 10th percentile) of a distribution to be potential outliers, and those smaller than the 5th percentile or larger than the 95th percentile to be possible extreme values. Many statistical packages highlight these values on box plots (as we do with stars and letters in Figure 8.1) to indicate that you should scrutinize them carefully. Some of these packages also have interactive tools, described in the next section, that allow for location of these values within the spreadsheet.

It is important to identify outliers because they can have dramatic effects on your statistical tests. Outliers and extreme values increase the variance in the data. Inflated variances decrease the power of the test and increase the chance of a Type II error (failure to reject a false null hypothesis; see Chapter 4). For this reason, some researchers automatically delete outliers and extreme values prior to conducting their analyses. This is bad practice! There are only two reasons for justifiably discarding data: (1) the data are in error (e.g., they were entered incorrectly in the field notebook); or (2) the data no longer represent valid observations from your original sample space (e.g., one of your dune plots was overrun by all-terrain vehicles). Deleting observations simply because they are "messy" is laundering or doctoring the results and legitimately could be considered scientific fraud.

Outliers are more than just noise. Outliers can reflect real biological processes, and careful thought about their meaning can lead to new ideas and hypotheses.[9] Moreover, some data points will appear as outliers only because the data are being forced into a normal distribution. Data from non-normal distributions such as the log-normal or exponential often seem to be extreme, but these data fall in place after they are appropriately transformed (see below).

Errors

Errors are recorded values that do not represent the original measurements or observations. Some, but not all, errors are also outliers. Conversely, not all outliers in your data are necessarily errors. Errors can enter the dataset in two ways: as errors in collection or as errors in transcription. Errors in collection are data that result from broken or miscalibrated instruments, or from mistaken data entry in the field. Errors in collection can be difficult to identify. If you or your field assistant were recording the data and wrote down an incorrect value, the error rarely can be corrected.[10] Unless the error is recognized immediately during data transcription, it will probably remain undetected and contribute to the variation of the data.

---

[9] Although biologists are trained to think about means and averages, and to use statistics to test for patterns in the central tendency of data, interesting ecology and evolution often happen in the statistical tails of a distribution. Indeed, there is a whole body of statistics dedicated to the study and analysis of extreme values (Gaines and Denny 1993), which may have long-lasting impacts on ecosystems (e.g., Foster et al. 1998). For example, isolated populations (peripheral isolates, *sensu* Mayr 1963) that result from a single founder event may be sufficiently distinct in their genetic makeup that they would be considered statistical outliers relative to the original population. Such peripheral isolates, if subject to novel selection pressures, could become reproductively isolated from their parent population and form new species (Schluter 1996). Phylogenetic reconstruction has suggested that peripheral isolates can result in new species (e.g., Green et al. 2002), but experimental tests of the theory have not supported the model of speciation by founder effect and peripheral isolates (Mooers et al.1999).

[10] To minimize collection errors in the field, we often engage in a call and response dialogue with each other and our field assistants:

> *Data Collector: "This is Plant 107. It has 4 leaves, no phyllodes, no flowers."*
> *Data Recorder: "Got it, 4 leaves no phyllodes or flowers for 107."*
> *Data Collector: "Where's 108?"*
> *Data Recorder: " Died last year. 109 should be about 10 meters to your left. Last year it had 7 leaves and an aborted flower. After 109, there should be only one more plant in this plot."*

Repeating the observations back to one another and keeping careful track of the replicates helps to minimize collection errors. This "data talk" also keeps us alert and focused during long field days.

Errors resulting from broken or miscalibrated instruments are easier to detect (once it is determined that the instrument is broken or after it has been recalibrated). Unfortunately, instrument error can result in the loss of a large amount of data. Experiments that rely on automated data collection need to have additional procedures for checking and maintaining equipment to minimize data loss. It may be possible to collect replacement values for data resulting from equipment failures, but only if the failures are detected very soon after the data are collected. Potential equipment failures are another reason to transcribe rapidly your raw data into spreadsheets and evaluate their accuracy.

Errors in transcription result primarily from mistyping values into spreadsheets. When these errors show up as outliers, they can be checked against the original field data sheets. Probably the most common transcription error is the misplaced decimal point, which changes the value by an order of magnitude. When errors in transcription do not show up as outliers, they can remain undetected unless spreadsheets are proofread very carefully. Errors in transcription are less common when data are transferred electronically from instruments directly into spreadsheets. However, transmission errors can occur and can result in "frame-shift" errors, in which one value placed in the wrong column results in all of the subsequent values being shifted into incorrect columns. Most instruments have built-in software to check for transmission errors and report them to the user in real-time. The original data files from automated data-collection instruments should not be discarded from the instrument's memory until after the spreadsheets have been checked carefully for errors.

Missing Data

A related, but equally important, issue is the treatment of missing data. Be very careful in your spreadsheets and in your metadata to distinguish between measured values of 0 and missing data (replicates for which no observations were recorded). Missing data in a spreadsheet should be given their own designation (such as the abbreviation "NA" for "not available") rather than just left as blank cells. Be careful because some software packages may treat blank cells as zeroes (or insert 0s in the blanks), which will ruin your analyses.

Detecting Outliers and Errors

We have found three techniques to be especially useful for detecting outliers and errors in datasets: calculating column statistics, checking ranges and precision of column values; and graphical exploratory data analysis. Other, more complex methods are discussed by Edwards (2000). We use as our example for outlier detection the measurement of heights of pitcher-leaves of *Darlingtonia californica* (the cobra lily, or California pitcher plant).[11] We collected these data, reported in Table 8.1, as part of a study of the growth, allometry, and photosynthesis of this species (Ellison and Farnsworth 2005). In this example of only 25 observations, the outliers are easy to spot simply by scanning the columns of numbers. Most ecological datasets, however, will have far more observations and it will be harder to find the unusual values in the data file.

COLUMN STATISTICS The calculation of simple column statistics within the spreadsheet is a straightforward way to identify unusually large or small values in the dataset. Measures of location and spread, such as the column mean, median, standard deviation, and variance, give a quick overview of the distribution of the values in the column. The minimum and maximum values of the column may indicate suspiciously large or small values. Most spreadsheet software packages have functions that calculate these values. If you enter these functions as the last six rows of the spreadsheet (Table 8.1), you can use them as a first check on the data. If you find an extreme value and determine that it is a true error, the spreadsheet will automatically update the calculations when you replace it with the correct number.

CHECKING RANGE AND PRECISION OF COLUMN VALUES Another way to check the data is to use spreadsheet functions to ensure that values in a given column are within reasonable boundaries or reflect the precision of measurement. For example, *Darlingtonia* pitchers rarely exceed 900 mm in height. Any measurement much above this value would be suspect. We could look over the dataset carefully to see if any values are extremely large, but for real datasets with hundreds or thousands of observations, manual checking is inefficient and inaccurate. Most spreadsheets have logical functions that can be used for value checking and that can automate this process. Thus, you could use the statement:

If (value to be checked) > (maximum value), record a "1"; otherwise, record a "0"

to quickly check to see if any of the values of height exceed an upper (or lower) limit.

Darlingtonia californica

[11] *Darlingtonia californica* is a carnivorous plant in the family Sarraceniaceae that grows only in serpentine fens of the Siskiyou Mountains of Oregon and California and the Sierra Nevada Mountains of California. Its pitcher-shaped leaves normally reach 800 mm in height, but occasionally exceed 1 m. These plants feed primarily on ants, flies, and wasps.

TABLE 8.1 **Ecological data set illustrating typical extreme values of *Darlingtonia californica***

	Plant #	Height (mm)	Mouth(mm)	Tube (mm)
	1	744	34.3	18.6
	2	700	34.4	20.9
	3	714	28.9	19.7
	4	667	32.4	19.5
	5	600	29.1	17.5
	6	777	33.4	21.1
	7	640	34.5	18.6
	8	440	29.4	18.4
	9	715	39.5	19.7
	10	573	33.0	15.8
A	**11**	**1500**	**33.8**	**19.1**
	12	650	36.3	20.2
	13	480	27.0	18.1
	14	545	30.3	17.3
	15	845	37.3	19.3
	16	560	42.1	14.6
	17	450	31.2	20.6
	18	600	34.6	17.1
	19	607	33.5	14.8
	20	675	31.4	16.3
	21	550	29.4	17.6
B	**22**	**5.1**	**0.3**	**0.1**
	23	534	30.2	16.5
	24	655	35.8	15.7
C	**25**	**65.5**	**3.52**	**1.77**
Mean		611.7	30.6	16.8
Median		607	33.0	18.1
Standard deviation		265.1	9.3	5.1
Variance		70271	86.8	26.1
Minimum		5.1	0.3	0.1
Maximum		1500	42.1	21.1

The data are morphological measurements of individual pitcher plants. Pitcher height, pitcher mouth diameter, and pitcher tube diameter were recorded for 25 individuals sampled at Days Gulch. Rows with extreme values are shown in boldface and designated as A, B, and C, and are noted with asterisks in Figure 8.1. Summary statistics for each variable are given at the bottom of the table. It is crucial to identify and evaluate extreme values in a dataset. Such values greatly inflate variance estimates, and it is important to ascertain whether those observations represent measurement error, atypical measurements or simply natural variation of the sample. Data from Ellison and Farnsworth (2005).

This same method can be used to check the precision of the values. We measured height using a measuring tape marked in millimeters, and routinely recorded measurements to the nearest whole millimeter. Thus, no value in the spreadsheet should be more precise than a single millimeter (i.e., no value should have fractions of millimeters entered). The logical statement to check this is:

If (value to be checked has any value other than 0 after the decimal point), record a "1"; otherwise, record a "0"

There are many similar checks on column values that you can imagine, and most can be automated using logical functions within spreadsheets.

Range checking on the data in Table 8.1 reveals that the recorded height of Plant 11 (1500 mm) is suspiciously large because this observation exceeds the 900-mm threshold. Moreover the recorded heights of Plant 22 (5.1 mm) and Plant 25 (65.5 mm) were misentered because the decimal values are greater than the precision of our measuring tapes. This audit would lead us to re-examine our field data sheets or perhaps return to the field to measure the plants again.

GRAPHICAL EXPLORATORY DATA ANALYSIS Scientific pictures—graphs—are one of the most powerful tools you can use to summarize the results of your study (Tufte 1986, 1990; Cleveland 1985, 1993). They also can be used to detect outliers and errors, and to illuminate unexpected patterns in your data. The use of graphics in advance of formal data analysis is called **graphical exploratory data analysis** or graphical EDA, or simply **EDA** (e.g., Tukey 1977; Ellison 2001). Here, we focus on the use of EDA for detecting outliers and extreme values. The use of EDA to hunt for new or unexpected patterns in the data ("data mining" or more pejoratively, "data dredging") is evolving into its own cottage industry (Smith and Ebrahim 2002).[12]

[12] The statistical objection to data dredging is that if we create enough graphs and plots from a large, complex dataset, we will surely find some relationships that turn out to be statistically significant. Thus, data dredging undermines our calculation of *P*-values and increases the chances of incorrectly rejecting the null hypothesis (Type I error). The philosophical objection to EDA is that you should already know in advance how your data are going to be plotted and analyzed. As we emphasized in Chapters 6 and 7, laying out the design and analysis ahead of time goes hand-in-hand with having a focused research question. On the other hand, EDA is essential for detecting outliers and errors. And there is no denying that you can uncover interesting patterns just by casually plotting your data. You can always return to the field to test these patterns with an appropriate experimental design. Good statistical packages facilitate EDA because graphs can be constructed and viewed quickly and easily.

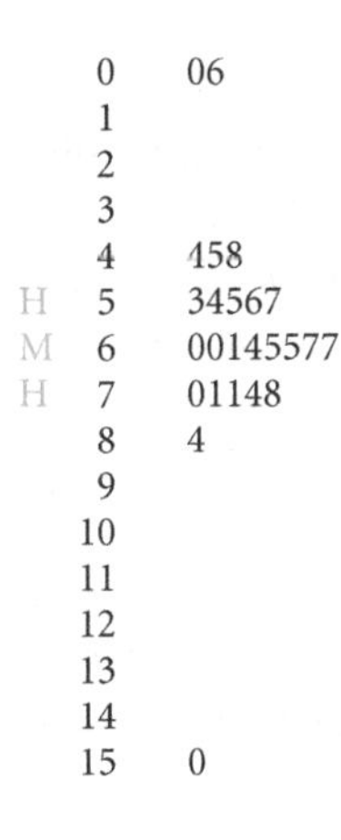

Figure 8.2 Stem-and-leaf plot. The data are measurements of the heights (in mm) of 25 individuals of *Darlingtonia californica* (see Table 8.1). In this plot, the first column of numbers on the left—the "stem"—represent the "hundreds" place, and each "leaf" to its right represents the "tens" place of each recorded value. For example, the first row illustrates the observations 00x and 06x, or an observation less than 10 (the value 5.1 of Plant 22 of Table 8.1), and another observation between 60 and 70 (the value 65.5 of Plant 25). The fifth row illustrates observations 44x, 45x, and 48x, which correspond to values 440 (Plant 8), 450 (Plant 17), and 480 (Plant 13). The last row illustrates observation 150x, corresponding to value 1500 (Plant 11). All observations in column 1 of Table 8.1 are included in this display. The median (607) is in row 7, which is labeled with the letter M. The lower quartile (545), also known as a **hinge**, is in row 6, which is labeled with the letter H. The upper quartile or hinge (700) is in row 8, also labeled H. Like histograms and box plots, stem-and-leaf plots are a diagnostic tool for visualizing your data and detecting extreme data points.

There are three indispensable types of graphs for detecting outliers and extreme values: **box plots, stem-and-leaf plots**, and **scatterplots**. The first two are best used for univariate data (plotting the distribution of a single variable), whereas scatterplots are used for bivariate or multivariate data (plotting relationships among two or more variables). You have already seen many examples of univariate data in this book: the 50 observations of spider tibial spine length (see Figure 2.6 and Table 3.1), the presence or absence of *Rhexia* in Massachusetts towns (see Chapter 2), and the lifetime reproductive output of orange-spotted whirligig beetles (see Chapter 1) are all sets of measurements of a single variable. To illustrate outlier detection and univariate plots for EDA, we use the first column of data in Table 8.1—pitcher height.

Our summary statistics in Table 8.1 suggest that there may be some problems with the data. The variance is large relative to the mean, and both the minimum and the maximum value look extreme. A box plot (top of Figure 8.1) shows that two points are unusually small and one point is unusually large. We can identify these unusual values in more detail in a stem-and-leaf plot (Figure 8.2), which is a variant of the more familiar histogram (see Chapter 1; Tukey 1977). The stem-and-leaf plot shows clearly that the two small values are an order of magnitude (i.e., approximately tenfold) smaller than the mean or median, and that the one large value is more than twice as large as the mean or median.

How unusual are these values? Some *Darlingtonia* pitchers are known to exceed a meter in height, and very small pitchers can be found as well. Perhaps these three observations simply reflect the inherent variability in the size of cobra lily pitchers. Without additional information, there is no reason to think that these values are in error or are not part of our sample space, and it would be inappropriate to remove them from the dataset prior to further statistical analysis.

We do have additional information, however. Table 8.1 shows two other measurements we took from these same pitchers: the diameter of the pitcher's tube, and the diameter of the opening of the pitcher's mouth. We would expect all of these variables to be related to each other, both because of previous research on *Darlingtonia* (Franck 1976) and because the growth of plant (and animal) parts tends to be correlated (Niklas 1994).[13] A quick way explore how these variables co-vary with one another is to plot all possible pair-wise relationships among them (Figure 8.3).

The **scatterplot matrix** shown in Figure 8.3 provides additional information with which to decide whether our unusual values are outliers. The two scatterplots in the top row of Figure 8.3 suggest that the unusually tall plant in Table

[13] Imagine a graph in which you plotted the logarithm of pitcher length on the *x*-axis and the logarithm of tube diameter on the *y*-axis. This kind of allometric plot reveals the way that the shape of an organism may change with its size. Suppose that small cobra lilies were perfect scaled-down replicas of large cobra lilies. In this case, increases in tube diameter are synchronous with increases in pitcher length, and the slope of the line (β_1) = 1 (see Chapter 9). The growth is isometric, and the small plants look like miniatures of the large plants. On the other hand, suppose that small pitchers had tube diameters that were relatively (but not absolutely) larger than the tube diameters of large plants. As the plants grow in size, tube diameters increase relatively slowly compared to pitcher lengths. This would be a case of negative allometry, and it would be reflected in a slope value $\beta_1 < 1$. Finally, if there is positive allometry, the tube diameter increases faster with increasing pitcher length ($\beta_1 > 1$). These patterns are illustrated by Franck (1976). A more familiar example is the human infant, in which the head starts out relatively large, but grows slowly, and exhibits negative allometry. Other body parts, such as fingers and toes, exhibit positive allometry and grow relatively rapidly.

Allometric growth (either positive or negative) is the rule in nature. Very few organisms grow isometrically, with juveniles appearing as miniature replicas of adults. Organisms change shape as they grow, in part because of basic physiological constraints on metabolism and uptake of materials. Imagine that (as a simplification) the shape of an organism is a cube with a side of length L. The problem is that, as organisms increase in size, volume increases as a cubic function (L^3) of length (L), but surface area increases only as a square function (L^2). Metabolic demands of an organism are proportional to volume (L^3), but the transfer of material (oxygen, nutrients, waste products) is proportional to surface area (L^2). Therefore, structures that function well for material transfer in small organisms will not get the job done in larger organisms. This is especially obvious when we compare species of very different body size. For example, the cell membrane of a tiny microorganism does a fine job transferring oxygen by simple diffusion, but a mouse or a human requires a vascularized lung, circulatory system, and a specialized transport molecule (hemoglobin) to accomplish the same thing. Both within and between species, patterns of shape, size, and morphology often reflect the necessity of boosting surface area to meet the physiological demands of increasing volume. See Gould (1977) for an extended discussion of the evolutionary significance of allometric growth.

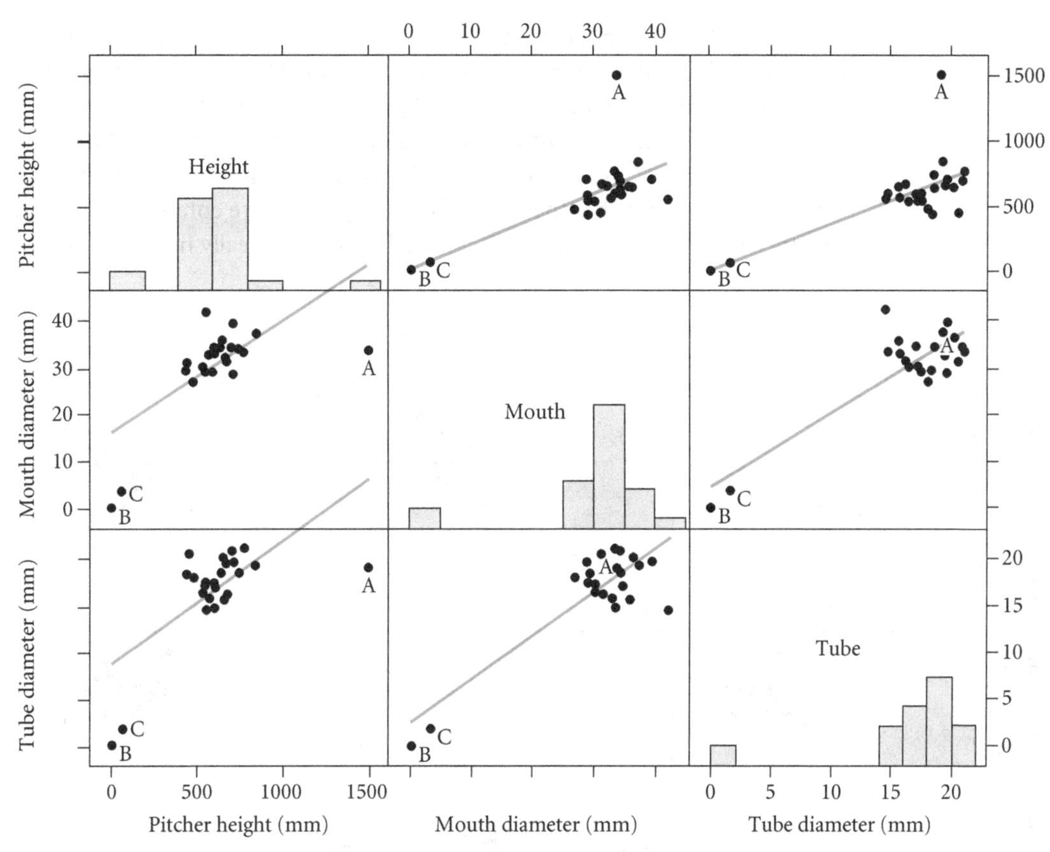

Figure 8.3 Scatterplot matrix illustrating the relationship between the height, mouth diameter, and tube diameter of 25 *D. californica* individuals (see Table 8.1). The extreme values labeled in Figure 8.1 and Table 8.1 are also labeled here. The diagonal panels in this figure illustrate the histograms for each of the 3 morphological variables. In the off-diagonal panels, scatterplots illustrate the relationship between the two variables indicated in the diagonal; the scatterplots above the diagonal are mirror images of those below the diagonal. For example, the scatterplot at the lower left illustrates the relationship between pitcher height (on the *x*-axis) and tube diameter (on the *y*-axis), whereas the scatterplot at the upper right illustrates the relationship between tube diameter (on the *x*-axis) and pitcher height (on the *y*-axis). The lines in each scatterplot are the best-fit linear regressions between the two variables (see Chapter 9 for more discussion of linear regression).

8.1 (Point A; Plant 11) is remarkably tall for its tube and mouth diameters. The plots of mouth diameter versus tube diameter, however, show that neither of these measurements are unusual for Plant 11. Collectively, these results suggest an error either in recording the plant's height in the field or in transcribing its height from the field datasheet to the computer spreadsheet.

In contrast, the short plants are not unusual in any regard except for being short. Not only are they short, but they also have small tubes and small mouths. If we remove Plant 11 (the unusually tall plant) from the dataset (Figure 8.4), the relationships among all three variables of the short plants are not unusual relative to the rest of the population, and there is no reason based on the data themselves to suspect that the values for these plants were entered incorrectly. Nonetheless, these two plants are very small (could we really have measured a

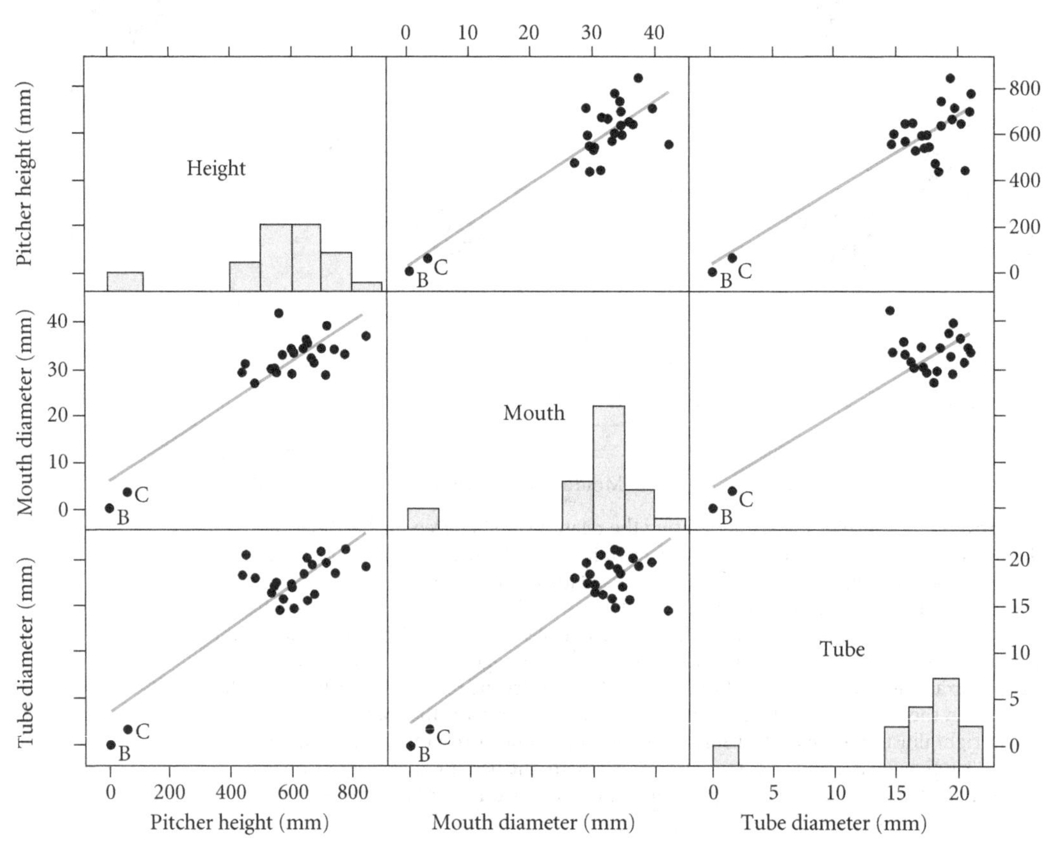

Figure 8.4 Scatterplot matrix illustrating the relationship between pitcher height, mouth diameter, and tube diameter. The data and figure layout are identical to Figure 8.3. However, the outlier Point A (see Table 8.1) has been eliminated, which changes some of these relationships. Because Point A was a large outlier for pitcher height (height = 1500 mm), the scale of this variable (x-axis in the left-hand column) now ranges from only 0 to 800 mm, compared to 0 to 1500 mm in Figure 8.3. It is not uncommon to see correlations and statistical significance change based on the inclusion or exclusion of a single extreme data value.

5mm-tall plant?) and we should re-check our datasheets and perhaps return to the field to re-measure these plants.

Creating an Audit Trail

The process of examining a dataset for outliers and errors is a special case of the generic **quality assurance and quality control** process (abbreviated **QA/QC**). As with all other operations on a spreadsheet or dataset, the methods used for outlier and error detection should be documented in a QA/QC section of the metadata. If it is necessary to remove data values from the spreadsheet, a description of the values removed should be included in the metadata, and a new spreadsheet should be created that contains the corrected data. This spreadsheet also should be stored on permanent media and given a unique identifier in the data catalog.

The process of subjecting data to QA/QC and modifying data files leads to the creation of an **audit trail.** An audit trail is a series of files and their associated documentation that enables other users to recreate the process by which the final, analyzed dataset was created. The audit trail is a necessary part of the metadata for any study. If you are preparing an environmental impact report or other document that is used in legal proceedings, the audit trail may be part of the legal evidence supporting the case.[14]

The Final Step: Transforming the Data

You will often read in a scientific paper that the data were "transformed prior to analysis." Miraculously, after the transformation, the data "met the assumptions" of the statistical test being used and the analysis proceeded apace. Little else is usually said about the data transformations, and you may wonder why the data were transformed at all.

But first, what is a transformation? By a transformation, we simply mean a mathematical function that is applied to all of the observations of a given variable:

$$Y^* = f(Y) \tag{8.1}$$

[14] Historians of science have reconstructed the development of scientific theories by careful study of inadvertent audit trails—marginal notes scrawled on successive drafts of handwritten manuscripts and different editions of monographs. The use of word processors and spreadsheets has all but eliminated marginalia and yellow pads, but does allow for the creation and maintenance of formal audit trails. Other users of your data and future historians of science will thank you for maintaining an audit trail of your data.

Y represents the original variable, Y^* is the transformed variable, and f is a mathematical function that is applied to the data. Most transformations are fairly simple algebraic functions, subject to the requirement that they are **continuous monotonic functions.**[15] Because they are monotonic, transformations do not change the rank order of the data, but they do change the relative spacing of the data points, and therefore affect the variance and shape of the probability distribution (see Chapter 2).

There are two legitimate reasons to transform your data before analysis. First, transformations can be useful (although not strictly necessary) because the patterns in the transformed data may be easier to understand and communicate than patterns in the raw data. Second, transformations may be necessary so that the analysis is valid—this is the "meeting the assumptions" reason that is used most frequently in scientific papers and discussed in biometry textbooks.

Data Transformations as a Cognitive Tool

Transformations often are useful for converting curves into straight lines. Linear relationships are easier to understand conceptually, and often have better statistical properties (see Chapter 9). When two variables are related to each other by multiplicative or exponential functions, the logarithmic transformation is one of the most useful data transformations (see Footnote 13). A classic ecological example is the species–area relationship: the relationship between species number and island or sample area (Preston 1962; MacArthur and Wilson 1967). If we measure the number of species on an island and plot it against the area of the island, the data often follow a simple **power function**:

$$S = cA^z \tag{8.2}$$

where S is the number of species, A is island area, and c and z are constants that are fitted to the data. For example, the number of species of plants recorded from each of the Galápagos Islands (Preston 1962) seems to follow a power relationship (Table 8.2). First, note that island area ranges over three orders of magnitude, from less than 1 to nearly 7500 km^2. Similarly, species richness spans two orders of magnitude, from 7 to 325.

[15] A continuous function is a function $f(X)$ such that for any two values of the random variable X, x_i and x_j, that differ by a very small number ($|x_i - x_j| < \delta$), then $|f(x_i) - f(x_j)| < \varepsilon$, another very small number. A monotonic function is a function $f(X)$ such that for any two values of the random variable X, x_i and x_j, if $x_i < x_j$ then $f(x_i) < f(x_j)$. A continuous monotonic function has both of these properties.

TABLE 8.2 **Species richness on 17 of the Galápagos Islands**

Island	Area (km^2)	Number of species	$\log_{10}$ (Area)	$\log_{10}$ (Species)
Albemarle (Isabela)	5824.9	325	3.765	2.512
Charles (Floreana)	165.8	319	2.219	2.504
Chatham (San Cristóbal)	505.1	306	2.703	2.486
James (Santiago)	525.8	224	2.721	2.350
Indefatigable (Santa Cruz)	1007.5	193	3.003	2.286
Abingdon (Pinta)	51.8	119	1.714	2.076
Duncan (Pinzón)	18.4	103	1.265	2.013
Narborough (Fernandina)	634.6	80	2.802	1.903
Hood (Española)	46.6	79	1.669	1.898
Seymour	2.6	52	0.413	1.716
Barringon (Santa Fé)	19.4	48	1.288	1.681
Gardner	0.5	48	–0.286	1.681
Bindloe (Marchena)	116.6	47	2.066	1.672
Jervis (Rábida)	4.8	42	0.685	1.623
Tower (Genovesa)	11.4	22	1.057	1.342
Wenman (Wolf)	4.7	14	0.669	1.146
Culpepper (Darwin)	2.3	7	0.368	0.845
Mean	526.0	119.3	1.654	1.867
Standard deviation	1396.9	110.7	1.113	0.481
Standard error of the mean	338.8	26.8	0.270	0.012

Area (originally given in square miles) has been converted to square kilometers, and island names from Preston (1962) are retained, with modern island names given in parentheses. The last two columns give the logarithm of area and species richness, respectively. Mathematical transformations such as logarithms are used to better meet the assumptions of parametric analyses (normality, linearity, and constant variances) and they are used as a diagnostic tool to aid in the identification of outliers and extreme data points. Because both the response variable (species richness) and the predictor variable (island area) are measured on a continuous scale, a regression model is used for analysis (see Table 7.1).

If we plot the raw data—the number of species S as a function of area A (Figure 8.5)—we see that most of the data points are clustered to the left of the figure (as most islands are small). As a first pass at the analysis, we could try fitting a straight line to this relationship:

$$S = \beta_0 + \beta_1 A \tag{8.3}$$

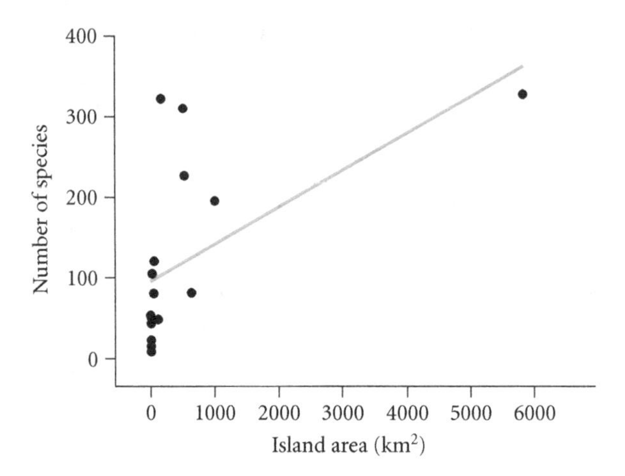

Figure 8.5 Plot of plant species richness as a function of Galápagos Island area using the raw data (Columns 2 and 3) of Table 8.2. The line shows the best-fit linear regression line (see Chapter 9). Although a linear regression can be fit to any pair of continuous variables, the linear fit to these data is not good: there are too many negative outliers at small island areas, and the slope of the line is dominated by Albermarle, the largest island in the dataset. In many cases, a mathematical transformation of the *X* variable, the *Y* variable, or both will improve the fit of a linear regression.

In this case, β_0 represents the intercept of the line, and β_1 is its slope (see Chapter 9). However, the line does not fit the data very well. In particular, notice that the slope of the regression line seems to be dominated by the datum for Albermarle, the largest island in the dataset. In Chapter 7, we warned of precisely this problem that can arise with outlier data points that dominate the fit of a regression line (see Figure 7.3). The line fit to the data does not capture well the relationship between species richness and island area.

If species richness and island area are related exponentially (see Equation 8.2), we can transform this equation by taking logarithms of both sides:

$$\log(S) = \log(cA^z) \tag{8.4}$$

$$\log(S) = \log(c) + z\log(A) \tag{8.5}$$

This transformation takes advantage of two properties of logarithms. First, the logarithm of a product of two numbers equals the sum of their logarithms:

$$\log(ab) = \log(a) + \log(b) \tag{8.6}$$

Second, the logarithm of a number raised to a power equals the power times the logarithm of the number:

$$\log(a^b) = b\log(a) \tag{8.7}$$

We can rewrite Equation 8.4, denoting logarithmically transformed values with an asterisk (*):

$$S^* = c^* + zA^* \tag{8.8}$$

Thus, we have taken an exponential equation, Equation 8.2, and transformed it into a linear equation, Equation 8.8. When we plot the logarithms of the data, the relationship between species richness and island area is now much clearer (Figure 8.6), and the coefficients have a simple interpretation. The value for z in Equation 8.2 and 8.8, which equals the slope of the line in Figure 8.6, is 0.331; this means that every time we increase A^* (the logarithm of island area) by one unit (that is, by a factor of 10 as we used $\log_{10}$ to transform our variables in Table 8.2, and $10^1 = 10$), we increase species richness by 0.331 units (that is, by a factor of approximately 2 because $10^{0.331}$ equals 2.14). Thus, we can say that a

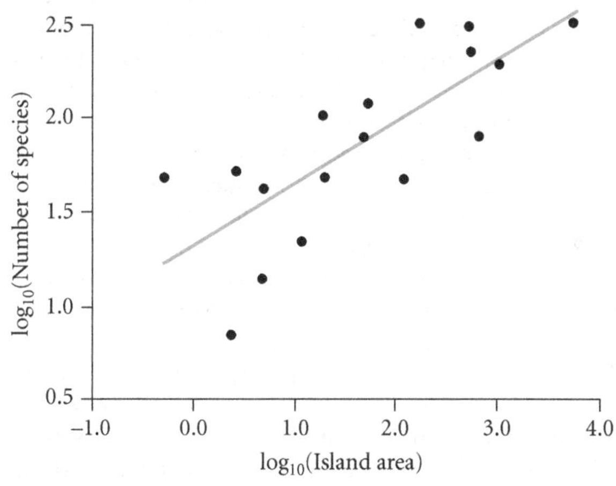

Figure 8.6 Plot of the logarithm of plant species richness as a function of the logarithm of Galápagos Island area (Columns 4 and 5 of Table 8.2). The line is the best-fit linear regression line (see Chapter 9). Compared to the linear plot in Figure 8.5, this regression fits the data considerably better: the largest island in the data set no longer looks like an outlier, and the linearity of the fit is improved.

10-fold increase in island area results in a doubling of the number of species present on the Galápagos Islands.[16]

Other transformations can be used to convert non-linear relationships to linear ones. For example, the cube-root transformation ($\sqrt[3]{Y}$) is appropriate for measures of mass or volume (Y^3) that are allometrically related to linear measures of body size or length (Y; see Footnote 13). In studies that examine relationships between two measures of masses or volumes (Y^3), such as comparisons of brain mass and body mass, both the X and Y variables are logarithmically transformed. The logarithmic transformation reduces variation in data that may range over several orders of magnitude. (For a detailed exposition of the brain-body mass relationship, see Allison and Cicchetti 1976 and Edwards 2000.)

[16] The history of the species-area relationship illustrates the dangers of **reification**: the conversion of an abstract concept into a material thing. The power function ($S = cA^z$) has formed the basis for several important theoretical models of the species-area relationship (Preston 1962; MacArthur and Wilson 1967; Harte et al. 1999). It also has been used to argue for making nature reserves as large as possible so that they contain the greatest possible number of species. This led to a long debate on whether a single large or several small preserves would protect the most species (e.g., Willis 1984; Simberloff and Abele 1984). However, Lomolino and Weiser (2001) have recently proposed that the species-area relationship has an asymptote, in which case a power function is not appropriate. But this proposal has itself been challenged on both theoretical and empirical grounds (Williamson et al. 2001). In other words, there is no consensus that the power function always forms the basis for the species-area relationship (Tjørve 2003, 2009; Martin and Goldenfeld 2006).

The power function has been popular because it appears to provide a good fit to many species-area datasets. However, a detailed statistical analysis of 100 published species-area relationships (Connor and McCoy 1979) found that the power function was the best-fitting model in only half of the datasets. Although island area is usually the single strongest predictor of species number, area typically accounts for only half of the variation in species richness (Boecklen and Gotelli 1984). As a consequence, the value of the species-area relationship for conservation planning is limited because there will be much uncertainty associated with species richness forecasts. Moreover, the species-area relationship can be used to predict only the number of species present, whereas most conservation management strategies will be concerned with the identity of the resident species. The moral of the story is that data can always be fit to a mathematical function, but we have to use statistical tools to evaluate whether the fit is reasonable or not. Even if the fit of the data is acceptable, this result, by itself, is rarely a strong test of a scientific hypothesis, because there are usually alternative mathematical models that can fit the data just as well.

Data Transformations because the Statistics Demand It

All statistical tests require that the data fit certain mathematical assumptions. For example, data to be analyzed using analysis of variance (see Chapters 7 and 10) must meet two assumptions:

1. The data must be **homoscedastic**—that is, the residual variances of all treatment groups need to be approximately equal to each other.
2. The **residuals**, or deviations from the means of each group, must be normal random variables.

Similarly, data to be analyzed using regression or correlation (see Chapter 9) also should have normally distributed residuals that are uncorrelated with the independent variable.

Mathematical transformations of the data can be used to meet these assumptions. It turns out that common data transformations often address both assumptions simultaneously. In other words, a transformation that equalizes variances (Assumption 1) often normalizes the residuals (Assumption 2).

Five transformations are used commonly with ecological and environmental data: the logarithmic transformation, the square-root transformation, the angular (or arcsine) transformation, the reciprocal transformation, and the Box-Cox transformation.

THE LOGARITHMIC TRANSFORMATION The **logarithmic transformation** (or **log transformation**) replaces the value of each observation with its logarithm:[17]

John Napier

[17] Logarithms were invented by the Scotsman John Napier (1550–1617). In his epic tome *Mirifici logarithmorum canonis descriptio* (1614), he sets forth the rationale for using logarithms (from the English translation, 1616):

Seeing there is nothing (right well-beloved Students of the Mathematics) that is so troublesome to mathematical practice, nor that doth more molest and hinder calculators, than the multiplications, divisions, square and cubical extractions of great numbers, which besides the tedious expense of time are for the most part subject to many slippery errors, I began therefore to consider in my mind by what certain and ready art I might remove those hindrances.

As illustrated in Equations 8.4–8.8, the use of logarithms allows multiplication and division to be carried out by simple addition and subtraction. Series of numbers that are multiplicative on a linear scale turn out to be additive on a logarithmic scale (see Footnote 2 in Chapter 3).

$$Y^* = \log(Y) \quad (8.9)$$

The logarithmic transformation (most commonly using the natural logarithm or the logarithm to the base e)[18] often equalizes variances for data in which the mean and the variance are positively correlated. A positive correlation means that groups with large averages will also have large variances (in ANOVA) or that the magnitude of the residuals is correlated with the magnitude of the independent variable (in regression). Univariate data that are positively skewed (skewed to the right) often have a few large outlier values. With a log-transformation, these outliers often are drawn into the mainstream of the distribution, which becomes more symmetrical. Datasets in which means and variances are positively correlated also tend to have outliers with positively skewed residuals. A logarithmic transformation often addresses both problems simultaneously.

Note that the logarithm of 0 is not defined: regardless of base b, there is no number for which $a^b = 0$. One way around this problem is to add 1 to each observation before taking its logarithm (as $\log(1) = 0$, again regardless of base b). However, this is not a useful or appropriate solution if the dataset (or especially if one treatment group) contains many zeros.[19]

THE SQUARE-ROOT TRANSFORMATION The **square-root transformation** replaces the value of each observation with its square root:

$$Y^* = \sqrt{Y} \quad (8.10)$$

[18] Logarithms can be taken with respect to many "bases," and Napier did not specify a particular base in his *Mirifici*. In general, for a value a and a base b we can write

$$\log_b a = X$$

which means that $b^X = a$. We have already used the logarithm "base 10" in our example of the species-area relationship in Table 8.2; for the first line in Table 8.2, the $\log_{10}(5825) = 3.765$ because $10^{3.765} = 5825$ (allowing for round-off error). A change in one $\log_{10}$ unit is a power of 10, or an order of magnitude. Other logarithmic bases used in the ecological literature are 2 (the octaves of Preston (1962) and used for data that increase by powers of 2), and e, the base of the natural logarithm, or approximately 2.71828... e is a transcendental number, whose value was proven by Leonhard Euler (1707–1783) to equal

$$\lim_{n \to \infty} (1 + \frac{1}{n})^n$$

The $\log_e$ was first referred to as a "natural logarithm" by the mathematician Nicolaus Mercator (1620–1687), who should not to be confused with the map-maker Gerardus Mercator (1512–1594). It is often written as ln instead of $\log_e$.

This transformation is used most frequently with count data, such as the number of caterpillars per milkweed or the number of *Rhexia* per town. In Chapter 2 we showed that such data often follow a Poisson distribution, and we pointed out that the mean and variance of a Poisson random variable are equal (both equal the Poisson rate parameter λ). Thus, for a Poisson random variable, the mean and variance vary identically. Taking square-roots of Poisson random variables yields a variance that is independent of the mean. Because the square root of 0 itself equals 0, the square-root transformation does not transform data values that are equal to 0. Thus, to complete the transformation, you should add some small number to the value prior to taking its square root. Adding 1/2 (0.5) to each value is suggested by Sokal and Rohlf (1995), whereas 3/8 (0.325) is suggested by Anscombe (1948).

THE ARCSINE OR ARCSINE-SQUARE ROOT TRANSFORMATION The **arcsine, arcsine-square root**, or **angular transformation** replaces the value of each observation with the arcsine of the square root of the value:

$$Y^* = \text{arcsine}\,\sqrt{Y} \quad \textbf{(8.11)}$$

This transformation is used principally for proportions (and percentages), which are distributed as binomial random variables. In Chapter 3, we noted that the mean of a binomial distribution = np and its variance = $np(1 - p)$, where p is the probability of success and n is the number of trials. Thus, the variance is a direct function of the mean (the variance = $(1 - p)$ times the mean). The arcsine transformation (which is simply the inverse of the sine function) removes this dependence. Because the sine function yields values only between –1 and +1, the inverse sine function can be applied only to data whose values fall

[19] Adding a constant before taking logarithms can cause problems for estimating population variability (McArdle et al. 1990), but it is not a serious issue for most parametric statistics. However, there are some subtle philosophical issues in how zeroes in the data are treated. By adding a constant to the data, you are implying that 0 represents a measurement error. Presumably, the true value is some very small number, but, by chance, you measured 0 for that replicate. However, if the value is a true zero, the most important source of variation may be between the absence and presence of the quantity measured. In this case, it would be more appropriate to use a discrete variable with two categories to describe the process. The more zeros in the dataset (and the more that are concentrated in one treatment group), the more likely it is that 0 represents a qualitatively different condition than a small positive measurement. In a population time series, a true zero would represent a population extinction rather than just an inaccurate measurement.

between –1 and +1: $-1 \leq Y_i \leq +1$. Therefore, this transformation is appropriate only for data that are expressed as proportions (such as p, the proportion or probability of success in a binomial trial). Two caveats to note with the arcsine transformation. First, if your data are percentages (0 to 100 scale), they must first be converted to proportions (0 to 1.0 scale). Second, in most software packages, the arcsine function gives transformed data in units of radians, not degrees.

THE RECIPROCAL TRANSFORMATION The **reciprocal transformation** replaces the value of each observation with its reciprocal:

$$Y^* = 1/Y \tag{8.12}$$

It is used most commonly for data that record rates, such as number of offspring per female. Rate data often appear hyperbolic when plotted as a function of the variable in the denominator. For example if you plot the number of offspring per female on the y-axis and number of females in the population on the x-axis, the resulting curve may resemble a hyperbola, which decreases steeply at first, then more gradually as X increases. These data are generally of the form

$$aXY = 1$$

(where X is number of females and Y is number of offspring per female), which can be re-written as a hyperbola

$$1/Y = aX$$

Transforming Y to its reciprocal $1/Y$ results in a new relationship

$$Y^* = 1/(1/Y) = aX$$

which is more amenable to linear regression.

THE BOX-COX TRANSFORMATION We use the logarithmic, square root, reciprocal and other transformations to reduce variance and skew in the data and to create a transformed data series that has an approximately normal distribution. The final transformation is the **Box-Cox transformation**, or generalized power transformation. This transformation is actually a family of transformations, expressed by the equation

$$\begin{aligned} Y^* &= (Y^\lambda - 1)/\lambda \quad (\text{for } \lambda \neq 0) \\ Y^* &= \log_e(Y) \quad (\text{for } \lambda = 0) \end{aligned} \tag{8.13}$$

where λ is the number that maximizes the **log-likelihood function:**

$$L = -\frac{\nu}{2}\log_e(s_T^2) + (\lambda - 1)\frac{\nu}{n}\sum_{i=1}^{n}\log_e Y \qquad \textbf{(8.14)}$$

where ν is the degrees of freedom, n is the sample size, and s_T^2 is the variance of the transformed values of Y (Box and Cox 1964). The value of λ that results when Equation 8.14 is maximized is used in Equation 8.13 to provide the closest fit of the transformed data to a normal distribution. Equation 8.14 must be solved iteratively (trying different values of λ until L is maximized) with computer software.

Certain values of λ correspond to the transformations we have already described. When $\lambda = 1$, Equation 8.13 results in a linear transformation (a shift operation; see Figure 2.7), when $\lambda = 1/2$, the result is the square-root transformation, when $\lambda = 0$, the result is the natural logarithmic transformation, and when $\lambda = -1$, the result is the reciprocal transformation. Before going to the trouble of maximizing Equation 8.14, you should try transforming your data using simple arithmetic transformations. If your data are right-skewed, try using the more familiar transformations from the series $1/\sqrt{Y}$, $\sqrt{Y}$, $\ln(Y)$, $1/Y$. If your data are left-skewed, try Y^2, Y^3, etc. (Sokal and Rohlf 1995).

Reporting Results: Transformed or Not?

Although you may transform the data for analysis, you should report the results in the original units. For example, the species–area data in Table 8.2 would be analyzed using the log-transformed values, but in describing island size or species richness, you would report them in their original units, **back-transformed**. Thus, the average island size is

$$\text{antilog}(\overline{\log A}) = \text{antilog}(1.654) = 45.110 \text{ km}^2$$

Note that this is very different from the arithmetic average of island size prior to transformation, which equals 526 km^2. Similarly, average species richness is

$$\text{antilog}(\overline{\log S}) = 10^{1.867} = 73.6 \text{ species}$$

which differs from the arithmetic average of species richness prior to transformation, 119.3.

You would construct confidence intervals (see Chapter 3) in a similar manner, by taking antilogarithms of the confidence limits constructed using the standard errors of the means of the transformed data. This will normally result in asymmetric confidence intervals. For the island area data in Table 8.2, the stan-

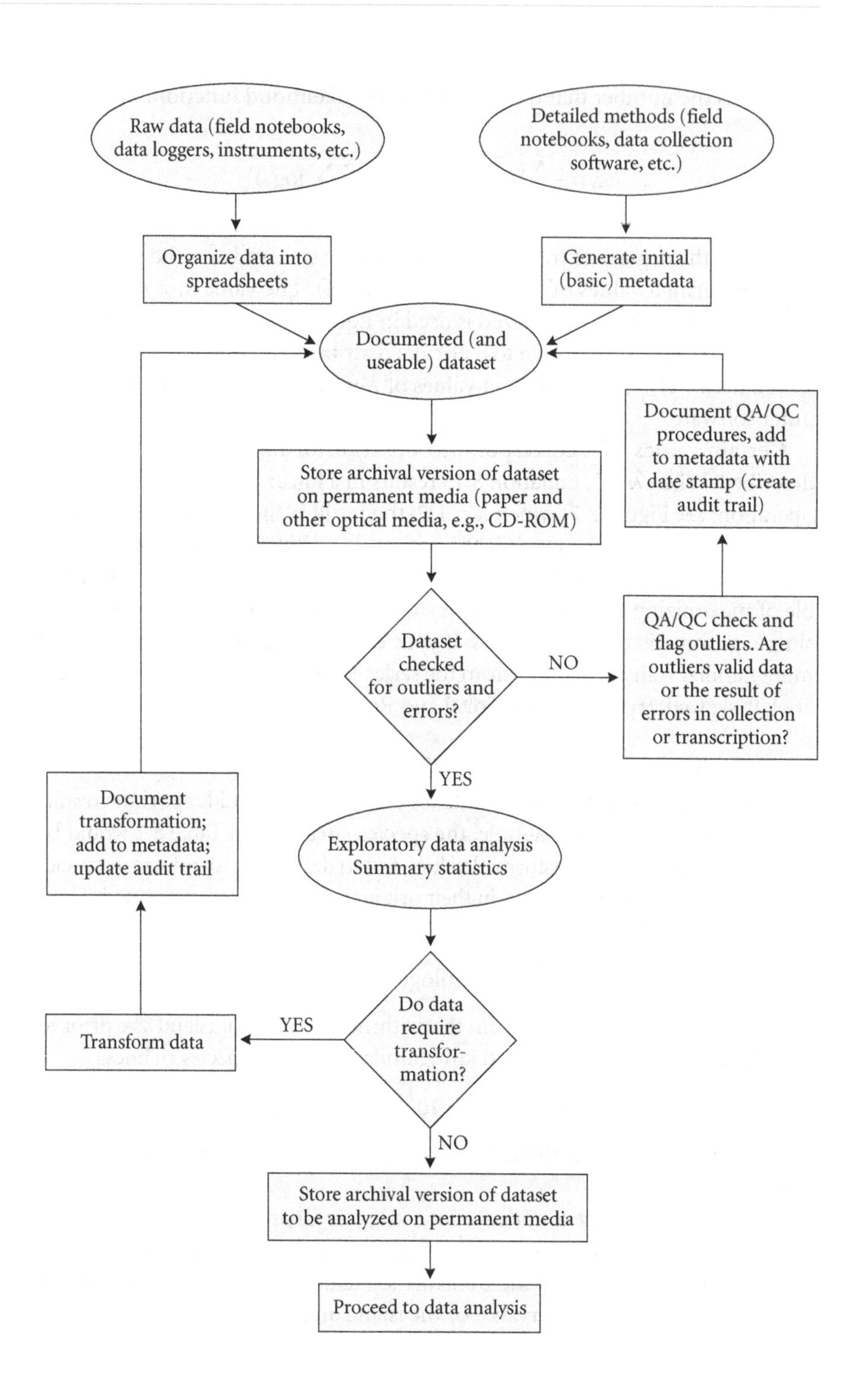
Raw data (field notebooks, data loggers, instruments, etc.)
Detailed methods (field notebooks, data collection software, etc.)
Organize data into spreadsheets
Generate initial (basic) metadata
Documented (and useable) dataset
Store archival version of dataset on permanent media (paper and other optical media, e.g., CD-ROM)
Document QA/QC procedures, add to metadata with date stamp (create audit trail)
Dataset checked for outliers and errors?
NO
QA/QC check and flag outliers. Are outliers valid data or the result of errors in collection or transcription?
YES
Document transformation; add to metadata; update audit trail
Exploratory data analysis Summary statistics
Do data require transformation?
YES
Transform data
NO
Store archival version of dataset to be analyzed on permanent media
Proceed to data analysis

◀ **Figure 8.7** The data management flowchart. This chart outlines the steps for managing, curating, and storing data. These steps should all be taken before formal data analysis begins. Documentation of metadata is especially important to ensure that the dataset will still be accessible and usable in the distant future.

dard error of the mean of log(A) = 0.270. For n = 17, the necessary value from the t distribution, $t_{0.025[16]} = 2.119$. Thus, the lower bound of the 95% confidence interval (using Equation 3.16) = 1.654 – 2.119 × 0.270 = 1.082. Similarly, the upper bound of the 95% confidence interval = 1.654 + 2.119 × 0.270 = 2.226. On a logarithmic scale these two values form a symmetrical interval around the mean of 1.654, but when they are back-transformed, the interval is no longer symmetrical. The antilog of 1.082 = $10^{1.082}$ = 12.08, whereas the antilog of 2.226 = $10^{2.226}$ = 168.27; the interval [12.08, 168.27] which is not symmetrical around the back-transformed mean of 45.11.

The Audit Trail Redux

The process of transforming data also should be added to your audit trail. Just as with data QA/QC, the methods used for data transformation should be documented in the metadata. Rather than writing over your data in your original spreadsheet with the transformed values, you should create a new spreadsheet that contains the transformed data. This spreadsheet also should be stored on permanent media, given a unique identifier in your data catalog, and added to the audit trail.

Summary: The Data Management Flow Chart

Organization, management, and curation of your dataset are essential parts of the scientific process, and must be done before you begin to analyze your data. This overall process is summarized in our data management flow chart (Figure 8.7).

Well-organized datasets are a lasting contribution to the broader scientific community, and their widespread sharing enhances collegiality and increases the pace of scientific progress. The free and public availability of data is a requirement of many granting agencies and scientific journals. Data collected with public funds distributed by organizations and agencies in the United States may be requested at any time by anyone under the provisions of the Freedom of Information Act. Adequate funds should be budgeted to allow for necessary and sufficient data management.

Data should be organized and computerized rapidly following collection, documented with sufficient metadata necessary to reconstruct the data collection, and archived on permanent media. Datasets should be checked carefully for errors in collection and transcription, and for outliers that are nonetheless valid data. Any changes or modifications to the original ("raw") data resulting from error- and outlier-detection procedures should be documented thoroughly in the metadata; the altered files should not replace the raw data file. Rather, they should be stored as new (modified) data files. An audit trail should be used to track subsequent versions of datasets and their associated metadata.

Graphical exploratory data analysis and calculation of basic summary statistics (e.g., measures of location and spread) should precede formal statistical analysis and hypothesis testing. A careful examination of preliminary plots and tables can indicate whether or not the data need to be transformed prior to further analysis. Data transformations also should be documented in the metadata. Once a final version of the dataset is ready for analysis, it too should be archived on permanent media along with its complete metadata and audit trail.

PART III

Data Analysis

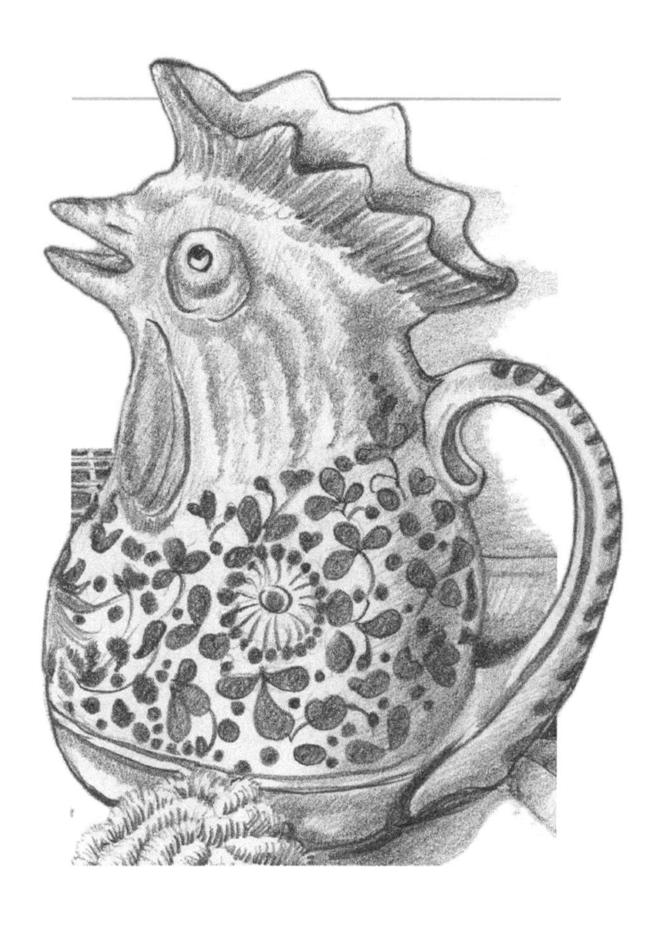

CHAPTER 9

Regression

Regression is used to analyze relationships between continuous variables. At its most basic, regression describes the linear relationship between a predictor variable, plotted on the x-axis, and a response variable, plotted on the y-axis. In this chapter, we explain how the method of least-squares is used to fit a regression line to data, and how to test hypotheses about the parameters of the fitted model. We highlight the assumptions of the regression model, describe diagnostic tests to evaluate the fit of the data to the model, and explain how to use the model to make predictions. We also provide a brief description of some more advanced topics: logistic regression, multiple regression, non-linear regression, robust regression, quantile regression, and path analysis. We close with a discussion of the problems of model selection: how to choose an appropriate subset of predictor variables, and how to compare the relative fit of different models to the same data set.

Defining the Straight Line and Its Two Parameters

We will start with the development of a linear model, because this is the heart of regression analysis. As we noted in Chapter 6, a regression model begins with a stated hypothesis about cause and effect: the value of the X variable causes, either directly or indirectly, the value of the Y variable.[1] In some cases, the direc-

[1] Many statistics texts emphasize a distinction between **correlation**, in which two variables are merely associated with one another, and **regression**, in which there is a direct cause-and-effect relationship. Although different kinds of statistics have been developed for both cases, we think the distinction is largely arbitrary and often just semantic. After all, investigators do not seek out correlations between variables unless they believe or suspect there is some underlying cause-and-effect relationship. In this chapter, we cover statistical methods that are sometimes treated separately in analyses of correlation and regression. All of these techniques can be used to estimate and test associations of two or more continuous variables.

tion of cause and effect is straightforward—we hypothesize that island area controls plant species number (see Chapter 8) and not the other way around. In other cases, the direction of cause and effect is not so obvious—do predators control the abundance of their prey, or does prey abundance dictate the number of predators (see Chapter 4)?

Once you have made the decision about the direction of cause and effect, the next step is to describe the relationship as a mathematical function:

$$Y = f(X) \tag{9.1}$$

In other words, we apply the function f to each value of the variable X (the input) to generate the corresponding value of the variable Y (the output). There are many interesting and complex functions that can describe the relationship between two variables, but the simplest one is that Y is a *linear* function of X:

$$Y = \beta_0 + \beta_1 X \tag{9.2}$$

In words, this function says "take the value of the variable X, multiply it by β_1, and add this number to β_0. The result is the value of the variable Y." This equation describes the graph of a straight line. This model has two parameters in it, β_0 and β_1, which are called the **intercept** and the **slope** of the line (Figure 9.1). The intercept (β_0) is the value of the function when $X = 0$. The intercept is measured in the same units as the Y variable. The slope (β_1) measures the change in the Y variable for each unit change in the X variable. The slope is therefore a rate and is measured in units of $\Delta Y/\Delta X$ (read as: "change in Y divided by change in X"). If the slope and intercept are known, Equation 9.2 can be used to predict the value of Y for any value of X. Conversely, Equation 9.2

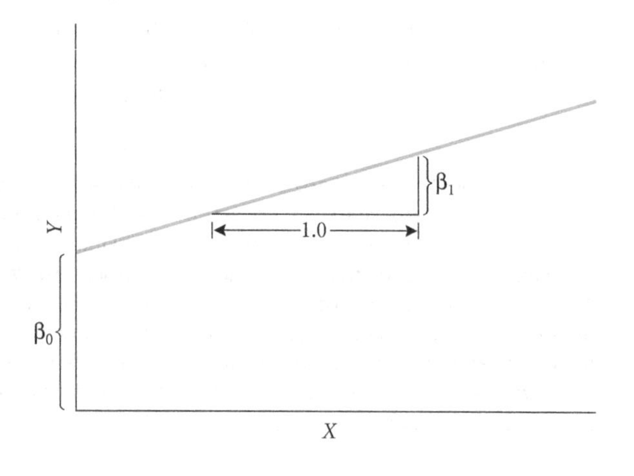

Figure 9.1 Linear relationship between variables X and Y. The line is described by the equation $y = \beta_0 + \beta_1 X$, where β_0 is the intercept, and β_1 is the slope of the line. The intercept β_0 is the predicted value from the regression equation when $X = 0$. The slope of the line β_1 is the increase in the Y variable associated with a unit increase in the X variable ($\Delta Y/\Delta X$). If the value of X is known, the predicted value of Y can be calculated by multiplying X by the slope (β_1) and adding the intercept (β_0).

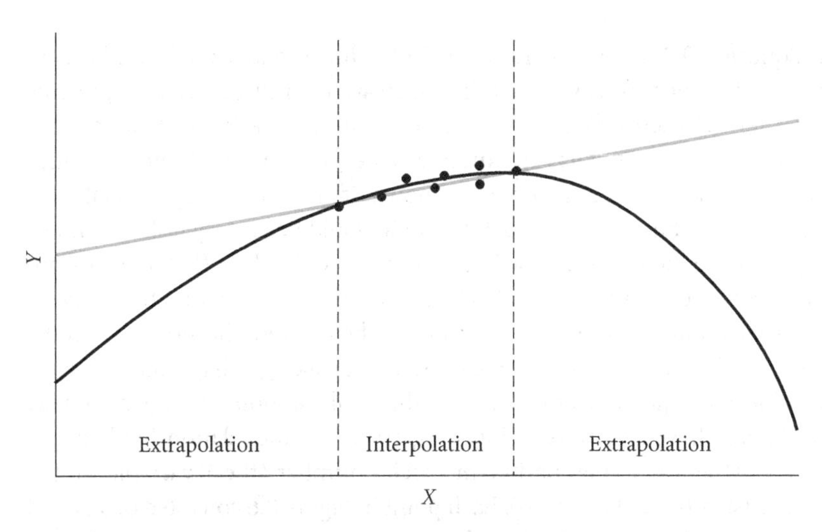

Figure 9.2 Linear models may approximate non-linear functions over a limited domain of the X variable. Interpolation within these limits may be acceptably accurate, even though the linear model (blue line) does not describe the true functional relationship between Y and X (black curve). Extrapolations will become increasingly inaccurate as the forecasts move further away from the range of collected data. A very important assumption of linear regression is that the relationship between X and Y (or transformations of these variables) is linear.

can be used to determine the value of X that would have generated a particular value of Y.

Of course, nothing says that nature has to obey a linear equation; many ecological relationships are inherently non-linear. However, the linear model is the simplest starting place for fitting functions to data. Moreover, even complex, non-linear (wavy) functions may be *approximately* linear *over a limited range of the X variable* (Figure 9.2). If we carefully restrict our conclusions to that range of the X variable, a linear model may be a valid approximation of the function.

Fitting Data to a Linear Model

Let's first recast Equation 9.2 in a form that matches the data we have collected. The data for a regression analysis consist of a series of paired observations. Each observation includes an X value (X_i) and a corresponding Y value (Y_i) that both have been measured for the same replicate. The subscript i indicates the replicate. If there is a total of n replicates in our data, the subscript i can take on any integer value from $i = 1$ to n. The model we will fit is

$$Y_i = \beta_0 + \beta_1 X_i + \varepsilon_i \tag{9.3}$$

As in Equation 9.2, the two parameters in the linear equation, β_0 and β_1, are unknowns. But there is also now a third unknown quantity, ε_i, which represents the error term. Whereas β_0 and β_1 are simple constants, ε_i is a normal random variable. This distribution has an expected value (or mean) of 0, and a variance equal to σ^2, which may be known or unknown. If all of the data points fall on a perfectly straight line, then σ^2 equals 0, and it would be an easy task to connect those points and then measure the intercept (β_0) and the slope (β_1) directly from the line.[2] However, most ecological datasets exhibit more variation than this—a single variable rarely will account for most of the variation in the data, and the data points will fall within a fuzzy band rather than along a sharp line. The larger the value of σ^2, the more noise, or error, there will be about the regression line.

In Chapter 8, we illustrated a linear regression for the relationship between island area (the X variable) and plant species number (the Y variable) in the Galápagos Islands (see Figure 8.6). Each point in Figure 8.6 consisted of a paired observation of the area of the island and its corresponding number of plant species (see Table 8.2). As we explained in Chapter 8, we used a logarithmic transformation of both the X variable (area) and the Y variable (species richness) in order to homogenize the variances and linearize the curve. A little later in this chapter, we will return to these data in their untransformed state.

Figure 8.6 showed a clear relationship between $\log_{10}$(area) and $\log_{10}$(species number), but the points did not fall in a perfect line. Where should the regression line be placed? Intuitively, it seems that the regression line should pass through the center of the cloud of data, defined by the point $(\bar{X}, \bar{Y})$. For the island data, the center corresponds to the point (1.654, 1.867) (remember, these are log-transformed values).

Now we can rotate the line through that center point until we arrive at the "best fit" position. But how should we define the best fit for the line? Let's first define the **squared residual** d_i^2 as the squared difference between the observed Y_i value and the Y value that is predicted by the regression equation ($\hat{Y}_i$). We use the small caret (or "hat") to distinguish between the observed value (Y_i) and the value predicted from the regression equation ($\hat{Y}_i$). The squared residual d_i^2 is calculated as

$$d_i^2 = (Y_i - \hat{Y}_i)^2 \tag{9.4}$$

[2] Of course, if there are only two observations, a straight line will fit them perfectly every time! But having more data is no guarantee that the straight line is a meaningful model. As we will see in this chapter, a straight line can be fit to any set of data and used to make forecasts, regardless of whether the fitted model is valid or not. However, with a large data set, we have diagnostic tools for assessing the fit of the data, and we can assign confidence intervals to forecasts that are derived from the model.

The residual deviation is squared because we are interested in the magnitude, and not the sign, of the difference between the observed and predicted value (see Footnote 8 in Chapter 2). For any particular observation Y_i, we could pass the regression through that point, so that its residual would be minimized ($d_i = 0$). But the regression line has to fit all of the data collectively, so we will define the sum of all of the residuals, also called the **residual sum of squares** and abbreviated as *RSS*, to be

$$RSS = \sum_{i=1}^{n}(Y_i - \hat{Y}_i)^2 \tag{9.5}$$

The "best fit" regression line is the one that minimizes the residual sum of squares.[3] By minimizing *RSS*, we are ensuring that the regression line results in the smallest average difference between each Y_i value and the $\hat{Y}_i$ value that is predicted by the regression model (Figure 9.3).[4]

Francis Galton

[3] Francis Galton (1822–1911)—British explorer, anthropologist, cousin of Charles Darwin, sometime statistician, and quintessential D.W.E.M. (Dead White European Male)—is best remembered for his interest in eugenics and his assertion that intelligence is inherited and little influenced by environmental factors. His writings on intelligence, race, and heredity led him to advocate breeding restrictions among people and undergirded early racist policies in colonial Australia. He was knighted in 1909.

In his 1866 article "Regression towards mediocrity in hereditary structure," Galton analyzed the heights of adult children and their parents with the least-squares linear regression model. Although Galton's data frequently have been used to illustrate linear regression and regression toward the mean, a recent re-analysis of the original data reveals they are not linear (Wachsmuth et al. 2003).

Galton also invented a device called the *quincunx*, a box with a glass face, rows of pins inside, and a funnel at the top. Lead shot of uniform size and mass dropped through the funnel was deflected by the pins into a modest number of compartments. Drop enough shot, and you get a normal (Gaussian) curve. Galton used this device to obtain empirical data showing that the normal curve can be expressed as a mixture of other normal curves.

[4] Throughout this chapter, we present formulas such as Equation 9.5 for sums of squares and other quantities that are used in statistics. However, we do not recommend that you use the formulas in this form to make your calculations. The reason is that these formulas are very susceptible to small round-off errors, which accumulate in a large sum (Press et al. 1986). Matrix multiplication is a much more reliable way to get the statistical solutions; in fact, it is the only way to solve more complex problems such as multiple regression. It is important for you to study these equations so that you understand how the statistics work, and it is a great idea to use spreadsheet software to try a few simple examples "by hand." However, for analysis and publication, you should let a dedicated statistics package do the heavy numerical lifting.

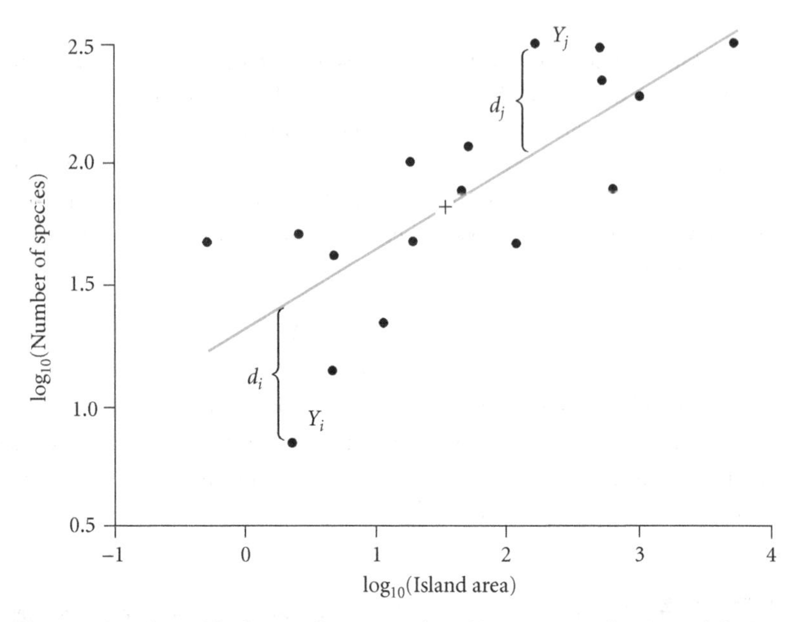

Figure 9.3 The residual sum of squares is found by summing the squared deviation (d_i's) each observation from the fitted regression line. The least-squares parameter estimates ensure that the fitted regression line minimizes this residual sum of squares. The + marks the midpoint of the data $(\bar{X}, \bar{Y})$. This regression line describes the relationship between the logarithm of island area and the logarithm of species number for plant species of the Galápagos Islands; data from Table 8.2. The regression equation is $\log_{10}(\text{Species}) = 1.320 + \log_{10}(\text{Area}) \times 0.331$; $r^2 = 0.584$.

We could fit the regression line through $(\bar{X}, \bar{Y})$ "by eye" and then tinker with it until we found the slope and intercept values that gave us the smallest value of *RSS*. Fortunately, there is an easier way to obtain the estimates of β_0 and β_1 that minimize *RSS*. But first we have to take a brief detour to discuss variances and covariances.

Variances and Covariances

In Chapter 3, we introduced you to the sum of squares (SS_Y) of a variable,

$$SS_Y = \sum_{i=1}^{n}(Y_i - \bar{Y})^2 \quad \textbf{(9.6)}$$

which measures the squared deviation of each observation from the mean of the observations. Dividing this sum by $(n-1)$ gives the familiar formula for the sample variance of a variable:

$$s_Y^2 = \frac{1}{n-1}\sum_{i=1}^{n}(Y_i - \bar{Y})^2 \quad \textbf{(9.7)}$$

By removing the exponent from Equation 9.6 and expanding the squared term, we can re-write it as

$$SS_Y = \sum_{i=1}^{n}(Y_i - \bar{Y})(Y_i - \bar{Y}) \tag{9.8}$$

Now let's consider the situation with two variables X and Y. Instead of the sum of squares for one variable, we can define the **sum of cross products** (SS_{XY}) as

$$SS_{XY} = \sum_{i=1}^{n}(X_i - \bar{X})(Y_i - \bar{Y}) \tag{9.9}$$

and the **sample covariance** (s_{XY}) as

$$s_{xy} = \frac{1}{n-1}\sum_{i=1}^{n}(X_i - \bar{X})(Y_i - \bar{Y}) \tag{9.10}$$

As we saw in Chapter 3, the sample variance is always a non-negative number. Because the square of the difference of each observation from its mean, $(Y_i - \bar{Y})^2$, is always greater than zero, the sum of all of these squared differences must also be greater than zero. But the same is not true for the sample covariance. Suppose that relatively large values of X are consistently paired with relatively small values of Y. The first term in Equation 9.9 or Equation 9.10 will be positive for the X_i values that are greater than $\bar{X}$. But the second term (and hence the product) will be negative because the relatively small Y_i values are less than $\bar{Y}$. Similarly, the relatively small X_i values (those less than $\bar{X}$) will be paired with the relatively large values of Y_i (those greater than $\bar{Y}$). If there are lots of pairs of data organized this way, the sample covariance will be a negative number.

On the other hand, if the large values of X are always associated with the large values of Y, the summation terms in Equations 9.9 and 9.10 all will be positive and will generate a large positive covariance term. Finally, suppose that X and Y are unrelated to one another, so that large or small values of X may sometimes be associated with large or small values of Y. This will generate a heterogeneous collection of covariance terms, with both positive and negative signs. The sum of these terms may be close to zero.

For the Galápagos plant data, $SS_{XY} = 6.558$, and $s_{XY} = 0.410$. All of the elements of this covariance term are positive, except for the contributions from the islands Duncan (−0.057) and Bindlow (−0.080). Intuitively, it would seem that this measure of covariance should be related to the slope of the regression line, because it describes the relationship (positive or negative) between variation in the X variable and variation in the Y variable.[5]

Least-Squares Parameter Estimates

Having defined the covariance, we can now estimate the regression parameters that minimize the residual sum of squares:

$$\hat{\beta}_1 = \frac{s_{XY}}{s_X^2} = \frac{SS_{XY}}{SS_X} \tag{9.11}$$

where the sums of squares of X is

$$SS_X = \sum_{i=1}^{n} \left(X_i - \overline{X}\right)^2$$

We use the symbol $\hat{\beta}_1$ to designate our estimate of the slope, and to distinguish it from β_1, the true value of the parameter. Keep in mind that β_1 has a true value only in a frequentist statistical framework; in a Bayesian analysis, the parameters themselves are viewed as a random sample from a distribution (see Chapter 5). Later in this chapter we discuss Bayesian parameter estimation in regression models.

Equation 9.11 illustrates the relationship between the slope of a regression model and the covariance between X and Y. Indeed, the slope is the covariance

[5] The covariance is a single number that expresses a relationship between a single pair of variables. Suppose we have a set of n variables. For each unique pair (X_i, X_j) of variables we could calculate each of the covariance terms s_{ij}. There are exactly $n(n - 1)/2$ such unique pairs, which can be determined using the binomial coefficient from Chapter 2 (see also Footnote 4 in Chapter 2). These covariance terms can then be arranged in a square $n \times n$ **variance-covariance matrix**, in which all of the variables are represented in each row and in each column of the matrix. The matrix entry row i, column j is simply s_{ij}. The diagonal elements of this matrix are the variances of each variable $s_{ii} = s_i^2$. The matrix elements above and below the diagonal will be mirror images of one another because, for any pair of variables X_i and X_j, Equation 9.10 shows that $s_{ij} = s_{ji}$. The variance-covariance matrix is a key piece of machinery that is used to obtain least-squares solutions for many statistical problems (see also Chapter 12 and the Appendix).

The variance-covariance matrix also makes an appearance in community ecology. Suppose that each variable represents the abundance of a species in a community. Intuitively, the covariance in abundance should reflect the kinds of interactions that occur between species pairs (negative for predation and competition, positive for mutualism, 0 for weak or neutral interactions). The **community matrix** (Levins 1968) and other measures of pairwise interaction coefficients can be computed from the variance-covariance matrix of abundances and used to predict the dynamics and stability of the entire **assemblage** (Laska and Wootton 1998).

of X and Y, scaled to the variance in X. Because the denominator $(n-1)$ is identical for the calculation of s_{XY} and s_X^2, $\hat{\beta}_1$ can also be expressed as a ratio of the sum of the cross products (SS_{XY}) to the sum of squares of X (SS_X). For the Galápagos plant data, $s_{XY} = 0.410$, and $s_X^2 = 1.240$, so $\hat{\beta}_1 = 0.410/1.240 = 0.331$. Remember that a slope is always expressed in units of $\Delta Y/\Delta X$, which in this case is the change in $\log_{10}$(species number) divided by the change in $\log_{10}$(island area).

To solve for the intercept of the equation ($\hat{\beta}_0$), we take advantage of the fact that the regression line passes through the point $(\bar{X}, \bar{Y})$. Combining this with our estimate of $\hat{\beta}_1$, we have

$$\hat{\beta}_0 = \bar{Y} - \hat{\beta}_1\bar{X} \tag{9.12}$$

For the Galápagos plant data, $\hat{\beta}_0 = 1.867 - (0.331)(1.654) = 1.319$.

The units of the intercept are the same as the units of the Y variable, which in this case is $\log_{10}$(species number). The intercept tells you the estimate of the response variable when the value of the X variable equals zero. For our island data, $X = 0$ corresponds to an area of 1 square kilometer (remember that $\log_{10}(1.0) = 0.0$), with an estimate of $10^{1.319} = 20.844$ species.

We still have one last parameter to estimate. Remember from Equation 9.3 that our regression model includes not only an intercept ($\hat{\beta}_0$) and a slope ($\hat{\beta}_1$), but also an error term (ε_i). This error term has a normal distribution with a mean of 0 and a variance of σ^2. How can we estimate σ^2? First, notice that if σ^2 is relatively large, the observed data should be widely scattered about the regression line. Sometimes the random variable will be positive, pushing the datum Y_i above the regression line, and sometimes it will be negative, pushing Y_i below the line. The smaller σ^2, the more tightly the data will cluster around the fitted regression line. Finally, if $\sigma^2 = 0$, there should be no scatter at all, and the data will "toe the line" perfectly.

This description sounds very similar to the explanation for the residual sum of squares (RSS), which measures the squared deviation of each observation from its fitted value (see Equation 9.5). In simple summary statistics (see Chapter 3), the sample variance of a variable measures the average deviation of each observation from the mean (see Equation 3.9). Similarly, our estimate of the variance of the regression error is the average deviation of each observation from the fitted value:

$$\hat{\sigma}^2 = \frac{RSS}{n-2} = \frac{\sum_{i=1}^{n}(Y_i - \hat{Y}_i)^2}{n-2} = \frac{\sum_{i=1}^{n}\left[Y_i - (\hat{\beta}_0 + \hat{\beta}_1 X_i)\right]^2}{n-2} \tag{9.13}$$

The expanded forms are shown to remind you of the calculation of *RSS* and $\hat{Y}_i$. As before, remember that $\hat{\beta}_0$ and $\hat{\beta}_1$ are the fitted regression parameters, and $\hat{Y}_i$ is the predicted value from the regression equation. The square root of Equation 9.13, $\hat{\sigma}$, is often called the **standard error of regression.** Notice that the denominator of the estimated variance is $(n - 2)$, whereas we previously used $(n - 1)$ as the denominator for the sample variance calculation (see Equation 3.9). The reason we use $(n - 2)$ is that the denominator is the degrees of freedom, the number of independent pieces of information that we have to estimate that variance (see Chapter 3). In this case, we have already used up two degrees of freedom to estimate the intercept and the slope of the regression line. For the Galápagos plant data, $\hat{\sigma} = 0.320$.

Variance Components and the Coefficient of Determination

A fundamental technique in parametric analysis is to partition a sum of squares into different components or sources. We will introduce the idea here, and return to it again in Chapter 10. Starting with the raw data, we will consider the sum of squares of the *Y* variable (SS_Y) to represent the total variation that we are trying to partition.

One component of that variation is pure or random error. This variation cannot be attributed to any particular source, other than random sampling from a normal distribution. In Equation 9.3 this source of variation is ε_i, and we have already seen how to estimate this residual variation by calculating the residual sums of squares (*RSS*). The remaining variation in Y_i is not random, but systematic. Some values of Y_i are large *because* they are associated with large values of X_i. The source of this variation is the regression relationship $Y_i = \beta_0 + \beta_1 X_i$. By subtraction, it follows that the remaining component of variation that can be attributed to the regression model (SS_{reg}) is

$$SS_{reg} = SS_Y - RSS \tag{9.14}$$

By rearranging Equation 9.14, the total variation in the data can be additively partitioned into components of the regression (SS_{reg}) and of the residuals (*RSS*):

$$SS_Y = SS_{reg} + RSS \tag{9.15}$$

For the Galápagos data, $SS_Y = 3.708$ and $RSS = 1.540$. Therefore, $SS_{reg} = 3.708 - 1.540 = 2.168$.

We can imagine two extremes in this "slicing of the variance pie." Suppose that all of the data points fell perfectly on the regression line, so that any value of Y_i could be predicted exactly knowing the value of X_i. In this case, $RSS = 0$, and $SS_Y = SS_{reg}$. In other words, all of the variation in the data can be attributed to the regression and there is no component of random error.

At the other extreme, suppose that the X variable had no effect on the resulting Y variable. If there is no influence of X on Y, then $\beta_1 = 0$, and there is no slope to the line:

$$Y_i = \beta_0 + \varepsilon_i \tag{9.16}$$

Remember that ε_i is a random variable with a mean of 0 and a variance of σ^2. Taking advantage of the shift operation (see Chapter 2) we have

$$Y \sim N(\beta_0, \sigma) \tag{9.17}$$

In words, Equation 9.17 says, "Y is a random normal variable, with a mean (or expectation) of β_0 and a standard deviation of σ." If none of the variation in Y_i can be attributed to the regression, the slope equals 0, and the Y_i's are drawn from a normal distribution with mean equal to the intercept of the regression (β_0), and a variance of σ^2. In this case $SS_Y = RSS$, and therefore $SS_{reg} = 0.0$.

Between the two extremes of $SS_{reg} = 0$ and $RSS = 0$ lies the reality of most data sets, which reflects both random and systematic variation. A natural index that describes the relative importance of regression versus residual variation is the familiar r^2, or **coefficient of determination:**

$$r^2 = \frac{SS_{reg}}{SS_Y} = \frac{SS_{reg}}{SS_{reg} + RSS} \tag{9.18}$$

The coefficient of determination tells you the proportion of the variation in the Y variable that can be attributed to variation in the X variable through a simple linear regression. This proportion varies from 0.0 to 1.0. The larger the value, the smaller the error variance and the more closely the data match the fitted regression line. For the Galápagos data, $r^2 = 0.585$, about midway between no correlation and a perfect fit. If you convert the r^2 value to a scale of 0 to 100, it is often described as the percentage of variation in Y "explained" by the regression on X. Remember, however, that the causal relationship between the X and Y variable is an hypothesis that is explicitly proposed by the investigator (see Equation 9.1). The coefficient of determination—no matter how large—does not by itself confirm a cause-and-effect relationship between two variables.

A related statistic is the **product-moment correlation coefficient**, r. As you might guess, r is just the square root of r^2. However, the sign of r (positive or negative) is determined by the sign of the regression slope; it is negative if $\beta_1 <$ 0 and positive if $\beta_1 > 0$. Equivalently, r can be calculated as

$$r = \frac{SS_{XY}}{\sqrt{(SS_X)(SS_Y)}} = \frac{s_{XY}}{s_X s_Y} \quad (9.19)$$

with the positive or negative sign coming from the sum of cross products term in the numerator.[6]

Hypothesis Tests with Regression

So far, we have learned how to fit a straight line to continuous X and Y data, and how to use the least-squares criterion to estimate the slope, the intercept, and the variance of the fitted regression line. The next step is to test hypotheses about the fitted regression line. Remember that the least squares calculations give us only *estimates* ($\hat{\beta}_0$, $\hat{\beta}_1$, $\hat{\sigma}^2$) of the true values of the parameters (β_0, β_1, σ^2). Because there is uncertainty in these estimates, we want to test whether some of these parameter estimates differ significantly from zero.

In particular, the underlying assumption of cause and effect is embodied in the slope parameter. Remember that, in setting up the regression model, we assume that X causes Y (see Figure 6.2 for other possibilities). The magnitude of β_1 measures the strength of the response of Y to changes in X. Our null hypothesis is that β_1 does not differ from zero. If we cannot reject this null hypothesis, there is no compelling evidence of a functional relationship between the X and the Y variables. Framing the null and alternative hypotheses in terms of our models, we have

$$Y_i = \beta_0 + \varepsilon_i \quad \text{(Null hypothesis)} \quad (9.20)$$

$$Y_i = \beta_0 + \beta_1 X_i + \varepsilon_i \quad \text{(Alternative hypothesis)} \quad (9.21)$$

[6] Rearranging Equation 9.19 reveals a close connection between r and β_1, the slope of the regression:

$$\beta_1 = r\left(\frac{s_Y}{s_X}\right)$$

Thus, the slope of the regression line turns out to be the correlation coefficient "rescaled" to the relative standard deviations of Y and X.

The Anatomy of an ANOVA Table

This null hypothesis can be tested by first organizing the data into an **analysis of variance** (ANOVA) table. Although an ANOVA table is naturally associated with an analysis of variance (see Chapter 10), the partitioning of the sum of squares is common to ANOVA, regression, and many other generalized linear models (McCullagh and Nelder 1989). Table 9.1 illustrates a complete ANOVA table with all of its components and equations. Table 9.2 illustrates the same ANOVA table with calculations for the Galápagos plant data. The abbreviated form of Table 9.2 is typical for a scientific publication.

The ANOVA table has a number of columns that summarize the partitioning of the sum of squares. The first column is usually labeled Source, meaning the component or source of variation. In the regression model, there are only two sources: the regression and the error. We have also added a third source, the total, to remind you that the total sum of squares equals the sum of the regression and the error sums of squares. However, the row giving the total sums of squares usually is omitted from published ANOVA tables. Our simple regression model has only two sources of variation, but more complex models may have several sources of variation listed.

TABLE 9.1 **Complete ANOVA table for single factor linear regression**

Source	Degrees of freedom (df)	Sum of squares (*SS*)	Mean square (*MS*)	Expected mean square	F-ratio	*P*-value
Regression	1	$SS_{reg} = \sum_{i=1}^{n}(\hat{Y}_i - \bar{Y})^2$	$\frac{SS_{reg}}{1}$	$\sigma^2 + \beta_1^2\sum_{i=1}^{n}X_i^2$	$\frac{SS_{reg}/1}{RSS/(n-2)}$	Tail of the F distribution with 1, $n-2$ degrees of freedom
Residual	$n-2$	$RSS = \sum_{i=1}^{n}(Y_i - \hat{Y}_i)^2$	$\frac{RSS}{(n-2)}$	σ^2		
Total	$n-1$	$SS_Y = \sum_{i=1}^{n}(Y_i - \bar{Y})^2$	$\frac{SS_Y}{(n-1)}$	σ_Y^2		

The first column gives the source of variation in the data. The second column gives the degrees of freedom (df) associated with each component. For a simple linear regression, there is only 1 degree of freedom associated with the regression, and $(n-2)$ associated with the residual. The degrees of freedom total to $(n-1)$ because 1 degree of freedom is always used to estimate the grand mean of the data. The single factor regression model partitions the total variation into a component explained by the regression and a remaining ("unexplained") residual. The sums of squares are calculated using the observed Y values (Y_i), the mean of the Y values ($\bar{Y}$), and the Y values predicted by the linear regression model ($\hat{Y}_i$). The expected mean squares are used to construct an F-ratio to test the null hypothesis that the variation associated with the slope term (β_1) equals 0.0. The *P*-value for this test is taken from a standard table of F-values with 1 degree of freedom for the numerator and $(n-2)$ degrees of freedom for the denominator. These basic elements (source, df, sum of squares, mean squares, expected mean square, F-ratio, and *P*-value) are common to all ANOVA tables in regression and analysis of variance (see Chapter 10).

TABLE 9.2 **Publication form of ANOVA table for regression analysis of Galápagos plants species-area data**

Source	df	*SS*	*MS*	F-ratio	*P*
Regression	1	2.168	2.168	21.048	0.000329
Residual	15	1.540	0.103		

The raw data are from Table 8.2; the formulas for these calculations are given in Table 9.1. The total sum of squares and degrees of freedom are rarely included in a published ANOVA table. In many publications, these results would be reported more compactly as "The linear regression of log-species richness against log-area was highly significant (model $F_{1,15} = 21.048, p < 0.001$). Thus, the best-fitting equation was $\log_{10}$(Species richness) = 1.867 + 0.331×log(island area); $r^2 = 0.585$." Published regression statistics should include the coefficient of determination (r^2) and the least-squares estimators of the slope (0.331) and the intercept (1.867).

The second column gives the degrees of freedom, usually abbreviated df. As we have discussed, the degrees of freedom depend on how many pieces of independent information are available to estimate the particular sum of squares. If the sample size is n, there is 1 degree of freedom associated with the regression model (specifically the slope), and $(n - 2)$ degrees of freedom associated with the error. The total degrees of freedom is $(1 + n - 2) = (n - 1)$. The total is only $(n - 1)$ because 1 degree of freedom was used in estimating the grand mean, $\overline{Y}$.

The third column gives the sum of squares (*SS*) associated with particular source of variation. In the expanded Table 9.1, we showed the formulas used, but in published tables (e.g., Table 9.2), only the numerical values for the sums of squares would be given.

The fourth column gives the mean square (*MS*), which is simply the sum of squares divided by its corresponding degrees of freedom. This division is analogous to calculating a simple variance by dividing SS_Y by $(n - 1)$.

The fifth column is the expected mean square. This column is not presented in published ANOVA tables, but it is very valuable because it shows exactly what is being estimated by each of the different mean squares. It is these expectations that are used to formulate hypothesis tests in ANOVA.

The sixth column is the calculated F-ratio. The F-ratio is a ratio of two different mean square values.

The final column gives the tail probability value corresponding to the particular F-ratio. Specifically, this is the probability of obtaining the observed F-ratio (or a larger one) if the null hypothesis is true. For the simple linear regression model, the null hypothesis is that $\beta_1 = 0$, implying no functional relationship between the X and Y variables. The probability value depends on the size of the F-ratio and on the number of degrees of freedom associated with the numerator and denominator mean squares. The probability values can be looked up in statistical tables, but they are usually printed as part of the standard regression output from statistics packages.

In order to understand how the F-ratio is constructed, we need to examine the expected mean square values. The expected value of the regression mean square is the sum of the regression error variance *and* a term that measures the regression slope effect:

$$E(MS_{reg}) = \sigma^2 + \beta_1^2 \sum_{i=1}^{n} X_i^2 \quad (9.22)$$

In contrast, the expected value of the residual mean square is simply the regression error variance:

$$E(MS_{resid}) = \sigma^2 \quad (9.23)$$

Now we can understand the logic behind the construction of the F-ratio. The F-ratio for the regression test uses the regression mean square in the numerator and the residual mean square in the denominator. If the true regression slope is zero, the second term in Equation 9.22 (the numerator of the F-ratio) also will equal zero. As a consequence, Equation 9.22 (the numerator of the F-ratio) and Equation 9.23 (the denominator of the F-ratio) will be equal. In other words, if the slope of the regression (β_1) equals zero, the expected value of the F-ratio will be 1.0. For a given amount of error variance, the steeper the regression slope, the larger the F-ratio. Also, for a given slope, the smaller the error variance, the larger the F-ratio. This also makes intuitive sense because the smaller the error variance, the more tightly clustered the data are around the fitted regression line.

The interpretation of the *P*-value follows along the lines developed in Chapter 4. The larger the F-ratio (for a given sample size and model), the smaller the *P*-value. The smaller the *P*-value, the more unlikely it is that the observed F-ratio would have been found if the null hypothesis were true. With *P*-values less than the standard 0.05 cutoff, we reject the null hypothesis and conclude that the regression model explains more variation than could be accounted for by chance alone. For the Galápagos data, the F-ratio = 21.048. The numerator of this F-ratio is 21 times larger than the denominator, so the variance accounted for by the regression is much larger than the residual variance. The corresponding *P*-value = 0.0003.

Other Tests and Confidence Intervals

As you might suspect, all hypothesis tests and confidence intervals for regression models depend on the variance of the regression ($\hat{\sigma}^2$). From this, we can calculate other variances and significance tests (Weisberg 1980). For example, the variance of the estimated intercept is

$$\hat{\sigma}^2_{\hat{\beta}_0} = \hat{\sigma}^2 \left(\frac{1}{n} + \frac{\bar{X}^2}{SS_X} \right) \quad (9.24)$$

An F-ratio also can be constructed from this variance to test the null hypothesis that $\beta_0 = 0.0$.

Notice that the intercept of the regression line is subtly different from the intercept that is calculated when the model has a slope of zero (Equation 9.16):

$$Y_i = \beta_0 + \varepsilon_i$$

If the model has a slope of 0, the expected value of the intercept is simply the average of the Y_i values:

$$E(\beta_0) = \bar{Y}$$

However, for the regression model with a slope term, the intercept is the expected value when $X_i = 0$, that is,

$$E(\beta_0) = \hat{Y}_i \mid (X_i = 0)$$

A simple 95% confidence interval can be calculated for the intercept as

$$\hat{\beta}_0 - t_{(\alpha,n-2)}\hat{\sigma}_{\hat{\beta}_0} \leq \beta_0 \leq \hat{\beta}_0 + t_{(\alpha,n-2)}\hat{\sigma}_{\hat{\beta}_0} \quad (9.25)$$

where α is the (two-tailed) probability level ($\alpha = 0.025$ for a 95% confidence interval), n is the sample size, t is the table value from the t-distribution for the specified α and n, and $\hat{\sigma}_{\hat{\beta}_0}$ is calculated as the square root of Equation 9.24.

Similarly, the variance of the slope estimator is

$$\hat{\sigma}^2_{\hat{\beta}_1} = \frac{\hat{\sigma}^2}{SS_X} \quad (9.26)$$

and the corresponding confidence interval for the slope is

$$\hat{\beta}_1 - t_{(\alpha,n-2)}\hat{\sigma}_{\hat{\beta}_1} \leq \beta_1 \leq \hat{\beta}_1 + t_{(\alpha,n-2)}\hat{\sigma}_{\hat{\beta}_1} \quad (9.27)$$

For the Galápagos data, the 95% confidence interval for the intercept (β_0) is 1.017 to 1.623, and the 95% confidence interval for the slope (β_1) is 0.177 to 0.484 (remember these are $\log_{10}$ values). Because neither of these confidence intervals bracket 0.0, the corresponding F-tests would cause us to reject the null hypothesis that both β_0 and β_1 equal 0. If β_1 equals 0, the regression line is horizontal, and the dependent variable [$\log_{10}$(species number)] does not systematically increase with changes in the independent variable [$\log_{10}$(island area)]. If β_0 equals 0, the dependent variable takes on a value of 0 when the independent variable is 0.

Although the null hypothesis has been rejected in this case, the observed data do not fall perfectly in a straight line, so there is going to be uncertainty associated with any particular value of the X variable, For example, if you were to repeatedly sample different islands that were identical in island area (X), there would be variation among them in the (log-transformed) number of species recorded.[7] The variance of the fitted value $\hat{Y}$ is

$$\hat{\sigma}^2_{(\hat{Y}|X)} = \hat{\sigma}^2\left(\frac{1}{n} + \frac{(X_i - \bar{X})^2}{SS_X}\right) \tag{9.28}$$

and a 95% confidence interval is

$$\hat{Y} - t_{(\alpha,n-2)}\hat{\sigma}_{(\hat{Y}|X)} \leq \hat{Y} \leq \hat{Y} + t_{(\alpha,n-2)}\hat{\sigma}_{(\hat{Y}|X)} \tag{9.29}$$

This confidence interval does not form a parallel band that brackets the regression line (Figure 9.4). Rather, the confidence interval flares out the farther away we move from $\bar{X}$ because of the term $(X_i - \bar{X})^2$ in the numerator of Equation 9.28. This widening confidence interval makes intuitive sense. The closer we are to the center of the cloud of points, the more confidence we have in estimating Y from repeated samples of X. In fact, if we choose $\bar{X}$ as the fitted

[7] Unfortunately, there is a mismatch between the statistical framework of random sampling and the nature of our data. After all, there is only one Galápagos archipelago, which has evolved a unique flora and fauna, and there are no multiple replicate islands that are identical in area. It seems dubious to treat these islands as samples from some larger sample space, which itself is not clearly defined—would it be volcanic islands? tropical Pacific islands? isolated oceanic islands? We could think of the data as a sample from the Galápagos, except that the sample is hardly random, as it consists of all the large major islands in the archipelago. There are a few additional islands from which we could collect data from, but these are considerably smaller, with very few species of plants and animals. Some islands are so small as to be essentially empty, and these 0 data (for which we cannot simply take logarithms) have important effects on the shape of the species–area relationship (Williams 1996). The fitted regression line for all islands may not necessarily be the same as for sets of large versus small islands. These problems are not unique to species–area data. In any sampling study, we must struggle with the fact that the sample space is not clearly defined and the replicates we collect may be neither random nor independent of one another, even though we use statistics that rely on these assumptions.

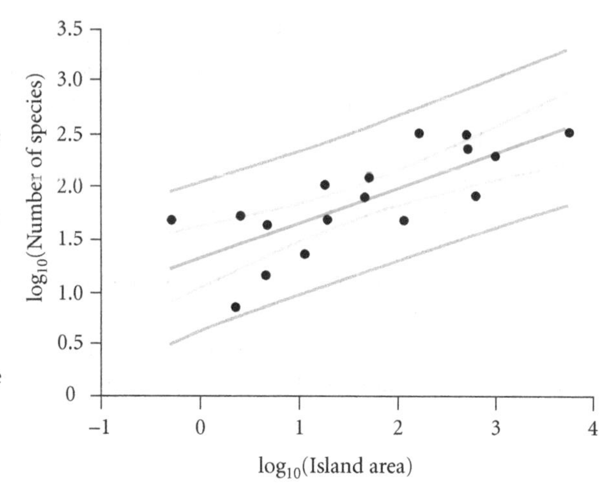

Figure 9.4 Regression line (dark blue), 95% confidence interval (light blue lines), and 95% prediction interval (gray lines) for a log-log regression of Galápagos plant species number against island area (see Table 8.2). The confidence interval describes the uncertainty in the collected data, whereas the prediction interval is used to evaluate new data not yet collected or included in the analysis. Notice that the confidence intervals flare outward as we move further away from the average island area in the collected data. The flaring confidence intervals reflect increasing uncertainty about predictions for X variables that are very different from those already collected. Notice also that these intervals are described on a logarithmic scale. Back-transformation (e^Y) is necessary to convert the units to numbers of species. Back-transformed confidence intervals are very wide and asymmetric around the predicted value.

value, the standard deviation of the fitted value is equivalent to a standard error of the average fitted value (see Chapter 3):

$$\hat{\sigma}_{(\hat{Y}|X)} = \hat{\sigma}/\sqrt{n}$$

The variance is minimized here because there are observed data both above and below the fitted value. However, as we move away from $\bar{X}$, there are fewer data in the neighborhood, so the prediction becomes more unreliable.

A useful distinction can be made here between **interpolation** and **extrapolation**. Interpolation is the estimation of new values that are within the range of the data we have collected, whereas extrapolation means estimating new values beyond the range of data (see Figure 9.2). Equation 9.29 ensures that confidence intervals for fitted values always will be smaller for interpolation than for extrapolation.

Finally, suppose we discover a new island in the archipelago, or sample an island that was not included in our original census. This new observation is designated

$$(\tilde{X}, \tilde{Y})$$

We will use the fitted regression line to construct a **prediction interval** for the new value, which is subtly different from constructing a confidence interval for a fitted value. The confidence interval brackets uncertainty about an estimate based on the available data, whereas the prediction interval brackets uncertainty about an estimate based on new data. The variance of the prediction for a single fitted value is

$$\sigma^2_{(\tilde{Y}|\tilde{X})} = \hat{\sigma}^2 \left(1 + \frac{1}{n} + \frac{(\tilde{X} - \bar{X})^2}{SS_X} \right) \tag{9.30}$$

and the corresponding prediction interval is

$$\tilde{Y} - t_{(\alpha,n-2)}\hat{\sigma}_{(\tilde{Y}|\tilde{X})} \leq \tilde{Y} \leq \tilde{Y} + t_{(\alpha,n-2)}\hat{\sigma}_{(\tilde{Y}|\tilde{X})} \quad (9.31)$$

This variance (Equation 9.30) is larger than the variance for the fitted value (Equation 9.28). Equation 9.30 includes variability both from the error associated with the new observation, and from the uncertainty in the estimates of the regression parameters from the earlier data.[8]

To close this section, we note that it is very easy (and seductive) to make a point prediction from a regression line. However, for most kinds of ecological data, the uncertainty associated with that prediction usually is too large to be useful. For example, if we were to propose a nature reserve of 10 square kilometers based only on the information contained in the Galápagos species–area relationship, the point prediction from the regression line is 45 species. However, the 95% prediction interval for this forecast is 9 to 229 species. If we increased the area 10-fold to 100 square kilometers, the point prediction is 96 species, with a range of 19 to 485 species (both ranges based on a back-transformed prediction interval from Equation 9.31).[9] These wide prediction intervals not only are typical for species–area data but also are typical for most ecological data.

Assumptions of Regression

The linear regression model that we have developed relies on four assumptions:

1. **The linear model correctly describes the functional relationship between *X* and *Y*.** This is the fundamental assumption. Even if the overall relationship is non-linear, a linear model may still be appropriate over a limited range of the *X* variable (see Figure 9.2). If the assumption of linearity is violated, the estimate of σ^2 will be inflated because it will include both a random error and a fixed error component; the latter represents the difference between the true function and the linear one that has been fit to the data. And, if the true relationship is not linear,

[8] A final refinement is that Equation 9.31 is appropriate only for a single prediction. If multiple predictions are made, the α value must be adjusted to create a slightly broader **simultaneous prediction interval**. Formulas also exist for **inverse prediction intervals**, in which a range of *X* values is obtained for a single value of the *Y* variable (see Weisberg 1980).

[9] To be picky, a back-transformed point estimates the median, not the mean, of the distribution. However, back-transformed confidence intervals are not biased because the quantiles (percentage points) translate across transformed scales.

predictions derived from the model will be misleading, particularly when extrapolated beyond the range of the data.

2. **The X variable is measured without error.** This assumption allows us to isolate the error component entirely as random variation associated with the response variable (Y). If there is error in the X variable, the estimates of the slope and intercept will be biased. By assuming there is no error in the X variable, we can use the least-squares estimators, which minimize the vertical distance between each observation and its predicted value (the d_i's in Figure 9.3). With errors in both the X and Y variable, one strategy would be to minimize the perpendicular distance between each observation and the regression line. This so-called Model II regression is commonly used in principle components and other multivariate techniques (see Chapter 12). However, because the least-squares solutions have proven so efficient and are used so commonly, this assumption often is quietly ignored.[10]
3. **For any given value of X, the sampled Y values are independent with normally distributed errors.** The assumption of normality allows us to use parametric theory to construct confidence intervals and hypothesis tests based on the F-ratio. Independence, of course, is the critical assumption for all sample data (see Chapter 6), even though it often is violated to an unknown degree in observational studies. If you suspect that the value Y_i influences the next observation you collect (Y_{i+1}), a time-series analysis may remove the correlated components of the error variation. Time-series analysis is discussed briefly in Chapter 6.
4. **Variances are constant along the regression line.** This assumption allows us to use a single constant σ^2 for the variance of regression line. If variances depend on X, then we would require a variance function, or an entire family of variances, each predicated on a particular value of X. Non-constant variances are a common problem in regression that can be recognized through diagnostic plots (see the following section) and sometimes remedied through transformations of the original X or Y variables (see Chapter 8). However, there is no guarantee that a transformation will linearize the relationship and generate constant variances. In such cases, other methods (such as non-linear regression, described

[10] Although it is rarely mentioned, we make a similar assumption of no errors for the categorical X variable in an analysis of variance (see Chapter 10). In ANOVA, we have to assume that individuals are correctly assigned to groups (e.g., no errors in species identification) and that all individuals within a group have received an identical treatment (e.g., all replicates in a "low pH" treatment received exactly the same pH level).

later in this chapter) or generalized linear models (McCullagh and Nelder 1989) should be used.

If all four of these assumptions are met, the least-squares method provides **unbiased estimators** of all the model parameters. They are unbiased because repeated samples from the same population will yield slope and intercept estimates that, on average, are the same as the true underlying slope and intercept values for the population.

Diagnostic Tests For Regression

We have now seen how to obtain least-squares estimates of linear regression parameters and how to test hypotheses about those parameters and construct appropriate confidence intervals. However, a regression line can be forced through any set of $\{X,Y\}$ data, regardless of whether the linear model is appropriate. In this section, we present some diagnostic tools for determining how well the estimated regression line fits the data. Indirectly, these diagnostics also help you to evaluate the extent to which the data meet the assumptions of the model.

The most important tool in diagnostic analysis is the set of residuals, $\{d_i\}$, which represents the differences between the observed values (Y_i) and the values predicted by the regression model ($\hat{Y}_i$) of Equation 9.4. The residuals are used to estimate the regression variance, but they also provide important information about the fit of the model to the data.

Plotting Residuals

Perhaps the single most important graph for diagnostic analysis of the regression model is the plot of the residuals (d_i) versus the fitted values ($\hat{Y}_i$). If the linear model fits the data well, this **residual plot** should exhibit a scatter of points that approximately follow a normal distribution and are completely uncorrelated with the fitted values (Figure 9.5A). In Chapter 11, we will explain the Kolmogorov-Smirnov test, which can be used to formally compare the residuals to a normal distribution (see Figure 11.4).

Two kinds of problems can be spotted in residual plots. First, if the residuals themselves are correlated with the fitted values, it means that the relationship is not really linear. The model may be systematically over-estimating or under-estimating $\hat{Y}_i$ for high values of the X variable (Figure 9.5B). This can happen, for example, when a straight line is forced through data that really represent an asymptotic, logarithmic, or other non-linear relationship. If the residuals first rise above the fitted values then fall below, and then rise above again, the data

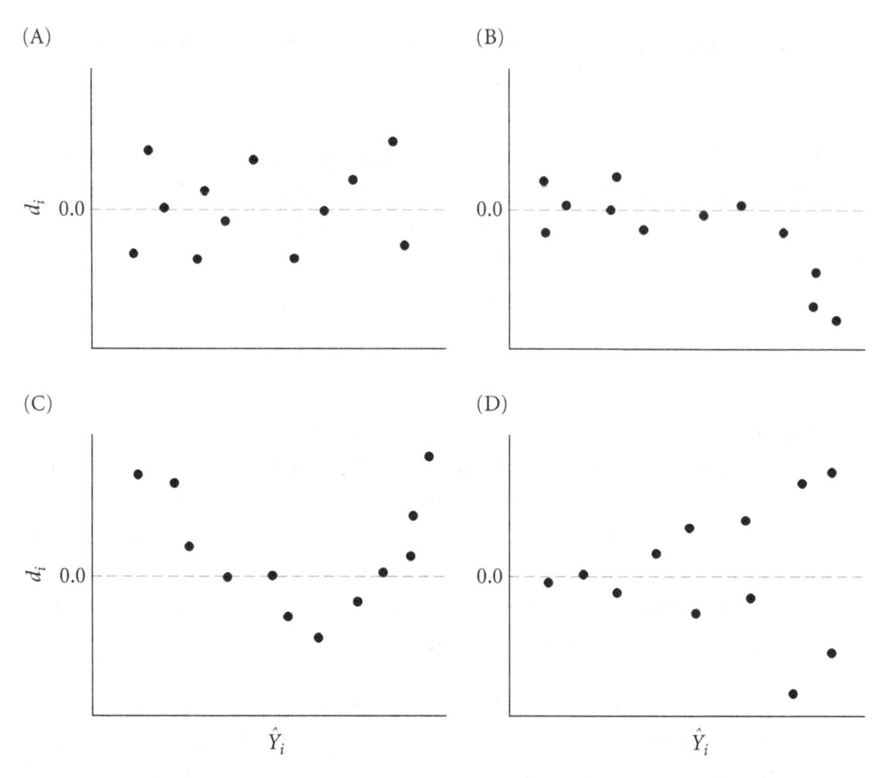

Figure 9.5 Hypothetical patterns for diagnostic plots of residuals (d_i) versus fitted values ($\hat{Y}_i$) in linear regression. (A) Expected distribution of residuals for a linear model with a normal distribution of errors. If the data are well fit by the linear model, this is the pattern that should be found in the residuals. (B) Residuals for a non-linear fit; here the model systemically overestimates the actual Y value as X increases. A mathematical transformation (e.g., logarithm, square root, or reciprocal) may yield a more linear relationship. (C) Residuals for a quadratic or polynomial relationship. In this case, large positive residuals occur for very small and very large values of the X variable. A polynomial transformation of the X variable (X^2 or some higher power of X) may yield a linear fit. (D) Residuals with heteroscedascity (increasing variance). In this case, the residuals are neither consistently positive nor negative, indicating that the fit of the model is linear. However, the average size of the residual increases with X (heteroscedascity), suggesting that measurement errors may be proportional to the size of the X variable. A logarithmic or square root transformation may correct this problem. Transformations are not a panacea for regression analyses and do not always result in linear relationships.

may indicate a quadratic rather than a linear relationship (Figure 9.5C). Finally, if the residuals display an increasing or decreasing funnel of points when plotted against the fitted values (Figure 9.5D), the variance is **heteroscedastic**: either increasing or decreasing with the fitted values. Appropriate data transformations, discussed in Chapter 8, can address some or all of these problems. Resid-

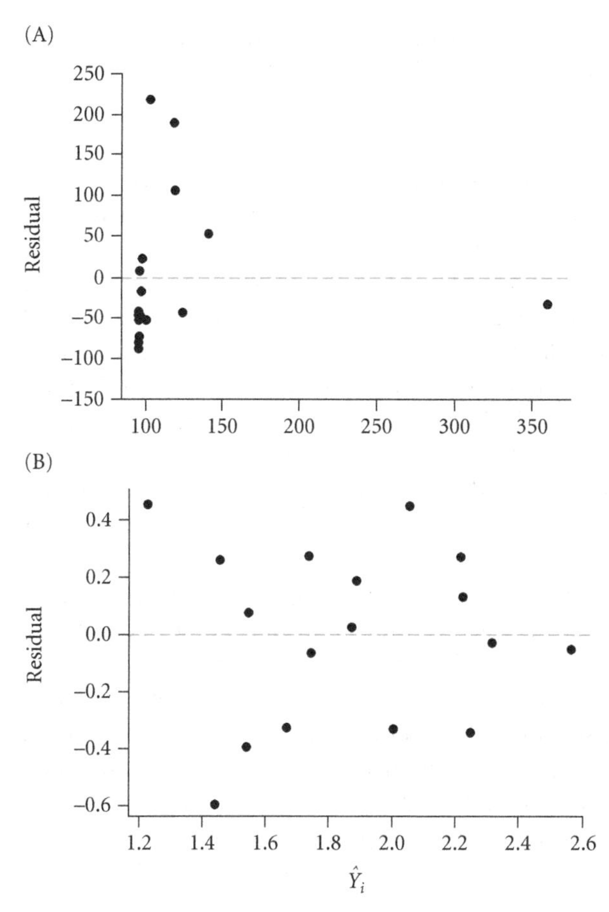

Figure 9.6 Residual plot for species–area relationship for plants of Galápagos Islands (see Table 8.2). In a residual plot, the *x*-axis is the predicted value from the regression equation ($\hat{Y}_i$), and the *y*-axis is the residual, which is the difference between the observed and fitted values. In a dataset that matches the predictions of the regression model, the residual plot should be a normally distributed cloud of points centered on the average value of 0 (see Figure 9.5a). (A) Residual plot for regression of untransformed data (see Figure 8.5). There are too many negative residuals at small fitted values, and the distribution of residuals does not appear normal. (B) Residual plot for regression of the same data after a $\log_{10}$ transformation of both the *x*- and *y*-axes (see Figures 8.6 and 9.3). The transformation has considerably improved the distribution of the residuals, which no longer deviate systematically at large or small fitted values. Chapter 11 describes the Kolmogorov-Smirnov test, which can be used to test the residuals for deviations from a normal distribution (see Figure 11.4).

ual plots also may highlight outliers, points that fall much further from the regression prediction than most of the other data.[11]

To see the effects of transformations, compare Figures 9.6A and 9.6B, which illustrate residual plots for the Galápagos data before and after $\log_{10}$ transformation. Without the transformation, there are too many negative residuals, and they are all clustered around very small values of $\hat{Y}_i$ (Figure 9.6A). After trans-

[11] **Standardized residuals** (also called **studentized residuals**) also are calculated by many statistics packages. Standardized residuals are scaled to account for the distance of each point from the center of the data. If the residuals are not standardized, data points far away from $\bar{X}$ appear to be relatively well-fit by the regression, whereas points closer to the center of the data cloud appear to not fit as well. Standardized residuals also can be used to test whether particular observations are statistically significant outliers. Other residual distance measures include Cook's distance and leverage values. See Sokal and Rohlf (1995) for details.

formation, the positive and negative residuals are evenly distributed and not associated with $\hat{Y}_i$ (Figure 9.6B).

Other Diagnostic Plots

Residuals can be plotted not only against fitted values, but also against other variables that might have been measured. The idea is to see whether there is any variation lurking in the errors that can be attributed to a systematic source. For example, we could plot the residuals from Figure 9.6B against some measure of habitat diversity on each island. If the residuals were positively correlated with habitat diversity, then islands that have more species than expected on the basis of area usually have higher habitat diversity. In fact, this additional source of systematic variation can be included in a more complex multiple regression model, in which we fit coefficients for two or more predictor variables. Plotting residuals against other predictor variables is often a more simple and reliable exploratory tactic than assuming a model with linear effects and linear interactions.

It also may be informative to plot the residuals against time or data collection order. This plot may indicate altered measurement conditions during the time period that the data were collected, such as increasing temperatures through the course of the day in an insect behavior study, or a pH meter with a weakening battery that gives biased results for later measurements. These plots again remind us of the surprising problems that can arise by not using randomization when we are collecting data. If we collect all of our behavior measurements on one insect species in the morning, or measure pH first in all of the control plots, we have introduced an unexpected confounding source of variation in our data.

The Influence Function

Residual plots do a good job of revealing non-linearity, heteroscedascity, and outliers. However, potentially more insidious are influential data points. These data may not show up as outliers, but they do have an inordinate influence on the slope and intercept estimates. In the worst situation, influential data points that are far removed from the cloud of typical data points can completely "swing" a regression line and dominate the slope estimate (see Figure 7.3).

The best way to detect such points is to plot an **influence function**. The idea is simple, and is actually a form of statistical jackknifing (see Footnote 2 in Chapter 5; see also Chapter 12). Take the first of the n replicates in your dataset and delete it from the analysis. Recompute the slope, intercept, and probability value. Now replace that first data point and remove the second data point. Again calculate the regression statistics, replace the datum, and continue through the data list. If you have n original data points, you will end up with n different regression analyses, each of which is based on a total of $(n - 1)$ points. Now take the slope and intercept estimates for each of those analyses and plot them together in a single scatter plot (Figure 9.7). On this same plot, graph the slope and intercept estimates from the complete data set. This graph of intercept versus slope estimates will itself always have a negative slope because as the regression line tips up, the slope increases and the intercept decreases. It may also be informative to create a histogram of r^2 values or tail probabilities that are associated with each of those regression models.

The influence function illustrates how much your estimated regression parameters would have changed just by the exclusion of a single datum. Ideally, the jackknifed parameter estimates should cluster tightly around the estimates from the full dataset. A cluster of points would suggest that the slope and intercept values are stable and would not change greatly with the deletion (or addition) of a single data point. On the other hand, if one of the points is very distant from the cluster, the slope and intercept are highly influenced by that sin-

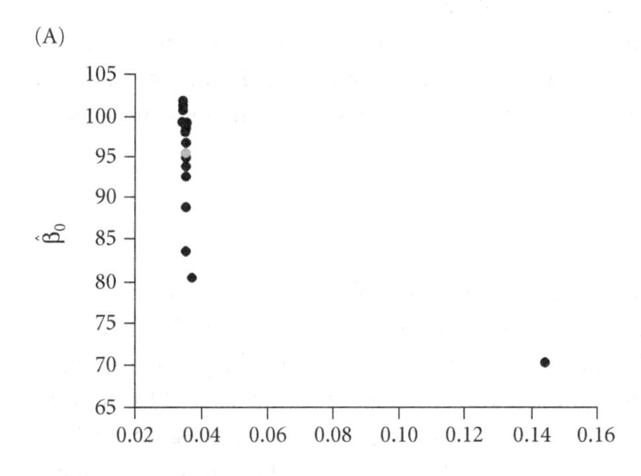

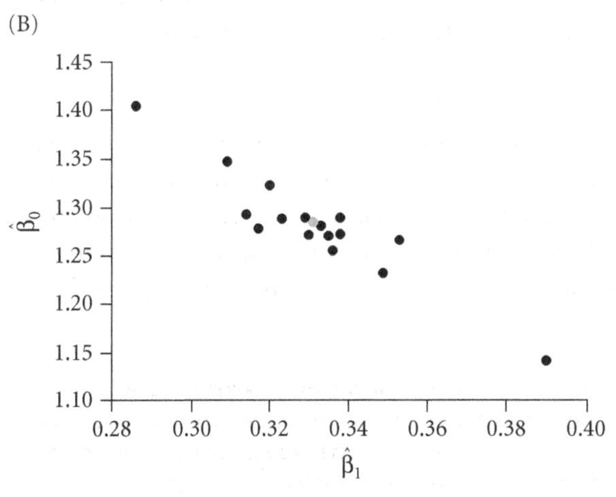

Figure 9.7 Influence function for species-area linear regression of Galápagos island plant data (see Table 8.2). In an influence function plot, each point represents the recomputed slope ($\hat{\beta}_1$) and intercept ($\hat{\beta}_0$) following deletion of a single observation from the dataset. The blue point is the slope and intercept estimate for the complete data set. (A) In the regression model with untransformed data, the observation for the large island Albemarle (see Figure 8.5) has a very large influence on the slope and intercept, generating the extreme point in the lower right-hand corner of the graph. (B) Influence function calculated for regression line fitted to $\log_{10}$ transformed data. After the logarithmic transformation (see Figures 8.6 and 9.3), there is now a more homogenous cloud of points in the influence function. Although the slope and intercept will change following the deletion of each observation, no single data point has an extreme influence on the calculated slope and intercept. In this dataset, the logarithmic transformation not only improves the linear fit of the data, but it also stabilizes the slope and intercept estimates, so they are not dominated by one or two influential data points.

gle datum, and we should be careful about what conclusions we draw from this analysis. Similarly, when we examine the histogram of probability values, we want to see all of the observations clustered nicely around the *P*-value that was estimated for the full dataset. Although we expect to see some variation, we will be especially troubled if one of the jackknifed samples yields a *P*-value very different from all the rest. In such a case, rejecting or failing to reject the null hypothesis hinges on a single observation, which is especially dangerous.

For the Galápagos data, the influence function highlights the importance of using an appropriate data transformation. For the untransformed data, the largest island in the dataset dominates the estimate of β_1. If this point is deleted, the slope increases from 0.035 to 0.114 (Figure 9.7A). In contrast, the influence function for the transformed data shows much more consistency, although $\hat{\beta}_1$ still ranged from 0.286 to 0.390 following deletion of a single datum (Figure 9.7B). For both data sets, probability values were stable and never increased above $P = 0.023$ for any of the jackknife analyses.

Using the jackknife in this way is not unique to regression. Any statistical analysis can be repeated with each datum systematically deleted. It can be a bit time-consuming,[12] but it is an excellent way to assess the stability and general validity of your conclusions.

Monte Carlo and Bayesian Analyses

The least-squares estimates and hypotheses tests are representative of classic frequentist parametric analysis, because they assume (in part) that the error term in the regression model is a normal random variable; that the parameters have true, fixed values to be estimated; and that the *P*-values are derived from probability distributions based on infinitely large samples. As we emphasized in Chapter 5, Monte Carlo and Bayesian approaches differ in philosophy or implementation (or both), but they also can be used for regression analysis.

Linear Regression Using Monte Carlo Methods

For the Monte Carlo approach, we still use all of the least-squares estimators of the model parameters. However, in a Monte Carlo analysis, we relax the assumption of normally distributed error terms, so that significance tests are no longer based on comparisons with F-ratios. Instead, we randomize and resample the data and simulate the parameters directly to estimate their distribution.

[12] Actually, there are clever matrix solutions to obtaining the jackknifed values, although these are not implemented in most computer packages. The influence function is most important for small data sets that can be analyzed quickly by using case selection options that are available in most statistical software.

For linear regression analysis, what is an appropriate randomization method? We can simply reshuffle the observed Y_i values and pair them randomly with one of the X_i values. For such a randomized dataset, we can then calculate the slope, intercept, r^2, or any other regression statistics using the methods described in the previous section. The randomization method effectively breaks up any covariation between the X and Y variables and represents the null hypothesis that the X variable (here, island area) has no effect on the Y variable (here, species richness). Intuitively, the expected slope and r^2 values should be approximately zero for the randomized data sets, although some values will differ by chance. For the Galápagos data, 5000 reshufflings of the data generated a set of slope values with a mean of –0.002 (very close to 0), and a range of –0.321 to 0.337 (Figure 9.8). However, only one of the simulated values (0.337) exceeded the observed slope of 0.331. Therefore, the estimated tail probability is 1/5000 =

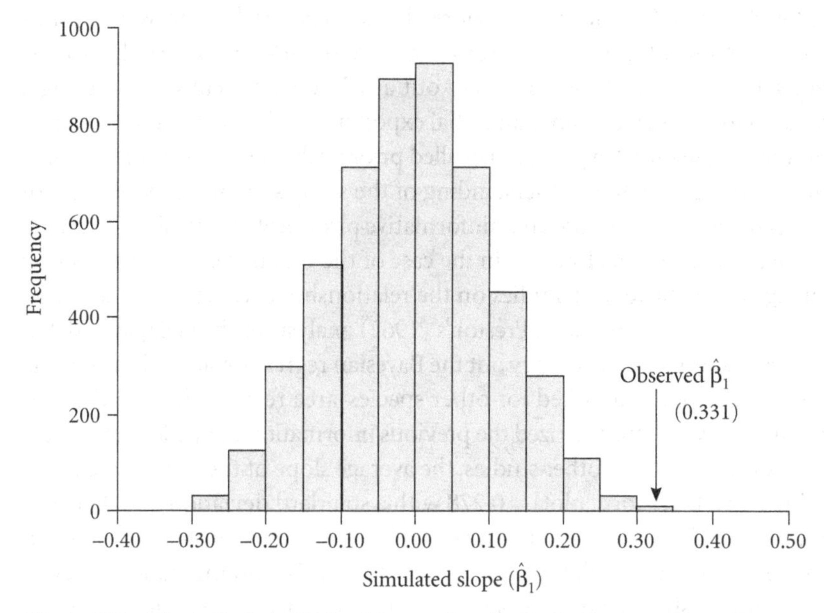

Figure 9.8 Monte Carlo analysis of slope values for the log-log transformed species–area relationship of plant species on the Galápagos Islands (see Table 8.2; see also Figures 8.6 and 9.3). Each observation in the histogram represents the least-squares slope of a simulated dataset in which the observed X and Y values were randomly reshuffled. The histogram illustrates the distribution of 5000 such random datasets. The observed slope from the original data is plotted as an arrow on the histogram. The observed slope of 0.331 exceeded all but 1 of the 5000 simulated values. Therefore, the estimated tail probability of obtaining a result this extreme under the null hypothesis is $P = 1/5000 = 0.0002$. This probability estimate is consistent with the probability calculated from the F-ratio for the standard regression test of these data ($P = 0.0003$; see Table 9.2).

0.0002. By comparison, the parametric analysis gave essentially the same result, with a *P*-value equal to 0.0003 for $F_{1,15} = 21.118$ (see Table 9.2).

One interesting result from the Monte Carlo analysis is that the *P*-values for the intercept, slope, and r^2 tests are all identical. Naturally, the observed parameter estimates and simulated distributions for each of these variables are all very different. But why should the tail probabilities be the same in each case? The reason is that β_0, β_1, and r^2 are not algebraically independent of one another; the randomized values reflect that interdependence. Consequently, the tail probabilities come out the same for these fitted values when they are compared to distributions based on the same randomized datasets. This doesn't happen in parametric analysis because *P*-values are not determined by reshuffling the observed data.

Linear Regression Using Bayesian Methods

Our Bayesian analysis begins with the assumption that the parameters of the regression equation do not have true values to be estimated, but rather are random variables (see Chapter 5). Therefore, the estimates of the regression parameters—the intercept, slope, and error term—are reported as probability distributions, not as point values. To carry out any Bayesian analysis, including a Bayesian regression, we require an initial expectation of the shapes of these distributions. These initial expectations, called prior probability distributions, come from previous research or understanding of the study system. In the absence of previous research, we can use an uninformative prior probability distribution, as described in Chapter 5. However, in the case of the Galápagos data, we can take advantage of the history of studies on the relationship between island area and species richness that pre-dated Preston's (1962) analysis of the Galápagos data.

For this example, we will carry out the Bayesian regression analysis using data that Preston (1962) compiled for other species-area relationships in the ecological literature. He summarized the previous information available (in his Table 8), and found that, for 6 other studies, the average slope of the species-area relationship (on a $\log_{10}$-$\log_{10}$ plot) = 0.278 with a standard deviation of 0.036. Preston (1962) did not report his estimates of the intercept, but we calculated it from the data he reported: the mean intercept = 0.854 and the standard deviation = 0.091. Lastly, we need an estimate of the variance of the regression error ε, which we calculated from his data as 0.234.

With this prior information, we use Bayes Theorem to calculate new distributions (the **posterior probability distributions**) on each of these parameters. Posterior probability distributions take into account the previous information plus the data from the Galápagos Islands. The results of this analysis are

$$\hat{\beta}_0 \sim N(1.863, 0.157)$$
$$\hat{\beta}_1 \sim N(0.328, 0.145)$$
$$\hat{\sigma}^2 = 0.420$$

In words, the intercept is a normal random variable with mean = 1.863 and standard deviation 0.157, the slope is a normal random variable with mean = 0.328 and standard deviation = 0.145, and the regression variance is a normal random variable with mean 0 and variance 0.420.[13]

How do we interpret these results? The first thing to re-emphasize is that in a Bayesian analysis, the parameters are considered to be random variables—they do not have fixed values. Thus, the linear model (see Equation 9.3) needs to be re-expressed as

$$Y_i \sim N(\beta_0 + \beta_1 X_i, \sigma^2) \quad (9.32)$$

In words, "each observation Y_i is drawn from a normal distribution with mean $= \beta_0 + \beta_1 X_i$ and variance $= \sigma^2$." Similarly, both the slope β_0 and the intercept β_1 are themselves normal random variables with estimated means $\hat{\beta}_0$ and $\hat{\beta}_1$ and corresponding estimated variances $\hat{s}_0^2$ and $\hat{s}_1^2$.

How do these estimates apply to our data? First, the regression line has a slope of 0.328 (the expected value of the β_1) and an intercept of 1.863. The slope differs only slightly from the least-squares regression line shown in Figure 9.4, which has a slope of 0.331. The intercept is substantially higher than the least-squares fit, suggesting that the larger number of small islands contributed more to this regression than the few large islands.

Second, we could produce a 95% credibility interval. The computation essentially is the same as that shown in Equations 9.25 and 9.27, but the interpretation of the Bayesian credibility interval is that the mean value would lie within the credibility interval 95% of the time. This differs from the interpretation of the 95% confidence interval, which is that the confidence interval will include the true value 95% of the time (see Chapter 3).

Third, if we wanted to predict the species richness for a given island size X_i, the *predicted* number of species would be found by first drawing a value for the slope β_1 from a normal distribution with mean = 0.328 and standard deviation = 0.145, and then multiplying this value by X_i. Next, we draw a value for the intercept β_0 from a normal distribution with mean = 1.863 and standard

[13] Alternatively, we could conduct this analysis without any prior information, as discussed in Chapter 4. This more "objective" Bayesian analysis uses uninformative prior probability distributions on the regression parameters. The results for the analysis using uninformative priors are $\hat{\beta}_0 \sim N(1.866, 0.084)$, $\hat{\beta}_1 \sim N(0.329, 0.077)$, and $\hat{\sigma}^2 = 0.119$. The lack of substantive difference between these results and those based on the informed prior distribution illustrates that good data overwhelm subjective opinion.

deviation = 0.157. Last, we add "noise" to this prediction by adding an error term drawn from a normal distribution with $X = 0$ and $\sigma^2 = 0.42$.

Finally, some words about hypothesis testing. There is no way to calculate a *P*-value from a Bayesian analysis. There is no *P*-value because we are not testing a hypothesis about how unlikely our data are given the hypothesis [as in the frequentist's estimate of $P(\text{data}|H_0)$] or how frequently we would expect to get these results given chance alone (as in the Monte Carlo analysis). Rather, we are estimating values for the parameters. Parametric and Monte Carlo analyses estimate parameter values too, but the primary goal often is to determine if the estimate is significantly different from 0. Frequentist analyses also can incorporate prior information, but it is still usually in the context of a binary (yes/no) hypothesis test. For example, if we have prior information that the slope of the species–area relationship, $\beta_1 = 0.26$, and our Galápagos data suggest that $\beta_1 = 0.33$, we can test whether the Galápagos data are "significantly different" from previously published data. If the answer is "yes," what do you conclude? And if the answer is "no"? A Bayesian analysis assumes that you're interested in using all the data you have to estimate the slope and intercept of the species area relationship.

Other Kinds of Regression Analyses

The basic regression model that we have developed just scratches the surface of the kinds of analyses that can be done with continuous predictor and response variables. We provide only a very brief overview to other kinds of regression analyses. Entire books have been written on each of these individual topics, so we cannot hope to give more than an introduction to them. Nevertheless, many of the same assumptions, restrictions, and problems that we have described for simple linear regression apply to these methods as well.

Robust Regression

The linear regression model estimates the slope, intercept, and variance by minimizing the residual sum of squares (*RSS*) (see Equation 9.5). If the errors follow a normal distribution, this residual sum of squares will provide unbiased estimates of the model parameters. The least-squares estimates are sensitive to outliers because they give heavier weights to large residuals. For example, a residual of 2 contributes $2^2 = 4$ to the *RSS*, but a residual of 3 contributes $3^2 = 9$ to the *RSS*. The penalty added to large residuals is appropriate, because these large values will be relatively rare if the errors are sampled from a normal distribution.

However, when true outliers are present—aberrant data points (including erroneous data) that were not sampled from the same distribution—they can seriously inflate the variance estimates. Chapter 8 discussed several methods for identifying and dealing with outliers in your data. But another strategy is to fit the model using a residual function other than least-squares, one that is not so sensitive to the presence of outliers. **Robust regression** techniques use different mathematical functions to quantify residual variation. For example, rather than squaring each residual, we could use its absolute value:

$$residual = \sum_{i=1}^{n} \left|(Y_i - \hat{Y}_i)\right| = \sum_{i=1}^{n} |d_i| \tag{9.33}$$

As with *RSS*, this *residual* is large if the individual derivations (d_i) are large. However, very large deviations are not penalized as heavily because their values are not squared. This weighting is appropriate if you believe the data are drawn from a distribution that has relatively fat tails (leptokurtic; see Chapter 3) compared to the normal distribution. Other measures can be used that give more or less weight to large deviations.

However, once we abandon *RSS*, we can no longer use the simple formulas to obtain estimates of the regression parameters. Instead, iterative computer techniques are necessary to find the combination of parameters that will minimize the residual (see Footnote 5 in Chapter 4).

To illustrate robust regression, Figure 9.9 shows the Galápagos data with an additional outlier point. A standard linear regression for this new dataset gives a slope of 0.233, 30% shallower than the actual slope of 0.331 shown in Figure 9.4. Two different robust regression techniques, regression using **least-trimmed squares** (Davies 1993) and regression using **M-estimators** (Huber 1981), were applied to the data shown in Figure 9.9. As in standard linear regression, these robust regression methods assume the X variable is measured without error.

Least-trimmed squares regression minimizes the residual sums of squares (see Equation 9.5) by trimming off some percentage of the extreme observations. For a 10% trimming—removing the largest decile of the residual sums of squares—the predicted slope of the species-area relationship for the data illustrated in Figure 9.9 is 0.283, a 17% improvement over the basic linear regression. The intercept for the least-trimmed squares regression is 1.432, somewhat higher than the actual intercept.

M-estimators minimize the residual:

$$residual = \sum_{i=1}^{n} \rho\left(\frac{Y_i - X_i b}{s}\right) + n\log s \tag{9.34}$$

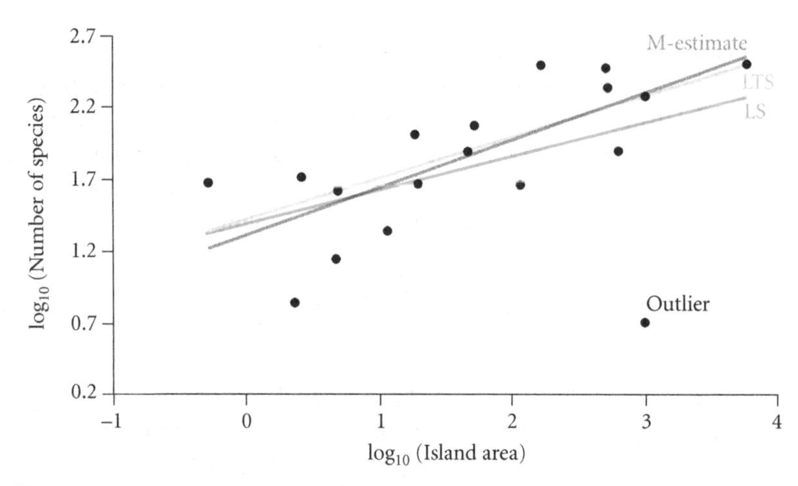

Figure 9.9 Three regressions for the species–area relationship of plant species on the Galápagos islands (see Table 8.2; see also Figures 8.6 and 9.3) with an added artificial outlier point. The standard linear regression (LS) is relatively sensitive to outliers. Whereas the original slope estimate was 0.331, the new slope is only 0.233, and has been dragged down by the outlier point. The least-trimmed squares (LTS) method discards the extreme 10% of the data (5% from the top and 5% from the bottom). The slope for the LTS regression is 0.283—somewhat closer to the original estimate of 0.331, although the intercept estimate is inflated to 1.432. The M-estimate weights the regression by the size of the residual, so that large outliers contribute less to the slope and intercept calculations. The M-estimator recovers the "correct" estimates for the slope (0.331) and the intercept (1.319), although the estimated variance is about double that of the linear regression that did not include the outlier. These robust regression methods are useful for fitting regression equations to highly variable data that include outliers.

where X_i and Y_i are values for predictor and response variables, respectively; $\rho = -\log f$, where f is a probability density function of the residuals ε that is scaled by a weight s, $[f(\varepsilon/s)]/s$, and b is an estimator of the slope β_1. The value for b that minimizes Equation 9.34 for a given value of s is a robust M-estimator of the slope of the regression line β_1. The M-estimator effectively weights the data points by their residuals, so that data points with large residuals contribute less to the slope estimate.

It is, of course, necessary to provide an estimate of s to solve this equation. Venables and Ripley (2002) suggest that s can be estimated by solving

$$\sum_{i=1}^{n} \psi\left(\frac{Y_i - X_i b}{s}\right)^2 = n \qquad (9.35)$$

where ψ is the derivative of ρ used in Equation 9.34. Fortunately, all these computations are handled easily in software such as S-Plus or R that have built-in robust regression routines (Venables and Ripley 2002).

M-estimate robust regression yields the correct slope of 0.331, but the variance around this estimate equals 0.25—more than three times as large as the estimate of the variance of the data lacking the outlier (0.07). The intercept is similarly correct at 1.319, but its variance is nearly twice as large (0.71 versus 0.43). Lastly, the overall estimate of the variance of the regression error equals 0.125, just over 30% larger than that of the simple linear regression on the original data.

A sensible compromise strategy is to use the ordinary least-squares methods for testing hypotheses about the statistical significance of the slope and intercept. With outliers in the data, such tests will be relatively conservative because the outliers will inflate the variance estimate. However, robust regression could then be used to construct prediction intervals. The only caveat is that if the errors are actually drawn from a normal distribution with a large variance, the robust regression is going to underestimate the variance, and you may be in for an occasional surprise if you make predictions from the robust regression model.

Quantile Regression

Simple linear regression fits a line through the center of a cloud of points, and is appropriate for describing a direct cause-and-effect relationship between the X and the Y variables. However, if the X variable acts as a limiting factor, it may impose a ceiling on the upper value of the Y variable. Thus, the X variable may control the maximum value of Y, but have no effect below this maximum. The result would be a triangle-shaped graph. For example, Figure 9.10 depicts the relationship between annual acorn biomass and an "acorn suitability index" measured for 43 0.2-hectare sample plots in Missouri (Schroeder and Vangilder 1997). The data suggest that low suitability constrains annual acorn biomass, but at high suitability, acorn biomass in a plot may be high or low, depending on other limiting factors.

A **quantile regression** minimizes deviations from the fitted regression line, but the minimization function is asymmetric: positive and negative deviations are weighted differently:

$$residual = \sum_{i=1}^{n} \left| Y_i - \hat{Y}_i \right| h \tag{9.36}$$

As in robust regression, the function minimizes the absolute value of the deviations, so it isn't as sensitive to outliers. However, the key feature is the multiplier h. The value of h is the quantile that is being estimated. If the deviation inside the absolute value sign is positive, it is multiplied by h. If the deviation is negative, it is multiplied by $(1.0 - h)$. This asymmetric minimization fits a regression line through the upper regions of the data for large h, and through the lower regions for small h. If $h = 0.50$, the regression line passes through the center of the cloud of data and is equivalent to a robust regression using Equation 9.5.

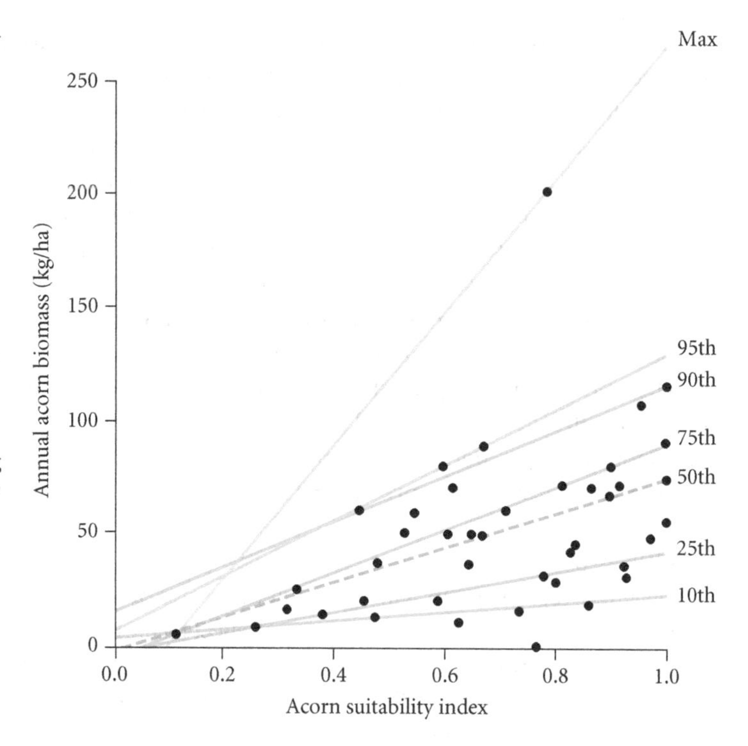

Figure 9.10 Illustration of quantile regression. The X variable is a measure of the acorn suitability of a forest plot, based on oak forest characteristics for $n = 43$ 0.2-ha sample plots in Missouri (data from Schroeder and Vangilder 1997). The Y variable is the annual acorn biomass in the plot. Solid lines correspond to quantile regression slopes. The dashed line (= 50th percentile) is the standard least-squares regression that passes through the center of the cloud of points. Quantile regression is appropriate when there is a limiting factor that sets an upper ceiling on a response variable. (After Cade et al. 1999.)

The result is a family of regression lines that characterize the upper and lower boundaries of the data set (Figure 9.10). If the X variable is the only factor affecting the Y variable, these quantile regressions will be roughly parallel to one another. However, if other variables come into play, the slopes of the regressions for the upper quantiles will be much steeper than for the standard regression line. This pattern would be indicative of an upper boundary or a limiting factor at work (Cade and Noon 2003).

Some cautions are necessary with quantile regression. The first is that the choice of which quantile to use is rather arbitrary. Moreover, the more extreme the quantile, the smaller the sample size, which limits the power of the test. Also, quantile regressions will be dominated by outliers, even though Equation 9.36 minimizes their effect relative to the least-squares solution. Indeed, quantile regression, by its very definition, is a regression line that passes through extreme data points, so you need to make very sure these extreme values do not merely represent errors.

Finally, quantile regression may not be necessary unless the hypothesis is really one of a ceiling or limiting factor. Quantile regression is often used with "triangular" plots of data. However, these data triangles may simply reflect heteroscedascity—a variance that increases at higher values of the X variable.

Data transformations and careful analysis of residuals may reveal that a standard linear regression is more appropriate. See Thompson et al. (1996), Garvey et al. (1998), Scharf et al. (1998), and Cade et al. (1999) for other regression methods for bivariate ecological data.

Logistic Regression

Logistic regression is a special form of regression in which the *Y* variable is categorical, rather than continuous. The simplest case is of a dichotomous *Y* variable. For example, we conducted timed censuses at 42 randomly-chosen leaves of the cobra lily (*Darlingtonia californica*), a carnivorous pitcher plant that captures insect prey (see Footnote 11 in Chapter 8). We recorded wasp visits at 10 of the 42 leaves (Figure 9.11). We can use logistic regression to test the hypothesis that visitation probability is related to leaf height.[14]

You might be tempted to force a regression line through these data, but even if the data were perfectly ordered, the relationship would not be linear. Instead, the best fitting curve is S-shaped or logistic, rising from some minimum value to a maximum asymptote. This kind of curve can be described by a function with two parameters, β_0 and β_1:

$$p = \frac{e^{\beta_0 + \beta_1 X}}{1 + e^{\beta_0 + \beta_1 X}} \tag{9.37}$$

The parameter β_0 is a type of intercept because it determines the probability of success ($Y_i = 1$) p when $X = 0$. If $\beta_0 = 0$, then $p = 0.5$. The parameter β_1 is similar to a slope parameter because it determines how steeply the curve rises to the maximum value of $p = 1.0$. Together, β_0 and β_1 specify the range of the *X* variable over which most of the rise occurs and determine how quickly the probability value rises from 0.0 to 1.0.

The reason for using Equation 9.37 is that, with a little bit of algebra, we can transform it as follows:

$$\ln\left(\frac{p}{1-p}\right) = \beta_0 + \beta_1 X \tag{9.38}$$

[14] Of course, we could use a much simpler *t*-test or ANOVA (see Chapter 10) to compare the heights of visited versus unvisited plants. However, an ANOVA for these data subtly turns cause and effect on its head: the hypothesis is that plant height influences capture probability, which is what is being tested with logistic regression. The ANOVA layout implies that the categorical variable *visitation* somehow causes or is responsible for variation in the continuous response variable leaf height.

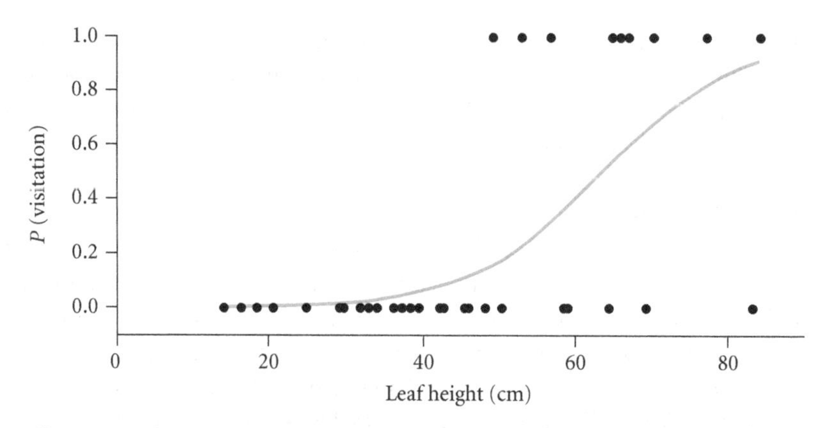

Figure 9.11 Relationship between leaf height and wasp visitation for the carnivorous plant *Darlingtonia californica*. Each point represents a different plant in a population in the Siskiyou Mountains of southern Oregon (Ellison and Gotelli, unpublished data). The x-axis is the height of the leaf, a continuous predictor variable. The y-axis is the probability of wasp visitation. Although this is a continuous variable, the actual data are discrete, because a plant is either visited (1) or not (0). Logistic regression fits an S-shaped (= logistic) curve to these data. Logistic regression is used here because the response variable is discrete, so the relationship has an upper and lower asymptote. The model is fit using the logit transformation (Equation 9.38). The best-fit parameters (using maximum likelihood by iterative fitting) are $\hat{\beta}_0 = -7.293$ and $\hat{\beta}_1 = 0.115$. The P-value for a test of the null hypothesis that $\beta_1 = 0$ is 0.002, suggesting that the probability of wasp visitation increases with increasing leaf height.

This transformation of the Y variable, called the **logit transformation**, converts the S-shaped logistic curve into a straight line. Although this transformation is indeed linear for the X variable, we cannot apply it directly to our data. If the data consist only of 0's and 1's for plants of different sizes, Equation 9.38 cannot be solved because $\ln[p/(1-p)]$ is undefined for $p = 1$ or $p = 0$. But even if the data consist of estimates of p based on multiple observations of plants of the same size, it still would not be appropriate to use the least-squares estimators because the error term follows a binomial distribution rather than a normal distribution.

Instead, we use a maximum likelihood approach. The maximum likelihood solution gives parameter estimates that make the observed values in the data set most probable (see Footnote 13 in Chapter 5). The maximum likelihood solution includes an estimate of the regression error variance, which can be used to test null hypotheses about parameter values and construct confidence intervals, just as in standard regression.

For the *Darlingtonia* data the maximum likelihood estimates for the parameters are $\hat{\beta}_0 = -7.293$ and $\hat{\beta}_1 = 0.115$. The test of the null hypothesis that $\beta_1 = 0$ generates a *P*-value of 0.002, suggesting that the probability of wasp visitation increases with leaf size.

Non-Linear Regression

Although the least squares method describes linear relationships between variables, it can be adapted for non-linear functions. For example, the double logarithmic transformation converts a power function for X and Y variables ($Y = aX^b$) into a linear relationship between log(X) and log(Y): $\log(Y) = \log(a) + b \times \log(X)$. However, not all functions can be transformed this way. For example, many non-linear functions have been proposed to describe the functional response of predators—that is the change in predator feeding rate as a function of prey density. If a predator forages randomly on a prey resource that is depleted through time, the relationship between the number eaten (N_e) and the initial number (N_0) is (Rogers 1972):

$$N_e = N_0\left(1 - e^{a(T_h N_e - T)}\right) \tag{9.39}$$

In this equation, there are three parameters to be estimated: a, the attack rate, T_h, the handling time per prey item, and T, the total prey handling time. The Y variable is N_e, the number eaten, and the X variable is N_0, the initial number of prey provided.

There is no algebraic transformation that will linearize Equation 9.39, so the least-squares method of regression cannot be used. Instead, **non-linear regression** is used to fit the model parameters in the untransformed function. As with logistic regression (a particular kind of non-linear regression), iterative methods are used to generate parameters that minimize the least-squares deviations and allow for hypothesis tests and confidence intervals.

Even when a transformation can produce a linear model, the least-squares analysis assumes that the error terms ε_i are normally distributed *for the transformed data*. But if the error terms are normally distributed for the original function, the transformation will not preserve normality. In such cases, non-linear regression is also necessary. Trexler and Travis (1993) and Juliano (2001) provide good introductions to ecological analyses using non-linear regression.

Multiple Regression

The linear regression model with a single predictor variable X can easily be extended to two or more predictor variables, or to higher-order polynomials of a single predictor variable. For example, suppose we suspected that species richness peaks at intermediate island sizes, perhaps because of gradients in disturbance frequency or intensity. We could fit a second-order polynomial that includes a squared term for island area:

$$Y_i = \beta_0 + \beta_1 X_i + \beta_2 X_i^2 + \varepsilon_i \tag{9.40}$$

This equation describes a function with a peak of species richness at an intermediate island size.[15] Equation 9.40 is an example of a **multiple regression**, because there are now actually two predictor variables, X and X^2, which contribute to variation in the Y variable. However, it is still considered a linear regression model (albeit a multiple linear regression) because the parameters β_i in Equation 9.40 are solvable using linear equations. If the data are modeled as a simple linear function, we ignore the polynomial term, and the systematic component of variation from X^2 is incorrectly pooled with the error term:

$$\varepsilon_i' = \beta_2 X_i^2 + \varepsilon_i \tag{9.41}$$

A more familiar example of multiple regression is when two or more distinct predictor variables are measured for each replicate. For example, in a study of variation in species richness of ants (S) in New England bogs and forests (Gotelli and Ellison 2002a,b), we measured the latitude and elevation of each study site and entered both variables in a multiple regression equation:

$$\log_{10}(\textit{forest } S) = 4.879 - 0.089(\textit{latitude}) - 0.001(\textit{elevation}) \tag{9.42}$$

Both slope parameters are negative because species richness declines at higher elevations and higher latitudes. The parameters in this model are called **partial regression parameters** because the residual sums of squares from the other variables in the model have already been accounted for statistically. For example, the parameter for latitude could be found by first regressing species richness on elevation, and then regressing the residuals on latitude. Conversely, the parameter for elevation could be found by regressing the residuals from the richness-latitude regression on elevation.

[15] The peak can be found by taking the derivative of Equation 9.40 and setting it equal to zero. Thus, the maximum species richness would occur at $X = \beta_1/\beta_2$. Although this equation produces a non-linear curve for the graph of Y versus X, the model is still a linear sum of the form $\Sigma\beta_i X_i$, where X_i is the measured predictor variable (which itself may be a transformed variable) and β_i's are the fitted regression parameters.

Because these partial regression parameters are based on residual variation not accounted for by other variables, they often will not be equivalent to parameters estimated in simple regression models. For example, if we regress species richness on latitude only, the result is

$$\log_{10}(\mathit{forest}\ S) = 5.447 - 0.105(\mathit{latitude}) \tag{9.43}$$

And, if we regress species richness on elevation only, we have

$$\log_{10}(\mathit{forest}\ S) = 1.087 - 0.001(\mathit{elevation}) \tag{9.44}$$

With the exception of the elevation term, all of the parameters differ among the simple linear regressions (Equations 9.43 and 9.44) and the multiple regression (Equation 9.42).

In linear regression with a single predictor variable, the function can be graphed as a line in two dimensions (Figure 9.12). With two predictor variables, the multiple regression equation can be graphed as a plane in a three-dimensional coordinate space (Figure 9.13). The "floor" of the space represents the two axes for the two predictor variables, and the vertical dimension represents the response variable. Each replicate consists of three measurements (Y variable, X_1 variable, X_2 variable), so the data can be plotted as a cloud of points in the three

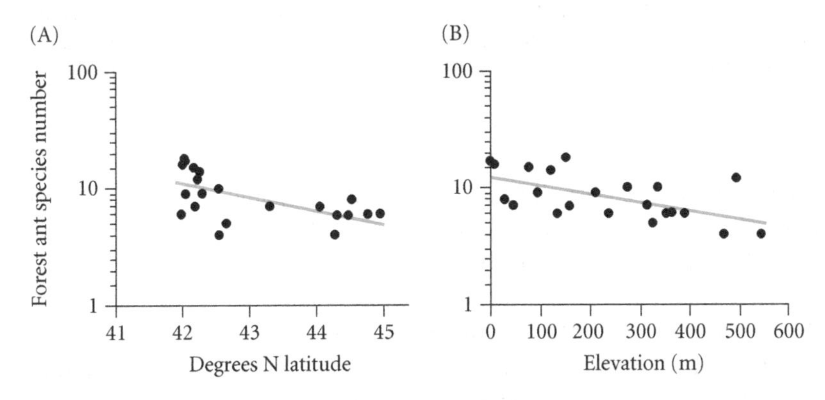

Figure 9.12 Forest ant species density as a function of (A) latitude and (B) elevation. Each point (n = 22) is the number of ant species recorded in a 64-m^2 forest plot in New England (Vermont, Massachusetts, and Connecticut). The least-squares regression line is shown for each variable. Note the $\log_{10}$ transformation of the y-axis. For the latitude regression, the equation is $\log_{10}$(ant species number) = 5.447 – 0.105 × (latitude); r^2 = 0.334. For the elevation regression, the equation is $\log_{10}$(ant species number) = 1.087 – 0.001 × (elevation); r^2 = 0.353. The negative slopes indicate that ant species richness declines at higher latitudes and higher elevations. (Data and sampling details in Gotelli and Ellison 2002a,b.)

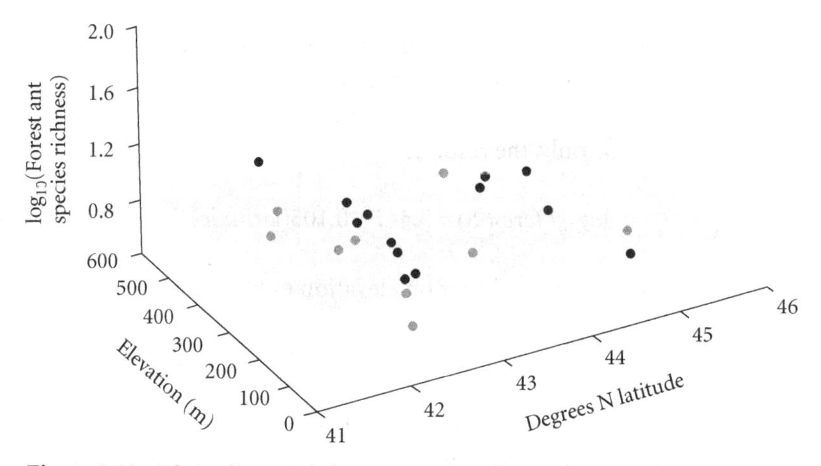

Figure 9.13 Three-dimensional representation of multiple regression data. The least squares solution is a plane that passes through the cloud of data. Each point ($n = 22$) is the number of ant species recorded in a 64-m^2 forest plot in northern New England (Vermont, Massachusetts, and Connecticut). The X variable is the plot latitude (the first predictor variable), the Y variable is the plot elevation (the second predictor variable), and the Z variable is the $\log_{10}$ of the number of ant species recorded (the response variable). A multiple regression equation has been fit to the equation: $\log_{10}$(ant species number) = 4.879 – 0.089×(latitude) – 0.001×(elevation); $r^2 = 0.583$. Notice that the r^2 value, which is a measure of the fit of the data to the model, is larger for the multiple regression model than it is for either of the simple regression models based on each predictor variable by itself (see Figure 9.12). The solution to a linear regression with one predictor variable is a straight line, whereas the solution to a multiple regression with two predictor variables is a plane, which is shown in this three-dimensional rendering. Residuals are calculated as the vertical distance from each datum to the predicted plane. (Data and sampling details in Gotelli and Ellison 2002a,b.)

dimensional space. The least-squares solution passes a plane through the cloud of points. The plane is positioned so that the sum of the squared vertical deviations of all of the points from the plane is minimized.

The matrix solutions for multiple regression are described in the Appendix, and their output is similar to that of simple linear regression: least-squares parameter estimates, a total r^2, an F-ratio to test the significance of the entire model, and error variances, confidence intervals, and hypothesis tests for each of the coefficients. Residual analysis and tests for outliers and influential points can be carried out as for single linear regression. Bayesian or Monte Carlo methods also can be used for multiple regression models.

However, a new problem arises in evaluating multiple regression models that was not a problem in simple linear regression: there may be correlations among the predictor variables themselves, referred to as **multicollinearity** (see Graham 2003). Ideally, the predictor variables are **orthogonal** to one another: all values of one predictor variable are found in combination with all values of the second predictor variable. When we discussed the design of two-way experiments in Chapter 7, we emphasized the importance of ensuring that all possible treatment combinations are represented in a fully crossed design.

The same principle holds in multiple regression: ideally, the predictor variables should not themselves be correlated with one another. Correlations among the predictor variables make it difficult to tease apart the unique contributions of each variable to the response variable. Mathematically, the least-squares estimates also start to become unstable and difficult to calculate if there is too much multicollinearity between the predictor variables. Whenever possible, design your study to avoid correlations between the predictor variables. However, in an observational study, you may not be able to break the covariation of your predictor variables, and you will have to accept a certain level of multicollinearity. Careful analysis of residuals and diagnostics is one strategy for dealing with covariation among predictor variables. Another strategy is to mathematically combine a set of intercorrelated predictor variables into a smaller number of orthogonal variables with multivariate methods such as principal components or discriminant analysis (see Chapter 12).

Is multicollinearity a problem in the ant study? Not really. There is very little correlation between elevation and latitude, the two predictor variables for ant species richness (Figure 9.14).

Path Analysis

All of the regression models that we have discussed so far (linear regression, robust regression, quantile regression, non-linear regression, and multiple regression) begin with the designation of a single response variable and one or more predictor variables that may account for variation in the response. But in reality, many models of ecological processes do not organize variables into a single

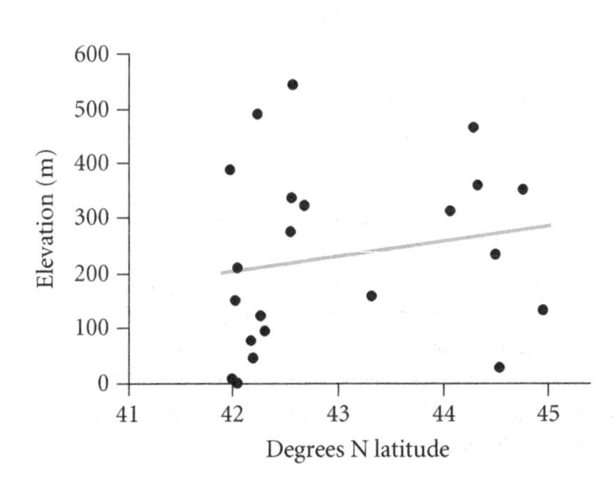

Figure 9.14 Lack of collinearity among the predictor variables strengthens multiple regression analyses. Each point represents the latitude and elevation in a set of 22 forest plots for which ant species richness had been measured (see Figure 9.12). Although elevation and latitude can be used simultaneously as predictors in a multiple regression model (see Figure 9.13), the fit of the model can be compromised if there are strong correlations among the predictor variables (collinearity). In this case, there is only a very weak correlation between the predictor variables latitude and elevation ($r^2 = 0.032$; $F_{1,20} = 0.66$, $P = 0.426$). It is always a good idea to test for correlations among predictor variables when using multiple regression. (Data and sampling details in Gotelli and Ellison 2002a,b.)

Figure 9.15 Path analysis for passive colonization and food web models of *Sarracenia* ▶ inquilines. Each oval represents a different taxon that can be found in the leaves of the northern pitcher plant, *Sarracenia purpurea*. Captured insect prey forms the basis for this complex food web, which includes several trophic levels. The underlying data consist of abundances of organisms measured in leaves that had different water levels in each leaf (Gotelli and Ellison 2006). Treatments were applied for one field season (May–August 2000) in a press experiment (see Chapter 6) to 50 plants (n = 118 leaves). The abundance data were fit to two different path models, which represent two different models of community organization. (A) In the passive colonization model, abundances of each taxon depend on the volume of water in each leaf and the level of prey in the leaf, but no interactions among taxa are invoked. (B) In the trophic model, the abundances are determined entirely by trophic interactions among the inquilines. The number associated with each arrow is the standardized path coefficient. Positive coefficients indicate that increased prey abundance leads to increased predator abundance. Negative coefficients indicate that increased predator abundance leads to decreased prey abundance. The thickness of each arrow is proportional to the size of the coefficient. Positive coefficients are blue lines, negative coefficients are gray lines.

response variable and multiple predictor variables. Instead, measured variables may act simultaneously on each other in cause-and-effect relationships. Rather than isolate the variation in a single variable, we should try to explain the overall pattern of covariation in a set of continuous variables.

This is the goal of **path analysis**. Path analysis forces the user to specify a path diagram that illustrates the hypothesized relationships among the variables. Variables are connected to one another by single- or double-headed arrows. Variables that do not interact directly are not connected by arrows. A path diagram of this sort represents a mechanistic hypothesis of interactions in a system of variables.[16] It also represents a statistical hypothesis about the structure of the variance-covariance matrix of these variables (see Footnote 5 in this chapter). Partial regression parameters can then be estimated for the individual paths in the diagram, and summary goodness-of-fit statistics can be derived for the overall evaluation of the model.

For example, path analysis can be used to test different models for community structure of the invertebrate species that co-occur in pitcher-plant leaves (Gotelli and Ellison 2006). The data consist of repeated censuses of entire communities that live in pitcher plant leaves (n = 118 leaves). The replicates are the

Sewall Wright

[16] Path analysis was first introduced by the population geneticist Sewall Wright (1889–1988) as a method for analyzing genetic covariation and patterns of trait inheritance. It has long been used in the social sciences, but has become popular with ecologists only recently. It has some of the same strengths and weaknesses as multiple regression analysis. See Kingsolver and Schemske (1991), Mitchell (1992), Petraitis et al. (1996), and Shipley (1997) for good discussions of path analysis in ecology and evolution.

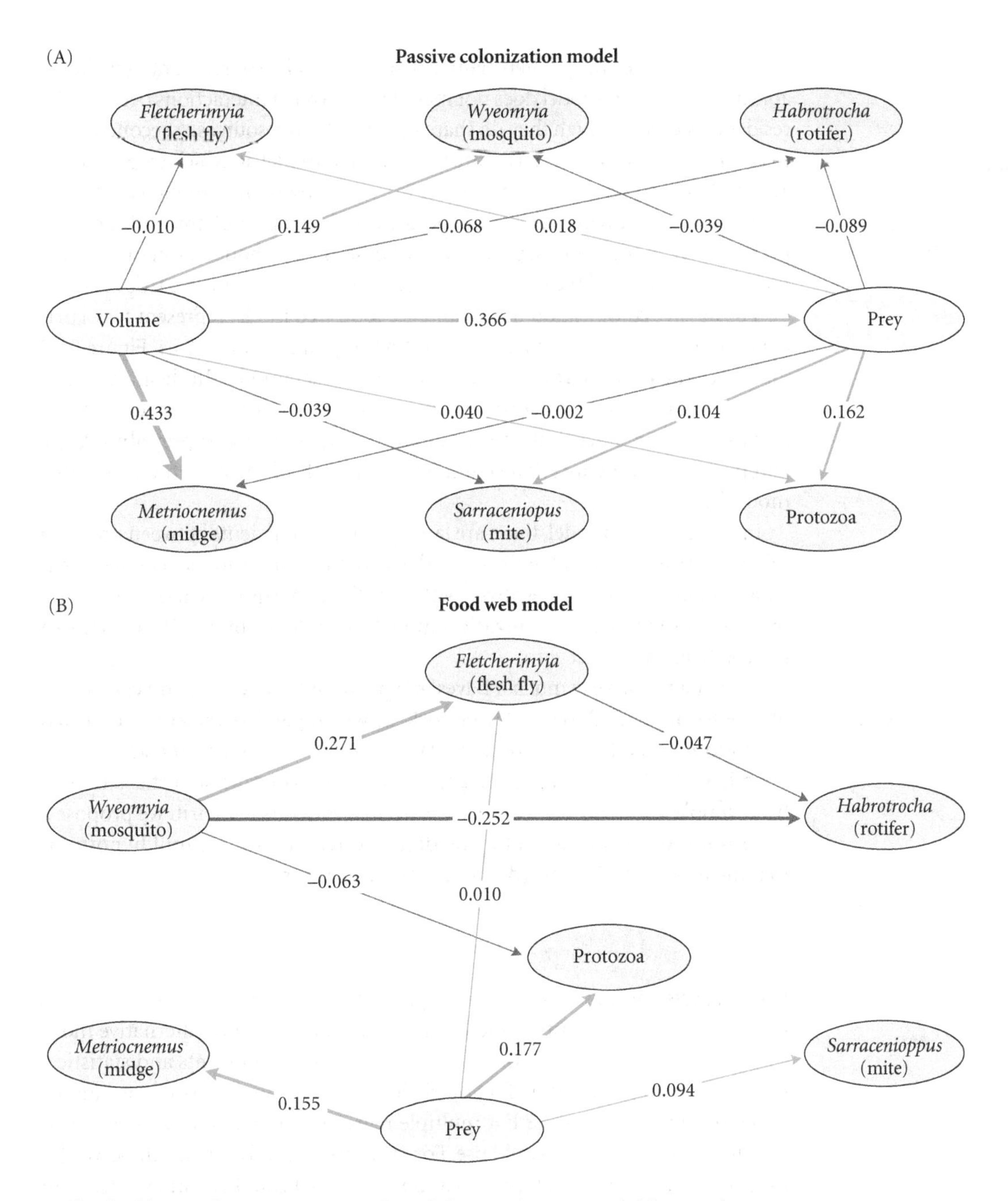

individual leaves that were censused, and the continuous variables are the average abundance of each invertebrate species.

One model to account for community structure is a passive colonization model (Figure 9.15A). In this model, the abundance of each species is determined largely by the volume of the pitcher-plant leaf and the food resources (ant prey) available in the leaf. This model does not include any explicit interactions between the resident species, although they do share the same food resources. A second model to account for community structure is a food web model, in which trophic interactions between species are important in regulating their abundances (Figure 9.15B). Prey resources occur at the bottom of the trophic chain, and these are processed by mites, midges, protozoa, and flesh flies. Although flesh flies are the top predator, some of the species in this community have multiple trophic roles.

The passive colonization model and the food web model represent two a priori hypotheses about the forces controlling inquiline abundances. Figure 9.15 illustrates the **path coefficients** associated with each model. In the passive colonization model, the largest coefficients are positive ones between leaf volume and midge abundance, leaf volume and prey abundance, and prey abundance and protozoa abundance. The confidence intervals of most coefficients in the model bracket zero.

In the food web model, there are large positive coefficients between prey and midges, between prey and protozoa, and between mosquito abundance and flesh-fly abundance. A strong negative coefficient links mosquitoes and their rotifer prey. As in the passive colonization model, confidence intervals for many of the coefficients bracket zero.

Path analysis is very much a Bayesian approach to modeling and can be implemented in a full Bayesian framework, in which path coefficients are treated as random variables (Congdon 2002). It requires the user to specify a priori hypotheses in the form of path diagrams with arrows of cause and effect. In contrast, simple and multiple regression is more frequentist in spirit: we propose a parsimonious linear model for the data and test the simple null hypothesis that the model coefficients (β_i) do not differ from zero.

Model Selection Criteria

Path analysis and multiple regression present a problem that has not arisen so far in our survey of statistical methods: how to choose among alternative models. For both multiple regression and path analysis, coefficients and statistical significance can be calculated for any single model. But how do we choose among different candidate models? For multiple regression, there may be many possible predictor variables we could use. For example, in the forest ant study, we also measured vegetation structure and canopy cover that could potentially affect ant species richness. Should we just use all of these variables, or is there some way to select a subset of predictor variables to most parsimoniously account for variation in the response variable? For path analysis, how do we decide which of two or more a priori models best fits the data?

Model Selection Methods for Multiple Regression

For a dataset that has one response variable and n predictor variables, there are $(2^n - 1)$ possible regression models that can be created. These range from simple linear models with only a single predictor variable to a fully saturated multiple regression model that has all n predictor variables present.[17]

It would seem that the best model would be the one that has the highest r^2 and explains most of the variation in the data. However, you would find that the fully saturated model always has the highest r^2. Adding variables to a regression model never increases the *RSS* and usually will decrease it, although the decrease can be pretty small when adding certain variables to the model.

Why not use only the variables for which the slope coefficients (β_i) in the saturated model are statistically different from zero? The problem here is that coefficients—and their statistical significance—depend on which other variables are included in the model. There is no guarantee that the reduced model containing only the significant regression coefficients is necessarily the best model. In fact, some of the coefficients in the reduced model may no longer be statistically significant, especially if there is multicollinearity among the variables.

Variable selection strategies include **forward selection**, **backward elimination**, and **stepwise methods**. In forward selection, we begin adding variables one at a time, and stop at a certain criterion. In backward elimination models, we start with the fully saturated model (all variables included) and begin eliminating variables one at a time. Stepwise models include both forward and backward steps to make comparisons by swapping variables in and out of the model and evaluating changes in the criterion.

Typically, two criteria are used to decide when to stop adding (or removing) variables. The first criterion is a change in the F-ratio of the fitted model. The F-ratio is a better choice than r^2 or the *RSS* because the change in the F-ratio depends both on the reduction in r^2 and on the number of parameters that are included in the model. In forward selection, we continue adding variables until

[17] There are actually even more possible models and variables than this. For example, if we have two predictor variables X_1 and X_2, we can create a composite variable X_1X_2. The regression model

$$Y_i = \beta_0 + \beta_1 X_{1i} + \beta_2 X_{2i} + \beta_3 X_{1i} X_{2i} + \varepsilon_i$$

has coefficients for both the main effects of X_1 and X_2, and a coefficient for the interaction between X_1 and X_2. This term is analogous to the interaction between discrete variables in a two-way ANOVA (see Chapter 10). As we have noted before, statistical software will cheerfully accept any linear model you care to construct; it is your responsibility to examine the residuals and consider the biological implications of different model structures.

the increase in the F-ratio falls below a specified threshold. In backwards elimination, we eliminate variables until there is too large a drop in the F-ratio. The second criterion for variable selection is **tolerance**. Variables are removed from a model or not added in if they create too much multicollinearity among the set of predictor variables. Typically, the software algorithm creates a multiple regression in which the candidate variable is the response variable, and the other variables already in the model are the predictor variables. The quantity $(1 - r^2)$ in this context is the tolerance. If the tolerance is too low, the new variable shows too much correlation with the existing set of predictor variables and is not included in the equation.

All computer packages contain default cutoff points for F-ratios and tolerances, and they will all generate a reasonable set of predictor variables in multiple regression. However, there are two problems with this approach. The first problem is that there is no theoretical reason why variable selection methods will necessarily find the best subset of predictor variables in multiple regression. The second problem is that these selection methods are based on optimization criteria, so they leave you with only a single best model. The algorithms don't give you alternative models that may have had very similar F-ratios and regression statistics. These alternative models include different variables, but are statistically indistinguishable from the single best-fitting model.

If the number of candidate variables is not too large (< 8), a good strategy is to calculate all possible regression models and then evaluate them on the basis of F-ratios or other least-squares criteria. This way, you can at least see if there is a large set of similar models that may be statistically equivalent. You may also find patterns in the variables—perhaps one or two variables always show up in the final set of models—that can help you choose a final subset. Our point here is that you cannot rely on automated or computerized statistical analyses to sift through a set of correlated variables and identify the correct model for your data. The existing selection methods are reasonable, but they are arbitrary.

Model Selection Methods in Path Analysis

Path analysis forces the investigator to propose specific models that are then evaluated and compared. The selection of the "best" model in path analysis presumes that the "correct" model is included among the alternatives you are evaluating. If none of the proposed path models is correct, you simply are choosing an incorrect model that fits relatively well compared to the alternative incorrect models.

In both path analysis and multiple regression, the problem is to choose a model that has enough variables to properly account for the variation in the data and minimize the residual sum of squares, but not so many variables that you are fitting equations to random noise in the data. **Information criterion statistics** balance a reduction in the sum of squares with the addition of parameters to the model. These methods penalize models with more parameters in them, even though such models inevitably reduce the sum of squares. For example, an information criterion index for multiple regression is the **adjusted r^2**:

$$r_{adj}^2 = 1 - \left(\frac{n-1}{n-p}\right)\left(1 - r^2\right) \quad \textbf{(9.45)}$$

where n is the sample size and p is the number of parameters in the model ($p = 2$ for simple linear regression with a slope parameter and an intercept parameter). Like r^2, this number increases as more of the residual variation is explained. However, r_{adj}^2 also decreases as more parameters are added to the model. Therefore, the model with the largest r_{adj}^2 is not necessarily the model with the most parameters. In fact, if the model has lots of parameters and fits the data poorly, r_{adj}^2 can even become negative.

For path analysis, the **Akaike information criterion**, or **AIC**, can be calculated as

$$\mathrm{AIC} = -2\log\left[L(\hat{\theta} \mid data)\right] + 2K \quad \textbf{(9.46)}$$

where $L(\hat{\theta} \mid y)$ is the likelihood of the estimated model parameter ($\hat{\theta}$) given the data, and K is the number of parameters in the model. In path analysis, the number of parameters in the model is not simply the number of arrows in the path diagram. The variances of each variable have to be estimated as well.

The AIC can be thought of as a "badness-of-fit" measure, because the larger the number, the more poorly the data fit the variance-covariance structure that is implied by the path diagram. For example, the random colonization model for the inquilines has 21 parameters and a cross-validation index (an AIC measure) of 0.702, whereas the food web model has 15 parameters and a cross-validation index of 0.466.

Bayesian Model Selection

Bayesian analysis can be used to compare different hypotheses or models (such as the regression models with and without the slope term β_1; see Equations 9.20 and 9.21). Two methods have been developed (Kass and Raftery 1995; Congdon 2002). The first, called **Bayes' factors**, estimates the relative likelihood of one hypothesis or model relative to another hypothesis or model. As in a full Bayesian analysis, we have prior probabilities $P(H_0)$ and $P(H_1)$ for our two hypothesis or models. From Bayes' Theorem (see Chapter 1), the posterior probabilities

$P(H_0|\text{data})$ and $P(H_1|\text{data})$ are proportional to the prior probabilities × the likelihoods $L(\text{data} | H_0)$ and $L(\text{data} | H_1)$, respectively:

$$P(H_i | \text{data}) \propto L(\text{data} | H_i) \times P(H_i) \quad (9.47)$$

We define the **prior odds ratio**, or the relative probability of one hypothesis versus the other hypothesis, as

$$P(H_0)/P(H_1) \quad (9.48)$$

If these are the only two alternatives, then $P(H_0) + P(H_1) = 1$ (by the First Axiom of Probability; see Chapter 1). Equation 9.48 expresses how much more or less likely one hypothesis is relative to the other before we conduct our experiment. At the start of the study, we expect the prior probabilities of each hypothesis or model to be roughly equal (otherwise, why waste our time gathering the data?), and so Equation 9.48 ≈ 1.

After the data are collected, we calculate posterior probabilities. The Bayes factor is the **posterior odds ratio**:

$$P(H_0 | \text{data}) \,/\, P(H_1 | \text{data}) \quad (9.49)$$

If Equation 9.49 ≫1, then we would have reason to believe that H_0 is favored over H_1, whereas if Equation 9.49 ≪ 1, then we would support H_1. We use Equations 9.47 and 9.48 together to solve for the posterior odds ratio:

$$P(H_0 | \text{data}) \,/\, P(H_1 | \text{data}) = [L(\text{data} | H_0) \,/\, L(\text{data} | H_1)] \times [P(H_0) \,/\, P(H_1)] \quad (9.50)$$

Equation 9.50 says that the posterior odds ratio equals the likelihood ratio times the prior odds ratio. Because the prior odds ratio equals one in an experiment with uninformative priors [$P(H_0) = P(H_1) = 0.5$], the likelihood ratio (also called a Bayes factor) can be used as an estimate of the posterior odds ratio.[18]

The second Bayesian method for comparing models is to approximate Bayes' factors when the prior probabilities are uninformative. This method, called the **Bayesian information criterion** (**BIC**) is used for model selection, such as choosing among regression models. Let λ = the likelihood ratio $L(\text{data} | M_0) \,/\, L(\text{data} | M_1)$ for two models M_0 and M_1, each with different numbers of param-

[18] Bayes' factors are most useful when there are informative priors. As most Bayesian software now uses uninformative priors, there is much less emphasis on Bayes' factors than there was when informative priors were the norm. See Kass and Raftery (1995) for review of Bayes factors.

eters p_1 and p_2 (for example, Equation 9.20 has one parameter, β_0, and Equation 9.21 has two parameters, β_0 and β_1). We define the BIC to be

$$BIC = 2\log_e\lambda - (p_1 - p_2)\log_e n \quad (9.51)$$

where n is the sample size. The best model is the one with the lowest BIC.[19]

In summary, selection of a reduced variable set is a common activity in multiple regression studies, but the criteria used are somewhat arbitrary. These analyses suffer when there is multicollinearity in the data and when the sample sizes are too small. In general, there should be at least 10 to 20 observations for each independent variable you are considering. Many regression studies do not really have adequate replication for multiple regression, and there is a danger that, even with stepwise methods, the resulting models are neither optimal nor parsimonious.

Regression and path analysis with AIC or BIC statistics are a step forward in this regard, because they force the investigator to pose explicit model structures. However, this kind of framework may not always be available, especially in the early stages of a study. Burnham and Anderson (2010) is an excellent introduction to the general topic of model fitting and information criteria.

Summary

Regression is a powerful statistical tool for assessing the relationship between two or more continuous variables. Least-squares solutions to parameters are appropriate and unbiased if the model is truly linear, the *X* variable is measured without error, the observations are independent samples, and the error terms have a normal distribution. Residual analysis is a key step to evaluate the validity of the regression assumptions. More advanced methods include robust regression for handling outliers, logistic and non-linear regression for functional relationships that are not linear, and multiple regression and path analysis for handling multiple predictor variables and complex a priori model structures.

[19] Link and Barker (2006) illustrated that AIC tends to select more complex models than BIC, but that model selection by BIC is strongly dependent on the choice of prior probability distributions on the parameters. But both methods focus on selecting an overall best model. Additional methods are available for identifying which variables are especially important in the model (e.g., Murray and Conner 2009; Doherty et al. 2012).

CHAPTER 10

The Analysis of Variance

The **analysis of variance**, or **ANOVA**, is Fisher's technique for partitioning the sum of squares, something we first encountered in our discussion of regression (see Chapter 9). More generally, ANOVA refers to a class of sampling or experimental designs in which the predictor variable is categorical and the response variable is continuous. Examples of such designs include the one-way layout, randomized block, and split-plot designs. In Chapter 7, we described the physical layout of these designs, the rationale for using them, and some of the advantages and disadvantages of each design. This chapter concentrates on the data analysis and hypothesis tests that are associated with each of those designs.

We first explain the mechanics of the partitioning of the sum of squares—a fundamental technique in ANOVA. We next outline the assumptions of ANOVA. If those assumptions are met, we can use Fisher's F-ratio to estimate *P*-values for the partitioned sum of squares. For each of the designs in Chapter 7, we explain the underlying model and the associated ANOVA table. We also describe how to plot data so you can understand interaction terms in two-way ANOVAs and ANCOVAs. We explain how to incorporate random and fixed factors in your analyses, and how to use these models to partition the variance in a dataset. We describe some procedures used after your basic ANOVA test for comparing means (a priori contrasts and a posteriori comparisons) and conclude with a discussion of how to interpret a set of *P*-values collected in multiple experiments.

It is easy to lose sight of the goal of ANOVA: the comparison of means among groups that have been sampled randomly. Chapter 5 gives details of one example, in which an investigator wishes to compare ant nest densities in forest and field habitats. In that sense, ANOVA is simply an extension of the familiar *t*-test that is used for comparing the means of two sampled groups. If you have not already done so, you should read Chapters 7 and 9 before tackling this chapter. Although Chapter 9 discusses regression, it also introduces the idea of a linear model, treatment effects, partitioning of the sum of squares, and the set-up of an ANOVA table. In fact, both regression and ANOVA are special cases of a more generalized linear model (McCullagh and Nelder 1989). Although statistical software will solve all of the equations presented in this chapter, it is important that you understand this material because the default settings used by many software packages will not generate the correct analysis for many common experimental designs.

Symbols and Labels in ANOVA

A major headache in interpreting ANOVA tables is understanding the symbols and labeling conventions. There are many variables to keep track of, and there is not a consistent set of notation in the literature.

We use a relatively simple system. First, the symbol Y is always reserved for the measured response variable, just as it has been used throughout this book. The symbol $\overline{Y}$ indicates the grand mean of the data. Any mean that is calculated for particular subgroup is indicated with a subscript, such as $\overline{Y}_i$. A subscripted variable that does not have a bar over it indicates a particular datum, such as Y_{ij}. The variable μ indicates the expected value of a variable in the model, and the residual error term is indicated by ε, usually with some subscripting to indicate the different treatment components.

Capital letters $A, B, C\ldots$ designate the different factors in the model. The different levels of the variables are indicated by subscripts $i, j, k\ldots$. For example, we would write A_i to indicate level i of Factor A, and B_j to indicate level j of Factor B. The maximum number of levels for a factor is indicated by the corresponding *lower case* of the letter. Thus, if A_i indicates treatment level i for Factor A, the number of levels of Factor A ranges from $i = 1$ to a. The exception to this pattern is the lower case n, which is reserved for the number of replicates used to estimate the sum of squares within groups (the residual sum of squares), the lowest level at which replicate samples are taken. We always use the standard variance symbol σ^2, regardless of whether the variance component is for a fixed or a random factor. We state clearly in the text or table legend whether each factor in the model is fixed or random.

ANOVA and Partitioning of the Sum of Squares

The analysis of variance is built on the concept of the partitioning of the sum of squares, which was introduced in Chapter 9. In brief, the total variation in a set of data can be expressed as a sum of the squares: the difference between each observation (Y_i) and the grand mean of the data ($\overline{Y}$) is squared and summed.

This total variation can be partitioned or divided into different components. Some components represent random or error variation that is not attributable to any specific cause; it may result from observation error and other unspecified forces (see Footnotes 10 and 11 in Chapter 1). Other components represent the effects of the experimental treatments applied to the replicates, or the differences among sampling categories. A statistical analysis involves specifying an underlying model for how the observations might be affected by different treatments, partitioning the sum of squares among the different components in the model, and then using the results to test statistical hypotheses for the strength of particular effects.

We illustrate the partitioning of the sum of squares with a one-way ANOVA design used to test the effects of early snowmelt on alpine plant growth (e.g., Price and Waser 1998; Dunne et al. 2003). Such an experiment might have 3 treatment groups and 4 replicates per treatment ($4 \times 3 = 12$ total observations). Four plots are unmanipulated: no changes are made to the plots other than those that occur during censusing. Four plots are warmed with permanent solar-powered heating coils that melt the spring snow pack earlier in the year than normal. Four additional plots serve as controls: they are fitted with heating coils that are never activated. After 3 years of treatment application, you measure the length of the flowering period for larkspur (*Delphinium nuttallianum*) in each plot.

The results are shown in Table 10.1, along with all of the calculations needed for this example. Although most ANOVA calculations are now done with a computer, it is worth your time to work through this pencil-and-paper example so that you have a solid understanding of how the sum of squares is partitioned.

As we described in Chapters 3 and 9, we begin our analysis by calculating the total sum of squares of the data, which is the sum of the squared deviations of each observation (Y_i) from the grand mean ($\overline{Y}$). In the one-way layout, there are $i = 1$ to a treatments and $j = 1$ to n replicates per treatment, with a total of $a \times n$ observations. In our example, there are $a = 3$ treatments (unmanipulated, control, and treatment) and $n = 4$ replicates per treatment, with a total sample size of $a \times n = 3 \times 4 = 12$. Thus, we can write

$$SS_{total} = \sum_{i=1}^{a}\sum_{j=1}^{n}(Y_{ij} - \overline{Y})^2 \qquad \textbf{(10.1)}$$

The total sum of squares for the data in Table 10.1 is 41.66.

This total sum of squares reflects the deviation of each observation from the grand mean. It can be decomposed (partitioned) into two different sources. The first **component of variation** is the **variation among groups**. The variation

TABLE 10.1 **Partitioning of the sum of squares in ANOVA**

Unmanipulated	Control	Treatment
10	9	12
12	11	13
12	11	15
13	12	16
$\bar{Y}_1 = 11.75$	$\bar{Y}_2 = 10.75$	$\bar{Y}_3 = 14.00$
$\sum_{j=1}^{n}(Y_{1j} - \bar{Y}_1)^2 = 4.75$	$\sum_{j=1}^{n}(Y_{2j} - \bar{Y}_2)^2 = 4.75$	$\sum_{j=1}^{n}(Y_{3j} - \bar{Y}_3)^2 = 10.00$
$\sum_{j=1}^{n}(\bar{Y}_1 - \bar{Y})^2 = 0.68$	$\sum_{j=1}^{n}(\bar{Y}_2 - \bar{Y})^2 = 8.08$	$\sum_{j=1}^{n}(\bar{Y}_3 - \bar{Y})^2 = 13.40$
$\sum_{j=1}^{n}(Y_{1j} - \bar{Y})^2 = 5.43$	$\sum_{j=1}^{n}(Y_{2j} - \bar{Y})^2 = 12.83$	$\sum_{j=1}^{n}(Y_{3j} - \bar{Y})^2 = 23.40$

among groups represents differences among the means or averages of each of the treatment groups. Thinking of each group mean as a single observation, this source of variation is

$$SS_{among\ groups} = \sum_{i=1}^{a}\sum_{j=1}^{n}(\bar{Y}_i - \bar{Y})^2 \tag{10.2}$$

This equation contains two summations, one over the a treatment groups, and one over the n_i observations in each treatment group. Doing the first summation is easy: take each treatment mean, subtract it from the grand mean, square it, and add the terms up for each of the a treatment groups. But how do we "sum over j" when there is no j subscript in the equation $(\bar{Y}_i - \bar{Y})^2$? Because there are n_i observations in each treatment group, you simply multiply the first summation by the constant n_i. Thus, Equation 10.2 is equivalent to

$$SS_{among\,groups} = \sum_{i=1}^{a} n_i(\bar{Y}_i - \bar{Y})^2$$

$$\bar{Y} = 12.17$$

$$\sum_{i=1}^{a}\sum_{j=1}^{n}(Y_{ij} - \bar{Y}_i)^2 = 19.50 = SS_{within\ groups}$$

$$\sum_{i=1}^{a}\sum_{j=1}^{n}(\bar{Y}_i - \bar{Y})^2 = 22.16 = SS_{among\ groups}$$

$$\sum_{i=1}^{a}\sum_{j=1}^{n}(Y_{ij} - \bar{Y})^2 = 41.66 = SS_{total}$$

The hypothetical raw data consist of observations of flowering period (in weeks) of larkspur in a set of 12 alpine meadow plots. Four of the plots receive an experimental warming treatment, four plots serve as controls (heating elements are set up, but not turned on), and four plots are unmanipulated. There are $i = 1$ to 3 groups (= treatments), with $n = 4$ replicates per group. This table illustrates basic ANOVA calculations and the partitioning of the sum of squares. In the first row of calculations, we calculate the grand mean $\bar{Y}$ for all 12 of the replicates (12.17), and the mean $\bar{Y}_i$ for each treatment group (11.75, 10.75, and 14.00). In the second row of calculations, we sum the squared deviations of each observation from its own group mean $(Y_{ij} - \bar{Y}_i)^2$ (4.75, 4.75, and 10.00), and add these to get the within-group sum of squares (19.50). In the third row of calculations, we determine the squared deviation of each group mean from the grand mean $(\bar{Y}_i - \bar{Y})^2$ multiplied by the sample size $n = 4$ (0.68, 8.08, and 13.40), and add these together to get the among-groups sum of squares (22.16). In the fourth row of calculations, we determine the squared deviation of each observation from the grand mean $(Y_{ij} - \bar{Y})^2$ (5.43, 12.83, and 23.40) and add these together to get the total sum of squares. Thus, the total sum of squares (41.66) can be additively partitioned into the within-group component (19.50) and the among-group component (22.16). This is a fundamental algebraic property that will hold for any set of numbers. Using such data to conduct a statistical test is meaningful only if the sampling meets the general assumptions of ANOVA and matches the particular design of the specific ANOVA model.

In the balanced designs we discuss in this chapter, n_i is the same in all the treatment groups ($n_i = n$), and this simplifies to

$$SS_{among\ groups} = n\sum_{i=1}^{a}(\bar{Y}_i - \bar{Y})^2$$

In the example shown in Table 10.1, the three sums of squares for the treatment groups are 0.17, 2.02, and 3.35. Because there are four replicates per treatment, the among-group sum of squares equals 4 × (0.17 + 2.02 + 3.35) = 22.16. In the one-way ANOVA model, the controlled factors represent processes that we hypothesize cause differences among the treatment groups. The effect of these factors is represented by the sum of squares among groups (Equation 10.2).

The remaining component is the **variation within groups**. Rather than calculate the deviation of each observation from the grand mean, we calculate the deviation of each observation from its own group mean and then sum across the groups and the replicates:

$$SS_{within\ groups} = \sum_{i=1}^{a}\sum_{j=1}^{n}(Y_{ij} - \bar{Y}_i)^2 \quad \textbf{(10.3)}$$

This component of variation in the example data set is 19.50. The within-group sum of squares is often called the **residual sum of squares**, the **residual variation**, or the **error variation.** As in regression, we refer to it as "residual" because it is variation that is not explained by controlled or experimental factors in our model. The within-group variation (Equation 10.3) is described as "error variation" because our statistical model incorporates this component as random sampling from a normal distribution. In more complex ANOVA models, we will partition the total sum of squares into multiple components of variation, each representing the contribution of a factor in the model. In all cases, though, what is left over is always the residual sum of squares.

One of Fisher's key contributions was to show that the components of variation are additive:

$$SS_{total} = SS_{among\ groups} + SS_{within\ groups} \quad \textbf{(10.4)}$$

In words, the total sum of squares equals the sum of squares among groups plus the sum of squares within groups. For the data in Table 10.1:

$$\begin{array}{ccccc} \sum_{i=1}^{a}\sum_{j=1}^{n}(Y_{ij}-\bar{Y})^2 & = & \sum_{i=1}^{a}\sum_{j=1}^{n}(\bar{Y}_i-\bar{Y})^2 & + & \sum_{i=1}^{a}\sum_{j=1}^{n}(Y_{ij}-\bar{Y}_i)^2 \\ 41.66 & = & 22.16 & + & 19.50 \end{array} \quad \textbf{(10.5)}$$

We emphasize that the partitioning of the sum of squares is a purely algebraic property: this result will hold for any set of numbers, regardless of what they represent or how they were collected.[1]

[1] A proof of the sum of squares partitioning can be quickly sketched. First, start with the total sum of squares:

$$SS_{total} = \sum_{i=1}^{a}\sum_{j=1}^{n}(Y_{ij}-\bar{Y})^2$$

Now let's add and subtract $\bar{Y}_i$, which does not change the total:

$$SS_{total} = \sum_{i=1}^{a}\sum_{j=1}^{n}(Y_{ij}-\bar{Y}+\bar{Y}_i-\bar{Y}_i)^2$$

Regrouping the elements gives us the two familiar components of the sum of squares:

$$SS_{total} = \sum_{i=1}^{a}\sum_{j=1}^{n}\left[(Y_{ij}-\bar{Y}_i)+(\bar{Y}_i-\bar{Y})\right]^2$$

Recalling the binomial expansion (see Chapter 2), $(a+b)^2 = a^2 + 2ab + b^2$:

Nevertheless, the sum of squares partitioning seems to be a natural measure of treatment effects. If the sum of squares among groups is relatively large compared to the sum of squares within groups, then differences among the treatments would seem to be important. On the other hand, if the within-group sum of squares is large relative to the sum of squares among groups, we conclude that differences among the groups are weak or inconsistent. We will see below how to quantify these ideas in the analysis of variance.

The Assumptions of ANOVA

Before we can use the sum of squares in a statistical model, however, the data have to meet the following set of assumptions.

1. *The samples are independent and identically distributed.* As always, this assumption forms the basis for any statistical sampling model. We assume the data represent a random sample of the sample space that you have defined, and that the observations within and between treatments are independent of one another (see Chapter 6). For all of the descriptions of ANOVA tables in this chapter, we have assumed the simplest case, in which sample sizes (n) are equal within all groups. See Sokal and Rohlf (1995) for an overview of ANOVA methods when sample sizes of the treatment groups are unequal.
2. *The variances are homogeneous among groups.* Although the means of the sampled groups may differ from one another, we assume that the variance within each group is approximately equal to the variance with-

---

$$SS_{total} = \sum_{i=1}^{a}\sum_{j=1}^{n}(Y_{ij} - \overline{Y}_i)^2 + 2\sum_{i=1}^{a}\sum_{j=1}^{n}(Y_{ij} - \overline{Y}_i)(\overline{Y}_i - \overline{Y}) + \sum_{i=1}^{a}\sum_{j=1}^{n}(\overline{Y}_i - \overline{Y})^2$$

Pleasingly, the second term of this expansion always equals zero (because the sum of the deviations from the mean = 0; see Chapter 9), and we are left with:

$$\sum_{i=1}^{a}\sum_{j=1}^{n}(Y_{ij} - \overline{Y})^2 = \sum_{i=1}^{a}\sum_{j=1}^{n}(\overline{Y}_i - \overline{Y})^2 + \sum_{i=1}^{a}\sum_{j=1}^{n}(Y_{ij} - \overline{Y}_i)^2$$

Equation 10.4 can be thought of as the Pythagorean Theorem (see Footnote 5 in Chapter 12) for distances of data from means and means from each other. Equation 10.4 ensures that $SS_{among\ groups}$ and $SS_{within\ groups}$ are orthogonal to one another. They do not "cast shadows" on each other and are therefore statistically independent. This independence is necessary for valid interpretation of F-ratios, and is yet another advantage to using squared deviations for expressing distance.

in all the other groups. Thus, each treatment group contributes roughly the same to the within group sum of squares. In linear regression, we make an analogous assumption that the variance is homogenous for different levels of the X variable (see Chapter 9). Also, as in linear regression, data transformations often will equalize variances (see Chapter 8).

3. *The residuals are normally distributed.* The residuals are assumed to follow a normal distribution, with a mean equal to zero. Thanks to the Central Limit Theorem (see Chapter 2), this assumption is not too restrictive, especially if samples sizes are large and approximately equal among treatments, or if the data themselves are means. Once again, suitably transformed data often can have normally distributed error terms.
4. *The samples are classified correctly.* For experimental studies, we assume that all individuals assigned to a particular treatment have been treated identically (e.g., all birds in a parasite-free treatment receive identical doses of antibiotic). For observational studies or natural experiments, we assume that all individuals grouped in a particular class actually belong in that class (e.g., in an environmental impact study, all study plots have been assigned correctly to impact or control groups). The violation of this assumption is potentially serious, and may compromise the estimate of P-values. In regression studies, the analogous assumption is no measurement errors in the X-variable (see Footnote 10 in Chapter 9). However, measurement errors in a regression design may not be as serious a problem as classification errors in an ANOVA design. A high-quality study that is conducted carefully is the best safeguard against errors of classification and measurement.
5. *The main effects are additive.* In certain ANOVA designs, such as the randomized block or split-plot, not all treatment factors are completely replicated. In such cases, it is necessary to assume that the main effects are strictly additive, and that there are no interactions among the different factors. We will address this assumption in more detail later in this chapter when we discuss these designs. Data transformations also can help ensure additivity, particularly when multiplicative factors are logarithmically transformed (see Chapter 3).

Hypothesis Tests with ANOVA

If the ANOVA assumptions are met (or not too severely violated), we can test hypotheses based on an underlying model that is fit to the data. For the one-way ANOVA, that model is

$$Y_{ij} = \mu + A_i + \varepsilon_{ij} \tag{10.6}$$

In this model, Y_{ij} is the replicate j associated with treatment level i, μ is the true grand mean or average [$(\overline{Y})$ is the estimate of μ], and ε_{ij} is the error term. Although each observation Y_{ij} has its own unique error ε_{ij} associated with it, remember that all of the ε_{ij}'s are drawn from a single normal distribution with a mean of 0. The most important element in the model is the term A_i. This term represents the additive linear component associated with level i of treatment A. There is a different coefficient A_i associated with each of the i treatment levels. If A_i is a positive number, treatment level i has an expectation that is greater than the grand average. If A_i is negative, the expectation is below the grand average. Because the A_i's represent deviations from the grand mean, by definition, their sum equals zero. The ANOVA allows us to estimate the A_i effects (the treatment mean minus the grand mean is an unbiased estimator of A_i) and to test hypotheses about the A_i's.

What is the null hypothesis? If there are no treatment effects, then $A_i = 0$ for all treatment levels. Therefore, the null hypothesis is

$$Y_{ij} = \mu + \varepsilon_{ij} \tag{10.7}$$

If the null hypothesis is true, any variation that occurs among the treatment groups (and there will always be some) reflects random error and nothing else.

The ANOVA table provides a general test of the null hypothesis of no treatment effects. Table 10.2 shows the basic components of the ANOVA table for a one-way design (see "The Anatomy of an ANOVA Table" in Chapter 9 for the details and abbreviations in a typical ANOVA table). We begin by calculating a mean square, which is simply a sum of squares divided by its corresponding degrees of freedom (see Chapter 9). Two mean squares are calculated in a one-way ANOVA.

The first mean square, for the variation among groups, has $(a - 1)$ degrees of freedom, where a is the number of treatments. The second mean square, for the variation within groups, has $a(n - 1)$ degrees of freedom. This makes intuitive sense, because, within each group, there should be $(n - 1)$ degrees of freedom. With a groups, that gives $a(n - 1)$ degrees of freedom for the mean square within groups. Also, notice that the total degrees of freedom = $(a - 1) + a(n - 1) = (an - 1)$, which is just one less than the total sample size. Why don't the degrees of freedom add up to the total sample size (an)? Because one degree of freedom is used in estimating the overall grand mean (μ).

Whereas the mean square within groups estimates the error variance σ^2, the mean square between groups estimates $\sigma^2 + \sigma_A^2$: the error variance *plus* the vari-

TABLE 10.2 **ANOVA table for one-way layout**

Source	Degrees of freedom (df)	Sum of squares (SS)	Mean square (MS)	Expected mean square	F-ratio	P-value
Among groups	$a-1$	$\sum_{i=1}^{a}\sum_{j=1}^{n}(\bar{Y}_i-\bar{Y})^2$	$\frac{SS_{among\ groups}}{(a-1)}$	$\sigma^2+n\sigma_A^2$	$\frac{MS_{among\ groups}}{MS_{within\ groups}}$	Tail of the F-distribution with $(a-1)$ $a(n-1)$ degrees of freedom
Within groups (residual)	$a(n-1)$	$\sum_{i=1}^{a}\sum_{j=1}^{n}(Y_{ij}-\bar{Y}_i)^2$	$\frac{SS_{within\ groups}}{a(n-1)}$	σ^2		
Total	$an-1$	$\sum_{i=1}^{a}\sum_{i=1}^{n}(Y_{ij}-\bar{Y})^2$	$\frac{SS_{total}}{(an-1)}$	σ_Y^2		

There are a groups, with n replicates per group. A_i is the treatment effect for group i. The grouping factor may be either a fixed or a random factor. In this simple design, there is only a single F-ratio that can be constructed, which tests the null hypothesis that the variation among sampled groups is not significantly different from 0. The F-ratio uses the among-groups mean square for the numerator and the within-groups mean square (residual) for the denominator. See Table 9.1 for further explanation of the elements of an ANOVA table.

ance due to the treatment groups. Therefore, the ratio of these two quantities (MS_{among}/MS_{within}) is an appropriate test of the treatment effect. The larger the treatment effect relative to the within-square error, the larger the F-ratio. If there is no effect of the treatments ($\sigma_A^2 = 0$) (the null hypothesis), then the mean square among groups $= \sigma^2 + 0 = \sigma^2$, which is the same as the mean square within groups, or the error variance σ^2. In this case, $MS_{among} = MS_{within}$, and their ratio, the F-ratio, = 1.0. The tail probability depends both on the size of the F-ratio and the degrees of freedom. For the one-way ANOVA, there are $(a - 1)$ degrees of freedom in the numerator and $a(n - 1)$ degrees of freedom in the denominator.

For the data in Table 10.1, the F-ratio is 5.11, with a corresponding *P*-value of 0.033 (Table 10.3). This *P*-value is small (it is less than 0.05), so we would reject the null hypothesis of no treatment effect. When we look at the raw data (Table 10.1), it seems appropriate that we should reject the null hypothesis: flowering periods were longer in the treatment group than in either the control or the unmanipulated group.

Constructing F-Ratios

Here are the general steps for constructing an F-ratio and using it to test hypotheses using ANOVA:

1. Use the mean squares associated with the particular ANOVA model that matches your sampling or experimental design. We introduced the basic

TABLE 10.3 **One-way ANOVA table for the hypothetical data in Table 10.1**

Source	Degrees of freedom (df)	Sum of squares (*SS*)	Mean square (*MS*)	F-ratio	*P*-value
Among groups	2	22.17	11.08	5.11	0.033
Within groups (residual)	9	19.50	2.17		
Total	11	41.67			

The formulas in Table 10.2 are used to calculate the sum of squares, mean square, and the F-ratio. The first column indicates the source of variation. The second column indicates the degrees of freedom, which are determined by the sample size and the number of groups being compared. The third column gives the sum of squares, which were calculated in Table 10.1. The fourth column is the mean square, which is calculated by dividing each sum of squares by its corresponding degrees of freedom. The next column gives the F-ratio(s), which is calculated as a ratio of the appropriate mean square values. For the simple one-way ANOVA, the only F-ratio that can be constructed tests for the variation among groups; it is calculated as (among-groups mean square) / (within-groups mean square). The *P*-value is determined from a set of statistical tables for an F-ratio with 2 and 9 degrees of freedom. The *P*-value indicates that there is only a 3.3% chance of obtaining these data (or other data even more extreme) if the null hypothesis were true. Because this *P*-value is less than the standard alpha level of 0.05, we reject the null hypothesis, and conclude that the variation among the groups is probably larger than can be explained by chance alone.

designs in Chapter 7, and in this chapter we will give you the ANOVA tables that go with them. We hope you took our advice in Chapters 4 and 6 and figured out the model *before* you collected your data!

2. Find the expected mean square that includes the particular effect you are trying to measure and use it as the numerator of the F-ratio.
3. Find a second expected mean square that includes all of the statistical terms in the numerator *except for the single term you are trying to estimate* and use it as the denominator of the F-ratio.
4. Divide the numerator by the denominator to get your F-ratio.
5. Using statistical tables or your computer's output, determine the probability value (*P*-value) associated with the F-ratio and its corresponding degrees of freedom. The null hypothesis is always that the effect of interest is zero. If the null hypothesis is true, the properly constructed F-ratio will usually have an expected value of 1.0. In contrast, if the effect is very large, the numerator will be much bigger than the denominator and will yield an F-ratio that is substantially larger than 1.0.[2]

[2] F-ratios less than 1.0 are also theoretically possible—such a result would indicate that the differences among group means were actually *smaller* than expected by chance. Very small F-ratios might reflect a failure to obtain random independent samples. For example, if the same replicate is measured mistakenly in more than one treatment group, the among-group sum of squares will be artificially small (see also Footnote 6 in Chapter 5).

TABLE 10.4 **ANOVA table for randomized block design**

Source	Degrees of freedom (df)	Sum of squares (SS)	Mean square (MS)
Among groups	$a-1$	$\sum_{i=1}^{a}\sum_{j=1}^{b}(\bar{Y}_i-\bar{Y})^2$	$\frac{SS_{among\ groups}}{(a-1)}$
Blocks	$b-1$	$\sum_{i=1}^{a}\sum_{j=1}^{b}(\bar{Y}_j-\bar{Y})^2$	$\frac{SS_{blocks}}{(b-1)}$
Within groups (residual)	$(a-1)(b-1)$	$\sum_{i=1}^{a}\sum_{j=1}^{b}(Y_{ij}-\bar{Y}_i-\bar{Y}_j+\bar{Y})^2$	$\frac{SS_{within\ groups}}{(a-1)(b-1)}$
Total	$ab-1$	$\sum_{i=1}^{a}\sum_{j=1}^{b}(Y_{ij}-\bar{Y})^2$	$\frac{SS_{total}}{(ab-1)}$

There are $i = 1$ to a treatment groups and $j = 1$ to b blocks, with each treatment group represented once within a block. The treatment effect is fixed and the blocks effect is random. Two hypothesis tests are possible, one for differences among treatments and one for differences among blocks. Both tests use the residual mean square as the denominator of the F-ratio. Because there is no replication within a block for a randomized block design, there is no test for the interaction between blocks and treatments, which is an important limitation of this model.

6. Repeat steps 2 through 5 for other factors that you are testing. A simple one-way ANOVA generates only a single F-ratio, but more complex models allow you to test multiple factors.

Most statistical software will handle all of these steps for you. However, you should take the time to examine the numerical values of the F-ratios and convince yourself that they were calculated the way they should have been. As we will explain, the default settings in many statistical packages may not generate the correct F-ratios for your particular model. Be careful!

A Bestiary of ANOVA Tables

Here we briefly present and describe the ANOVA tables and hypothesis tests for the other ANOVA designs discussed in Chapter 7. We also introduce a new design, ANCOVA, now that you have learned about regression (in Chapter 9).

Randomized Block

In the randomized block design, each set of a treatments is physically (or spatially) grouped in a block (see Figure 7.6), with each treatment represented exact-

Expected mean square	F-ratio	P-value
$\sigma^2 + b\sigma_A^2$	$\frac{MS_{among\ groups}}{MS_{within\ groups}}$	Tail of the F-distribution with $(a-1)$, $(a-1)(b-1)$ degrees of freedom
$\sigma^2 + a\sigma_B^2$	$\frac{MS_{blocks}}{MS_{within\ groups}}$	Tail of the F-distribution with $(b-1)$, $(a-1)(b-1)$ degrees of freedom
σ^2		
σ_Y^2		

ly once in each block. There are $a = 1$ to i treatment groups $j = 1$ to b blocks, so the total sample size is ba observations. The model we are testing is:

$$Y_{ij} = \mu + A_i + B_j + \varepsilon_{ij} \tag{10.8}$$

In addition to the random error term ε_{ij} and the treatment effect A_i, there is now a block effect B_j: values measured in some blocks are consistently higher or lower than in other blocks, above and beyond the effect of the treatment A_i. Notice that there is no interaction term included for blocks and treatments. Such an interaction may, of course, exist, but we cannot estimate it; it lurks hidden in both the error sum of squares and the treatment sum of squares.

The ANOVA table for the randomized block design (Table 10.4) contains the usual sum of squares for differences among treatment means, but also contains a sum of squares for the differences among blocks, which has $(b-1)$ degrees of freedom. This sum of squares is calculated by first obtaining the average of all the treatments within each block and then measuring the variation among the blocks. The error sum of squares now contains $(a-1)(b-1)$ degrees of freedom. For the corresponding one-way ANOVA with $n = b$ replicates per treatment, there would have been $(a-1)(b)$ degrees of freedom. These numbers differ by $(a-1)$ degrees of freedom, which are used to estimate the block effect.

Two null hypotheses can be tested with the randomized block design. The first null hypothesis is that there are no differences among blocks. The F-ratio used to test this hypothesis is $MS_{among\ blocks}/MS_{within\ groups}$. Testing for block effects usually is not of interest; the primary reason for using block designs is that we expect there to be differences among blocks, and we want to adjust for those differences in our comparison of treatments. The second null hypothesis, which is the one which we normally care most about, is that there are no differences among treatments. The F-ratio used to test this hypothesis is calculated in the usual way as $MS_{among\ groups}/MS_{within\ groups}$. However, as we noted above, the mean square within groups has fewer degrees of freedom than the error mean square in the simple one-way ANOVA. The reason is that some of the original error degrees of freedom have been used to estimate the block effect. If the differences among the blocks are large, the reduction in the among-group sum of squares will be substantial, and the test for the treatment effect will be more powerful, even with fewer degrees of freedom. However, if the differences among blocks are small, the reduction in the among-group sum of squares will be small, and the test for the treatment effect will be less powerful. If the block effects are weak, we have squandered the $(a - 1)$ degrees of freedom that are necessary to estimate them.

Nested ANOVA

In a nested design, the data are organized hierarchically, with one class of objects nested within another (see Figure 7.8). A familiar example is a taxonomic classification, in which species are grouped within genera and genera are grouped within families. The key feature to recognizing a nested design is that the subgroupings are not repeated in the higher level categories. For example, the ant genera *Myrmica*, *Aphaenogaster*, and *Pheidole* occur only within the ant subfamily Myrmicinae; they are not found in the subfamily Dolichoderinae or in the subfamily Formicinae. Similarly, the ant genera *Formica* and *Camponotus* are only found in the subfamily Formicinae. Nested designs superficially may resemble crossed or orthogonal designs. However, in a proper orthogonal design, every level of one factor is represented with every level of another factor. In Chapter 7, our example of a crossed design was a nitrogen and phosphorus addition experiment in which every level of nitrogen was paired with every level of phosphorus.

It is important to recognize the difference in these designs because they call for different kinds of analyses. Although there are many variations of nested designs, we will use the simplest one possible, in which an investigator takes two or more subsamples from a single replicate of a one-way ANOVA. Thus, there are $i = 1$ to a treatment levels, $j = 1$ to b replicates within each treatment level, and $k = 1$ to n subsamples within each replicate. The total sample size (for the balanced design) is $a \times b \times n$. The model being tested is

$$Y_{ijk} = \mu + A_i + B_{j(i)} + \varepsilon_{ijk} \quad \textbf{(10.9)}$$

A_i is the treatment effect, and $B_{j(i)}$, is the variation among replicates, which are nested within treatments. The symbol $j(i)$ reminds us that replicate level j is nested within treatment level i. Finally, ε_{ijk} is the random error term, indicating the error associated with subsample k, replicate j, treatment i. Corresponding to these three levels of variation, there are three mean squares in the ANOVA table: variation among treatments, variation among replicates within a treatment, and error variation. The error variation is calculated for subsamples within each replicate (Table 10.5).

The most important feature in the ANOVA table for nested designs is the F-ratio for the treatment effect. The denominator of this F-ratio is the mean square for replicates within a treatment—not the usual denominator of error variance. The reason is that the individual subsamples are nested within replicates, so they are not independent of one another. This $MS_{within\ groups}$ is appropriate for testing for differences among replicates within a treatment, but *not* for testing for differences among treatments. The correct calculation for the treatment effect has only $a(b - 1)$ degrees of freedom in the denominator, representing the independent variation among the replicates.

The result of the nested ANOVA test for treatment differences would be algebraically identical to a simple one-way ANOVA in which you first calculated the average of the subsamples within a replicate. This one-way ANOVA would also have $a(b - 1)$ degrees of freedom in the denominator, which corresponds to the number of truly independent replicates.

In contrast, if you mistakenly use the $MS_{within\ groups}$ to test treatment effects in a nested design, you have $ab(n - 1)$ degrees of freedom, which is considerably larger and more likely to cause you to incorrectly reject the null hypothesis (a Type I error). The choice of the correct denominator for the F-ratio becomes clear when you examine the expected mean squares in the ANOVA table.[3]

[3] Some authors have suggested that the within-treatment variance can be pooled under certain circumstances. Suppose that the test for variation among replicates within a treatment is non-significant. Then it is tempting to pool the variation and treat each subsample as a truly independent replicate. This will certainly increase the power of the test for treatment effect, but it will also increase the chance of a Type I error if there really is variation among replicates. The critical level for making the decision to pool is generally set much higher (e.g., $\alpha = 0.25$) to reduce the probability of incorrectly accepting the null hypothesis of no variation among replicates (a Type II error).

Our advice is to not pool data. The choice of an α level for the pooling decision is reasonable, but arbitrary (Underwood 1997). We think you should stick with the nested ANOVA that truly reflects the structure of your data. As we pointed out in Chapter 7, the nested design doesn't give you any added power for detecting treatment effects, and it seems dangerous to try and squeeze additional degrees of freedom out of the analysis after the fact. If it is logistically possible, a better strategy is to reduce or eliminate subsampling in your design and increase the number of truly independent replicates for estimating the treatment effect. Once again, this is an issue that should be resolved before you collect your data.

TABLE 10.5 **ANOVA table for nested design**

Source	Degrees of freedom (df)	Sum of squares (*SS*)	Mean square (*MS*)
Among groups	$a-1$	$\sum_{i=1}^{a}\sum_{j=1}^{b}\sum_{k=1}^{n}(\overline{Y}_i-\overline{Y})^2$	$\frac{SS_{among\ groups}}{(a-1)}$
Among replicates within groups	$a(b-1)$	$\sum_{i=1}^{a}\sum_{j=1}^{b}\sum_{k=1}^{n}(\overline{Y}_{j(i)}-\overline{Y}_i)^2$	$\frac{SS_{replicates(groups)}}{a(b-1)}$
Subsamples within replicates (residual)	$ab(n-1)$	$\sum_{i=1}^{a}\sum_{j=1}^{b}\sum_{k=1}^{n}(Y_{ijk}-\overline{Y}_{j(i)})^2$	$\frac{SS_{subsamples}}{ab(n-1)}$
Total	$abn-1$	$\sum_{i=1}^{a}\sum_{j=1}^{b}\sum_{k=1}^{n}(Y_{ijk}-\overline{Y})^2$	$\frac{SS_{total}}{(abn-1)}$

There are $i = 1$ to a treatment groups, $j = 1$ to b replicates nested within each treatment group, and $k = 1$ to n subsamples nested within each replicate. The among-groups factor is fixed, and the replicates within groups are treated as a random factor. Notice that the test for treatment effect uses the among-replicates mean square, not the residual mean square. The residual mean square (subsamples) is used as the denominator in the test for variation among replicates within a treatment. A common error in the analysis of nested designs is to treat the subsamples as independent (which they are not) and to use the residual mean square in the denominator of the F-ratio to test for the treatment effect.

The analysis we have presented here represents the simplest possible nested design. Many other possible designs include mixtures of nested and crossed factors, and designs such as the split-plot and repeated measures can be interpreted as special forms of nested designs. Our advice is to avoid complicated designs with several nested and crossed factors. In some cases, it may not even be possible to construct a valid ANOVA model for these designs. If your data are organized in a complicated nested design, you can always analyze the averages of the non-independent subsamples; often this will collapse the design into a simpler model. As we discussed in Chapter 7, averaging the subsamples reduces your apparent sample size, but it preserves the true degrees of freedom and independent replication that is necessary for valid hypothesis tests.

Two-Way ANOVA

We return to the example of the two-way design from Chapter 7: a study of barnacle recruitment in which one factor is substrate (3 levels: cement, slate, granite) and the second factor is predation (4 levels: unmanipulated, control, predator exclusion, predator inclusion). There are $i = 1$ to a levels of the first factor, $j = 1$ to b levels of the second factor, and $k = 1$ to n replicates for each unique ij

Expected mean square	F-ratio	P-value
$\sigma^2 + bn\sigma_A^2 + n\sigma_{B(A)}^2$	$\frac{MS_{among\ groups}}{MS_{among\ replicates(groups)}}$	Tail of the F-distribution with $(a-1), a(b-1)$ degrees of freedom
$\sigma^2 + n\sigma_{B(A)}^2$	$\frac{MS_{among\ replicates(groups)}}{MS_{subsamples}}$	Tail of the F-distribution with $a(b-1), ab(n-1)$ degrees of freedom
σ^2		
σ_Y^2		

treatment combination. There are *ab* unique treatment combinations, and $a \times b \times n$ total replicates. In the example from Chapter 7, $a = 3$ substrate levels, $b = 4$ predation levels, and $n = 10$ replicates per treatment combination. There are $ab = (3)(4) = 12$ unique treatment combinations, and a total sample size of $a \times b \times n = (3)(4)(10) = 120$ (see Figure 7.9).

Instead of a single mean square to represent the treatment effect as in the one-way ANOVA, there are now three mean squares for the treatment factors. There is a sum of squares and mean square for each **main effect**, or treatment group: one for substrate with $(a - 1) = 2$ degrees of freedom and one for predation with $(b - 1) = 3$ degrees of freedom. These sums of squares are used to test for differences in means for each factor, just as in a one-way ANOVA.

However, a subtle difference is that the sum of squares associated with the main effect of substrate in the two-way ANOVA is calculated by *averaging* over all the levels of predation. Similarly, the sum of squares associated with the main effect for predation is calculated by averaging over all the levels of substrate. In contrast, the main effect for the one-way ANOVA simply averages the replicates within each treatment, because there is no second factor present in the experimental design.[4]

[4] Of course, this second factor is still present in nature, but it hasn't been incorporated into the design. The one-way design may explicitly restrict the experiment to only one level of the other factor. For example, a one-way design for the predation experiment might be conducted only on natural substrate. Alternatively, the one-way design may not explicitly control for a second factor. However if the design includes proper randomization and replication (see Chapter 6), the second factor will simply contribute to the residual variation. For example, the one-way design for the substrate experiment ignores predation, but if the replicates are independent of one another and their placement is random, the predation effect will be a component of the unexplained residual variation.

In addition to the mean squares for the two main effects, there is a third effect that is estimated for the two-way ANOVA: the interaction between the two factors. The **interaction effect** measures differences in the means of treatment groups that cannot be predicted on the additive basis of the two main effects. Quantifying the interaction effect for two or more treatments is often the key reason for conducting a multi-factor experiment or sampling study. Later in this chapter, we will study the interaction effect in more detail. For now, we simply note that the interaction effect has $(a - 1)(b - 1)$ degrees of freedom. In the barnacle example, the interaction effect has $(3 - 1)(4 - 1) = 6$ degrees of freedom.

The two main effects and the interaction term each show up as elements in our model:

$$Y_{ijk} = \mu + A_i + B_j + AB_{ij} + \varepsilon_{ijk} \tag{10.10}$$

TABLE 10.6 **ANOVA table for two-way design**

Source	Degrees of freedom (df)	Sum of squares (*SS*)	Mean square (*MS*)
Factor *A*	$a - 1$	$\sum_{i=1}^{a}\sum_{j=1}^{b}\sum_{k=1}^{n}(\overline{Y}_i - \overline{Y})^2$	$\frac{SS_A}{(a-1)}$
Factor *B*	$b - 1$	$\sum_{i=1}^{a}\sum_{j=1}^{b}\sum_{k=1}^{n}(\overline{Y}_j - \overline{Y})^2$	$\frac{SS_B}{(b-1)}$
Interaction ($A \times B$)	$(a - 1)(b - 1)$	$\sum_{i=1}^{a}\sum_{j=1}^{b}\sum_{k=1}^{n}(\overline{Y}_{ij} - \overline{Y}_i - \overline{Y}_j + \overline{Y})^2$	$\frac{SS_{AB}}{(a-1)(b-1)}$
Within groups (residual)	$ab(n - 1)$	$\sum_{i=1}^{a}\sum_{j=1}^{b}\sum_{k=1}^{n}(Y_{ijk} - \overline{Y}_{ij})^2$	$\frac{SS_{within\ groups}}{ab(n-1)}$
Total	$abn - 1$	$\sum_{i=1}^{a}\sum_{j=1}^{b}\sum_{k=1}^{n}(Y_{ijk} - \overline{Y})^2$	$\frac{SS_{total}}{(abn-1)}$

There are $i = 1$ to a levels of Factor A, $j = 1$ to b levels of Factor B, and n replicates of each unique ij treatment combination. Because Factors A and B are both fixed, the main effects and the interaction effect are all tested against the residual mean square. This model is the standard "default" output for nearly all statistical packages, even though in many designs, Factors A or B may be random, not fixed.

Following our procedure for constructing F-ratios, each of the corresponding mean squares will be used in the numerator, and the error term will always be used in the denominator (Table 10.6). For the two-way design, the denominator degrees of freedom for the error term are $ab(n-1) = (4)(3)(9) = 108$ degrees of freedom. However this standard procedure for a two-way ANOVA is only valid if the two factors are what is known as **fixed effects**. If the two factors are **random effects**, the expected mean squares change, and we have to construct the F-ratios differently. We will first work through the typical ANOVA tables for fixed factors, and then return to this problem later in the chapter when we discuss random and fixed factors.

Finally, it is instructive to contrast the two-way design for these data (2 factors with 4 and 3 treatment levels, respectively) with the corresponding one-way design (1 factor with 12 treatment levels). In the one-way design, the error term has $a(n-1) = 12(10-1) = 108$ degrees of freedom. For the two-way design, the denominator degrees of freedom for the error term are $ab(n-1) = (4)(3)(9) =$

Expected mean square	F-ratio	P-value
$\sigma^2 + nb\sigma_A^2$	$\frac{MS_A}{MS_{within\ groups}}$	Tail of the F-distribution with $(a-1)$, $ab(n-1)$ degrees of freedom
$\sigma^2 + na\sigma_B^2$	$\frac{MS_B}{MS_{within\ groups}}$	Tail of the F-distribution with $(b-1)$, $ab(n-1)$ degrees of freedom
$\sigma^2 + n\sigma_{AB}^2$	$\frac{MS_{AB}}{MS_{within\ groups}}$	Tail of the F-distribution with $(a-1)(b-1)$, $ab(n-1)$ degrees of freedom
σ^2		
σ_Y^2		

108 degrees of freedom. Thus, the calculation of degrees of freedom and the mean squares are identical, whether we analyze the data as a one-way layout, or as a proper two-way layout.

Compare carefully the degrees of freedom for the treatment effects in these two models. For the two-way design, if you add the degrees of freedom from the two main effects and the interaction you get: 2 (substrate main effect) + 3 (predation main effect) + 6 (predation × substrate interaction) = 11. This is the same number of degrees of freedom that we had in the simple one-way design with 12 treatment levels. Moreover, you would find that the sum of squares for these terms adds up the same way as well. We have effectively partitioned the treatment degrees of freedom from the one-way ANOVA into components that reflect the logical structure of the two-way layout.

ANOVA for Three-Way and *n*-Way Designs

In theory, the two-factor design can be extended to any number of factors. Each factor has different treatment levels within it, and all treatments are completely crossed. Each level of one treatment is applied with each level of every other treatment, so that all combinations are represented. For example, a three-factor experiment that manipulates the presence or absence of herbivores, carnivores, and predators (see Table 7.3) has two levels for each of the three factors. In the three-factor model, there is the grand mean (μ), three main effects (A, B, C), three pairwise interactions (AB, AC, BC), a single three-way interaction term (ABC), and an error term (ε). The model is:

$$Y_{ijkl} = \mu + A_i + B_j + C_k + AB_{ij} + AC_{ik} + BC_{jk} + ABC_{ijk} + \varepsilon_{ijkl} \qquad (10.11)$$

The main effects of herbivore, carnivore, and predator each have $(a - 1)$ degrees of freedom (1 in this example). The pair-wise interaction terms represent the non-additive effect for each pair of possible trophic factors. Each of these interaction terms has $(a - 1)(b - 1)$ degrees of freedom (this expression also equals 1 in this example). Finally, there is a single three-way interaction term also with $(a - 1)(b - 1)(c - 1)$ degrees of freedom (coincidentally also equal to 1 in this example). As before (with a fixed effects model), all of the main effects and interaction terms are tested using the error term as the mean square in the denominator (Table 10.7).

Split-Plot ANOVA

In the split-plot design, treatments of one factor are spatially grouped together, as in a randomized block design. A second treatment is then applied to the entire block or plot (see Figure 7.12). In the barnacle example of Chapter 7, the whole plot factor is predation, because it is the entire block that is caged or manipulated. The within-plot factor is substrate, because each substrate type is represented within each block. The model is

$$Y_{ijk} = \mu + A_i + B_{j(i)} + C_k + AC_{ik} + CB_{kj(i)} \, [+\varepsilon_{ijkl}] \quad (10.12)$$

The whole-plot treatment is A_i, the different plots (nested within factor A) are $B_{j(i)}$, the within-plot factor is C_k, and the error term is ε_{ijkl}. We show this error term in brackets for completeness, but it cannot be isolated in this model because there is no replication of factor C within blocks (each level of C is represented only once within a block).

Two different error terms are used for the hypothesis tests in the split-plot design. To test for effects of the whole-plot treatment A, we use the mean square for blocks $B_{j(i)}$ as the denominator of the F-ratio. This is because entire blocks serve as the independent replicates with respect to the whole-plot treatments.

The within-plot treatment C is tested against the $C \times B$ interaction term, as is the $A \times C$ interaction (Table 10.8). As in a standard two-way ANOVA, this model generates F-ratios and hypothesis tests for the main effects of A, C, and the interaction between them ($A \times C$). However, because the residual error term (ε_{ijkl}) cannot be completely isolated, the split-plot model assumes there is no interaction between factor C and the subplots ($CB_{kj(i)} = 0$; see Underwood 1997, page 393*ff* for a full discussion).

Repeated Measures ANOVA

A repeated measures design is one in which multiple observations are taken on a single individual or replicate. If there is a lot of variability from one replicate to the next, this technique controls for that source of variation. However, the repeated observations on a single individual or replicate are not statistically independent of one another, and care must be taken to ensure that the analysis reflects this structure of dependence in the data.

Two kinds of designs are included in repeated measures analysis. The first kind of design is one in which a single replicate is exposed to different experimental treatments, each applied at a different time and in a randomized order. For example, individual plants might be exposed to a series of different CO_2 concentrations. At each concentration, photosynthetic rates are measured (e.g., Potvin et al. 1986). This design can also be used when the replicates are repeatedly censused, but no manipulations are applied. In this case, the treatments are simply the effects of time.

For both variations, the model is

$$Y_{ij} = \mu + A_i + B_j + \varepsilon_{ij} \quad (10.13)$$

TABLE 10.7 **ANOVA table for three-way factorial design**

Source	Degrees of freedom (df)	Sum of squares (*SS*)
Factor *A*	$a-1$	$\sum_{i=1}^{a}\sum_{j=1}^{b}\sum_{k=1}^{c}\sum_{l=1}^{n}(\bar{Y}_i-\bar{Y})^2$
Factor *B*	$b-1$	$\sum_{i=1}^{a}\sum_{j=1}^{b}\sum_{k=1}^{c}\sum_{l=1}^{n}(\bar{Y}_j-\bar{Y})^2$
Factor *C*	$c-1$	$\sum_{i=1}^{a}\sum_{j=1}^{b}\sum_{k=1}^{c}\sum_{l=1}^{n}(\bar{Y}_k-\bar{Y})^2$
$A\times B$ Interaction	$(a-1)(b-1)$	$\sum_{i=1}^{a}\sum_{j=1}^{b}\sum_{k=1}^{c}\sum_{l=1}^{n}(\bar{Y}_{ij}-\bar{Y}_i-\bar{Y}_j+\bar{Y})^2$
$A\times C$ Interaction	$(a-1)(c-1)$	$\sum_{i=1}^{a}\sum_{j=1}^{b}\sum_{k=1}^{c}\sum_{l=1}^{n}(\bar{Y}_{ik}-\bar{Y}_i-\bar{Y}_k+\bar{Y})^2$
$B\times C$ Interaction	$(b-1)(c-1)$	$\sum_{i=1}^{a}\sum_{j=1}^{b}\sum_{k=1}^{c}\sum_{l=1}^{n}(\bar{Y}_{jk}-\bar{Y}_j-\bar{Y}_k+\bar{Y})^2$
$A\times B\times C$ Interaction	$(a-1)(b-1)(c-1)$	$\sum_{i=1}^{a}\sum_{j=1}^{b}\sum_{k=1}^{c}\sum_{l=1}^{n}\begin{pmatrix}\bar{Y}_{ijk}-\bar{Y}_{ij}-\bar{Y}_{ik}-\bar{Y}_{jk}+\\ \bar{Y}_i+\bar{Y}_j+\bar{Y}_k-\bar{Y}\end{pmatrix}^2$
Within groups (residual)	$abc(n-1)$	$\sum_{i=1}^{a}\sum_{j=1}^{b}\sum_{k=1}^{c}\sum_{l=1}^{n}(Y_{ijkl}-\bar{Y}_{ijk})^2$
Total	$abcn-1$	$\sum_{i=1}^{a}\sum_{j=1}^{b}\sum_{k=1}^{c}\sum_{l=1}^{n}(Y_{ijkl}-\bar{Y})^2$

There are $i = 1$ to a levels of Factor A, $j = 1$ to b levels of Factor B, $k = 1$ to c levels of Factor C, and n replicates of each unique ijk treatment combination. Factors A, B, and C are fixed. This model generates three main effects (A, B, C), three pairwise interaction terms (AB, AC, BC) and one three-way interaction term (ABC). Because Factors A, B, and C are all fixed, all the main effects and interaction terms are tested against the residual mean square. In order to analyze such a design, all possible treatment combinations of the different factors must be established with replication for each.

in which there are $i = 1$ to a treatments (or times) and $j = 1$ to n replicates (or individuals). If you compare Equations 10.13 and 10.8, you will see that the repeated measures model (and corresponding ANOVA table) is the same one

Mean square (*MS*)	Expected mean square	F-ratio	*P*-value
$\frac{SS_A}{(a-1)}$	$\sigma^2 + bcn\sigma_A^2$	$\frac{MS_A}{MS_{within\ groups}}$	Tail of the F-distribution with $(a-1)$, $ab(n-1)$ degrees of freedom
$\frac{SS_B}{(b-1)}$	$\sigma^2 + acn\sigma_B^2$	$\frac{MS_B}{MS_{within\ groups}}$	Tail of the F-distribution with $(b-1)$, $ab(n-1)$ degrees of freedom
$\frac{SS_C}{(c-1)}$	$\sigma^2 + abn\sigma_C^2$	$\frac{MS_C}{MS_{within\ groups}}$	Tail of the F-distribution with $(c-1)$, $abc(n-1)$ degrees of freedom
$\frac{SS_{AB}}{(a-1)(b-1)}$	$\sigma^2 + cn\sigma_{AB}^2$	$\frac{MS_{AB}}{MS_{within\ groups}}$	Tail of F-distribution with $(a-1)(b-1)$, $abc(n-1)$ degrees of freedom
$\frac{SS_{AC}}{(a-1)(c-1)}$	$\sigma^2 + bn\sigma_{AC}^2$	$\frac{MS_{AC}}{MS_{within\ groups}}$	Tail of the F-distribution with $(a-1)(c-1)$, $abc(n-1)$ degrees of freedom
$\frac{SS_{BC}}{(b-1)(c-1)}$	$\sigma^2 + an\sigma_{BC}^2$	$\frac{MS_{BC}}{MS_{within\ groups}}$	Tail of the F-distribution with $(b-1)(c-1)$, $abc(n-1)$ degrees of freedom
$\frac{SS_{ABC}}{(a-1)(b-1)(c-1)}$	$\sigma^2 + n\sigma_{ABC}^2$	$\frac{MS_{ABC}}{MS_{within\ groups}}$	Tail of the F-distribution with $(a-1)(b-1)(c-1)$, $abc(n-1)$ degrees of freedom
$\frac{SS_{within\ groups}}{abc(n-1)}$	σ^2		
$\frac{SS_{total}}{(abcn-1)}$	σ_Y^2		

we used for a randomized block design (see Table 10.4). Each individual serves as its own block, and the error degrees of freedom are accordingly adjusted. As in a randomized block design, this analysis assumes there is no interaction between replicates (individuals) and treatments—only simple, additive differences among individuals in their response to the treatments. Moreover, the treatments have to be applied in a random order, and there must be enough elapsed time between treatment applications to ensure that the responses are independent. We have to assume there are no carry-over effects from one treat-

TABLE 10.8 **ANOVA table for split-plot design**

Source	Degrees of freedom (df)	Sum of squares (*SS*)	Mean square (*MS*)
Factor *A* (whole-plot treatment)	$a-1$	$\sum_{i=1}^{a}\sum_{j=1}^{b}\sum_{k=1}^{c}(\bar{Y}_i-\bar{Y})^2$	$\frac{SS_A}{(a-1)}$
Factor *B*(*A*) (plots nested within *A*)	$a(b-1)$	$\sum_{i=1}^{a}\sum_{j=1}^{b}\sum_{k=1}^{c}(\bar{Y}_j-\bar{Y})^2$	$\frac{SS_B}{a(b-1)}$
Factor *C* (within-plot treatment)	$(c-1)$	$\sum_{i=1}^{a}\sum_{j=1}^{b}\sum_{k=1}^{c}(\bar{Y}_k-\bar{Y})^2$	$\frac{SS_C}{(c-1)}$
$A \times C$ Interaction	$(a-1)(c-1)$	$\sum_{i=1}^{a}\sum_{j=1}^{b}\sum_{k=1}^{c}(\bar{Y}_{ik}-\bar{Y}_i-\bar{Y}_k+\bar{Y})^2$	$\frac{SS_{AC}}{(a-1)(c-1)}$
$B(A) \times C$ Interaction	$a(b-1)(c-1)$	$\sum_{i=1}^{a}\sum_{j=1}^{b}\sum_{k=1}^{c}(Y_{ijk}-\bar{Y}_{ik})^2$	$\frac{SS_{BC}}{a(b-1)(c-1)}$
Total	$abc-1$	$\sum_{i=1}^{a}\sum_{j=1}^{b}\sum_{k=1}^{c}(Y_{ijk}-\bar{Y})^2$	$\frac{SS_{total}}{(abc-1)}$

There are $i = 1$ to a levels of Factor *A* (applied to entire plots), $j = 1$ to b plots nested within Factor *A*, and $k = 1$ to c levels of factor *C* (applied within plots). Factors *A* and *C* are fixed, Factor *B* (plots) is random. In this design, the F-ratio for factor *A* uses the plot factor (*B*, nested within *A*) as the error term for the F-ratio denominator because it is the plots that are the independent replicates for Factor *A*. The F-ratio test for Factor *C* (the within-plot factor) and for the $A \times C$ interaction both use the $B \times C$ interaction as the denominator. This model assumes that there is no interaction between plots and Factor *A*, because this interaction cannot be estimated in the split-plot design.

ment to the next, such as nutrient accumulation, habitat alteration, physiological adaptation, or (in behavioral studies) habituation or learning responses. Also, notice that the treatment order is confounded with time: because treatments are not applied simultaneously, an individual receives certain treatments early in the experimental period and other treatments late in the experimental period. This is another reason why randomizing the order of treatment application is essential.

The second kind of repeated measures design is one in which different treatments are applied, but each replicate (or individual) receives only a single treatment, and then is measured at different times. You can think of this design as an

Expected mean square	F-ratio	P-value
$\sigma^2 + c\sigma_B^2 + bc\sigma_A^2$	$\frac{MS_A}{MS_{B(A)}}$	Tail of the F-distribution with $(a-1)$, $a(b-1)$ degrees of freedom
$\sigma^2 + c\sigma_B^2$		
$\sigma^2 + \sigma_{BC}^2 + ba\sigma_C^2$	$\frac{MS_C}{MS_{B(A)xC}}$	Tail of the F-distribution with $(c-1)$, $a(b-1)(c-1)$ degrees of freedom
$\sigma^2 + \sigma_{BC}^2 + b\sigma_{AC}^2$	$\frac{MS_{AC}}{MS_{B(A)xC}}$	Tail of the F-distribution with $(a-1)(c-1)$, $a(b-1)(c-1)$ degrees of freedom
$\sigma^2 + \sigma_{BC}^2$		
σ_Y^2		

extension of the one-way ANOVA: instead of taking a single observation per replicate, we apply the treatment and take multiple observations on each replicate at different times. The model is

$$Y_{ijk} = \mu + A_i + B_{j(i)} + C_k + AC_{ik} + CB_{kj(i)} \quad (10.14)$$

In this model, there are i = 1 to a treatments, j = 1 to b replicates (or individuals) nested within treatments, and c = 1 to k times. If you compare Equation 10.14 with Equation 10.12 you will see that this repeated measures model is structurally equivalent to the split-plot model. In the repeated measures design, each replicate (or individual) is equivalent to an entire plot. The whole plot factor is simply the treatment that has been applied in the one-way design. The within-plot factor is time, with each sampling period the equivalent of a different treatment level. In the terminology of repeated measures designs, the whole-plot treatment effect corresponds to the between-subjects effect, and the within-plot treatment effect corresponds to the within-subjects effect.

As in the split-plot design, we have to assume there is no interaction between time and replicate: in other words, we must assume that the profile of the temporal trend of the response is the same for all replicates within a given treatment. Because each individual is, by definition, unique, there is no additional sampling we can perform to estimate this interaction term. One final variant of the repeated measures design is one in which the "time" measurements are made before and after the treatment application. In this case, we have one form of a BACI (Before-After-Control-Impact) design, with before and after measurements in control and impact plots (see Chapter 7).

As we noted in Chapter 7, all of these designs assume statistical circularity—the variances of the difference between any pair of times is the same. This assumption is rarely met with time-series data, and other kinds of analyses may be more appropriate for incorporating temporal variation.[5] As in the split-plot analysis, you must avoid the common mistake of analyzing these data as a pure two-way design. The repeated observations collected on a single replicate are not statistically independent, so the mean square for variation among replicates within a treatment is the correct denominator to use in your test of treatment effects.

ANCOVA

Now that you understand how to use linear regression for continuous variables (see Chapter 9), we can develop a new model that is a hybrid of regression and analysis of variance. **ANCOVA (analysis of covariance)** is used for ANOVA designs in which an additional continuous variable (the **covariate**) is measured for each replicate. The hypothesis is that the covariate also contributes to variation in the response variable. If the covariate had not been measured, that source of variation would have been lumped with pure error in the residual. In the analysis of covariance, we can statistically remove that source of variation from the residual. If the covariate has an important effect, the size of the residual is much smaller, and our test for treatment differences will be more powerful. A useful way to think about ANCOVA is that it is an analysis of variance performed on residuals from the regression of the response variable on the covariate.

Returning to the barnacle example of Chapter 7, we know that barnacle settlement is different on horizontal versus vertical surfaces, and we suspect that

[5] We have discussed the univariate analysis of repeated measures data. A different strategy is to analyze the repeated measures data as a multivariate analysis of variance (MANOVA; see Chapter 12), in which the response vector is the set of repeated measures for each replicate. The MANOVA has different assumptions from the univariate analysis, and it may be less powerful and require more replication. See Gurevitch and Chester (1986) and Potvin et al. (1990) for additional discussion.

variation in the placement angle of each replicate might contribute to the pattern in our data. Of course we would try to place our treatment substrates in an approximately horizontal orientation, but there is bound to be variation from one replicate to the next. Therefore, for each replicate, we measure the angle of orientation on a continuous scale from 0 to 90 degrees.[6] This continuous variable will be used as a covariate in the analysis.

The model is

$$Y_{ij} = \mu + A_i + \beta_i(X_{ij} - \overline{X}_i) + \varepsilon_{ij} \tag{10.15}$$

where A_i is the treatment effect ($i = 1$ to a treatments), X_{ij} is the covariate measured for observation Y_{ij}, $\overline{X}_i$ is the average value of the covariate for treatment group i, and ε_{ij} is (as usual) the error term. Notice that this model contains a term A_i for the treatment effect (as in ANOVA) and a slope term β_i for the covariate effect (as in a regression). Equation 10.15 represents the most complex case. It implies that each treatment group is described by its own unique regression line, each with a distinct slope and intercept.

A number of simpler models are embedded in Equation 10.15. For example, suppose that all of the treatment groups have a common regression slope. In this case, we can use the same slope term (β_C) for each treatment group, rather than a separate term for each treatment group (β_i):

$$Y_{ij} = \mu + A_i + \beta_C(X_{ij} - \overline{X}_i) + \varepsilon_{ij} \tag{10.16}$$

If the slope terms representing the covariate effect do not differ from zero, the model reduces to

$$Y_{ij} = \mu + A_i + \varepsilon_{ij}$$

which is just a one-way ANOVA (see Equation 10.6).

[6] Be very careful when using angular data of any sort. The barnacle example we describe here can be analyzed with conventional statistics. But if the angular data are measured on a complete circular scale, there is trouble. For example, in a study of ant foraging activity, suppose we measure daily foraging activity on a scale of 0 to 24 hours. If activity peaks at midnight, we might record two foraging observations at 2300 hours (= 11:00 pm) and 0100 hours (= 1:00 am). But if you take the mean of these numbers, the average is noon (1200 hours) not midnight (0000 hours)! Because the scale of these measurements is circular, rather than linear, we cannot use standard formulas to calculate even basic measures such as means and variances. An entirely different set of procedures is necessary for these circular statistics. See Fischer (1993) for an introduction to circular statistics in biology.

Conversely, if there is no effect of the treatment and no interaction between the treatment and the covariate, the model reduces to a linear regression on the covariate:

$$Y_{ij} = \mu + \beta_T(X_{ij} - \overline{X}) + \varepsilon_{ij} \quad \textbf{(10.17)}$$

The slope in this model (β_T) is fit to all the data, and the treatment designations are ignored.

The analysis of covariance usually involves first a test for differences among the β_i's. This test asks whether or not the slopes are heterogeneous for the group as a whole. If the slopes are heterogeneous, the effect of the treatment depends on the value of the covariate (the X variable) and we use the full model for analysis (see Equation 10.15). In extreme cases, if the regression lines cross, the ordering of the expected values for each treatment will differ between small and large values of the covariate. This "interaction" is similar to an interaction term between two categorical variables in a factorial ANOVA: the differences among treatment means for Factor A depend on the level of Factor B (and vice versa).

If the test for heterogeneous slopes is not significant, a common slope β_C can be fit to all of the treatments (Equation 10.16), and the adjusted means can be calculated. The **adjusted mean** is the expected value of the treatment group that would be found if all groups had the same mean covariate score:

$$\overline{Y}_i = \mu + A_i + \beta_C \overline{X} \quad \textbf{(10.18)}$$

Ideally, in an analysis of covariance, the covariate measurements for the different treatment groups will cover the same range of values and will not differ in their averages (see Figure 10.1A). The ANCOVA in this case is meaningful because the data in the treatment groups are being compared over the same range of the covariate. This distribution of the covariate scores is exactly what we would expect if proper randomization methods have been used to determine the spatial placement of the replicates and the assignment of the treatments. But suppose in the barnacle experiment you inadvertently placed the cement substrates in a flat area of shore, the granite substrates in a steep area, and the slate substrates in an area with intermediate slope (Figure 10.1B). Now the ANCOVA is on shaky ground because the treatments are confounded with differences in the covariate. Statistically, we can still carry out the ANCOVA in the same way, but we have to extrapolate the regression lines beyond the range of data for each treatment, and then assume (and hope) that the same linear relationship still holds. If linearity holds, ANCOVA will statistically partition the variance that is associated with the covariate. But it will not remedy the fact that the design

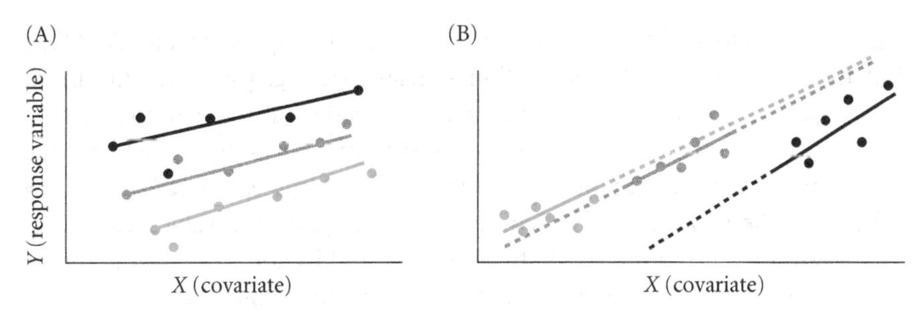

Figure 10.1 Safe and dangerous data sets for analysis of covariance. The covariate is shown on the x-axis and the response variable is shown on the y-axis. The three colors represent three treatment groups. Each point is an independent replicate. (A) In this case, measurements of the covariate occur over the same range for each treatment group. The ANCOVA comparisons are safe because the regression relationships do not have to be extrapolated. (B) In this case, the covariate measurements for each of the three groups are non-overlapping. The design is essentially confounded, because the blue treatment is associated with only small values of the covariate and the black treatment with only large values. The analysis assumes that the regression relationships remain linear when they are extrapolated beyond the observed range of data (indicated by the dashed regression lines). The statistical analysis is the same in both cases, but the conclusions are much less reliable for the data in (B).

is inherently confounded. As we have emphasized throughout this text, randomization is the best safeguard against surprises as in Figure 10.1B.

Random versus Fixed Factors in ANOVA

The tests that we have presented so far are the sort of results that you would obtain from a standard analysis with any computer package. Unfortunately, for two-factor and more complicated designs, these analyses may not be correct.

An important issue that arises in multi-factor designs is whether each factor should be analyzed as a **fixed factor** or a **random factor**. In a fixed factor analysis, the different treatment levels that are used are the only ones that are of interest, and the inferences are restricted to those particular levels. In a random factor analysis, the particular treatment levels represent a random sample of all possible levels that might have been established; the inferences are intended to be valid not just for the particular treatment levels tested, but for other levels of the treatment that were not included in the design. In a **mixed model analysis**, some of the factors are fixed and others are random. The default output for ANOVA in nearly all statistical software packages is for a fixed factor ANOVA.

Why is the distinction between fixed and random factors important? The reason is that the expected mean square values are different for random and fixed factor ANOVAs. Consequently, the F-ratios used to test hypotheses about random versus fixed factors will differ. For example, in a two-way fixed factor ANOVA, both the main effects and the interaction term are tested against the residual mean square (see Table 10.6). But in a random effects model, the main effects are tested using the interaction mean square in the denominator of the F-ratio, not the residual mean square. Both the mean squares and the degrees of freedom are very different for these two terms, which means that your significance values will change completely. In a mixed model, Factor *A* is fixed, and Factor *B* is random. In this case, Factor *A* is tested against interaction mean square, but Factor *B* is tested against the residual mean square. Tables 10.9 and 10.10 give the expected mean squares and proper F-ratios for both random effects and mixed models for the two-way ANOVA. Note that fixed factor, random factor, and mixed model ANOVAs apply only to models with two or more factors. One-way ANOVAs are calculated the same way, whether the treatment factor is fixed or random.

The correct calculation of F-ratios for random and mixed model ANOVAs is very important, but equally important is the fact that these different models dic-

TABLE 10.9 **ANOVA table for two-way ANOVA with random effects**

Source	Degrees of freedom (df)	Sum of squares (*SS*)	Mean square (*MS*)
Factor *A*	$a-1$	$\sum_{i=1}^{a}\sum_{j=1}^{b}\sum_{k=1}^{n}(\bar{Y}_i-\bar{Y})^2$	$\frac{SS_A}{(a-1)}$
Factor *B*	$b-1$	$\sum_{i=1}^{a}\sum_{j=1}^{b}\sum_{k=1}^{n}(\bar{Y}_j-\bar{Y})^2$	$\frac{SS_B}{(b-1)}$
Interaction ($A \times B$)	$(a-1)(b-1)$	$\sum_{i=1}^{a}\sum_{j=1}^{b}\sum_{k=1}^{n}(\bar{Y}_{ij}-\bar{Y}_i-\bar{Y}_j+\bar{Y})^2$	$\frac{SS_{AB}}{(a-1)(b-1)}$
Within groups (residual)	$ab(n-1)$	$\sum_{i=1}^{a}\sum_{j=1}^{b}\sum_{k=1}^{n}(Y_{ijk}-\bar{Y}_{ij})^2$	$\frac{SS_{within\ groups}}{ab(n-1)}$
Total	$abn-1$	$\sum_{i=1}^{a}\sum_{j=1}^{b}\sum_{k=1}^{n}(Y_{ijk}-\bar{Y})^2$	$\frac{SS_{total}}{(abn-1)}$

There are $i = 1$ to a levels of Factor *A*, $j = 1$ to b levels of Factor *B*, and n replicates of each unique *ij* treatment combination. The F-ratio for both main effects are tested using the interaction mean square in the denominator. In contrast, the residual mean square is used in the denominator of the standard F-ratio (see Table 10.6), in which both treatments are fixed factors.

tate different sampling strategies. In the fixed effects model, our goal should be to replicate as much as we can for each treatment combination. All of those extra replicates will give us more degrees of freedom in the residual mean square, which will boost our statistical power. But for the random effects or mixed model, we should use a different strategy. Instead of trying to increase the replication within each treatment level, we should try instead to increase the number of treatment levels that we establish, and not worry so much about the replication within each treatment level. The number of treatment levels determines the number of degrees of freedom in the interaction mean square, and this is where we want to increase the sampling. In the worst case scenario, if we run a two-way random effects ANOVA with only two treatment levels for each factor, we have only 1 degree of freedom for the interaction sum of squares—no matter how much replication we have used (see Footnote 10 in Chapter 7)!

The proper sampling effort for a fixed versus a random effects model should make some intuitive sense. If the domain of inference is restricted to the treat-

Expected mean square	F-ratio	P-value
$\sigma^2 + n\sigma^2_{AB} + nb\sigma^2_A$	$\frac{MS_A}{MS_{AB}}$	Tail of the F-distribution with $(a-1)$, $(a-1)(b-1)$ degrees of freedom
$\sigma^2 + n\sigma^2_{AB} + na\sigma^2_B$	$\frac{MS_B}{MS_{AB}}$	Tail of the F-distribution with $(b-1)$, $(a-1)(b-1)$ degrees of freedom
$\sigma^2 + n\sigma^2_{AB}$	$\frac{MS_{AB}}{MS_{within\ groups}}$	Tail of the F-distribution with $(a-1)(b-1)$, $ab(n-1)$ degrees of freedom
σ^2		
σ^2_Y		

TABLE 10.10 **ANOVA table for two-way mixed model**

Source	Degrees of freedom (df)	Sum of squares (*SS*)	Mean square (*MS*)
Factor *A*	$a-1$	$\sum_{i=1}^{a}\sum_{j=1}^{b}\sum_{k=1}^{n}(\bar{Y}_i-\bar{Y})^2$	$\frac{SS_A}{(a-1)}$
Factor *B*	$b-1$	$\sum_{i=1}^{a}\sum_{j=1}^{b}\sum_{k=1}^{n}(\bar{Y}_j-\bar{Y})^2$	$\frac{SS_B}{(b-1)}$
Interaction ($A \times B$)	$(a-1)(b-1)$	$\sum_{i=1}^{a}\sum_{j=1}^{b}\sum_{k=1}^{n}(\bar{Y}_{ij}-\bar{Y}_i-\bar{Y}_j+\bar{Y})^2$	$\frac{SS_{AB}}{(a-1)(b-1)}$
Within groups (residual)	$ab(n-1)$	$\sum_{i=1}^{a}\sum_{j=1}^{b}\sum_{k=1}^{n}(Y_{ijk}-\bar{Y}_{ij})^2$	$\frac{SS_{within\ groups}}{ab(n-1)}$
Total	$abn-1$	$\sum_{i=1}^{a}\sum_{j=1}^{b}\sum_{k=1}^{n}(Y_{ijk}-\bar{Y})^2$	$\frac{SS_{total}}{(abn-1)}$

Factor *A* is fixed, Factor *B* is random. There are $i = 1$ to a levels of Factor *A*, $j = 1$ to b levels of Factor *B*, and n replicates of each unique ij treatment combination. Notice that the F-ratio for Factor *A* uses the interaction mean square in the denominator, but the F-ratio for Factor *B* and for the interaction effect ($A \times B$) uses the residual mean square in the denominator. Degrees of freedom for these two effects also are different.

ments we have established, then extra replication will boost our power in the fixed effects model. But if the treatment levels are just a random subset of many possible treatment levels, we should be trying to increase our coverage of those different levels for the random effects model. If the treatment levels represent a continuous variable that has been converted to a discrete factor for ANOVA, the best strategy may be to abandon the ANOVA design altogether, and treat the problem as an experimental regression or response surface design, as we discussed in Chapter 7.

It is not always easy to decide whether a factor should be analyzed as fixed or random. If the factor is a set of randomly chosen sites or times, it almost always should be treated as a random factor, which is how the blocks are analyzed in a randomized block design and in a repeated measures design. If the factor represents a well-defined set of categories that is limited in number and does not represent continuous variation, such as species or sexes, it probably should be treated as a fixed factor.

Expected mean square	F-ratio	P-value
$\sigma^2 + n\sigma^2_{AB} + nb\sigma^2_A$	$\dfrac{MS_A}{MS_{AB}}$	Tail of the F-distribution with $(a-1)$, $(a-1)(b-1)$ degrees of freedom
$\sigma^2 + na\sigma^2_B$	$\dfrac{MS_B}{MS_{within\ groups}}$	Tail of the F-distribution with $(b-1)$, $ab(n-1)$ degrees of freedom
$\sigma^2 + n\sigma^2_{AB}$	$\dfrac{MS_{AB}}{MS_{within\ groups}}$	Tail of the F-distribution with $(a-1)(b-1)$, $ab(n-1)$ degrees of freedom
σ^2		
σ^2_Y		

One approach is to consider the ratio x/X where x is the number of treatments in the study and X is the number of possible treatment levels. If this ratio is close to 0, you probably have a random factor, whereas if it is close to 1.0, you probably have a fixed factor.

Many naturally continuous variables, such as nutrient concentration or population density, probably should be analyzed as random factors if you establish discrete levels for your ANOVA design. However, if there is special significance to the particular treatment levels you have used, you might be better off using a fixed design. For example, in a density study, you might establish only two treatments: ambient density and a density of zero. Although there are many other possible levels, ambient density might be viewed as equilibrium conditions, whereas the absence of a species represents the pre-invasion condition. A fixed design might be appropriate here.

You should state clearly in the methods section of your paper or report which factors were fixed and which factors were random, and then use the appropriate design for your ANOVA. There is no excuse for relying on the default fixed factor analysis, particularly since most statistical software allows you to specify which error terms will be used. Tables 10.9 and 10.10 illustrate the proper expected mean squares for only the two-way ANOVA model. For more complex designs, expected mean squares for mixed and random effects can be found in other ANOVA texts (e.g., Winer et al. 1991). However, in some complex designs it may not be possible to construct valid F-ratios for mixed models. This is another reason to use simple ANOVA designs and to avoid designs that use many factors with nesting or repeated measures in your analyses.

Partitioning the Variance in ANOVA

Many uses of ANOVA in ecology and environmental science are aimed at testing hypotheses regarding main effects and interaction terms. However, as we pointed out in Chapter 4, hypothesis testing is only the first step in analyzing data. We are also interested in estimating model parameters or in determining how much variation in the data can be attributed to a particular source (see Figure 4.6). This notion of partitioning the variation is familiar from regression analysis (see Chapter 9), in which the value of the coefficient of determination (r^2) may be more important than whether or not the null hypothesis is rejected.

We can partition the variance in an ANOVA in a similar way, but the procedure is mathematically less straightforward than it is for regression. In the past, investigators incorrectly tried to infer the relative importance of a particular factor from the size of the mean square or the calculated *P*-value. Both approaches are incorrect. The mean square almost always estimates more than one variance component, so it cannot be used by itself. And the *P*-value will be sensitive to the sample size used in the study.

The correct approach is to begin with the expected mean squares for your particular ANOVA table, and then transform them algebraically to isolate the variance component of interest. For example, in a one-way fixed factor ANOVA, we know that the residual mean square estimates the within-group variation:

$$\sigma_e^2 = MS_{within\ groups} \tag{10.19}$$

The among-group mean square estimates the treatment effect plus the residual:

$$\sigma_e^2 + n\,\sigma_A^2 = MS_{among\ groups} \tag{10.20}$$

If we rearrange Equation 10.20, we can isolate the treatment effect:

$$\sigma_A^2 = \frac{MS_{among\ groups} - MS_{within\ groups}}{n} \tag{10.21}$$

One further refinement is necessary. Because this is a fixed-factor ANOVA, there are a finite number of treatment levels of Factor A. Therefore, the estimate of the variance due to the treatment effect (which, strictly speaking is not a true "variance") must be adjusted by the factor $(a - 1)/a$. This adjustment accounts for the difference in the degrees of freedom we have for a true variance measure (a) and a variance estimate from a finite sample ($a - 1$). Notice that, as the number of treatment groups gets larger and larger, the correction factor $[(a - 1)/a]$ approaches 1.0. This makes intuitive sense, because a fixed effects model with a large number of treatment levels effectively is a random effects model.

With this bit of bookkeeping for the fixed effects components, we have

$$\sigma_A^2 = \frac{(MS_{among\ groups} - MS_{within\ groups})(a-1)}{na} \quad (10.22)$$

Once variance components have been isolated in this way, their contribution to the total can be calculated as a **proportion of explained variance** (**PEV**):

$$PEV_A = \frac{\sigma_A^2}{\sigma_A^2 + \sigma_e^2} \quad (10.23)$$

and

$$PEV_e = \frac{\sigma_e^2}{\sigma_A^2 + \sigma_e^2} \quad (10.24)$$

Algebraically, this partitioning is equivalent to the calculation of r^2 in linear regression (see Chapter 9).

As an example, for the data in Table 10.1, $a = 3$ groups, $n = 4$ replicates per group, $MS_{among\ groups} = 11.085$, and $MS_{within\ groups} = 2.167$. From Equations 10.21 and 10.22, the variance estimate for the treatment effect is 1.486, and the variance estimate for the residual is 2.167. From Equation 10.23, $PEV_A = 41\%$: $(1.486)/(1.486 + 2.167) = 0.406$.

For two-way ANOVAs, we follow the same principle of algebraically isolating the term of interest, although the calculations are slightly more complex. As you might guess, the calculations in the two-way ANOVA depend on whether we are using fixed effects, random effects, or mixed models (Table 10.11).

Even with the correct calculations, the technique still has some important limitations. Perhaps the most serious conceptual problem is that the partitioning applies only to those factors that were actually measured and included in the model. Any variation in unmeasured factors, regardless of its impor-

TABLE 10.11 **Variance components in two-way ANOVA models with fixed, mixed, or random factors**

Component of variance	Fixed effects model (*A* fixed, *B* fixed)	Random effects model (*A* random, *B* random)
A	$\frac{(MS_A - MS_{residual})(a-1)}{abn}$	$\frac{(MS_A - MS_{A\times B})}{bn}$
B	$\frac{(MS_B - MS_{residual})(b-1)}{abn}$	$\frac{(MS_B - MS_{A\times B})}{an}$
$A \times B$	$\frac{(MS_{A\times B} - MS_{residual})(a-1)(b-1)}{abn}$	$\frac{(MS_{A\times B} - MS_{residual})}{n}$
Residual	$MS_{residual}$	$MS_{residual}$

tance, is smuggled into the residual mean square, or into the measured treatment effects (if there are interactions between the measured and unmeasured factors).

Even in a one-way ANOVA, the amount of variation attributable to the treatment effect versus the residual or error will depend not only on the number of treatment groups that are used; but also on the particular levels that are chosen (Underwood and Petraitis 1993; Petraitis 1998). Similarly, in a regression study with a continuous predictor variable, the amount of variation accounted for will depend on the range of X-values chosen (see Figures 7.2 and 7.3).

This limitation means that it is difficult, if not impossible, to compare the strength of relative forces in different studies.[7] However, the same caveats apply to the interpretation of P-values calculated from ANOVA. When it is calculated correctly and interpreted conservatively, the partitioning of variance is a useful complement to standard hypothesis testing with ANOVA.

[7] An entirely different approach to evaluating effects across studies is meta-analysis (Arnqvist and Wooster 1995). In meta-analysis, each individual study is treated as an independent replicate for a broader hypothesis test (e.g., trophic cascades are stronger in aquatic versus terrestrial food webs; Shurin et al. 2002). From the published study, the meta-analysis extracts an effect size, usually by comparing the average of control and treatment groups, standardized by the estimated variance. Meta-analysis has recently become popular, but also controversial, in ecology and evolution (e.g., Whittaker 2010 and the associated set of articles in the September, 2010 issue of *Ecology* discussed by Strong 2010). Properly quantifying the effect size (Osenberg et al. 1999) and controlling for publication bias (Jennions and Møller 2002) are two of the current challenges in meta-analysis. See Gurevitch et al. (2001) for an introduction to the methods.

Mixed effects model (*A* fixed, *B* random)
$\frac{(MS_A - MS_{A \times B})(a-1)}{bna}$
$\frac{(MS_B - MS_{A \times B})}{an}$
$\frac{(MS_{A \times B} - MS_{residual})}{n}$
$MS_{residual}$

Here a is the number of treatment levels in Factor A, b is the number of treatment levels in Factor B, and n is the number of replicates per treatment (in a balanced design). *MS* indicates the particular mean square value that is used from the ANOVA table. Because the expected mean squares are different for fixed, random, and mixed model ANOVAs, the estimates of the variance components differ for these models. Once the variance components have been estimated, they can be used to measure the percentage of variation that can be attributed to each factor in the analysis.

After ANOVA: Plotting and Understanding Interaction Terms

At this point, you have posed your scientific question (see Chapter 4), designed and executed your experiment (see Chapters 6 and 7), collected and curated your data (see Chapter 8), and analyzed your data with ANOVA using frequentist, Monte Carlo, or Bayesian methods (see Chapter 5).

The next critical step is to plot the results of the analysis. It is impossible to statistically or biologically interpret the results of an ANOVA without careful reference to graphs of the data, and you should never publish an ANOVA table alone without also publishing some measure of the effect sizes (e.g., group means and variances), either as a table or a graph. We have already discussed exploratory data analysis (EDA) as a means of checking your data for outliers and errors (see Chapter 8) and examining general patterns in your data. Here, we are emphasizing publication-style plots that highlight the patterns of means and variances and allow you to interpret meaningfully the results of the statistical tests with ANOVA.

Plotting Results from One-Way ANOVAs

Let's consider the simplest one-way design for a generic experimental study: three treatment groups, consisting of a set of unmanipulated plots (U) that are not altered in any way except for the census or measurements that are taken, a set of control plots (C) that incorporate potential handling effects, but do not actually implement the treatment of interest, and the treatment plots (T), which implement the treatment of interest and (by necessity) also include the handling effects.

We can plot these data in a simple bar graph. The y-axis of the graph represents the response variable, plotted in its units of measurement. On the x-axis, there is a label for each of the treatment groups being plotted (3 groups in this case). For each group, plot a simple bar (or a box-and-whisker plot; see Chapters 3 and 8). The height of the bar represents the average of the treatment group. Add a vertical line above the bar to indicate the standard deviation around the mean. Ideally, the sample mean and the sample standard deviation would be based on at least 10 observations in each group (see Chapter 6, "The Rule of 10"). For more complicated designs, you may wish to use shading to indicate treatments that may represent important subgroups in your data. These subgroups can then be compared with a priori contrasts, discussed later in this chapter.

With this plot in hand, we can now interpret our ANOVA results. If the F-ratio test for differences among groups is not significant, it means there is no evidence that the population means are any more different than would be expected by chance, due to random sampling error (Figure 10.2A). The caveat is that our experiment or design has followed the assumptions of ANOVA and that our sample size is sufficiently large to have reasonable power for detecting an effect (see Chapter 4). Note that if the within-group variance is large and the sample size is small, the sample means may be quite different from one another, even though the null hypothesis hasn't been rejected. This is why it is so important to plot the standard deviations along with the sample means, and to carefully consider the sample size of the experiment.

What if the F-ratio test for differences among groups is significant? There are three general patterns in the means that lead to three rather different interpretations. First, suppose the treatment plot mean is elevated (or reduced) compared to the control and unmanipulated plots ($T > C = U$; Figure 10.2B). This result suggests that the treatment has a significant effect on the response variable, and that the result does not represent an artifact of the design. Note that the treatment effect would have been inferred erroneously if we had (incorrectly) set the experiment up without the controls and only compared the treatment and unmanipulated groups. Alternatively, suppose that the treatment and control groups have similar means, but they are both elevated above the unmanipulated plots ($T = C > U$; Figure 10.2C). This pattern suggests that the treatment effect is not important and that the difference among the means reflects a handling effect or other artifact of the manipulation. Finally, suppose that the three means differ from one another, with the treatment group mean being the highest and the unmanipulated group mean being the lowest ($T > C > U$; Figure 10.2D). In this case, there is evidence of a handling effect because $C > U$. However, the fact that $T > C$ means that the treatment effect is also real and does not represent just handling artifacts. Once you have established the pattern in your data using ANOVA and the graphical output, the next step is to determine whether that pattern tends to support or refute the scientific hypothesis you are evaluating (see Chapter 4). Always keep in mind the difference between statistical significance and biological significance!

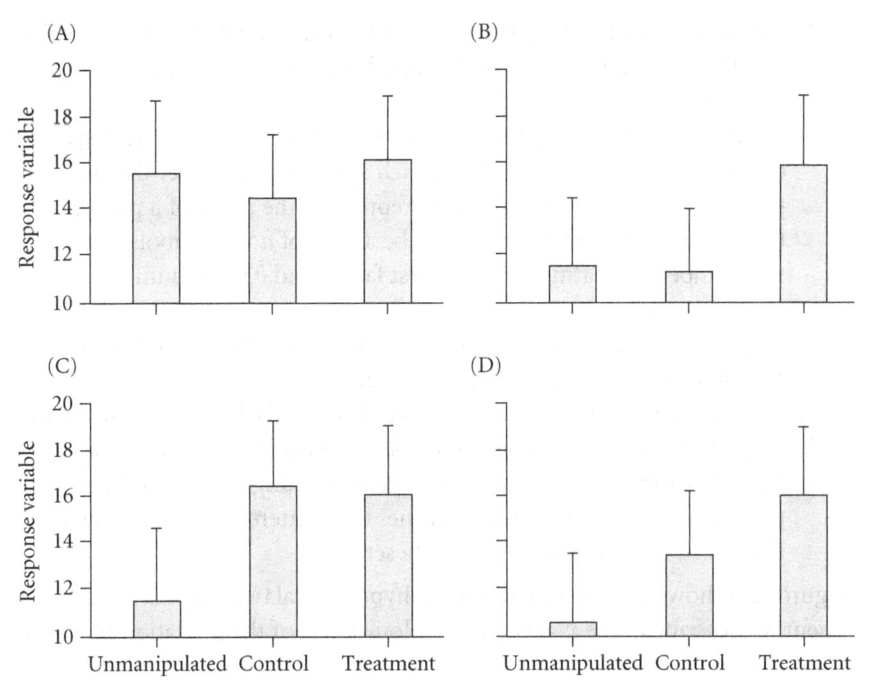

Figure 10.2 Possible outcomes of a hypothetical experiment with three treatments and a one-way ANOVA layout. Unmanipulated plots are not altered in any way except for effects that occur during sampling. Control plots do not receive the treatment, but may receive a sham treatment to mimic handling effects. The treatment group has the manipulation of interest, but also potentially includes handling effects. The height of each bar represents the mean response for the group, and the vertical line indicates 1 standard deviation around the mean. (A) The ANOVA test for treatment effects is non-significant, and the differences among groups in the mean response can be attributed to pure error. (B–D) The ANOVA test is always significant, but the pattern—and hence the interpretation—differs in each case. (B) The treatment mean is elevated compared to the control and unmanipulated replicates, indicating a true treatment effect. (C) Both the control and the treatment group means are elevated relative to the unmanipulated plot, indicating a handling effect, but no distinct treatment effect. (D) There is evidence of a handling effect because the mean of the control group exceeds that of the unmanipulated plot, even though the desired (biological) treatment was not applied. However, above and beyond this handling effect, there appears to be a treatment effect because the mean of the treatment group is elevated relative to the mean of the control group. A priori contrasts or a posteriori comparisons can be used to pinpoint which groups are distinctly different from another. Our point here is that the results of the ANOVA might be identical, but the interpretation depends on the particular pattern of the group means. ANOVA results cannot be interpreted meaningfully without a careful examination of the pattern of means and variances of the treatment groups.

Plotting Results from Two-Way ANOVAs

For plotting the results of two-way ANOVAs, simple bar graphs also can be used, but they are harder to interpret graphically and do not display the main effects and interaction effects very well. We suggest the following protocol for plotting the means in a two-way ANOVA:

1. Establish a plot in which the y-axis is the continuous response variable, and the x-axis represents the different levels for the first factor in the experiment.
2. To represent the second factor in the experiment, use a different symbol or color for each treatment level. Each symbol, placed over the appropriate category label on the x-axis, represents the mean of a particular treatment combination. There will be a total of $a \times b$ symbols where a is the number of treatments in the first factor and b is the number of treatments in the second factor.
3. Align each symbol above the appropriate factor label to establish the combination of treatments represented.
4. Use a (colored) line to connect the symbols across the levels of the first factor.
5. To plot standard deviations use vertical lines pointing upward for the higher treatment levels and vertical lines pointing downward for lower treatment levels. If the figure becomes too cluttered, error bars can be illustrated for only some of the data series.

Figure 10.3 shows just such a plot for the hypothetical two-way barnacle experiment we described. The x-axis gives the four levels of the predation treatment

Figure 10.3 Possible outcomes for a hypothetical experiment testing for the effects of substrate and predation treatment on barnacle recruitment. Each symbol represents a different treatment combination mean. Predation treatments are indicated by the x-axis label and substrate treatments are indicated by the different colors. Vertical bars indicate standard errors or standard deviations (which must be designated in the figure legend). Each panel represents the pattern associated with a particular statistical outcome. (A) Neither treatment effects nor the interaction are significant, and the means of all the treatment combinations are statistically indistinguishable. (B) The predation effect is significant, but the substrate effect and interaction effect are not. In this graph, treatment means are highest for the predator exclusion and lowest for the predator inclusion, with a similar pattern for all substrates. (C) The substrate effect is significant, but the predator effect and the interaction effect are not. There are no differences in the means of the predation treatment, but recruitment is always highest on the granite substrates and lowest on the cement substrates, regardless of the predation treatment. (D) Both predation and substrate effects are significant, but the interaction effect is not. Means depend on both the substrate and the predation treatment, but the effect is strictly additive, and the profiles of the means are parallel across substrates and treatments. (E) Significant interaction effect. Means of the treatments differ significantly, but there is no longer a simple additive effect of either predation or substrate. The ranking of the substrate means depends on the predation treatment, and the ranking of the predation means depends on the substrate. Main effects may not be significant in this case because treatment averages across substrates or predation treatments do not necessarily differ significantly. (F) The interaction effect is significant, which means that the effect of the predation treatment depends on the substrate treatment and vice versa. In spite of this interaction, it is still possible to talk about the general effects of substrate on recruitment. Regardless of the predation treatment, recruitment is always highest on granite substrates and lowest on cement substrates. The interaction effect is statistically significant, but the profiles for the means do not actually cross.

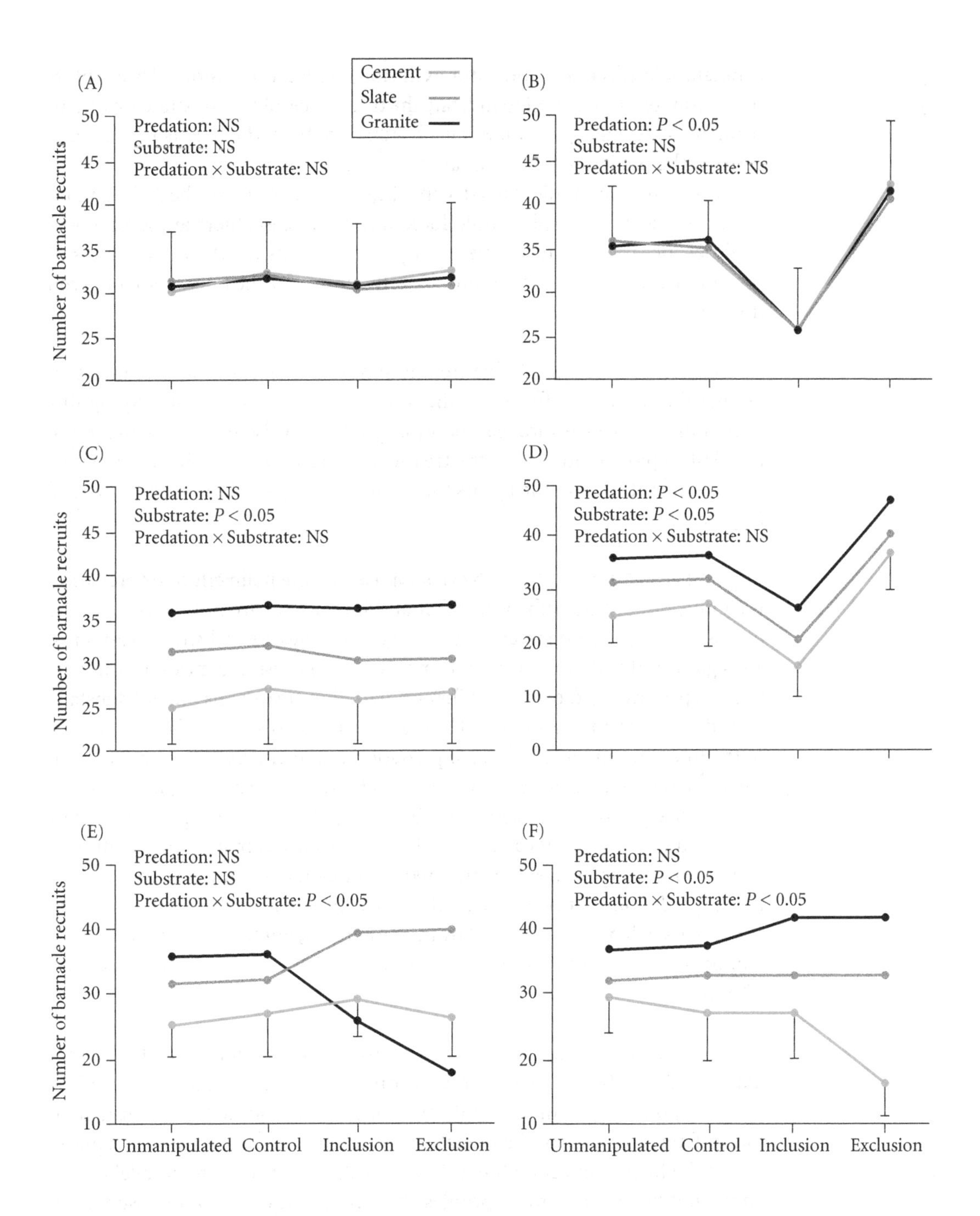
Cement
Slate
Granite
(A)
Predation: NS
Substrate: NS
Predation × Substrate: NS
(B)
Predation: P < 0.05
Substrate: NS
Predation × Substrate: NS
(C)
Predation: NS
Substrate: P < 0.05
Predation × Substrate: NS
(D)
Predation: P < 0.05
Substrate: P < 0.05
Predation × Substrate: NS
(E)
Predation: NS
Substrate: NS
Predation × Substrate: P < 0.05
(F)
Predation: NS
Substrate: P < 0.05
Predation × Substrate: P < 0.05
Number of barnacle recruits
Unmanipulated
Control
Inclusion
Exclusion

(unmanipulated, control, predator exclusion, predator inclusion). Three colors (blue, gray, black) are used to indicate the three levels of the substrate treatment (cement, slate, and granite). For each substrate type, the means for the four levels of the predation treatment are connected by a line.

Once again, we should consider the different outcomes of the ANOVA and what the associated graphs would look like. In this case, there are several possibilities, because we now have two hypothesis tests for the main effects of predation and substrate and a third hypothesis test for the interaction between these two.

NO SIGNIFICANT EFFECTS As before, the simplest scenario is the one in which neither the two main effects nor the interaction term are statistically significant. If the sample sizes are reasonably large, this will also be the case that is the messiest to plot because all of the treatment group averages will be very similar to one another, and the points for the means may be closely superimposed (Figure 10.3A).

ONE SIGNIFICANT MAIN EFFECT Next, suppose that the main effect for predation is significant, but the substrate effect and interaction are not. Therefore, the means of each predation treatment, averaged over the three substrate treatments, are significantly different from one another. In contrast, the means of the substrate types, averaged over the different predation treatments are not substantially different from one another. The graph will show distinctly different clumps of treatment means at each level of predation treatment, but the means of each substrate type will be nearly identical at each predation level (Figure 10.3B).

Conversely, suppose the substrate effect is significant, but the predation effect is not. Now the means of each of the three substrate treatments are significantly different, averaging over the four predation treatments. However, the averages for the predation treatments do not differ. The three connected lines for the substrate types will be well separated from one another, but the slopes of those lines will be basically flat because there is no effect of the four treatments (Figure 10.3C).

TWO SIGNIFICANT MAIN EFFECTS The next possibility is a significant effect of predation and of substrate, but no interaction term. In this case, the profile of the mean responses is again different for each of the substrate types, but it is no longer flat across each of the predation treatments. A key feature of this graph is that the lines connecting the different treatment groups are parallel to one another. When the treatment profiles are parallel, the effects of the two factors are strictly additive: the particular treatment combination can be predicted knowing the average effects of each of the two individual treatments. Additivity of treatment effects and a parallel treatment profile are diagnostic for an ANOVA in which both of the main effects are significant and the interaction term is not significant (Figure 10.3D).

SIGNIFICANT INTERACTION EFFECT The final possibility we will consider is that the interaction term is significant, but neither of the two main effects are. If the interactions are strong, the lines of the profile plot may cross one another (Figure 10.3E). If the interactions are weak, the lines may not cross one another, although they are no longer (statistically) parallel (Figure 10.3F).

Understanding the Interaction Term

With a significant interaction term, the means of the treatment groups are significantly different from one another, but we can no longer describe a simple additive effect for either of the two factors in the design. Instead, the effect of the first factor (e.g., predation) depends on the level of the second factor (e.g., substrate type). Thus, the differences among the substrate types depend on which predation treatment is being considered. In control and unmanipulated plots, recruitment was highest on granite substrates, whereas in predator inclusion and exclusion plots, recruitment was highest on slate. Equivalently, we can say that the effect of the predator treatment depends on the substrate. For the granite substrate, abundance is highest in the controls, whereas for the slate substrate, abundance is highest in the predator inclusion and exclusion treatments.

The graphic portrayal of the interaction term as non-parallel plots of treatment means also has an algebraic interpretation. From Table 10.6, the sum of squares for the interaction term in the two-way ANOVA is

$$SS_{AB} = \sum_{i=1}^{a}\sum_{j=1}^{b}\sum_{k=1}^{n}(\overline{Y}_{ij} - \overline{Y}_i - \overline{Y}_j + \overline{Y})^2 \tag{10.25}$$

Equivalently, we can add and subtract a term for $\overline{Y}$, giving

$$SS_{AB} = \sum_{i=1}^{a}\sum_{j=1}^{b}\sum_{k=1}^{n}\left[(\overline{Y}_{ij} - \overline{Y}) - (\overline{Y}_i - \overline{Y}) - (\overline{Y}_j - \overline{Y})\right]^2 \tag{10.26}$$

The first term $(\overline{Y}_{ij} - \overline{Y})$ in this expanded expression represents the deviation of each treatment group mean from the grand average. The second term $(\overline{Y}_i - \overline{Y})$ represents the deviation from the additive effect of Factor *A*, and the third term $(\overline{Y}_j - \overline{Y})$ represents the additive effect of Factor *B*. If the additive effects of Factors *A* and *B* together account for all of the deviations of the treatment means from the grand mean, then the interaction effect is zero. Thus, the interaction term measures the extent to which the treatment means differ from the strictly additive effects of the two main factors. If there is no interaction term, then knowing the substrate type and knowing the predation treatment would allow us to predict perfectly the response when these factors are combined. But if there is a strong interaction, we can no longer predict the combined effect, even if we understand the behavior of the solitary factors.[8]

It is clear why the interaction term is significant in Figure 10.3E, but why should the main effects be non-significant in this case? The reason is that, if you average the means across each predation treatment or across each substrate type, they would be approximately equal, and there would not be consistent differences for each factor considered in isolation. For this reason, it is sometimes claimed that nothing can be said about main effects when interaction terms are significant. This statement is only true for very strong interactions, in which the profile curves cross one another. However, in many cases, there may be overall trends for single factors even when the interaction term is significant.

For example, consider Figure 10.3F, which would certainly generate a statistically significant interaction term in a two-way ANOVA. The significant interaction tells us that the differences between the means for the substrate types depend on the predation treatment. However, in this case the interaction arises mostly because the cement substrate × predator exclusion treatment has a mean that is very small relative to all other treatments. In all of the predation treatments, the granite substrate (black line) has the highest recruitment and the cement substrate (blue line) has the lowest recruitment. In the control treatment the differences among the substrate means are relatively small, whereas in the predator exclusion treatment, the differences among the substrate means are relatively large. Again, we emphasize that ANOVA results cannot be interpreted properly without reference to the patterns of means and variances in the data.

Finally, we note that data transformations may sometimes eliminate significant interaction terms. In particular, relationships that are multiplicative on a

[8] A morbid example of a statistical interaction is the effect of alcohol and sedatives on human blood pressure. Suppose that alcohol lowers blood pressure by 20 points and sedatives lower blood pressure by 15 points. In a simple additive world, the combination of alcohol and sedatives should lower blood pressure by 35 points. Instead, the interaction of alcohol and sedatives can lower blood pressure by 50 points or more and is often lethal. This result could not have been predicted simply by understanding their effects separately. Interactions are a serious problem in both medicine and environmental science. Simple experimental studies may quantify the effects of single-factor environmental stressors such as elevated CO_2 or increased temperature, but there may be strong interactions between these factors that can cause unexpected outcomes.

linear scale are additive on a logarithmic scale (see discussions in Chapters 3 and 8), and the logarithmic transformation can often eliminate a significant interaction term. Certain ANOVA designs do not include interaction terms, and we have to assume strict additivity in these models.

Plotting Results from ANCOVAs

We will conclude this section by considering the possible outcomes for an ANCOVA. The ANCOVA plot should use the continuous covariate variable plotted on the *x*-axis, and the *Y* variable plotted on the *y*-axis. Each point represents an independent replicate, and different symbols or colors should be used to indicate replicates in different treatments. Initially, fit a linear regression line to each treatment group for a general plot of the results. We will present a single example in which there is one covariate and three treatment levels being compared. Here are the possibilities:

COVARIATE, TREATMENT, AND INTERACTION NON-SIGNIFICANT In this case, the three regression lines do not differ from one another, and the slope of those lines is not different from zero. The fitted intercept of each regression effectively estimates the average of the *Y* values (Figure 10.4A).

COVARIATE SIGNIFICANT, TREATMENT AND INTERACTION NON-SIGNIFICANT In this case, the three regression lines do not differ from one another, but now the slope of the common regression line is significantly different from zero. The results suggest that the covariate accounts for variation in the data, but there are no differences among treatments after the effect of the covariate is (statistically) removed (Figure 10.4B).

TREATMENT SIGNIFICANT, COVARIATE AND INTERACTION TERM NON-SIGNIFICANT In this case, the regression lines again have equivalent zero slopes, but now the intercepts for the three levels are different, indicating a significant treatment effect. Because the covariate does not account for much of the variation in the data, the result is qualitatively the same as if a one-way ANOVA were used and the covariate measures ignored (Figure 10.4C).

TREATMENT AND COVARIATE SIGNIFICANT, INTERACTION TERM NON-SIGNIFICANT In this case, the regression lines have equivalent non-zero slopes and the intercepts are significantly different. The covariate does account for some of the variation in the data, but there is still residual variation that can be attributed to the treatment effect. This result is equivalent to fitting a single regression line to the entire data set, and then using a one-way ANOVA on the residuals to test for treatment effects. When this result is obtained, it is appropriate to fit a common regression

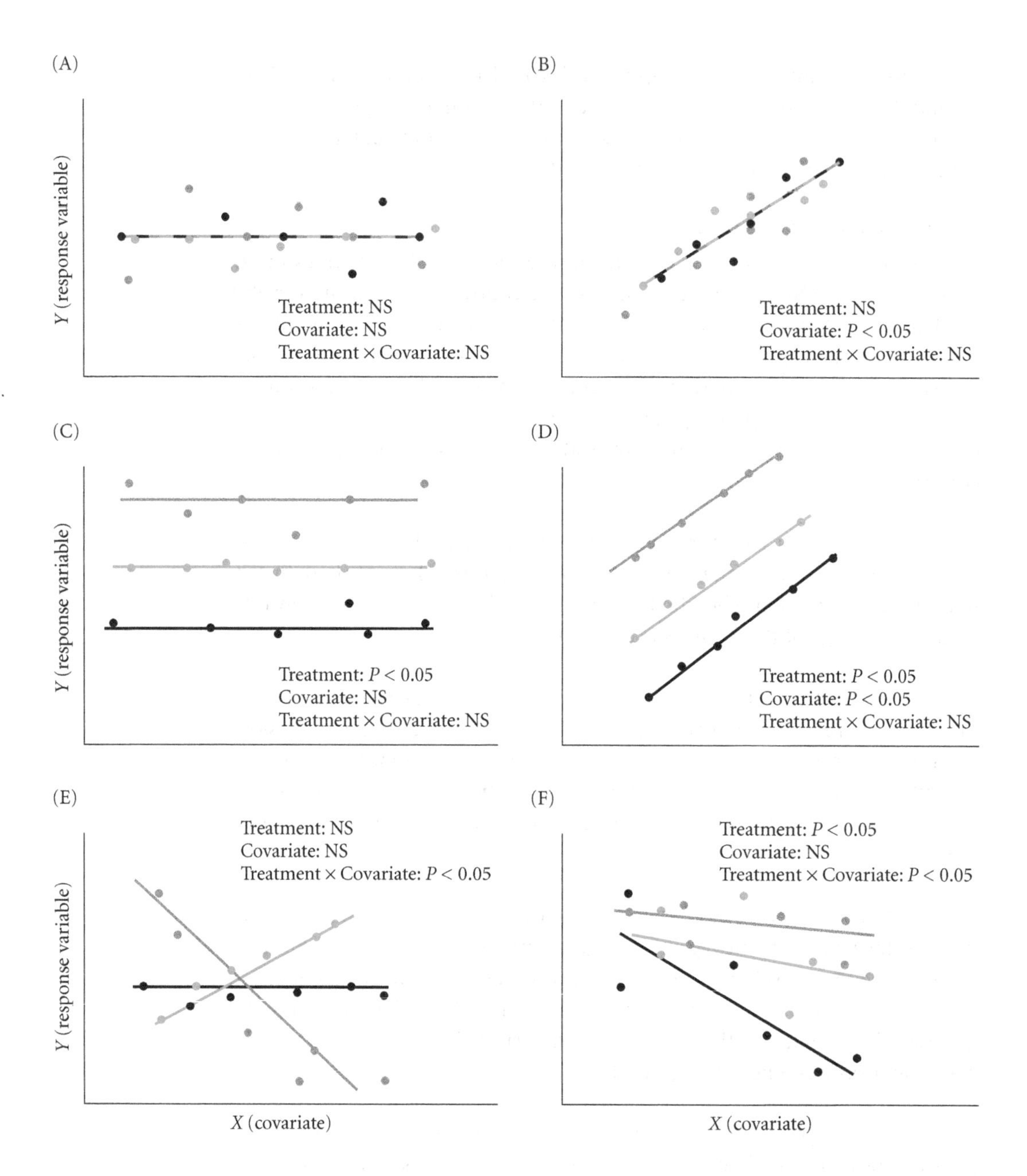

slope and use Equation 10.18 to estimate the adjusted treatment means (Figure 10.4D).

INTERACTION TERM SIGNIFICANT This case represents heterogeneous regression slopes, in which it is necessary to fit a separate regression line, with its own slope

◀**Figure 10.4** Possible outcomes for experiments with ANCOVA designs. In each panel, different symbols indicate replicates in each of three treatment groups. Each panel indicates a different possible experimental outcome with the associated ANCOVA result. (A) No significant treatment or covariate effects. The data are best fit by a single grand average with sampling error. (B) Covariate effect significant, no significant treatment effect. The data are best fit by a single regression line with no differences among treatments. (C) Treatment effect significant, no significant covariate. The covariate term is not significant (regression slope = 0), and the data are best fit by a model with different group means, as in a simple one-way ANOVA. (D) Treatment and covariate are both significant. The data are best fit by a regression model with a common slope, but different intercepts for each treatment group. This model is used to calculate adjusted means, which are estimated for the grand mean of the covariate. (E) A treatment × covariate interaction in which the order of the treatment means differs for different values of the covariate. The regression slopes are heterogeneous, and each treatment group has a different slope and intercept. (F) A treatment × covariate interaction in which the order of the treatment groups does not differ for different values of the covariate. Both the treatment and the treatment × covariate interaction are significant.

and intercept, to each treatment group (Figure 10.4E,F). When the treatment × covariate interaction is significant, it may not be possible to discuss a general treatment effect because the difference among the treatments may depend on the value of the covariate. If the interaction is strong and the regression lines cross, the ordering of the treatment means will be reversed at high and low values of the covariate (Figure 10.4E). If the treatment effect is strong, the rank order of the treatment means may remain the same for different values of the covariate, even though the interaction term is significant (Figure 10.4F).

Comparing Means

In the previous section, we emphasized that comparing the means of different treatment groups is essential for correctly interpreting the analysis of variance results. But how do we decide which means are truly different from one another? The ANOVA only tests the null hypothesis that the treatment means were all sampled from the same distribution. If we reject this null hypothesis, the ANOVA results do not specify which particular means differ from one another.

To compare different means, there are two general approaches. With a posteriori ("after the fact") comparisons, we use a test to compare all possible pairs of treatment means to determine which ones are different from one another. With a priori ("before the fact") contrasts, we specify ahead of time particular combinations of means that we want to test. These combinations often reflect specific hypotheses that we are interested in assessing.

Although a priori contrasts are not used often by ecologists and environmental scientists, we favor them for two reasons: first, because they are more specific, they are usually more powerful than generalized tests of pairwise differences of means. Second, the use of a priori contrasts forces an investigator to think clearly about which particular treatment differences are of interest, and how those relate to the hypotheses being addressed.

We illustrate both a priori and a posteriori approaches with the analysis of a simple ANOVA design. The data come from Ellison et al. (1996), who studied the interaction between red mangrove (*Rhizophora mangle*) roots and epibenthic sponges. Mangroves are among the few vascular plants that can grow in full-strength salt water; in protected tropical swamps, they form dense forests fringing the shore. The mangrove prop roots extend down to the substrate and are colonized by a number of species of sponges, barnacles, algae, and smaller invertebrates and microbes. The animal assemblage obviously benefits from this available hard substrate, but are there any effects on the plant?

Ellison et al. (1996) wanted to determine experimentally whether there were positive effects of two common sponge species on the root growth of *Rhizophora mangle*. They established four treatments, with 14 to 21 replicates per treatment.[9] The treatments were as follows: (1) unmanipulated; (2) foam attached to bare mangrove roots (the foam is a control "fake sponge" that mimics the hydrodynamic and other physical effects of a living sponge, but is biologically inert[10]); (3) *Tedania ignis* (red fire sponge) living colonies transplanted to bare mangrove roots; (4) *Haliclona implexiformis* (purple sponge) living colonies transplanted to bare mangrove roots. Replicates were established by randomly pre-selecting bare mangrove roots. Treatments were then randomly assigned to the replicates. The response variable was mangrove root growth, measured as mm/day for each

[9] The complete design was actually a randomized block with unequal sample sizes, but we will treat it as a simple one-way ANOVA to illustrate the comparisons of means. For our simplified example, we analyze only 14 randomly chosen replicates for each treatment, so our analysis is completely balanced.

[10] This particular foam control deserves special mention. In this book, we have discussed experimental controls and sham treatments that account for handling effects, and other experimental artifacts that we want to distinguish from meaningful biological effects. In this case, the foam control is used in a subtly different way. It controls for the physical (e.g., hydrodynamic) effects of a sponge body that disrupts current flow and potentially affects mangrove root growth. Therefore, the foam control does not really control for any handling artifacts because replicates in all four treatments were measured and handled in the same way. Instead, this treatment allows us to partition the effect of living sponge colonies (which secrete and take up a variety of minerals and organic compounds) from the effect of an attached sponge-shaped structure (which is biologically inert, but nevertheless alters hydrodynamics).

TABLE 10.12 **Analysis of variance table for experiment testing effects of sponges on growth of red mangrove roots**

Source	Degrees of freedom (df)	Sum of squares (*SS*)	Mean square (*MS*)	F-ratio	*P*-value
Treatments	3	2.602	0.867	5.286	0.003
Residual	52	8.551	0.164		
Total	55	11.153			

Four treatments were established (unmanipulated, foam, and two living sponge treatments) with 14 replicates in each treatment used in this analysis; total sample size = 14 × 4 = 56. The F-ratio for the treatment effect is highly significant, indicating that mean growth rates of mangrove roots differed significantly among the four treatments. Means and standard deviations of root growth (mm/day) for each treatment are illustrated in Figure 10.6. (Data from Ellison et al. 1996.)

replicate. Table 10.12 gives the analysis of variance for these data and Figure 10.5 illustrates the patterns among the means. The *P*-value for the F-ratio is highly significant ($F_{3,52}$ = 5.286, P = 0.001), and the treatments account for 19% of the variation in the data (computed using Equation 10.23). The next step is to focus on which particular treatment means are different from one another.

A Posteriori Comparisons

We will use Tukey's "honestly significant difference" (HSD) test to compare the four treatment means in the mangrove growth experiment. This is one of many

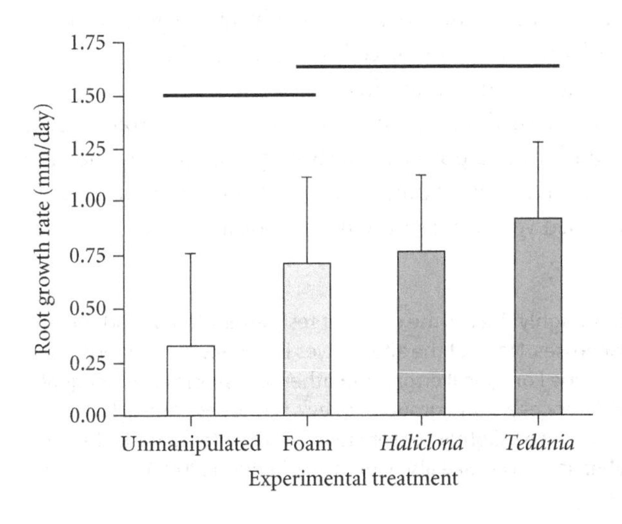

Figure 10.5 Growth rate (mm/day) of mangrove roots in four experimental treatments. Four treatments were established (unmanipulated, foam, and two living sponge treatments), with 14 replicates in each treatment. The height of the bar is the average growth rate for the treatment, and the vertical error bar is one standard deviation around the mean. The horizontal lines join treatment groups that do not differ significantly by Tukey's HSD test (see Table 10.13). (Data from Ellison et al. 1996. Statistical analyses of these data are presented in Tables 10.12–10.15.)

a posteriori procedures available for comparing pairs of means after ANOVA.[11] Tukey's HSD test statistically controls for the fact that we are carrying out many simultaneous comparisons. Therefore, the *P*-value must be adjusted downward for each individual test to achieve an experiment-wise error rate of $\alpha = 0.05$. In the next section, we will discuss in more detail the general problem of how to deal with multiple *P*-values in a study. The first step is to calculate the HSD:

$$HSD = q\sqrt{\left(\frac{1}{n_i} + \frac{1}{n_j}\right)MS_{residual}} \quad \textbf{(10.27)}$$

where q is the value from a statistical table of the studentized range distribution, n_i and n_j are the sample sizes for the means of group i and group j that are being compared, and $MS_{residual}$ is the familiar residual mean square from the one-way ANOVA table. For the data in Table 10.12, $q = 3.42$ (taken from a set of statistical tables), n_i and $n_j = 14$ replicates for all of the means, and $MS_{residual} = 0.164$. The calculated HSD = 0.523.

Therefore, any pair of means among the four treatments that differs by at least 0.523 in average daily root growth is significantly different at $P = 0.05$. Table 10.13 shows the matrix of pair-wise differences between each of the means, and the corresponding *P*-value calculated by Tukey's test.

The analysis suggests that the unmanipulated roots differed significantly from growth of the roots with sponges attached. The two living sponge treatments did not differ significantly from each other or from the foam treatment. However, the foam treatment was marginally non-significant ($P = 0.07$) in comparison with the unmanipulated roots. These patterns are illustrated graphically in Figure 10.5, in which horizontal lines are placed over sets of treatment means that do not differ significantly from one another.

Although these pair-wise tests did reveal particular pairs of treatments that differ, Tukey's HSD test and other a posteriori tests may occasionally indicate that none of the pairs of means are significantly different from one another, even though the overall F-ratio led you to reject the null hypothesis! This inconsis-

[11] Day and Quinn (1989) thoroughly discuss the different tests after ANOVA and their relative strengths and weaknesses. None of the alternatives is entirely satisfactory. Some suffer from excessive Type I or Type II errors, and others are sensitive to unequal sample sizes or variance differences among groups. Tukey's HSD does control for multiple comparisons, although it may be slightly conservative with a potential risk of a Type II error (not rejecting H_0 when it is false). See also Quinn and Keough (2002) for a discussion of the different choices.

TABLE 10.13 **Mean differences between all possible pairs of treatments in an experimental study of effects of sponges on root growth of red mangrove**

	Unmanipulated	Foam	*Haliclona*	*Tedania*
Unmanipulated	0.000			
Foam	0.383 (0.072)	0.000		
Haliclona	0.536 (0.031)	0.053 (0.986)	0.000	
Tedania	0.584 (0.002)	0.202 (0.557)	0.149 (0.766)	0.000

Four treatments were established (unmanipulated, foam, two living sponge treatments) with 14 replicates in each treatment used in this analysis; total sample size = $14 \times 4 = 56$. The ANOVA is given in Table 10.12. This table illustrates the differences among the treatment means. Each entry is the difference in mean root growth (mm/day) of a pair of treatments. The diagonal of the matrix compares each treatment to itself, so the difference is always zero. The value in parentheses is the associated tail probability based on Tukey's HSD test. This test controls the α level to account for the fact that six pairwise tests have been conducted. The larger the difference between the pairs of means, the lower the P-value. The unmanipulated roots differ significantly from the two treatments with living sponges (*Tedania, Haliclona*). Differences between all other pairs of means are non-significant. However, the pairwise comparison of the unmanipulated and foam treatments is only marginally above the $P = 0.05$ level, whereas the foam treatment does not differ significantly from either of the living sponge treatments. These patterns are illustrated graphically in Figure 10.5. (Data from Ellison et al. 1996.)

tent result can arise because the pairwise tests are not as powerful as the overall F-ratio itself.

A Priori Contrasts

A priori contrasts are much more powerful, both statistically, and logically, than pairwise a posteriori tests. The idea is to establish **contrasts**, or specified comparisons between particular sets of means that test specific hypotheses. If we follow certain mathematical rules, a set of contrasts will be orthogonal or independent of one another, and they will actually represent a mathematical partitioning of the among-groups sum of squares.

To create a contrast, we first assign an integer (positive, negative, or 0) to each treatment group. This set of integer coefficients is the contrast that we are testing. Here are the rules for building contrasts:

1. The sum of the coefficients for a particular contrast must equal 0.
2. Groups of means that are to be averaged together are assigned the same coefficient.
3. Means that are not included in the comparison of a particular contrast are assigned a coefficient of 0.

Let's try this for the mangrove experiment. If living sponge tissue enhances root growth, then the average growth of the two living sponge treatments should be greater than the growth of roots in the inert foam treatment.

Our contrast values are

Contrast I control (0) foam (2) *Tedania* (–1) *Haliclona* (–1)

Haliclona and *Tedania* receive the same coefficient (–1) because we want to compare the average of those two treatments with the foam treatment. The control receives a coefficient of 0 because we are testing an hypothesis about the effect of *living* sponge, so the relevant comparison is only with the foam. Foam receives a coefficient of 2 so that it is balanced against the average of the two living sponges and the sum of the coefficients is 0. Equivalently, you could have assigned coefficients of control (0), foam (6), *Tedania* (–3) and *Haliclona* (–3) to achieve the same contrast.

Once the contrast is established, we use it to construct a new mean square, which has one degree of freedom associated with it. You can think of this as a weighted mean square, for which the weights are the coefficients that reflect the hypothesis. The mean square for each contrast is calculated as

$$MS_{contrast} = \frac{n\left(\sum_{i=1}^{a} c_i \overline{Y}_i\right)^2}{\sum_{i=1}^{a} c_i^2} \quad \textbf{(10.28)}$$

The term in parentheses is just the sum of each coefficient (c_i) multiplied by the corresponding group mean ($\overline{Y}_i$). For our first contrast, the mean square for foam versus living sponge is

$$MS_{foam\,vs\,living} = \frac{\left((0)(0.329)+(2)(0.712)+(-1)(0.765)+(-1)(0.914)\right)^2 \times 14}{0^2+2^2+(-1)^2+(-1)^2} = 0.151 \quad \textbf{(10.29)}$$

This mean square has 1 degree of freedom, and we test it against the error mean square. The F-ratio is 0.151/0.164 = 0.921, with an associated *P*-value of 0.342 (Table 10.14). This contrast suggests that the growth rate of mangrove roots covered with foam was comparable to growth rates of mangrove roots covered with living sponges. This result makes sense because the means for these three treatment groups are fairly similar.

There is no limit to the number of potential contrasts we could create. However, we would like our contrasts to be orthogonal to one another, so that the results are logically independent. Orthogonal contrasts ensure that the P-values are not excessively inflated or correlated with one another. In order to create orthogonal contrasts, two additional rules must be followed:

4. If there are a treatment groups, at most there can be $(a - 1)$ orthogonal contrasts created (although there are many possible sets of such orthogonal contrasts).
5. All of the pair-wise cross products must sum to zero (see Appendix). In other words, a pair of contrasts Q and R is independent if the sum of the products of their coefficients c_{Qi} and c_{Ri} equals zero:

$$\sum_{i=1}^{a} c_{Qi}c_{Ri} = 0 \tag{10.30}$$

Building orthogonal contrasts is a bit like solving a crossword puzzle. Once the first contrast is established, it constrains the possibilities for those that are remaining. For the mangrove root data, we can build two additional orthogonal contrasts to go with the first one. The second contrast is:

Contrast II control (3) foam (–1) *Tedania* (–1) *Haliclona* (–1)

This contrast sums to zero: it is orthogonal with the first contrast because it satisfies the cross products rule (Equation 10.30):

$$(0)(3)+(2)(-1)+(-1)(-1)+(-1)(-1)=0 \tag{10.31}$$

This second contrast compares root growth in the control with the average of root growth of the foam and the two living sponges. This contrast tests whether enhanced root growth reflects properties of living sponges or the physical consequences of an attached structure per se. This contrast gives a very large F-ratio (14.00) that is highly significant (see Table 10.14). Again, this result makes sense because the average growth of the unmanipulated roots (0.329) is substantially lower than that of the foam treatment (0.712) or either of the two living sponge treatments (0.765 and 0.914).

One final orthogonal contrast can be created:

Contrast III control (0) foam (0) *Tedania* (1) *Haliclona* (–1)

The coefficients of this contrast again sum to zero, and the contrast is orthogonal to the first two because the sum of the cross products is zero (Equation 10.30). This third contrast tests for differences in growth rate of the two living sponge species but does not assess the mutualism hypothesis. It is of less interest than the other two contrasts, but it is the only contrast that can be created that is orthogonal to the first two. This contrast yields an F-ratio of only 0.947, with a corresponding *P*-value of 0.335 (see Table 10.14). Once again, this result is expected because the average growth rates of roots in the two living sponge treatments are similar (0.765 and 0.914).

Notice that the sum of squares for each of these three contrasts add up to the total for the treatment sum of squares:

$$\begin{aligned} SS_{treatment} &= SS_{contrast\ I} + SS_{contrast\ II} + SS_{contrast\ III} \\ 2.602 &= 0.151 + 2.296 + 0.155 \end{aligned} \tag{10.32}$$

In other words, we have now partitioned the total among-group sum of squares into three orthogonal, independent contrasts. This is similar to what happens in a two-way ANOVA, in which we decompose the overall treatment sum of squares into two main effects and an interaction term.

This set of three contrasts is not the only possibility. We could have also specified this set of orthogonal contrasts:

Contrast I	control (1)	foam (1)	*Tedania* (–1)	*Haliclona* (–1)
Contrast II	control (1)	foam (–1)	*Tedania* (0)	*Haliclona* (0)
Contrast III	control (0)	foam (0)	*Tedania* (1)	*Haliclona* (–1)

Contrast I compares the average growth of roots covered with the living sponges with the average growth of roots in the two treatments with no living sponges. Contrast II specifically compares the growth of control roots versus foam-covered roots, and Contrast III compares growth rates of roots covered with *Haliclona* versus roots covered with *Tedania*. These results (Table 10.15) suggest that living sponges enhance root growth compared to unmanipulated roots or those with attached foam. However, the Contrast II reveals that root growth with foam was enhanced compared to unmanipulated controls, suggesting that mangrove root growth is responsive to hydrodynamics or some other physical property of attached sponges, rather than the presence of living sponge tissue.

Either set of contrasts is acceptable, although we think the first set more directly addresses the effects of hydrodynamics versus living sponges. Finally, we note that there are at least two non-orthogonal contrasts that may be of interest. These would be

control (0)	foam (1)	*Tedania* (–1)	*Haliclona* (0)
control (0)	foam (1)	*Tedania* (0)	*Haliclona* (–1)

These contrasts compare root growth with each of the living sponges versus root growth with the foam control. Both contrasts are non-significant (*Tedania* $F_{1,52} = 0.12$, $P = 0.73$; *Haliclona* $F_{1,52} = 0.95$, $P = 0.33$), again suggesting that the effect of living sponges on mangrove root growth is virtually the same as the effect of biologically inert foam. However, these two contrasts are not orthogonal to one another (or to our other contrasts) because their cross products do not sum to zero. Therefore, these calculated *P*-values are not completely independent of the *P*-values we calculated in the other two sets of contrasts.

In this example, the a posteriori comparisons using Tukey's HSD test gave a somewhat ambiguous result, because it wasn't possible to separate cleanly the mean values on a pairwise basis (see Table 10.13). The a priori contrasts (in contrast) gave a much cleaner result, with a clear rejection of the null hypothesis and confirmation that the sponge effect could be attributed to the physical effects of the sponge colony, and not necessarily to any biological effects of living sponge (see Tables 10.14 and 10.15). The a priori contrasts were more successful because they are more specific and more powerful than the pairwise tests. Notice also that the pairwise analysis required six tests, whereas the contrasts used only three specified comparisons.

One cautionary note: a priori contrasts really do have to be established a priori—that is, *before* you have examined the patterns of the treatment means! If the contrasts are set up by grouping together treatments with similar means, the resulting *P*-values are entirely bogus, and the chance of a Type I error (falsely

TABLE 10.14 **A priori contrasts for analysis of variance for effects of sponges on mangrove root growth**

Source	Degrees of freedom (df)	Sum of squares (*SS*)	Mean square (*MS*)	F-ratio	*P*-value
Treatments	3	2.602	0.867	5.287	0.026
Foam vs. living	1	0.151	0.151	0.921	0.342
Unmanipulated vs. living, foam	1	2.296	2.296	14.000	<0.001
Tedania vs. *Haliclona*	1	0.155	0.155	0.945	0.335
Residual	52	8.539	0.164		
Total	56	11.141			

Treatments were established as in Tables 10.12 and 10.13. The simple one-way ANOVA test for treatment effects (Table 10.12) has now been partitioned into three independent contrasts, each with 1 degree of freedom. Each contrast is tested against the residual mean square. The analysis reveals that the growth of unmanipulated roots is significantly different from the average growth of roots covered with living sponges or foam. There was no significant difference between growth of roots with foam and growth of roots with living sponges. There was also no significant difference between growth of roots covered with the two different sponge species (*Tedania* and *Haliclona*). Note that 3 degrees of freedom and the sum of squares for the total treatment effect has now been partitioned into 3 additive components, each of which tests a more specific a priori hypothesis about mean differences. (Data from Ellison et al. 1996.)

TABLE 10.15 **An alternative set of a priori contrasts for analysis of variance for effects of sponges on mangrove root growth**

Source	Degrees of freedom (df)	Sum of squares (*SS*)	Mean square (*MS*)	F-ratio	*P*-value
Treatments	3	2.602	0.867	5.275	0.003
Unmanipulated, foam versus living	1	1.425	1.425	8.687	0.005
Unmanipulated versus foam	1	1.027	1.027	6.261	0.016
Tedania versus *Haliclona*	1	0.155	0.155	0.947	0.335
Residual	52	8.551	0.164		
Total	56	11.153			

Treatments were established as in Tables 10.12 and 10.13. The simple one-way ANOVA test for treatment effects (Table 10.12) again has been partitioned into three independent contrasts, each with 1 degree of freedom. Each contrast is tested against the residual mean square. These contrasts are slightly different from those of Table 10.14. The first contrast indicates that mangrove root growth rates were higher on treatments with living sponges than on treatments with foam or without sponges. The second contrast indicates that root growth was also enhanced by the foam treatment compared to the controls. There was no significant difference between the growth of roots covered with the two different sponge species (*Tedania* and *Haliclona*). Note that 3 degrees of freedom and the sum of squares for the total treatment effect has now been partitioned into 3 additive components, each of which tests a more specific a priori hypothesis about mean differences. (Data from Ellison et al. 1996.)

rejecting a true null hypothesis) is greatly inflated. If you do not have true a priori contrasts, and simply want to determine which means are different, you should use one of the a posteriori methods.

In summary, a priori contrasts are a powerful, but under-utilized, tool that allows you to compare particular sets of means that are relevant to your hypotheses. They are easy to calculate, and they can also be readily applied to two-way ANOVAS and more complex designs. In fact, you can even use a priori contrasts to partition main effects and interaction terms, as in a standard two-way ANOVA. We greatly favor thoughtful a priori contrasts in lieu of the numerous a posteriori comparison tests that are available.

Bonferroni Corrections and the Problem of Multiple Tests

In our discussion of a priori contrasts and a posteriori comparisons, we alluded to the problem of multiple comparisons. Both the a priori contrasts and

Tukey's HSD test do control internally for the number of different statistical comparisons that are being made. But such control is not present when we are comparing the results of multiple tests.

The difficulty is that the more statistical tests that are conducted the more likely it is that a P-value less than 0.05 will be obtained for one or more of those tests, even though the patterns reflect only random error. For example, if you conducted 20 independent statistical tests, and the null hypothesis were true in all 20 of them, you would still expect to reject the null hypothesis 5% of the time which would be approximately $(5\%) \times 20 = 1$ statistical test. Therefore, many authors advocate that you adjust the acceptable level for a Type I error (α) when conducting multiple tests.

The straightforward **Bonferroni method** is to establish the **experiment-wide error rate**, and divide it by the number of tests to give an adjusted α level (α') for each individual test:

$$\alpha' = \alpha/k \tag{10.33}$$

where α is the experiment-wide error rate, k is the number of tests being compared, and α' is the adjusted alpha level for each individual test. Thus, with 20 statistical tests, and an experiment-wide error rate of 0.05, we only would reject the null hypothesis for any particular test if $P < \alpha'$, which by Equation 10.33 = $0.05/20 = 0.0025$.

The Bonferroni procedure is unnecessarily conservative because it does not account for cases in which more than one test is simultaneously rejected. A better correction is the **Dunn-Sidak method**, which is the correct calculation of the probability of rejecting the null hypothesis at least once, based on the simple rules of probability and combinatorics (see Chapter 2). If α' is the adjusted alpha level, then the probability of making no Type I errors in k tests is

$$P(\text{no Type I error}) = (1 - \alpha')^k \tag{10.34}$$

The probability of making at least one Type I error, which is the experiment-wide error rate, therefore is

$$P(\text{at least one Type I error}) = \alpha = 1 - (1 - \alpha')^k \tag{10.35}$$

Rearranging in terms of the adjusted alpha level for each individual test gives

$$\alpha' = 1 - (1 - \alpha)^{\frac{1}{k}} \tag{10.36}$$

For the example of 20 independent tests, with an experiment-wide error rate of 0.05, we have a critical value of $\alpha' = 1 - (1 - 0.05)^{1/20} = 0.00256$, slightly higher than the simple Bonferroni correction of 0.00250.

These kinds of corrections are very popular with journal editors and reviewers, and frequently appear in journal articles. However, we do not favor any such adjustments to alpha levels, and encourage you to not use them unless your hand is forced by an editor.

Why are we arguing against the accepted standard? Here are the problems with adjusting α. First, the analyses assume that the tests are all independent of one another, which they rarely are. If the tests are not independent, the procedure is excessively conservative (leading to higher probability of a Type II error). Second, the analyses are predicated on the idea that all of the null hypotheses are true, which again represents an extreme condition.

But the most important objection is that the adjustment of α goes against common sense. An important rationale for using a standard α for rejecting or accepting null hypotheses is that the same standard of evidence is then applied to all hypotheses tests in the scientific literature (see Chapter 4). But if α is adjusted in each study, that standard suddenly doesn't hold anymore. And, what is the precise sample-space against which these adjustments should be made: the number of tests in the particular paper? in the journal issue as a whole? or perhaps the lifetime number of tests that investigators have conducted throughout their careers? Legitimate arguments could be made for each one of these.

Adjusting α's effectively penalizes investigators for conducting multiple tests, because the standard for rejection of the null hypothesis goes up the more tests that are conducted. Yet it is often the pattern of which particular tests are rejected that is the key to distinguishing between different scientific hypotheses.[12] Adjusting α's causes us to throw out precisely this key piece of information.

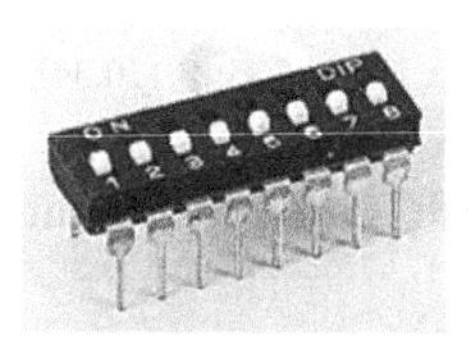

DIP Switch

[12] Rosenzweig and Abramsky (1997) refer to the evaluation of multiple pieces of evidence as the "DIP switch test." Each test represents a different DIP switch, and the collective pattern of which switches are flipped provides the cumulative evidence for or against a particular hypothesis. A DIP switch, or more precisely a Dual-Inline Package switch, is a small, two-position (on or off) switch that was used in early personal computers and consumer electronics. The photo illustrates eight DIP switches, which can be arranged in $2^8 = 256$ combinations of on and off. DIP switches replaced the jumpers that had to be moved around manually on older circuit boards. Since the late 1990s, DIP switches in computers have been largely replaced by non-volatile memory, which can store the information contained in the physical configuration of a DIP switch even when the computer is turned off. However, DIP switches are still widely used in industrial equipment.

Indeed, if the tests are truly independent, one could argue that adjustment of α's should actually be made in other direction! For example, suppose we conduct three independent experiments to test a particular hypothesis, and the α level for rejecting the null hypothesis in each case is 0.11. For any individual test, this is somewhat close to 0.05. But with the Bonferroni or Dunn-Sidak methods, we are not even close to the necessary rejection level ($\alpha = 0.0167$ and $\alpha = 0.0169$).

But shouldn't we be suspicious? After all, if the null hypothesis were always true, what are the chances of obtaining $P = 0.11$ three times in a row? **Fisher's combined probabilities** is a test that assesses a sequence of independent probability values.[13] The test statistic for combined probabilities (CP) is calculated as

$$CP = -2\sum_{i=1}^{k} \ln(p_i) \tag{10.37}$$

where k is the number of independent tests, and $\ln(p_i)$ is the natural logarithm of the tail probability for test i. The CP statistic follows a chi-square distribution (see Chapter 11 for a detailed discussion of a chi-square distribution) with $2n$ degrees of freedom. For three independent tests, each with a P of 0.11, $CP = 13.24$, with 6 degrees of freedom. From tabled values of the chi-square distribution, the combined tail probability is 0.039, which we would declare statistically significant. This makes intuitive sense because a set of three experiments from a truly random distribution would be unlikely to generate three P-values as small as 0.11. In contrast to the Bonferroni and Dunn-Sidak procedures, Fisher's test is vulnerable to Type I error if the tests are not truly independent. With non-independent tests, you are much more likely to reject H_0 incorrectly because each test is being counted as an independent assessment of the overall probability, which it is not.

Certainly it is important to avoid excessive statistical tests that are not independent of one another. For example, you should never use a series of t-tests when your data really fit the structure of a one-way ANOVA with a single P-value. For this same reason, we prefer orthogonal a priori contrasts to a posteriori comparisons, which involve many pairwise tests that are not independent of one another. As we will see in Chapter 12, multivariate ANOVA (MANOVA) often is preferable to a series of univariate tests when the response variables are all highly correlated and measured on the same replicates. Similarly, principal components analysis and discriminate functions are often useful for collapsing a set of intercorrelated variables into a smaller number of orthogonal predictor variables (see Chapter 12).

[13] These different lines of argument also reflect the distinction between Fisher's P and Neyman-Pearson's α (see Footnote 19 in Chapter 4).

But once you have reduced the analysis to the appropriate number of tests, we prefer to let the raw *P*-values stand and interpret them with some common sense. Most reviewers and authors would be properly skeptical of a single significant *P*-value out of a set of 20 independent tests, whereas six or seven significant results would indicate something interesting is going on. We hope you can convince journal editors and reviewers to let the raw *P*-values stand and pay attention to how you interpret them, rather than constantly downgrading all of your hard work by using Bonferroni adjustments.

Summary

The analysis of variance is a powerful statistical tool that is based on the algebraic partitioning of the sum of squares. This method assumes independent random samples, homogenous variances, normal error distributions, error-free classifications, and (for certain designs) additivity of main effects. If these assumptions are met, F-ratios of mean square terms can be calculated that test for additive or interaction effects of different treatments. The probability of obtaining different F-ratios can be used to test statistically the null hypothesis of no treatment effect. The ANOVA mean squares also can be used to partition the variance in the data into distinct components. Different experimental designs, such as the randomized block, nested ANOVA, multi-factor ANOVA, split-plot, and repeated measures ANOVA each have their own distinct ANOVA tables, in which different mean squares are used to construct appropriate F-ratios. The analysis of covariance (ANCOVA) is a special form of ANOVA that is a hybrid between ANOVA and regression. The correct construction of F-ratios is sensitive to whether the treatments are fixed factors, in which there are only a few distinct treatment levels, or random factors, in which the treatment levels are a random subset from a larger population. It is important to recognize which ANOVA design matches your sampling or experimental design. Because many ANOVA designs have equivalent numbers of factors, it is easy to generate an incorrect analysis by relying on the default settings of statistical software packages. After ANOVA, means and variances of treatments can be plotted to reveal the pattern of main effects and interactions between factors. A posteriori tests can be used to make pair-wise comparisons among different means, but a priori contrasts are a more powerful method for testing hypotheses and decomposing the treatment effects into distinct components. Bonferroni corrections and other methods are often used to correct *P*-values for multiple testing, but good arguments can be made that *P*-values should be interpreted without adjustment.

CHAPTER 11

The Analysis of Categorical Data

Many ecological and environmental studies generate response variables that are categorical, rather than continuous. For example, plants may be present or absent in a sampled area; beetles may be red, orange, or black; and foraging tigers may turn right more often than they turn left. Similarly, predictor variables may be categorical, rather than continuous. For example, treatment groups in an experimental or observational study represent different categories, such as different kinds of sponges on mangrove roots (see Chapter 10) or different species of beetles. These are all examples of **categorical variables**. The values that categorical variables can take are called **levels**, which may be ordered (e.g., low light, intermediate light, high light) or unordered (e.g., beetles, flies, wasps).

We use logistic regression (see Chapter 9) to analyze datasets with continuous predictor variables and response variables that have two levels (e.g., present, absent). We use ANOVA (see Chapter 10) to analyze datasets with categorical predictor variables and continuous response variables. However, when *both* predictor and response variables are categorical—data that result from a tabular design (see Chapter 7)—neither ANOVA nor regression are appropriate analytical tools.

In this chapter, we focus on data sets that have a single response variable with two or more levels (such as the presence or absence of an organism, or the coat color of an animal). The data in such a study represent counts—or frequencies—of observations in each category. When there is a single categorical predictor variable, the data are organized as two-way contingency tables and hypotheses are tested with the chi-square test, the *G*-test, or Fisher's Exact Test (for the special case of a 2×2 contingency table). When there are multiple predictor variables, the data are organized as multi-way contingency tables that can be analyzed using either log-linear models or classification trees. Bayesian analysis of contingency tables provides probabilistic estimates of expected cell frequencies in two-way contingency tables or of the parameters in log-linear models and classification trees. Analysis of multiple categorical response variables is discussed in Chapter 12.

We conclude this chapter with a discussion of goodness-of-fit tests. We can test whether a sample of data conforms to, is consistent with, or **fits**, a known distribution, such as the binomial, Poisson, or normal. Goodness-of-fit tests—the chi-square, *G*-, and Kolmogorov-Smirnov tests—can be used to test whether the residuals of raw or transformed data follow a normal distribution, which is a requirement for many parametric tests.

Two-Way Contingency Tables

Organizing the Data

Categorical data typically are organized as two-way contingency tables in which the rows represent levels of the predictor variable, and the columns represent levels of the response variable. The entries in such a table are the counts—or **frequencies**—of observations in each category. These are the fundamental data for contingency table analysis.[1] Analysis of contingency tables is done correctly only on the raw counts, not on the percentages, proportions, or relative frequencies of the data.

We illustrate tests on two-way contingency tables using data from a study examining factors associated with the status of 73 populations of rare plant species in New England (Farnsworth 2004). Each population was scored for whether or not it was declining in size; whether or not it was legally protected; the presence or absence of invasive species; and five qualitative levels of light (from 0 for deep shade to 4 for bright sun). The first three lines of this dataset are illustrated in Table 11.1. A summary of the association between two of the categorical variables in this dataset is presented in Table 11.2 as a **contingency table**.

This contingency table illustrates the relationship between the protection status of the rare plant populations and whether or not the populations are declining. For this analysis, we will use the protection status as the predictor variable

[1] The terminology associated with contingency table data is a bit inconsistent. Technically speaking, **frequencies** are the raw counts of observations in the cells of a contingency table. **Relative frequencies** are **proportions**, obtained by dividing the number of observations in one of the cells by row, column, or grand totals of the table. Finally, **percentages** are proportions that have been multiplied by 100, so that they range from 0 to 100. However, many scientists use the term "frequencies" to mean relative frequencies and use the terms "percentage" and "proportion" interchangeably. The distinction among these terms is critical, however, as statistical tests on data in contingency tables must be done on the raw counts (i.e., frequencies), not on relative frequencies or proportions.

TABLE 11.1 Three rows from a dataset measuring factors associated with population status of rare plant species in New England

Species	Invasive species present?	Population declining?	Legal protection?	Light level
Aristolochia	No	No	No	2
Hydrastis	No	Yes	No	0
Liatris	Yes	Yes	No	4
...	...	...	...	...

Each observation in this study is a particular population of a species. For each population, the data table records the presence or absence of invasive species, whether the population is declining or not, whether it is legally protected or not, and the understory light level, scored as a rank from 0 (lowest light level) to 4 (highest light level). The total sample size was 73 species. (Data from Farnsworth 2004.)

and the population status as the response variable. We set the model up this way because we believe that protection status potentially affects population status, and not the other way around. Note, however, that in setting up a contingency table, it does not matter whether you put predictors in rows and responses in columns or vice-versa. Statistical tests applied to contingency tables give the same result either way. Interpreting cause-and-effect relationships from the results

TABLE 11.2 Two-way contingency table summarizing the relationship between protection and population status

	Protection status		
Population status	**Not protected**	**Protected**	*Row total*
Declining	$Y_{1,1} = 18$	$Y_{1,2} = 8$	$\sum_{j=1}^{m} Y_{1,j} = 26$
Stable or increasing	$Y_{2,1} = 15$	$Y_{2,2} = 32$	$\sum_{j=1}^{m} Y_{2,j} = 47$
Column total	$\sum_{i=1}^{n} Y_{i,1} = 33$	$\sum_{i=1}^{n} Y_{i,2} = 40$	$\sum_{i=1}^{n}\sum_{j=1}^{m} Y_{i,j} = 73$

Each cell in the table specifies a particular combination of protection and population status. The numbers indicate the number of observed populations within a particular classification. For example, there were 18 unprotected populations that were declining and 32 protected populations that were stable or increasing. Some contingency tables may have cell values of 0 if there are no observations for a particular combination of factors. In any two-way contingency table, the grand total of the cell values (73) equals the sum of the row totals (26 + 47), which also equals the sum of the column totals (33 + 40). The row and column sums are sometimes called the marginal totals of the data. (Data from Farnsworth 2004.)

of these tests, however, does require distinguishing between predictors and responses.

Two-way contingency tables have rows and columns. By convention, rows are indexed using the letter i and columns are indexed using the letter j. Row indices range from 1 to n, where n is the number of row categories (two in Table 11.2). Column indices range from 1 to m, where m is the number of column categories (two in Table 11.2). The value $Y_{i,j}$ in each cell represents the frequency or number of observations that are in both the level represented by the row and the level represented by the column. For example, the value 18 in the upper left-hand cell ($Y_{1,1}$) of Table 11.2 means that 18 populations were sampled that were both declining and unprotected.[2]

Three other quantities that are derived from contingency tables are **row totals**, **column totals**, and the **grand total**. The row totals are the sums of the observations in each row; in Table 11.2, there are a total of 26 populations that are declining and 47 that are stable or increasing. The column totals are the sums of the observations in each column; in Table 11.2, there are 33 populations that are unprotected and 40 populations that have some degree of legal protection. The grand total (which also equals the total sample size) can be calculated in three ways: as the sum of the row totals (26 + 47 = 73), the sum of the column totals (33 + 40 = 73), or the sum of the cell frequencies (18 + 8 + 15 + 32 = 73). If all three of these sums are not equal, re-check your arithmetic!

A convenient way to visualize the data in contingency tables is to use a **mosaic plot** (Friendly 1994). Figure 11.1 illustrates a mosaic plot of the data shown in Table 11.2. In a mosaic plot, the two "axes" are the two variables—protection status on the x-axis (the predictor variable) and population status on the y-axis (the response variable). The size of the "tiles" is scaled to the proportion of observations in each cell of the contingency table. Therefore, the total area of each tile is proportional to the relative frequency of that combination.

Are the Variables Independent?

SPECIFYING THE NULL HYPOTHESIS Contingency tables are used to test the null hypothesis that the predictor and response variables are not associated with

[2] Unfortunately, there are neither standard methods for organizing contingency tables in spreadsheets, nor standard ways by which statistical software handles categorical data. Some packages import contingency tables directly, others generate contingency tables from datasets in which each row represents a single observation, and still others can work with data in either format. If you have a favorite statistical package, and know you will be working with categorical data, check the software's formatting requirements before you begin to code your data.

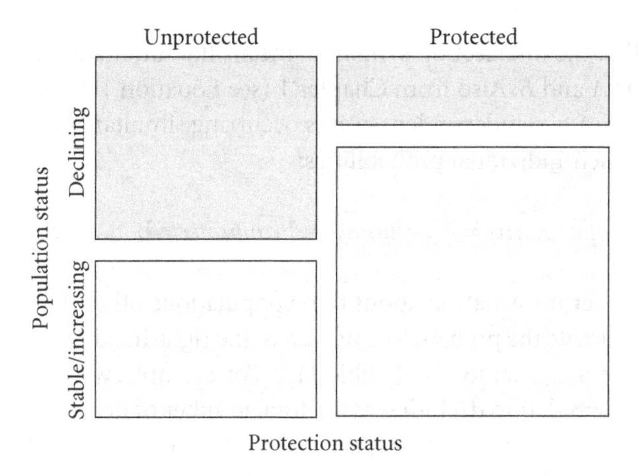

Figure 11.1 Illustration of a mosaic plot. This plot portrays the relationship between the status of rare plant populations and whether or not the land on which they occur is protected (Farnsworth 2004). The cell frequencies (from Table 11.2) are represented by "tiles" whose size is proportional to their relative frequency in the dataset. The width of the columns is proportional to the column totals. For example, because 40 of the 73 populations are protected, the protected column occupies 55% (or 40/73) of the width of the plot. The height of each tile is proportional to the cell frequency. Thus, 32 of the 40 protected populations are stable or increasing, so this tile occupies 32/40, or 80%, of the right-hand column.

each other. In the example shown in Table 11.2, we ask if there is an association or relationship between whether or not a population is protected and whether or not it is declining. Specifically, we want to know whether unprotected populations are more likely to be declining than are protected populations. If this pattern can be established, it would suggest that protection has a measurable effect in preventing population decline. To test this scientific hypothesis, we first must specify an appropriate statistical null hypothesis. In this case, the simplest null hypothesis is that the two variables are independent of each other, and the observed degree of association is no stronger than we would expect by chance or random sampling. Our discussion in Chapter 2 of the probability of independent events gives us a framework for testing this null hypothesis.

CALCULATING THE EXPECTED VALUES Our null hypothesis asserts that the values in each of the four cells in Table 11.2 are independent of each other. We ask, therefore, what the **expected values** would be in each of these four cells if stability of the population and its protection status were independent of one another? As an example, take the top left cell, which gives the number of populations that are declining and not protected. What is the expected number of populations in this category under the null hypothesis of independence? The expected value is simply the total number of observations (N) times the probability (P) of a population being both unprotected *and* declining:

$$\hat{Y}_{declining,\ unprotected} = N \times P(declining \cap unprotected) \quad (11.1)$$

Recall from Chapter 1 that the intersection symbol $\cap$ means the simultaneous occurrence of two events A and B. Also from Chapter 1 (see Equation 1.3), we know that the probability of two independent events occurring simultaneously equals the product of their individual probabilities:

$$P(declining \cap unprotected) = P(declining) \times P(unprotected) \quad (11.2)$$

Because we have no other information about these populations other than the data themselves, we estimate the probabilities of each of the right-hand terms in Equation 11.2 from the marginal totals of Table 11.2. For example, we estimate the probability of a population declining as the total number of declining populations (26 = the row total of the first row) divided by the total number of populations (73). This probability = 0.356. Similarly, we estimate the probability of a population being unprotected using the column total of the first column (33 = the total number of unprotected populations) divided by the total number of populations (73). This probability = 0.452. Substituting these values back into Equation 11.1, we have

$$\hat{Y}_{declining,\ unprotected} = 73 \times 0.356 \times 0.452 = 11.75 \quad (11.3)$$

Thus, if protection status and decline status are independent of one another, we expect approximately 12 of the 73 sampled populations would be unprotected *and* declining.

The general formula for computing the expected value of a cell in a two-way contingency table with $i = 1$ to n rows and $j = 1$ to m columns is

$$\hat{Y}_{i,j} = \frac{row\ total \times column\ total}{sample\ size} = \frac{\sum_{j=1}^{m} Y_{i,j} \times \sum_{i=1}^{n} Y_{i,j}}{N} \quad (11.4)$$

By applying Equation 11.4 to each of the cells in our example (see Table 11.2), we obtain the expected values for each cell (Table 11.3). Because we used the data to generate these probabilities, the row totals, column totals, and grand total of the expected values are always identical to the observed totals.

Testing the Hypothesis: Pearson's Chi-square Test

The chi-square test is the standard method for testing the hypothesis that the variables in a two-way contingency table are independent.

TABLE 11.3 **Expected values for data in Table 11.2**

	Protection status		
Population status	**Not protected**	**Protected**	*Row total*
Declining	11.75	14.25	26
Stable or increasing	21.25	25.75	47
Column total	33	40	*Grand total* = 73

From the data in Table 11.2, these expected values summarize the relationship between status of rare plant populations and whether or not the land on which they occur is protected. The expected values are calculated under the null hypothesis that there is no association between protection status and population status. If the two factors are independent, then the expected cell frequency is simply the row total times the column total divided by the grand total (Equation 11.4). Notice that the sums of the expected values yield the same row and column totals as do the original data (Table 11.2). (Data from Farnsworth 2004.)

THE TEST STATISTIC The test statistic, referred to as the **Pearson chi-square statistic** (Pearson 1900),[3] is calculated as

$$X^2_{\text{Pearson}} = \sum_{\textit{all cells}} \frac{(\textit{Observed} - \textit{Expected})^2}{\textit{Expected}} \tag{11.5}$$

In other words, for each cell, we subtract the expected value from the observed value, square this difference, and then divide this squared difference by the expected value. We then sum these quotients. For the data in Table 11.2, and the expected values in Table 11.3, we have

Karl Pearson

[3] Karl Pearson (1857–1936) was one of the founders of modern statistics. After graduating from Cambridge University in 1879, he moved to University College, London, where in 1911 he was named Galton Professor of Eugenics. In addition to developing the chi-square test, he came up with the term "standard deviation" to describe $\sqrt{\sigma^2}$, invented the correlation coefficient (r), made significant advances in regression analysis, and co-founded the journal *Biometrika*, which publishes statistical contributions with biological applications. Pearson was particularly interested in large-sample (asymptotic) statistics. He disagreed with Fisher on the relative importance of large-sample versus small-sample statistics. This disagreement was significant enough that Fisher declined the job offer of chief statistician at the Galton Laboratory for National Eugenics, which was run by Pearson.

$$X^2_{\text{Pearson}} = \frac{(18-11.75)^2}{11.75} + \frac{(8-14.25)^2}{14.25} + \frac{(15-21.25)^2}{21.25} + \frac{(32-25.75)^2}{25.75}$$

$$= 3.32 + 2.74 + 1.83 + 1.52$$

$$= 9.42$$

Intuitively, you can see that the chi-square statistic measures the extent to which the observed frequencies differ from the expected frequencies under the null hypothesis of independence. Suppose that the observed frequencies in each cell exactly matched the expected frequencies. The chi-square value would equal 0. The more the observed values deviate from the expected frequencies, the larger the chi-square value. Also, notice that the chi-square statistic is a squared measure, so it is only the size of the deviation that matters, not whether it is positive or negative. In this sense, the chi-square statistic is very similar to the residual sum of squares that we described for linear regression (see Chapter 9).

The expected distribution of the Pearson's chi-square statistic is the chi-square (χ^2) distribution.[4] As with the F-distribution (see Chapters 5, 10) and *t*-distribution (see Chapter 3), the shape of the chi-square distribution varies with the number of degrees of freedom.

DEGREES OF FREEDOM What are the degrees of freedom for Table 11.2? For a given sample size (in this case 73), and for the two row totals (here, 26 and 47) and the two column totals (here, 33 and 40), as soon as one cell value is specified, all the others are fixed. Thus, there is only 1 degree of freedom in this 2×2 table.

If the contingency table has more than two columns, all but one of each row's entries can vary (once all but one have been specified, the last is determined by

[4] The chi-square (χ^2) distribution is a probability density function that takes on values from 0 to $+\infty$. Its probability density function, evaluated only for positive values of x, is

$$f(x) = \frac{(\frac{1}{2})^{\frac{\nu}{2}}}{\Gamma(\frac{\nu}{2})} x^{(\frac{\nu}{2}-1)} e^{-\frac{1}{2}x}$$

where $\Gamma(x)$ is the Gamma function (see Footnote 11 in Chapter 5) and ν is the number of degrees of freedom. In fact, a χ^2 distribution with ν degrees of freedom is identical to a Gamma distribution evaluated with $a = \nu/2$ and $b = 1/2$. The χ^2 distribution has only a single parameter, ν. Its expected value = ν, and its expected variance = 2ν. The F distribution used for ANOVA (see Chapters 5 and 10) also can be expressed as the ratio of two χ^2 distributions.

the row total). Similarly, if the contingency table has more than two rows, all but one of each column's entries can vary. In general, the number of degrees of freedom for a contingency table, indicated by the Greek letter ν, is equal to

$$df = \nu = (\text{number of rows} - 1) \times (\text{number of columns} - 1) \quad (11.6)$$

This calculation for degrees of freedom differs from our experience with ANOVA (see Chapter 10) and regression (see Chapter 9). In those analyses, the degrees of freedom depended, in part, on the total sample size. But in the contingency table analysis, the degrees of freedom depend only on the number of categories or cells in the table. Thus, in Table 11.2, if we had data for 730 populations, rather than 73, the chi-square test still would have only 1 degree of freedom, because we are still analyzing only a 2 × 2 table. However, the statistical test would be much more precise with 730 populations than with only 73. This gain in precision would be lost if the data had been transformed into proportions or percentages, which is one reason that the analysis of categorical data is done with the raw frequencies.

CALCULATING THE *P*-VALUE As with all hypothesis tests, we are interested in finding the probability of the data, given the null hypothesis [$P(\text{data}|H_0)$], where the null hypothesis is that the two variables are independent. The *P*-value is the tail probability: the probability of obtaining our data or data more extreme than we did, given the null hypothesis (see Chapter 4). For the Pearson chi-square test applied to Table 11.2, we find the probability of obtaining a value of X^2_{Pearson} as large or larger than 9.42 (from Equation 11.5) relative to a χ^2 distribution with 1 degree of freedom (from Equation 11.6). This value can be looked up in a table (e.g., Rohlf and Sokal 1995), but more likely it will be provided by a statistical software package. The probability of obtaining this X^2_{Pearson} value = 0.0022. As this value is much less than the standard critical value (= probability of committing a Type I error) of 0.05, we reject the null hypothesis that the two variables, population status and protection status, are independent.

As we emphasized in Chapter 4, the *P*-value is used to make the decision to reject or not to reject the null hypothesis. The next step is to examine a graph of the data and decide whether the statistical pattern in the data supports or contradicts the scientific hypothesis. In this case, the scientific hypothesis is that protection status enhances population stability, and that is indeed the pattern seen in Figure 11.1: of the 40 protected populations, 80% (= 32) were stable or increasing. In contrast, of the 33 unprotected populations, only 45% (= 15) were stable or increasing. The chi-square probability test tells us that this difference is unlikely to have occurred by chance ($P = 0.0022$), and we conclude from this simple analysis that protection status is associated with population stability.

The chi-square statistic is the most familiar and basic test for independence in a contingency table. We now explore two alternatives to the chi-square test: the *G*-test and Fisher's Exact Test.

An Alternative to Pearson's Chi-Square: The *G*-Test

Like the chi-square test, the *G*-test for independence also compares the observed cell frequencies with their expected frequencies. It asks how well the distribution of observed values compares with the distribution of expected values based on the **multinomial probability distribution.**[5]

THE TEST STATISTIC The test statistic is the ratio of the probability of the observed frequencies to the probability of the expected frequencies (and thus the *G*-test is often referred to as a **likelihood ratio test**). This ratio can be computed in two ways. The first way, which is the most straightforward, uses the expected values that were calculated with Equation 11.4 and illustrated in Table 11.3:

$$G = 2 \times \sum_{all\,cells} \left[Observed \times \ln\left(\frac{Observed}{Expected} \right) \right] \quad (11.7)$$

Alternatively, *G* can be calculated directly from the observed frequencies, without first calculating the expected cell values:

[5] The multinomial distribution is a simple extension of the binomial distribution (see Chapter 2, Equations 2.2 and 2.3) to variables that can take on more than two values. For a random variable Y that can take on j values, $Y = \{Y_1, \ldots, Y_j\}$, each with probability p_j where by the first axiom of probability,

$$\sum_{i=1}^{j} p_i = 1$$

and for N trials

$$\sum_{i=1}^{j} Y_i = N$$

the multinomial distribution is defined as

$$P(Y_1, \ldots, Y_j) = N! \prod_{i=1}^{j} \frac{p_i^{Y_i}}{Y_i!}$$

For the data in Table 11.2, using this formula, the probability of observing the four cell frequencies $Y_{1,1}$, $Y_{1,2}$, $Y_{2,1}$, and $Y_{2,2}$ (which equal 18, 8, 15, and 32, respectively) is

$$\frac{N!}{Y_{1,1}! \times Y_{1,2}! \times Y_{2,1}! \times Y_{2,2}!} \times \left(\frac{Y_{1,1}}{N} \right)^{Y_{1,1}} \times \left(\frac{Y_{1,2}}{N} \right)^{Y_{1,2}} \times \left(\frac{Y_{2,1}}{N} \right)^{Y_{2,1}} \times \left(\frac{Y_{2,2}}{N} \right)^{Y_{2,2}} = 0.00202$$

Compare this result to that given in the text for the *G*-test.

$$G = 2 \times \left(\sum_{i=1}^{n}\sum_{j=1}^{m}[Y_{i,j}\ln(Y_{i,j})] - \sum_{i=1}^{n}\left[(\sum_{j=1}^{m}Y_j)\ln(\sum_{j=1}^{m}Y_j)\right] - \sum_{j=1}^{m}\left[(\sum_{i=1}^{n}Y_i)\ln(\sum_{i=1}^{n}Y_i)\right] + \left[(\sum_{i=1}^{n}\sum_{j=1}^{m}Y_{i,j})\ln(\sum_{i=1}^{n}\sum_{j=1}^{m}Y_{i,j})\right] \right) \quad (11.8)$$

In words, Equation 11.8 says to take the sum of the product of each cell frequency times its natural logarithm, subtract from that the sum of the product of the row totals times their natural logarithms, subtract from that the sum of the product of the column totals times their natural logarithms, and then add the product of the grand total times its natural logarithm. Last, multiply this entire mess times two. Equation 11.8 is particularly useful when there are a large number of cells and it is troublesome to compute expected values. Equations 11.7 and 11.8 are algebraically equivalent, and both generate a G-value of 9.575 for the data in Table 11.2.

DEGREES OF FREEDOM The degrees of freedom for the G-test are the same as for Pearson's chi-square test, and equal the product of the (number of rows −1) and the (number of columns −1) (see Equation 11.6). For the data in Table 11.2, there is only one degree of freedom.

CALCULATING THE P-VALUE Like Pearson's chi-square statistic, the G-statistic is distributed as a χ^2 random variable with ν degrees of freedom. The P-value for $G = 9.575$ with 1 degree of freedom = 0.00197 (within the round-off error of the calculation in Footnote 5).

The Chi-square Test and the G-Test for $R \times C$ tables

The chi-square test and the G-test are not restricted to simple 2×2 tables such as Table 11.2. They can be used for any two-way contingency table, regardless of the number of levels or categories for each variable. Such tables often are referred to as $R \times C$ tables, because they can have an arbitrary number of *r*ows and *c*olumns. Table 11.4 presents another subset of Farnsworth's (2004) data as a 2×5 table—the relationship between population status (two categories) and light level (five categories). For these data, the Pearson's X^2 statistic = 2.83, and with 4 degrees of freedom [(5–1 columns) × (2–1 rows)], the P-value = 0.59. This value is much greater than 0.05, so we cannot reject the null hypothesis that these two variables are independent. Similarly, the G-statistic = 3.41, and with 4 degrees of freedom the P-value = 0.49, also causing us to not reject the null hypothesis of independence. Both analyses suggest that light level at a site does not affect whether a population is stable or declining. Although frequency data theoretically can be fit to an $R \times C$ table of any size, the more categories there are in the table, the larger the total sample size should be in order for the test to perform properly.

TABLE 11.4 **Observed (in bold) and expected (in parentheses) values used to test the independence of population status and light level**

	Light level					
Population status	**0**	**1**	**2**	**3**	**4**	*Row total*
Declining	**5** (3.2)	**0** (0.7)	**3** (3.6)	**12** (12.5)	**6** (6.0)	26
Stable or increasing	**4** (5.8)	**2** (1.3)	**7** (6.4)	**23** (22.5)	**11** (11.0)	47
Column total	9	2	10	35	17	*Grand total* = 73

In this two-way table, the association between light level and population status is illustrated for 73 monitored plant populations (see Table 11.1). Sites were classified according to 5 light levels (0 = lowest, 4 = highest). Although there are 10 cells in the table, this is still considered a two-way contingency table because there are only two factors (light level and population status). As in the simpler two-way table, the marginal totals and the grand total of the observed data equal the marginal totals and the grand total of the expected values. The expected values were calculated under the null hypothesis of no association between light level and population status. (Data from Farnsworth 2004.)

CORRECTIONS FOR SMALL SAMPLE SIZES AND SMALL EXPECTED VALUES When the total sample size is large (Legendre and Legendre 1998 consider "large" to be greater than 10 times the number of cells in the contingency table), the distributions of both Pearson's chi-square statistic and the G-statistic are equally close to that of a χ^2 random variable. However, when the total sample size is small—less than five times the number of cells—there may be many observed values equal to zero or many low expected values. Low expected values are a potential problem because they are used in the denominator of the chi-square statistic (see Equation 11.5). Therefore, if the expected value of a cell is very small, even modest deviations in the observed value can increase greatly the overall chi-square statistic. When the sample size is small, the G-statistic should be adjusted using a method proposed by Williams (1976):

$$G_{\text{adjusted}} = G / q_{\min} \tag{11.9}$$

where

$$q_{\min} = 1 + \frac{\left[N \sum_{i=1}^{n} \frac{1}{\sum_{j=1}^{m} Y_{i,j}} - 1 \right] \times \left[N \sum_{j=1}^{m} \frac{1}{\sum_{i=1}^{n} Y_{i,j}} - 1 \right]}{6 \nu N} \tag{11.10}$$

In Equation 11.10, m is the number of columns, n is the number of rows, N is the total sample size, and ν is the degrees of freedom. As before, $Y_{i,j}$ represents the frequency of observations in (row i, column j) of the contingency table. The first term in the numerator is one less than the total sample size multiplied by the sum of the reciprocals of the column totals. The second term in the numerator is one less than the total sample size multiplied by the sum of the reciprocals of the row totals. For Table 11.4,

$$q_{\min} = 1 + \frac{[73\times(\frac{1}{9}+\frac{1}{2}+\frac{1}{10}+\frac{1}{35}+\frac{1}{17})-1]\times[73\times(\frac{1}{26}+\frac{1}{47})-1]}{6\times 4\times 73} = 1.109$$

and the adjusted G-statistic = 3.41/1.109 = 3.072, which has a P-value of 0.55. Note that the adjusted G-statistic is more conservative (generates larger P-values) than the unadjusted G-statistic ($P = 0.49$).

Standard statistical texts (e.g., Sokal and Rohlf 1995) recommend the use of G_{adjusted} whenever any expected values are less than 5, and some statistical software packages produce warning messages about the validity of the chi-square or G-test when expected values are small. On the other hand, Fienberg (1980) showed in a simulation study that the correction is not needed as long as all expected values are greater than 1. In practice, however, the use of G_{adjusted} vs. G only makes a difference when sample sizes are small and P-values are marginally close to 0.05. Despite the warning messages, most statistical software packages do not incorporate G_{adjusted} in their menus of options, although many include another correction in which 0.5 is added to or subtracted from each observed value in the table. This correction, known as Yates' continuity correction (Sokal and Rohlf 1995), also results in smaller values of G or X^2_{Pearson}, leading, like G_{adjusted}, to a more conservative test (i.e., a test less likely to reject the null hypothesis).

ASYMPTOTIC VERSUS EXACT TESTS Pearson's chi-square test and the G-test are asymptotic tests—that is, the distributions of both G and X^2_{Pearson} approach that of a χ^2 random variable as sample size gets infinitely large. Therefore, the computed P-values are based on an approximation, albeit a reasonable one. Using one key assumption, however, it is possible to compute an exact P-value for an $R \times C$ contingency table. That assumption is that the row and column totals are fixed by the investigator a priori.

For example, in Table 11.2, an exact test would be appropriate if the intent of the study had been to survey exactly 26 declining populations, 47 stable or increasing populations, 33 unprotected populations, and 40 protected populations. Although this assumption is unlikely in this example, it is not difficult to imagine experiments in which the number of individuals in each treatment group is fixed, and the intensity of treatment applied to the individuals is also fixed. More typically, in sampling studies, either the row or the column totals are fixed by the investigator, but not both. Thus, we might have chosen to survey 33 unprotected and 40 protected populations, but placed no restriction on their population status. Or, we might have selected 26 declining and 47 stable populations, without regard to their protection status.

When both the row and column totals are fixed, the calculation of an exact *P*-value is conceptually simple but computationally intensive. The exact probability (or *P*-value) is the probability of obtaining the observed cell frequencies and all other possible cell frequencies that are more extreme than the expected values, given the fixed row and column totals. For a 2 × 2 table, this procedure is called Fisher's Exact Test. As before, the null hypothesis is that the row and column variables are independent.

First we calculate the total number of permutations (see Chapter 2) that can result in 2 × 2 table with fixed row and column totals. This value is

$$\binom{N}{Y_{1,1}+Y_{1,2}}\binom{N}{Y_{1,1}+Y_{2,1}} = \frac{N!}{(Y_{1,1}+Y_{1,2})!(Y_{2,1}+Y_{2,2})!} \times \frac{N!}{(Y_{1,1}+Y_{2,1})!(Y_{1,2}+Y_{2,2})!} \quad \textbf{(11.11)}$$

Second, we calculate from the multinomial distribution (see Footnote 5 in this chapter) the number of ways that we can obtain the exact combination of cell values that we observed:

$$\frac{N!}{Y_{1,1}!\times Y_{1,2}!\times Y_{2,1}!\times Y_{2,2}!} \quad \textbf{(11.12)}$$

Dividing Equation 11.12 by Equation 11.11 results in the probability of the precise combination of cell values that we observed (equivalent to the calculating the number of successes divided by the number of trials):

$$P_{observed} = \frac{(Y_{1,1}+Y_{1,2})!\times(Y_{2,1}+Y_{2,2})!\times(Y_{1,1}+Y_{2,1})!\times(Y_{1,2}+Y_{2,2})!}{Y_{1,1}!\times Y_{1,2}!\times Y_{2,1}!\times Y_{2,2}!\times N!} \quad \textbf{(11.13)}$$

The tail probability equals Equation 11.13 *plus* the probabilities of all the more extreme cases. For a 2 × 2 table, the tail probability is calculated by simply enumerating all the more extreme cases, reapplying Equation 11.13 to each case, and summing the results. Note that the marginal totals (the values in the numerator), and the grand total (the value of *N* in the denominator) remain the same for all the more extreme cases; only the individual $Y_{i,j}$ values in the denominator of Equation 11.13 change in the other cases. The exact *P*-value is the sum of all the iterations of Equation 11.13; for Table 11.2, the exact P-value = 0.0031. For the 2 × 5 contingency table illustrated in Table 11.4, the exact P-value = 0.60.

Which Test To Choose?

All three of these alternatives—the asymptotic Pearson's chi-square test, the asymptotic G-test, and their exact-test counterparts—give somewhat different results, especially for small sample sizes. A common but misguided way to choose which test to use is to run all three of them and pick the one that gives the best result—the smallest P-value. A more appropriate way to choose among them is to match the sampling design of the study with the assumptions of the test.

Three different designs can generate identical contingency tables, and each of these has an associated test. First, our study could have been defined by the total sample size (73 in the case of Table 11.2) without any a priori assignment of number of populations (or individuals) to each category. This is the most common assumption in field studies: choose the total sample size and let the chips fall where they may. Sokal and Rohlf (1995) refer to this as a Model I design.

In a Model II design, either the row totals or the column totals, but not both, could be fixed in advance. In the Farnsworth (2004) study, the number of protected populations was fixed in advance at 40 and the number of unprotected ones was fixed at 33 (perhaps because of regulatory decisions made in advance), but the fraction of these populations that was declining was free to vary.

Finally, if the investigator fixes both row and column totals a priori, we have a Model III design. Model II and Model III designs are more likely in experimental studies, whereas Model I designs are more common in observational studies.

The G-test was developed explicitly for Model I designs (total sample size fixed), but it can be used for either Model I or Model II designs (row or column totals fixed). Pearson's chi-square test was not developed with any particular design in mind; as long as the sample size is large, it will give results that are virtually identical to the G-test. When sample sizes in Model I or Model II designs are small, the G-test with either Williams' or Yates' correction is appropriate. Exact tests really are appropriate only for Model III designs (both row and column totals fixed).

Finally, we note that there are Monte Carlo or computer simulation analogs to each of the three sampling designs. In a Monte Carlo simulation, we do not rely on statistical distributions to determine P-values, but instead estimate them directly from sampling (see Chapter 5). For example, a Monte Carlo simulation of a Model I design randomly assigns 73 populations to each of the four cells in Table 11.2. However, the probability of placement in each of those cells is proportional to the expected cell value (see Equation 11.4). In this way, the marginal totals of the simulated data are, on average, the same as the observed marginal totals. For each simulated data set, the standard chi-square statistic is calculated. One thousand (or some other large number) of simulated values are created, and we directly compare the observed chi-square value to this distribution of simulated values. For the data in Table 11.2, the observed chi-square value was larger than 990 out of 1000 simulated values. Therefore, the estimated tail probability from the Monte Carlo analysis is 0.010, qualitatively similar, although not identical to, the results of the *G*-test and chi-square test.

Multi-Way Contingency Tables

Most categorical ecological data include more than two variables (see Table 11.1), and so we are often testing for independence among multiple predictor variables. As in two-way tables, the data are best entered into a spreadsheet (see Table 11.1) but organized as a contingency table (Table 11.5).

Organizing the Data

Once we get beyond two factors, data are organized into multi-dimensional contingency tables that cannot be visualized easily in a single two-dimensional table. One way to display multidimensional categorical data is with multiple two-way tables, one for each level of the each higher-dimensional factor. For exam-

TABLE 11.5 **Complete classification of data for 73 rare plant populations**

Invasive species		**Absent**					**Present**				
Light level		**0**	**1**	**2**	**3**	**4**	**0**	**1**	**2**	**3**	**4**
Population status	**Protected**										
Declining	No	2	0	2	2	0	1	0	0	7	4
	Yes	1	0	1	1	0	1	0	0	2	2
Stable or increasing	No	0	0	1	1	4	1	1	2	2	3
	Yes	3	1	0	14	2	0	0	4	6	2

The populations are classified according to whether population status (declining or stable), protection status (yes or no), presence of invasive species (yes or no) and light level (5 levels; 0 = lowest, 4 = highest). Each cell in the table represents the number of populations found in a particular combination of conditions. For example, there were 14 protected populations that were stable or increasing in sites with no invasive species and a light level of 3. There were 7 unprotected populations that were declining in sites with invasive species present and a light level of 3. These data can be analyzed as a four-way contingency table. The population status is treated as the response variable, and the light level, protection status, and presence of invasive species are treated as potential predictor variables. (Data from Farnsworth 2004.)

ple, to visualize the Farnsworth (2004) data for 5 light levels, we could assemble 5 two-way tables. Each two-way table would show the counts for protection and population status under each different light level. Finally, if we wanted to include invasive species status as a fourth factor, the entire set of two-way tables would have to be duplicated for populations with and without invasive species.

An alternative layout of a four-way contingency table is shown in Table 11.5. This layout illustrates the association among population status, presence of invasive species, protection status, and light level. Each cell gives the number of populations recorded in a particular combination of categories. For example, 14 populations that were protected and not declining were recorded at sites with light level 3 and no invasive species present. Seven populations that were not protected and were declining were recorded at sites with light level 3 and invasive species absent. This table is more difficult to interpret than is a two-way table, but a mosaic plot (Figure 11.2) again provides a convenient and rapid way to visualize such multidimensional data.

You can see that the more factors that are included in the table, the more specific and detailed the characterization of each population. The trade-off is that the more factors included, the more the sample size is reduced in each cell, because the total number of observations is divided among a larger number of cells.

ARE THE VARIABLES INDEPENDENT? As in the analysis of two-way tables, we want to know whether the variables in a multi-way table are associated with one another. However, there are many more ways for multi-way data to be associated with each other than there are for two-way data. Before we turn to a rigorous specification of the appropriate null hypothesis, we must first understand how to calculate the expected values for a multi-way contingency table.

REPRISE: EXPECTED VALUES FOR A TWO-WAY CONTINGENCY TABLE We can calculate the expected values for multi-way contingency tables by expanding on an alternative method for calculating expected values for a two-way contingency table. Bear with us as we describe these calculations, because they will provide you not only with a new way to test hypotheses in two-way tables, but also with a firm base for testing hypotheses in multi-way tables.

For a two-way table with $i = 1$ to n rows and $j = 1$ to m columns, we calculated the expected value for each cell using Equation 11.4:

$$\hat{Y}_{i,j} = \frac{row\ total \times column\ total}{sample\ size} = \frac{\sum_{j=1}^{m} Y_{i,j} \times \sum_{i=1}^{n} Y_{i,j}}{N}$$

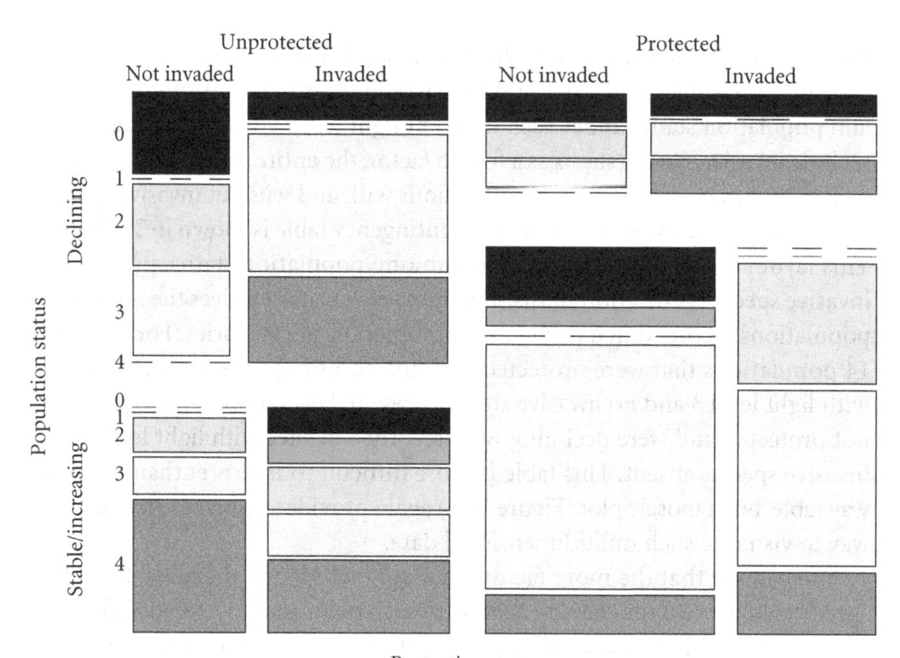

Figure 11.2 A mosaic plot illustrating the four-way frequency data from Table 11.5. These data are for 73 rare plant populations, classified according to their legal protection status (protected or not), population status (declining versus stable or increasing), presence of invasive species (yes or no), and light intensity (five increasing levels). As in Figure 11.1, the sizes of the tiles are scaled to the relative cell frequencies. The main categories are population status (Declining or Stable/Increasing on the y-axis) and legal protection (Protection Status on the x-axis), and these larger tiles are the same relative size as those shown in Figure 11.1. These main categories then are subdivided into whether or not they are affected by invasive species (divisions Not Invaded and Invaded on the x-axis), and into five light levels on the y-axis. Each light level is indicated with a different shading (black = 0; deepest blue = 1; middle blue = 2; lightest blue = 3; gray = 4). Cells with frequencies = 0 are indicated by the dashed lines; for example, for there are no sampled populations that are protected but declining, with no invading species, in light level 1 or light level 4. Data from Farnsworth (2004).

If we take the natural logarithm of both sides of Equation 11.4, we get

$$\ln(\hat{Y}_{i,j}) = \ln(\sum_{j=1}^{m} Y_{i,j}) + \ln(\sum_{i=1}^{n} Y_{i,j}) - \ln(N) \quad \textbf{(11.14)}$$

Thus, the natural logarithm of the expected cell value is the sum of the natural logarithm of the row total plus the natural logarithm of the column total minus the natural logarithm of the grand total (which equals the sample size).

If we re-write Equation 11.14 symbolically as

$$\ln(\hat{Y}_{i,j}) = [\theta] + [A]_{row} + [B]_{column} \tag{11.15}$$

it should remind you of the basic equation for the model in a two-way analysis of variance (see Chapter 10, Equation 10.10),

$$\hat{Y}_{ij} = \mu + A_i + B_j + AB_{ij}$$

although the interaction term (AB_{ij}) is missing from our contingency table equation. By analogy to ANOVA, $[\theta]$ in Equation 11.15 is the grand mean of the predicted values (μ in the ANOVA equation) and $[A]$ and $[B]$ are the main effects due to the row and column variables (A_i and B_j in the ANOVA equation).

The terms in Equation 11.15 are calculated as follows. The grand mean $[\theta]$ is the mean of the natural logarithms of the expected cell frequencies:

$$[\theta] = \frac{1}{nm}\sum_{i=1}^{n}\sum_{j=1}^{m}\ln(\hat{Y}_{i,j}) \tag{11.16}$$

For each row, the main effect $[A]_{row}$ is the mean of the logarithms of the expected row sum, minus the grand mean:

$$[A]_{row} = \left[\frac{1}{m}\sum_{j=1}^{m}\ln(\hat{Y}_{i,j})\right] - [\theta] \tag{11.17}$$

and for each column, the main effect $[B]_{column}$ is the mean of the logarithms of the expected column sum, minus the grand mean:

$$[B]_{column} = \left[\frac{1}{n}\sum_{i=1}^{n}\ln(\hat{Y}_{i,j})\right] - [\theta] \tag{11.18}$$

Note that both $[A]_{row}$ and $[B]_{column}$ are residuals: they represent deviations from the grand mean $[\theta]$. Therefore, if we took the sums over all rows of $[A]$ and the sums over all columns of $[B]$, these sums should equal 0:

$$\sum_{i=1}^{n}[A]_{row(i)} = \sum_{j=1}^{m}[B]_{column(j)} = 0 \tag{11.19}$$

Equation 11.15 does not include the interaction term $[AB]$ (equivalent to the AB_{ij} term in the ANOVA equation). Why not? In a two-way table, we are testing the null hypothesis that the two variables are *independent* and the interaction term $[AB] = 0$; we hypothesize that there is no interaction between independent variables. Therefore, we compute the expected values for each cell using Equation 11.14, because it corresponds to our null hypothesis of no interaction.

On to Multi-Way Tables!

We can easily extend Equations 11.14–11.19 to contingency tables of any dimension. Instead of rows and columns, we'll generalize to d dimensions, which we will index using subscripts $i, j, k, \ldots$ Each dimension can assume an arbitrary number of levels, which we will index using subscripts n, m, p, and $q\ldots$ So for our four-dimensional data in Table 11.5, we must calculate the expected values for four dimensions i, j, k, and l. Applying the notation of Equation 11.15, we use the model

$$\ln(\hat{Y}_{i,j,k,l}) = [\theta]+[A]+[B]+[C]+[D]+[AB]+[AC]+[AD]+[BC]+ \\ [BD]+[CD]+[ABC]+[ABD]+[ACD]+[BCD] \quad (11.20)$$

(We have omitted the subscripts on the terms in square brackets for clarity.) This equation is called a **log-linear model** because the logarithm of the expected frequencies is a linear function of the predictor variables. You have already seen another example of a log-linear model when we discussed logistic regression in Chapter 9.

You may also notice that the full interaction term, $[ABCD]$, is missing from Equation 11.20. Why is this? If we added in this interaction term into Equation 11.20, the expected values would equal exactly the observed values, and so the fit would be perfect! But of course we are not interested in a perfect fit; we are interested in how well the data can be fit by a model with as few parameters as possible. We can use Equation 11.20, therefore, to test the hypothesis that the full interaction term $[ABCD]$ equals zero. We calculate expected values using Equation 11.20, and use Equation 11.5 or Equation 11.7 to compute a chi-square or G-statistic. Unfortunately, solving Equation 11.20 to get expected values is difficult (Fienberg 1970).

Using R software, we computed expected values for Table 11.5, calculated the chi-square statistic (see Equation 11.5) and its associated P-value, and tested the hypothesis that the four variables are independent. The chi-square statistic = 53.32, which for 32 degrees of freedom has a P-value of 0.01. We therefore reject the null hypothesis and conclude that the four variables are not independent.

However, this overall P-value tells us only that at least two of the variables are not independent, but it does not tell us which ones. Similarly, when we obtain a small P-value for a one-way ANOVA, we know that the means of the treatment groups differ overall, but we do not know which particular means are different and which are similar. In ANOVA, we used a priori contrasts and a posteriori comparisons to determine which means are different (see Chapter 10). In contingency table analysis, two methods are available to us to determine which variables are independent and which are not: hierarchical log-linear models and classification trees.

HIERARCHICAL LOG-LINEAR MODELING Equation 11.20 is used to test the hypothesis that the four-way interaction term $[ABCD] = 0$. More specifically we are interested in determining which factors are associated with population status (population declining or stable). The variable *Declining* is the response variable and the other three variables (*Protected, Invaded,* and *Light*) are predictor variables. This model or initial hypothesis is equivalent to the hypothesis that each outcome of the variable *Declining* (Yes or No) is independent of the three predictor variables, *Protected, Invaded,* or *Light*. Such a model would look like this:

$$\ln \hat{Y}_{i,j,k,l} = [\theta]+[A]+[B]+[C]+[D]+[AB]+[AC]+[BC]+[ABC] \quad (11.21)$$

where $[A]$ is the effect due to *Protected,* $[B]$ is the effect due to *Invaded,* $[C]$ is the effect due to *Light,* and $[D]$ is the effect due to *Declining.* This model includes all the interaction terms of the three predictor variables, but *does not* include any interaction terms with *Declining* $[D]$. Why not? Because our null hypothesis is that the value of *Declining* is independent of the predictor variables—that is, there is no interaction between *Declining* and any of the other variables.

For the data in Table 11.5, the chi-square statistic for the model described by Equation 11.21 has a *P*-value of 0.004 with 19 degrees of freedom. We therefore reject the hypothesis that *Declining* is independent of the three predictor variables. We must now examine models that incorporate some interaction between *Declining* and the other predictor variables. Our strategy is to add the minimum number of interaction terms necessary to fit the data. We seek the simplest model that provides a good fit to the data and does not cause us to reject the null hypothesis.

Let's start by adding in one term for the interaction between *Declining* and *Protected* ($[AD]$):

$$\ln \hat{Y}_{i,j,k,l} = [\theta]+[A]+[B]+[C]+[D]+[AB]+[AC]+[BC]+[ABC]+[AD] \quad (11.22)$$

This model has a chi-square statistic = 29.6 with 18 degrees of freedom, yielding a marginally significant *P*-value of 0.04. This model is "less significant" than that of Equation 11.21, which is what we're interested in here (a model that is "not significantly different" from the hypothesized model that "fits" the data).

How do we decide which interaction terms to include in the model and which to exclude? The problem is exactly analogous to the one we faced in multiple regression and path analysis (see Chapter 9): choosing a subset of important variables from a much larger set of potential predictor variables. The trick is to find a balance between a model that does a good job of reducing the deviations between the observed values and the model predictions, but does not include so many variables that it is overparamaterized. Statistics based on the Akaike Information Criterion (AIC) are examples of statistics that provide a "badness-of-fit" index to take into account residual deviations and the number of model parameters (see "Model Selection Criteria" in Chapter 9).

For the plant population data, the fit of the model did, in fact, improve with the addition of a [*Declining* × *Protected*] interaction term. The improved fit is reflected in a decrease in AIC from 81.81 to 73.61. In contrast, the addition of the other two-way interaction terms in the model, [*Declining* × *Light*] or [*Declining* × *Invaded*] did not improve the fit of the model.

Lastly, we considered the addition of three-way interaction terms. Addition of any three-way term (other than the [*ABC*] term in Equation 11.21) resulted in models with poorer fits, as indicated by an increase in AIC. These analyses suggest that population declines are reduced by protection status, but they are not affected by light levels or the presence of invasive species in a site.

The models that we have tested in this analysis are **hierarchical**—the terms of the simpler models (which exclude certain interactions) are nested subsets of the more complex models (which include them). Considering all the possible combinations and interactions among four variables, there are 74 alternative models or hypotheses that could be fit! All of these models include the main effects, but each model also includes a different set of interaction terms between the main effect variables. The hypotheses of interest are whether or not these interaction terms equal zero. These models are hierarchical because if there is a higher-order interaction term (e.g., a three-way interaction term) then the model also includes all the lower-order (e.g., two-way) interaction terms.

DEGREES OF FREEDOM FOR LOG-LINEAR MODELS You may have noticed that the different models we fit to the data in Table 11.5 had different degrees of freedom. For a two-way table, the degrees of freedom equaled the product of one less than the number of rows × one less than the number of columns. For multiway tables, the number of degrees of freedom is based on a summation of the degrees of freedom for the main effects and for all of the interaction terms. Degrees of freedom for the interaction terms are calculated in exactly the same way that they are for ANOVA (see Chapter 10). If Factor A has a category levels, and Factor B has b category levels, the degrees of freedom for the interaction term $A \times B = (a-1)(b-1)$. For the full model (all main effects and interactions except the four-way interaction $ABCD$), the number of degrees of freedom ν is calculated as:

$$\begin{aligned}\nu = nmpq - [1 + \\ &(n-1)+(m-1)+(p-1)+(q-1)+ \\ &(n-1)(m-1)+(n-1)(p-1)+(n-1)(q-1)+(m-1)(p-1)+(m-1)(q-1)+(p-1)(q-1)+ \\ &(n-1)(m-1)(p-1)+(n-1)(m-1)(q-1)+(n-1)(p-1)(q-1)+(m-1)(p-1)(q-1)]\end{aligned} \quad (11.23)$$

where n, m, p, and q are the number of possible levels of variables A, B, C, and D, respectively. In our example, Factors A, B, and D all have 2 levels (yes or no), whereas Factor C (light) has 5 levels. If you look closely at all of these terms, you should be able to recognize the components for the 4 main effects (the second row of Equation 11.23), the 6 two-way interaction terms (the third row of Equation 11.23), and the 4 three-way interactions terms (the last row of Equation 11.23). This full model, therefore, would have 4 degrees of freedom. Notice that the more terms we include in the model, the fewer the residual degrees of freedom are left to evaluate the fit of the data to the model. When we test a hierarchical model, we lose degrees of freedom because the test degrees of freedom actually reflect the *difference* between degrees of freedom in this model and the next model in the hierarchy.

The first model we evaluated, Equation 11.21, would have

$$\begin{aligned}\nu = nmpq - [1 + \\ &(n-1)+(m-1)+(p-1)+(q-1)+ \\ &(n-1)(m-1)+(n-1)(p-1)+(m-1)(p-1)+ \\ &(n-1)(m-1)(p-1)]\end{aligned} \quad (11.24)$$

degrees of freedom. The four terms on the second line are for the main effects, the next three terms are for the pairwise interactions, ([AB], [AC], and [BC], respectively), and the last term is for the three-way interaction term [ABC]. Solving Equation 11.24 for our four variables = 19 degrees of freedom, as we reported above. The model in Equation 11.22, which added in the interaction term of *Declining* × *Protected* ([AD]), has one fewer degree of freedom [$(n-1)(q-1) = 1$], or 18 degrees of freedom. You may not be able to solve hierarchical models by hand calculations, but you should at least be able to reconstruct the degrees of freedom and to understand and interpret the results of the model-fitting statistics.

CLASSIFICATION TREES **Classification trees** are useful for both exploring and modeling dependencies among tabular or categorical variables. Although classification trees have not been used widely in the statistical analysis of ecological data, most ecologists and environmental biologists are already quite familiar with one type of classification tree—the taxonomic key, which is used to identify an organism to the species level.

In a taxonomic key, a series of binary decisions are made about morphological features of the specimen (e.g., leaves versus needles). These forks in the key lead to finer and finer distinctions (e.g., simple leaves versus compound leaves) and the key finally ends when there are no further branching decisions and we have hopefully arrived at the correct species identification (e.g., red maple). Tabular data can be analyzed statistically in the same way, with the classification tree indicating dichotomous data forks that reflect partitions in the categorical variables. We give only a brief introduction to this technique here; additional ecological examples are provided by De'ath and Fabricius (2000), Cutler et al. (2007), and De'ath (2007); statistical details can be found in Breiman et al. (1984) and Ripley (1996).

Classification trees are constructed by splitting a categorical dataset into smaller and smaller groups, where each split depends on a single variable. Each split results in two groups, and each group is again split based on a single variable. In a taxonomic key, the splitting proceeds until there is only one species left at the end of each "twig." In classification trees, splitting proceeds only until no further improvement in the fit of the model is obtained relative to the number of splits used.

Rarely does one of the splits uniquely describe a group. Rather, the splitting results in a group characterized by a relatively high probability of occurrence in one or more classes (outcomes of a particular variable) relative to the others. Splitting ends when the endpoint is as homogeneous or consistent as possible, given the uncertainty in the data. A node that predicts all replicates to be in one class is completely homogeneous (or pure) and has an **impurity** value equal to zero. For classification trees, impurity is based on the proportion p of replicates in each of n classes.

Impurity increases as predictive ability decreases. Impurity can be calculated either as the information (entropy) index familiar to ecologists as the Shannon index (see Footnote 25 in Chapter 13):

$$\text{Impurity} = -\sum_{i=1}^{n} p_i \ln(p_i) \qquad (11.25)$$

or as the Gini index:[6]

$$\text{Impurity} = 1 - \sum_{i=1}^{n} p_i^2 \tag{11.26}$$

Tree-splitting proceeds by minimizing either Equation 11.25 or 11.26 at each split. Minimizing these indices is roughly equivalent to calculating conditional probability distributions of a variable A for each class of variable B, as described by Legendre and Legendre (1998: 230*ff*). With either of these indices, a completely pure sample ($p = 1.0$, with all replicates classified in one group) generates an index value of zero.

We again use the data in Table 11.5 to illustrate this procedure. As in the log-linear model analysis, we are trying to determine which factors best predict whether or not a population is declining.[7]

Our dataset contains 73 observations: 26 populations are declining and 47 are stable or increasing. The first split divided the data into two groups, one group of 40 populations and another of 33 populations (Figure 11.3). This split is based on protection status. As we discussed earlier, populations that are protected are much more likely to be stable or increasing (32 of 40 populations)

[6] Equation 11.25, which is also referred to as the Shannon-Weaver or Shannon-Wiener index, is named after the mathematician and cryptographer, Claude E. Shannon (1916–2001). The Shannon index represents the amount of uncertainty or entropy in coded strings of text. The more different letters in the text and the more even their frequencies, the harder it is to predict the next letter in the sequence. Shannon was a researcher at Bell Labs who made fundamental contributions to information theory. He applied Boolean algebra and binary arithmetic to the arrangement of relay switches, which provided the logical foundation for digital circuit design. Shannon wrote a paper on the design of a computer program to play chess, and he applied principles of information theory to win at blackjack in the casinos of Las Vegas. In 1943, Shannon met fellow computer scientist and cryptographer Alan Turing (see Footnote 21 in Chapter 13), with whom he shared many interests.

Claude E. Shannon

The Gini index is more familiar to economists than ecologists, but is actually on firmer statistical ground as a diversity measure. In a modified form, this index is equivalent to Hurlbert's (1971) *p*robability of an *i*nterspecific *e*ncounter (*PIE*). *PIE* measures the probability that two individuals randomly chosen from a community represent two different species. Unlike the Shannon index, *PIE* has a straightforward statistical interpretation and is relatively insensitive to sample size (Gotelli and Graves 1996). See Equation 13.15 and Footnote 25 in Chapter 13 for more discussion of the Gini index and PIE as diversity measures.

[7] For this example, tree construction was done using the rpart library in R version 2.14.0. Details of computations for producing classification trees can be found in Ripley (1996) and Venables and Ripley (2002).

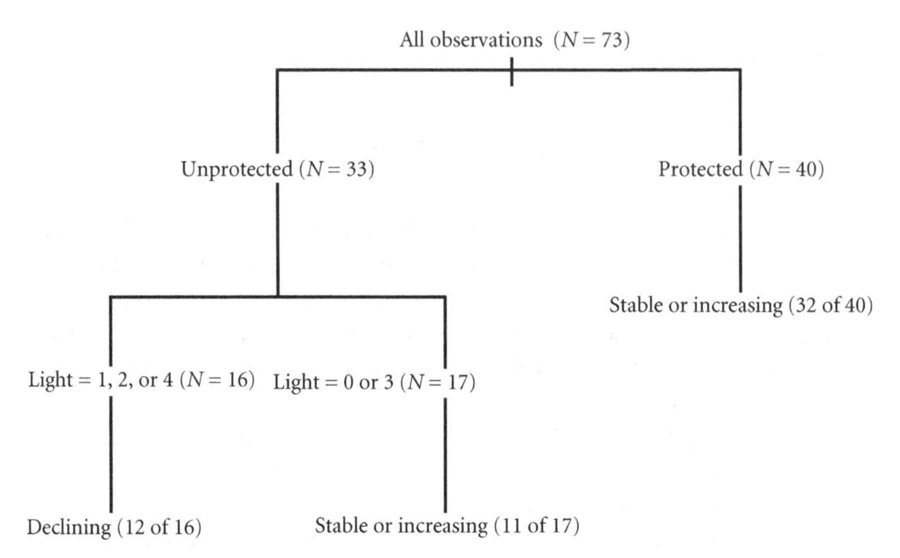

Figure 11.3 Classification tree for the rare plant population data in Table 11.5. This tree predicts population status (Declining versus Stable or Increasing) as a function of protection status, light level, and whether or not invasive species are present. The best-fitting model included protection status as the primary fork, with light level being used to split the unprotected populations in two groups. Notice that the final sample size at the three branch tips (16 + 17 + 40) accounts for all 73 of the populations in the dataset (Table 11.5). (Data from Farnsworth 2004.)

than they are to be declining (8 of 40 populations; see Figure 11.1). You should not be surprised that the total number of observations in each of these two groups matches the column totals in Table 11.2.

The remaining group of 33 observations (unprotected populations) consists of 18 populations that are declining and 15 that are stable or increasing—almost even odds. We split this group again into two groups, this time based on light availability (see Figure 11.3). Within the group of unprotected populations, those that are either in light level 0 or light level 3 are more likely to be stable or increasing (11 of 17 populations), whereas those in light levels 1, 2, or 4 are more likely to be declining (12 of 16 populations).

No further significant improvement in fit is obtained by further splitting of these groups, so tree construction stops here. Notice that the first split in this decision tree corresponds to the addition of the *Declining* × *Protected* interaction term ([*AD*]) in Equation 11.22. The second split corresponds to the addition of a *Declining* × *Protected* × *Light* term [*ACD*] into this model. In the log-linear analysis, the inclusion of this three-way interaction term did improve the fit of the model relative to Equation 11.22 ($P = 0.11$ with this term versus $P = 0.04$ without it). However, the penalty for adding in this interaction (it reduces the degrees of freedom from 18 to 14) increased the AIC from 73.61 to 75.50. By the "rules" of log-lincar model selection, this three-way interaction term was not retained, although it did emerge as a significant fork in the classification tree.

Don't be troubled by these apparent inconsistencies between the results of different kinds of statistical analyses. Because the log-linear models and classification trees are based on different selection criteria and different stopping rules they may give different answers when used with the same data set. Neither analysis is "best" or "correct," although in the context of management decisions, the classification tree may be easier to interpret than the log-linear analysis. Classification trees also can be used when the predictors are continuous, quantitative variables. Such trees are called **regression trees** (De'ath and Fabricus 2000).

Bayesian Approaches to Contingency Tables

Ecologists often refer to the chi-square and *G*-tests as non-parametric tests. In contrast, ANOVA and regression are recognized as parametric tests because they assume the data and residuals are normally distributed (see Chapter 5). Because chi-square and *G*-tests generate expected values from the data themselves, some researchers mistakenly believe that there is no underlying probability distribution required by the test. In fact, this is not the case. Log-linear modeling, which underlies the *G*-test, assumes that a Poisson or multinomial distribution describes the data. This assumption also allows us to use Bayesian techniques to analyze contingency tables.

Bayesian methods for contingency tables fall into two broad categories. One set of methods estimates expected frequencies as distributions (with modes or medians equal to the expected values). For a given cell, if the credibility interval of the expected value includes the observed value, the associated rows and columns are considered to be independent (Lindley 1964; Leonard 1975; Gelman et al. 1995).

The other set of methods estimates parameter values (with credibility intervals) for the terms in hierarchical log-linear models; parameters for which the credibility intervals include zero are dropped from the final model. Bayesian methods for choosing among log-linear models use selection criteria akin to AIC (Madigan and Raftery 1994; Albert 1996, 1997; Gelman et al. 1995). The application of Bayesian methods to classification and regression trees is an active area of current statistical research (Denison et al. 2002; Chipman et al. 2010), but unfortunately it is well beyond the scope of this primer.

Tests for Goodness of Fit

The statistical tests that we have encountered in this book, whether parametric or Bayesian, assume that our set of samples or observations accurately reflects a random variable with some underlying distribution. Examples include the binomial, Poisson, normal, lognormal, or exponential (see Chapter 2). Meaningful analyses also should result in a set of residuals that are distributed according to a known distribution, usually the normal distribution. We use **goodness-of-fit tests**, which are similar in structure to chi-square and *G*-tests, to determine if our observed data fit an hypothesized distribution or if our residuals conform to a normal (or other) distribution. Goodness-of-fit tests also are important for testing of specific ecological models. If the models make quantitative predictions about the expected values for different categories of data, we can use goodness-of-fit tests to see how well our data match the predictions of different models (Hilborn and Mangel 1997).

A goodness-of-fit test works in much the same way as the tests for independence presented earlier in this chapter: a set of observations of a single variable is compared with the expected values given a particular distribution. The mechanics are identical—use Equation 11.5 for a chi-square test for goodness-of-fit or Equation 11.7 for a *G*-test for goodness-of-fit. These two tests work well for discrete distributions and random variables, but do not work well for continuous distributions such as the normal distribution. For continuous distributions, we will introduce a different procedure, the Kolmogorov-Smirnov test, to evaluate the goodness-of-fit.

Goodness-of-Fit Tests for Discrete Distributions

As a simple example, we return to the controversy over the fairness of the Belgian Euro coin (see Footnote 7 in Chapter 1). Two statisticians with too much time on their hands spun a Belgian Euro coin 250 times and observed 140 heads and 110 tails. Is the Belgian Euro a fair coin? In other words, do these data fit the expected frequencies of heads and tails from a binomial distribution with $p = 0.50$?

CHI-SQUARE AND *G*-TESTS FOR BINOMIAL DATA The probability of heads on a fair coin = 0.5, so the expected frequencies of heads and tails are 125 and 125. We now have observed and expected values, and can calculate either of the two test statistics: $X^2_{Pearson}$ (see Equation 11.5) or G (see Equation 11.7). The $X^2_{Pearson}$ statistic = $(140 - 125)^2/125 + (110 - 125)^2/125 = 3.60$ and the *G*-statistic = $2[140 \times \ln(140/125) + 110 \times \ln(110/125)] = 3.61$. Both are distributed as χ^2 random variables and have tail probability values of 0.0577 and 0.0574, respectively. These *P*-values are suspiciously small, but as they are > 0.05, they are not considered significant by conventional hypothesis testing in the sciences. Thus, we would conclude that our observed data fit a binomial distribution, and there is (just barely) insufficient evidence to reject the null hypothesis that the Belgian Euro is a fair coin.[8]

BAYESIAN ALTERNATIVE The classical, asymptotic tests and the exact test fail to reject only marginally the null hypothesis that the Euro coin is fair, testing the hypothesis $P(\text{data}|\text{H}_0)$. A Bayesian analysis instead tests $P(\text{H}_1|\text{ data})$ by asking: do the data actually provide evidence that the Belgian Euro is biased? To assess this evidence, we must calculate the posterior odds ratio, as described in Chapter 9 (using Equation 9.49). MacKay (2002) and Hamaker (2002) conducted these analyses using several different prior probability distributions. First, if we have no reason to initially prefer one hypothesis over another (H_0: the coin is fair; H_1: the coin is biased), then the ratio of their priors = 1 and the posterior odds ratio is equal to the likelihood ratio

$$\frac{P(\text{data}\,|\,\text{H}_0)}{P(\text{data}\,|\,\text{H}_1)}$$

The likelihoods (numerator and denominator) of this ratio are calculated as:

$$P(D\,|\,H_i) = \int_0^1 P(D\,|\,p, H_i)P(p\,|\,H_i)dp \tag{11.27}$$

where D is data, H_i is either hypothesis ($i = 0$ or 1), and p is the probability of success in a binomial trial. For the null hypothesis (fair coin), our prior probability is that there is no bias; thus $p = 0.5$, and the likelihood $P(\text{data}\,|\text{H}_0)$ is the probability of obtaining exactly 140 heads in 250 trials, = 0.0084 (see Footnote 8). For the alternative hypothesis (biased coin), however, there are at least three ways to set the prior.

[8] This probability also can be calculated exactly using the binomial distribution. From Equation 2.3, the probability of getting *exactly* 140 heads in 250 spins with a fair coin is

$$\binom{250}{140}0.5^{140}0.5^{250-140} = 0.0084$$

But for a two-tailed significance test, we need the probability of obtaining 140 or more heads in our sample plus the probability of obtaining 110 or fewer tails in our sample. By analogy with the Fisher's Exact Test (see Equation 11.13) we simply add up the probabilities of obtaining 140, 141, ..., 250 tails and 0, 1, ..., 110 tails in a sample of 250, with expected probability of 0.50. This sum = 0.0581, a value almost identical to that obtained with the chi-square and *G*-tests.

First, we could use a uniform prior, which means that we have no initial knowledge as to how much bias there is and thus all biases are equally likely, and $P(p \mid H_1) = 1$. The integral then reduces to the sum of the probabilities over all choices of p (from 0 to 1) of the binomial expansion for 140 heads out of 250 trials:

$$P(D \mid H_1) = \sum_{p=0}^{1} \binom{250}{140} p^{140}(1-p)^{110} = 0.003 \tag{11.28}$$

The likelihood ratio is therefore 0.0084/0.00398 = 2.1, or approximately 2:1 odds in favor of the Belgian Euro being a *fair coin*!

Alternatively, we could use a more informative prior. The conjugate prior (see Footnote 11 in Chapter 5) for the binomial distribution is the beta distribution,

$$P(p \mid H_1, \alpha) = \frac{\Gamma(2\alpha)}{\Gamma(\alpha)^2} p^{\alpha-1}(1-p)^{\alpha-1} \tag{11.29}$$

where p is the probability of success, $\Gamma(\alpha)$ is the gamma distribution , and α is a variable parameter that expresses our prior belief in the bias of the coin. As α increases, our prior belief in the bias also increases. The likelihood of H_1 is now

$$P(D \mid H_1) = \frac{\Gamma(2\alpha)}{\Gamma(\alpha)^2} \binom{250}{140} \int_0^1 p^{140+\alpha-1}(1-p)^{110+\alpha-1} dp \tag{11.30}$$

For $\alpha = 1$ we have the uniform prior (and use Equation 11.28). By iteratively solving Equation 11.30, we can obtain likelihood ratios for a wide range of values of α. The most extreme odds-ratio obtainable using Equation 11.30 (remember, the numerator is fixed at 0.0084) is 0.52 (when $\alpha = 47.9$). For this informative prior, the odds are approximately 2:1 (= 1/0.52) in favor of the Belgian Euro being a *biased coin*.

Lastly, we could specify a very sharp prior that exactly matches the data: p = 140/250 = 0.56. Now, the likelihood of H_1 = 0.05078 (the binomial expansion in Equation 11.28 with $p = 0.56$) and the likelihood ratio = 0.0084/0.05078 = 0.16 or 6:1 odds (= 1/0.16) in favor of the Belgian Euro having exactly the bias observed.

Although our more informative priors suggest that the coin is biased, in order to reach that conclusion we had to specify priors that suggest strong advance knowledge that the Euro is indeed biased. An "objective" Bayesian analysis lets the data "speak for themselves" by using less informative priors, such as the uniform prior and Equation 11.28. That analysis provides little support for the hypothesis that the Belgian Euro is biased.

In sum, the asymptotic, exact, and Bayesian analyses all agree: the Belgian Euro is most likely a fair coin. The Bayesian analysis is very powerful in this context, because it provides a measure of the probability of the hypothesis of interest. In contrast, the conclusion of the frequentist analysis rests on P-values that are only slightly larger than the critical value of 0.05.

GENERALIZING THE CHI-SQUARE AND G-TESTS FOR MULTINOMIAL DISCRETE DISTRIBUTIONS Both the chi-square and G-tests can be used to test the fit of data that have more than two levels to discrete distributions, such as the Poisson or multinomial. The procedure is the same: calculate expected values, apply Equation 11.5 or 11.7, and compute the tail probability of the test statistic relative to a χ^2 distribution with ν degrees of freedom. The only tricky part is determining the degrees of freedom.

We distinguish two types of distributions that we use in applying goodness-of-fit tests: distributions with parameters estimated independently of the data and distributions with parameters estimated from the data themselves. Sokal and Rohlf (1995) call these **extrinsic hypotheses** and **intrinsic hypotheses**, respectively. For an extrinsic hypothesis, the expectations are generated by a model or prediction that does not depend on the data themselves. Examples of extrinsic hypotheses include the 50:50 expectation for an unbiased Euro coin, and the 3:1 ratio of dominant to recessive phenotypes in a simple Mendelian cross of two heterozygotes. For extrinsic hypotheses, the degrees of freedom are always one less than the number of levels the variable can assume.

Intrinsic hypotheses are those in which the parameters have to be estimated from the data themselves. An example of an intrinsic hypothesis is fitting data of counts of rare events to a Poisson distribution (see Chapter 2). In this case, the parameter of the Poisson distribution is not known, and must be estimated from the data themselves. We therefore use a degree of freedom for the total number of individuals sampled and use another degree of freedom for each estimated parameter. Because the Poisson distribution has one additional parameter (λ, the rate parameter), the degrees of freedom for that test would be two less than the number of levels.

It also is possible to use a chi-square or G-test to test the fit of a dataset to a normal distribution, but in order to do that, the data first have to be grouped into discrete levels. The normal distribution has two parameters to be estimated from the data (the mean μ and the standard deviation σ). Therefore, the degrees of freedom for a chi-square or G-test for goodness-of-fit to a normal distribution equals three less than the number of levels. However, the normal distribution is continuous, not discrete, and we would have to divide the data up into an arbitrary number of discrete levels in order to conduct the test. The results of such a test will differ depending on the number of levels chosen. To assess the goodness-of-fit for continuous distributions, the Kolmogorov-Smirnov test is a more appropriate and powerful test.

Testing Goodness-of-Fit for Continuous Distributions: The Kolmogorov-Smirnov Test

Many statistical tests require that either the data or the residuals fit a normal distribution. In Chapter 9, for example, we illustrated diagnostic plots for examining the residuals from a linear regression (see Figures 9.5 and 9.6). Although we asserted that Figure 9.5A illustrated the expected distribution of residuals for a linear model with a normal distribution of errors, we would like to have a quantitative test for that assertion. The Kolmogorov-Smirnov test is one such test.

The Kolmogorov-Smirnov test compares the distribution of an empirical sample of data with an hypothesized distribution, such as the normal distribution. Specifically, this test compares the empirical and expected cumulative distribution functions (CDF). The CDF is defined as the function $F(Y) = P(X < Y)$ for a random variable X. In words, if X is a random variable with a probability density function (PDF) $f(X)$, then the cumulative distribution function $F(Y)$ equals the area under $f(X)$ in the interval $X < Y$ (see Chapter 2 for further discussion of PDF's and CDF's). Throughout this book we have used the cumulative distribution function extensively—the tail probability, or P-value, is the area under the curve beyond the point Y, which is equal to $1 - F(Y)$.

Figure 11.4 illustrates two CDF's. The first is the empirical CDF of the residual values from the linear regression described in Chapter 9 of $\log_{10}$(plant species richness) on $\log_{10}$(island area). The second is the hypothetical CDF that would be expected if these residuals were normally distributed with mean = 0 and standard deviation = 0.31 (these values are estimated from the residuals themselves). The CDF of the empirical data is not a smooth curve because we have only 17 points, whereas the expected CDF is a smooth curve because it comes from an underlying normal distribution with infinitely many points.

The Kolmogorov-Smirnov test is a two-sided test. The null hypothesis is that the observed CDF $[F_{obs}(Y)]$ = the CDF of the hypothesized distribution $[F_{dist}(Y)]$. The test statistic is the absolute value of the maximum vertical difference between the observed and expected CDF's (arrow in Figure 11.4). Basically, one measures this distance at each observed point and takes the largest one.

For Figure 11.4, the maximum distance is 0.148. This maximum difference is compared with a table of critical values (e.g., Lilliefors 1967) for a particular sample size and P-value. If the maximum difference is less than the associated critical value, we cannot reject the null hypothesis that our data do not differ

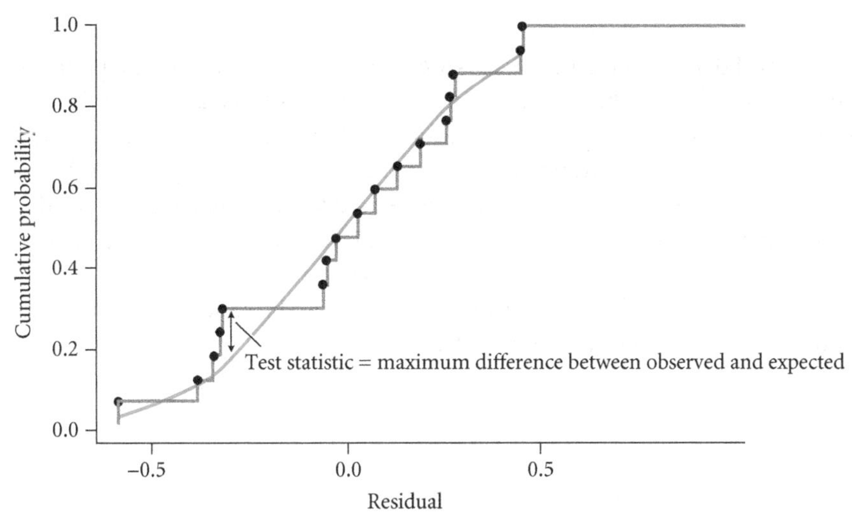

Figure 11.4 Goodness-of-fit test for a continuous distribution. Comparison of the cumulative distribution function of the observed residuals (black points, gray line) and the expected distribution if they were from a normal distribution (blue line). Residuals are calculated from the species-area regression of $\log_{10}$(species number) on $\log_{10}$(island area) for plant species of the Galápagos (data in Table 8.2; see Figure 9.6). The Kolmogorov-Smirnov test for goodness-of-fit compares these two distributions. The test statistic is the maximum difference between these two distributions, indicated by the double-headed arrow. The maximum difference for these data is 0.148, compared to a critical value of 0.206 for a test at $P = 0.05$. Because the null hypothesis cannot be rejected, the residuals appear to follow a normal distribution, which is one of the assumptions of linear regression (see Chapter 9). The Kolmogorov-Smirnov test can be used to compare a sample of continuous data to any continuous distribution, such as the normal, exponential, or log-normal (see Chapter 2).

from the expected distribution. In our example, the critical value for $P = 0.05$ is 0.206. Because the observed maximum difference was only 0.148, we conclude that the distribution of residuals from our regression is consistent with the hypothesis that the true errors come from a normal distribution (we do not reject H_0). On the other hand, if the maximum difference had been greater than the 0.206, we would reject the null hypothesis, and we should question whether a linear regression model, which assumes that the errors have a normal distribution, is the appropriate one to use for analyzing the data.

The Kolmogorov-Smirnov test is not limited to the normal distribution; it can be used for any continuous distribution. If you want to test whether your data are log-normally or exponentially distributed, for example, you can use the Kolmogorov-Smirnov test to compare the empirical CDF with a log-normal or exponential CDF (Lilliefors 1969).

Summary

Categorical data from tabular designs are a common outcome of ecological and environmental research. Such data can be analyzed easily using the familiar chi-square or *G*-tests. Hierarchical log-linear models or classification trees can test detailed hypotheses regarding associations among and between categorical variables. A subset of these methods also can be used to test whether or not the distribution of the data and residuals are consistent with, or are fit by, particular theoretical distributions. Bayesian alternatives to these tests have been developed, and these methods allow for estimation of expected frequency distributions and parameters of the log-linear models.

CHAPTER 12

The Analysis of Multivariate Data

All of the analytical methods that we have described so far apply only to single response variables, or univariate data. Many ecological and environmental studies, however, generate two or more response variables. In particular, we often analyze how multiple response variables—**multivariate data**—are related simultaneously to one or more predictor variables. For example, a univariate analysis of pitcher plant size might be based on a single variable: pitcher height. A multivariate analysis would be based on multiple variables: pitcher height, mouth opening, tube and keel widths, and wing lengths and spread (see Table 12.1). Because these response variables are all measured on the same individual, they are not independent of one another. Statistical methods for analyzing univariate data—regression, ANOVA, chi-square tests, and the like—may be inappropriate for analyzing multivariate data. In this chapter, we introduce several classes of methods that ecologists and environmental scientists use to describe and analyze multivariate data.

As with the univariate methods we discussed in Chapters 9–11, we can only scratch the surface and describe the important elements of the most common forms of multivariate analysis. Gauch (1982) and Manly (1991) present classical multivariate analysis techniques used commonly by ecologists and environmental scientists, whereas Legendre and Legendre (1998) provide a more thorough treatment of developments in multivariate analysis through the mid-1990s. Development of new multivariate methods as well as assessment of existing techniques is an active area of statistical research. Throughout this chapter, we highlight some of these newer methods and the debates surrounding them.

Approaching Multivariate Data

Multivariate data look very much like univariate data: they consist of one or more independent (predictor) variables and two or more dependent (response) variables. The distinction between univariate and multivariate data largely lies in how the data are organized and analyzed, not in how they are collected.[1]

Most ecological and environmental studies yield multivariate data. Examples include a set of different morphological, allometric, and physiological measurements taken on each of several plants assigned to different treatment groups; a list of species and their abundances recorded at multiple sampling stations along a river; a set of environmental variables (e.g., temperature, humidity, rainfall, etc.) and a corresponding set of organismal abundances or traits measured at several sites across a geographic gradient. In some cases, however, the explicit goal of a multivariate analysis can alter the sampling design. For example, a mark–recapture study of multiple species at multiple sites would not be designed the same way as a mark–recapture study of a single species because different methods of analysis might have to be used for rare versus common species (see Chapter 14).

Multivariate data can be quantitative or qualitative, continuous, ordered, or categorical. We focus our attention in this chapter on continuous quantitative variables. Extensions to qualitative multivariate variables are discussed by Gifi (1990).

The Need for Matrix Algebra

The mathematics of multivariate analysis is expressed best using **matrix algebra**. Matrix algebra allows us to use equations for multivariate analysis that look like those we have used earlier for univariate analysis. The similarity is more than skin deep; the interpretations of these equations are close, and many of the equations used for univariate statistical methods also can be written using matrix notation. We summarize basic matrix notation and matrix algebra in the Appendix.

A multivariate random variable $\mathbf{Y}$ is a set of n univariate variables $\{Y_1, Y_2, Y_3, \ldots, Y_n\}$ that are all measured for the same observational or experimental unit such as a single island, a single plant, or a single sampling station. We designate a multivariate random variable as a boldfaced, capital letter, whereas we con-

[1] In fact, you have already been exposed to multivariate data in Chapters 6, 7, and 10. In a repeated-measures design, multiple observations are taken on the same individual at different times. In a randomized block or split-plot design, certain treatments are physically grouped together in the same block. Because the response variables measured within a block or on a single individual are not independent of one another, the analysis of variance has to be modified to take account of this data structure (see Chapter 10). In this chapter, we will describe some multivariate methods in which there is not a corresponding univariate analysis.

tinue to write univariate random variables as italicized, capital letters. Each observation of **Y** is written as $\mathbf{y}_i$, which is a **row vector**. The elements $y_{i,j}$ of this row vector correspond to the measurement of the *j*th variable Y_j (*j* ranges from 1 to *n* variables) measured for individual *i* (*i* ranges from 1 to *m* individual observations or experimental units):

$$\mathbf{y}_i = [y_{i,1}, y_{i,2}, y_{i,3}, \ldots y_{i,j}, \ldots, y_{i,n}] \quad (12.1)$$

For convenience, we usually omit the subscript *i* in Equation 12.1. The row vector **y** is an example of a **matrix** (plural *matrices*) with one row and *n* columns. Each observation $y_{i,j}$ within the square brackets is called an **element** of the matrix (or row vector) **y**.

An ecologist would organize typical multivariate data in a simple spreadsheet with *m* rows, each representing a different individual, and *n* columns, each representing a different measured variable. Each row of the matrix corresponds to the row vector $\mathbf{y}_i$, and the entire table is the matrix **Y**. An example of a multivariate dataset is presented in Table 12.1. Each of the $m = 89$ rows represents a different plant, and each of the $n = 10$ columns represents a different morphological variable that was measured for each plant.

Several matrices will recur throughout this chapter. Equation 12.1 is the basic *n*-dimensional vector describing a single observation: $\mathbf{y} = [y_1, y_2, y_3, \ldots, y_n]$. We also define a **vector of means** as

$$\overline{\mathbf{Y}} = [\overline{Y}_1, \overline{Y}_2, \overline{Y}_3, \ldots, \overline{Y}_n] \quad (12.2)$$

TABLE 12.1 **Multivariate data for a carnivorous plant, *Darlingtonia californica***

Site	Plant	Height	Mouth	Tube	Keel	Wing1	Wing2	Wspread	Hoodmass	Tubemass	Wingmass
TJH	1	654	38	17	6	85	76	55	1.38	3.54	0.29
TJH	2	413	22	17	6	55	26	60	0.49	1.48	0.06
⋮	⋮	⋮	⋮	⋮	⋮	⋮	⋮	⋮	⋮	⋮	⋮

Each row represents a single observation of a cobra lily (*Darlingtonia californica*) plant measured at a particular site (T. J. Howell's fen, TJH) in the Siskiyou Mountains of southern Oregon. Each column represents a different morphological variable. Measurements for each plant included seven morphological variables (all in mm): height (height), mouth opening (mouth), tube diameter (tube), keel diameter (keel), length of each of the two "wings" (wing1, wing2) making up the fishtail appendage, and distance between the two wing tips (wspread). The plants were dried and the hood (hoodmass), tube (tubemass), and fishtail appendage (wingmass) were weighed (±0.01 g). The total dataset, collected in 2000 by A. Ellison, R. Emerson, and H. Steinhoff, consists of 89 plants measured at 4 sites. This is a typical multivariate dataset: the different variables measured on the same individual are not independent of one another.

which is the n-dimensional vector of sample means (calculated as described in Chapter 3) of each Y_j. The elements of this row vector are the means of each of the n variables, Y_j. In other words, $\bar{\mathbf{Y}}$ is equal to the set of sample means of each column of our original data matrix **Y**. In a univariate analysis, the corresponding measurement is the sample mean of the single response variable, which is an estimate of the true population mean μ. Similarly, the vector $\bar{\mathbf{Y}}$ is an estimate of the vector $\boldsymbol{\mu} = [\mu_1, \mu_2, \mu_3, \ldots, \mu_n]$ of the true population means of each of the n variables.

We need three matrices to describe the variation among the $y_{i,j}$ observations. The first is the **variance-covariance matrix, C** (see Footnote 5 in Chapter 9). The variance-covariance matrix is a **square matrix**, in which the diagonal elements are the sample variances of each variable (calculated in the usual way; see Chapter 3, Equation 3.9). The off-diagonal elements are the sample covariances between all possible pairs of variables (calculated as we did for regression and ANOVA; see Chapter 9, Equation 9.10).

$$\mathbf{C} = \begin{bmatrix} s_1^2 & c_{12} & \cdots & c_{1,n} \\ c_{2,1} & s_2^2 & \cdots & c_{2,n} \\ \vdots & \vdots & \ddots & \vdots \\ c_{n,1} & c_{n,2} & \cdots & s_n^2 \end{bmatrix} \quad \textbf{(12.3)}$$

In Equation 12.3, s_j^2 is the sample variance of variable Y_j, and $c_{j,k}$ is the sample covariance between variables Y_j and Y_k. Note that the sample covariance between variables Y_j and Y_k is the same as the sample covariance between variables Y_k and Y_j. Therefore, the elements of the variance-covariance matrix are **symmetric**, and are mirror images of one another across the diagonal ($c_{j,k} = c_{k,j}$).

The second matrix we use to describe the variances is the matrix of sample standard deviations:

$$\mathbf{D(s)} = \begin{bmatrix} \sqrt{s_1^2} & 0 & \cdots & 0 \\ 0 & \sqrt{s_2^2} & \cdots & 0 \\ \vdots & \vdots & \ddots & \vdots \\ 0 & 0 & \cdots & \sqrt{s_n^2} \end{bmatrix} \quad \textbf{(12.4)}$$

D(s) is a **diagonal matrix**: a matrix in which elements are all 0 except for those on the main diagonal. The non-zero elements are the sample standard deviations of each Y_j.

Lastly, we need a matrix of sample correlations:

$$\mathbf{P} = \begin{bmatrix} 1 & r_{12} & \cdots & r_{1,n} \\ r_{2,1} & 1 & \cdots & r_{2,n} \\ \vdots & \vdots & \ddots & \vdots \\ r_{n,1} & r_{n,2} & \cdots & 1 \end{bmatrix} \quad (12.5)$$

In this matrix, each element $r_{j,k}$ is the sample correlation (see Chapter 9, Equation 9.19) between variables Y_j and Y_k. Because the correlation of a variable with itself = 1, the diagonal elements of **P** all equal 1. Like the variance-covariance matrix, the matrix of correlations is symmetric ($r_{j,k} = r_{k,j}$).

Comparing Multivariate Means

Comparing Multivariate Means of Two Samples: Hotelling's T^2 Test

We introduce the concepts of multivariate analysis through a straightforward extension of the univariate *t*-test to the multivariate case. The classical *t*-test is used to test the null hypothesis that the mean value of a single variable does not differ between two groups or populations. For example, we could test the simple hypothesis that pitcher height of *Darlingtonia californica* (see Table 12.1) does not differ between populations sampled at Day's Gulch (DG) and T.J. Howell's fen (TJH). At each site, 25 plants were randomly sampled, from which we first calculated the mean height (DG = 618.8 mm; TJH = 610.4 mm) and the standard deviation (DG = 100.6 mm; TJH = 83.7 mm). Based on a standard *t*-test, the plants from these two populations do not differ significantly in height ($t = 0.34$, with 48 degrees of freedom, $P = 0.74$). We could conduct additional *t*-tests for each of the variables in Table 12.1 with or without corrections for multiple tests (see Chapter 10). However, we are more interested in asking whether overall plant morphology—quantified as the vector of means of all the morphological variables taken together—differs between the two sites. In other words, we wish to test the null hypothesis that the vectors of means of the two groups are equal.

Hotelling's T^2 test (Hotelling 1931) is a generalization of the univariate *t*-test to multivariate data. We first compute for each group the means of each of the variables Y_j of interest and then assemble them into two mean **column vectors** (see Equation 12.2), $\bar{\mathbf{Y}}_1$ (for TJH) and $\bar{\mathbf{Y}}_2$ (for DG). Note that a column vector is simply a row vector that has been **transposed**—the columns and rows are interchanged. The rules of matrix algebra, explained in the Appendix, sometimes require vectors to be single columns, and sometimes require vectors to be single rows. But the data are the same regardless of whether they are represented as row vectors or column vectors.

For this example, we have seven morphological variables: pitcher height, mouth opening, tube and keel widths, and lengths and spread of the wings of the fishtail appendage at the mouth of the pitcher (see Table 12.1), each measured at two sites.

$$\overline{\mathbf{Y}}_1 = \begin{bmatrix} 610.0 \\ 31.2 \\ 19.9 \\ 6.7 \\ 61.5 \\ 59.9 \\ 77.9 \end{bmatrix} \qquad \overline{\mathbf{Y}}_2 = \begin{bmatrix} 618.8 \\ 33.1 \\ 17.9 \\ 5.6 \\ 82.4 \\ 79.4 \\ 84.2 \end{bmatrix} \tag{12.6}$$

For example, average pitcher height at TJH was 610.0 mm (the first element of $\overline{\mathbf{Y}}_1$), whereas average pitcher mouth opening at DG was 33.1 mm (the second element of $\overline{\mathbf{Y}}_2$). Incidentally, the variables used in a multivariate analysis do not necessarily have to be measured in the same units, although many multivariate analyses rescale variables into standardized units, as described later in this chapter.

Now we need some measure of variance. For the *t*-test, we used the sample variances (s^2) from each group. For Hotelling's T^2 test, we use the sample variance-covariance matrix **C** (see Equation 12.3) for each group. The sample variance-covariance matrix for site TJH is

$$\mathbf{C}_1 = \begin{bmatrix} 7011.5 & 284.5 & 32.3 & -12.5 & 137.4 & 691.7 & 76.5 \\ 284.5 & 31.8 & -0.7 & -1.9 & 55.8 & 77.7 & 66.3 \\ 32.3 & -0.7 & 5.9 & 0.6 & -19.9 & -6.9 & -13.9 \\ -12.5 & -1.9 & 0.6 & 1.1 & -0.9 & -3.0 & -5.6 \\ 137.4 & 55.8 & -19.9 & -0.9 & 356.7 & 305.6 & 397.4 \\ 691.7 & 77.7 & -6.9 & -3.0 & 305.6 & 482.1 & 511.6 \\ 76.5 & 66.3 & -13.9 & -5.6 & 397.4 & 511.6 & 973.3 \end{bmatrix} \tag{12.7}$$

This matrix summarizes all of the sample variances and covariances of the variables. For example, at TJH, the sample variance of pitcher height was 7011.5 mm^2, whereas the sample covariance between pitcher height and pitcher mouth opening was 284.5 mm^2. Although a sample variance-covariance matrix can be constructed for each group ($\mathbf{C}_1$ and $\mathbf{C}_2$), Hotelling's T^2 test assumes that these two matrices are approximately equal, and uses a matrix $\mathbf{C_P}$ that is the pooled estimate of the covariance of the two groups:

$$\mathbf{C_P} = \frac{[(m_1 - 1)\mathbf{C}_1 + (m_2 - 1)\mathbf{C}_2]}{(m_1 + m_2 - 2)} \tag{12.8}$$

where m_1 and m_2 are the sample sizes for each group (here, $m_1 = m_2 = 25$).

Hotelling's T^2 statistic is calculated as

$$T^2 = \frac{m_1 m_2 (\overline{\mathbf{Y}}_1 - \overline{\mathbf{Y}}_2)^{\mathrm{T}} \mathbf{C_P}^{-1} (\overline{\mathbf{Y}}_1 - \overline{\mathbf{Y}}_2)}{(m_1 + m_2)} \tag{12.9}$$

where $\overline{\mathbf{Y}}_1$ and $\overline{\mathbf{Y}}_2$ are the vectors of means (see Equation 12.2), $\mathbf{C_P}$ is the pooled sample variance-covariance matrix (see Equation 12.8), superscript T denotes matrix transpose, and superscript (–1) denotes matrix inversion (see Appendix). Note that in this equation, there are two types of multiplication. The first is matrix multiplication of the vectors of differences of means and the inverse of the covariance matrix:

$$(\overline{\mathbf{Y}}_1 - \overline{\mathbf{Y}}_2)^{\mathrm{T}} \mathbf{C_P}^{-1} (\overline{\mathbf{Y}}_1 - \overline{\mathbf{Y}}_2)$$

The second is scalar multiplication of this product by the product of the sample sizes, $m_1 m_2$.

Following the rules of matrix multiplication, Equation 12.9 gives T^2 as a scalar, or a single number. A linear transformation of T^2 yields the familiar test statistic F:

$$\mathrm{F} = \frac{(m_1 + m_2 - n - 1)T^2}{(m_1 + m_2 - 2)n} \tag{12.10}$$

where n is the number of variables. Under the null hypothesis that the population mean vectors of each group are equal (i.e., $\mu_1 = \mu_2$), F is distributed as an F random variable with n numerator and $(m_1 + m_2 - n - 1)$ denominator degrees of freedom. Hypothesis testing then proceeds in the usual way (see Chapter 5). For the seven morphological variables, T^2 equals 84.62, with 7 numerator and 42 denominator degrees of freedom. Applying Equation 12.10 gives an F-value of 10.58. The critical value of F with 7 numerator and 42 denominator degrees of freedom equals 2.23, which is much less than the observed value of 10.58.

Therefore, we can reject the null hypothesis (with a *P*-value of 1.3×10^{-7}) that the two population mean vectors, $\boldsymbol{\mu}_1$ and $\boldsymbol{\mu}_2$, are equal.

Comparing Multivariate Means of More Than Two Samples: A Simple MANOVA

If we wish to compare univariate means of more than two samples of treatment groups, we use ANOVA in lieu of the standard *t*-test. Similarly, if we are comparing multivariate means of more than two groups, we use a **multivariate ANOVA**, or **MANOVA**, in lieu of Hotelling's T^2 test. As in Hotelling's T^2 test, we have $j = 1, \ldots, n$ variables or measurements taken for each individual and $i = 1, \ldots, m$ total observations. In the MANOVA, however, we have $k = 1, \ldots, g$ groups, and $l = 1, \ldots, q$ observations within each group. The multivariate observations (Equation 12.1) are denoted as $\mathbf{y}_{k,l}$, the lth vector of observations within the kth group. If our design is balanced, so that there are the same number q of observations in each of the g groups, then our total sample size $m = gq$.

By analogy with a one-way ANOVA (see Chapter 10, Equation 10.6), we analyze the model

$$\mathbf{Y} = \boldsymbol{\mu} + \mathbf{A}_k + \mathbf{e}_{kl} \tag{12.11}$$

where $\mathbf{Y}$ is the matrix of measurements (with m rows of observations and n columns of variables), $\boldsymbol{\mu}$ is the population (grand) mean, $\mathbf{A}_k$ is the matrix of deviations of the kth treatment from the sample grand mean, and $\mathbf{e}_{kl}$ is the error term: the difference between the lth individual in the kth treatment group and the mean of the kth treatment. The $\mathbf{e}_{kl}$'s are assumed to come from a multivariate normal distribution (see next section). The procedure is basically the same as a one-way ANOVA (Chapter 10), except that instead of comparing the group means, we compare the group **centroids**—the multivariate means.[2] The null hypothesis is that the means of the treatment groups are all the same: $\boldsymbol{\mu}_1 = \boldsymbol{\mu}_2 = \boldsymbol{\mu}_3 = \ldots = \boldsymbol{\mu}_g$. As with ANOVA, the test statistic is the ratio of the among-group

[2] In a one-dimensional space, two means can be compared as the arithmetic difference between them, which is what an ANOVA does for univariate data. In a two-dimensional space the mean vector of each group can be plotted as a point $(\bar{x}, \bar{y})$ in a Cartesian graph. These points represent the centroids (often thought of as the center of gravity of the cloud of points), and we can calculate the geometric distance between them. Finally, in a space with three (or more) dimensions, the centroid can again be located by a set of Cartesian coordinates with one coordinate for each dimension in the space. MANOVA compares the distances among those centroids, and tests the null hypothesis that the distances among the group centroids are no more different from those expected due to random sampling.

sums of squares divided by the within-group sums of squares, but these are calculated differently for MANOVA, as described below. This ratio has an F-distribution (see Chapter 10).

THE SSCP MATRICES In an ANOVA, the sums of squares are simple numbers; in a MANOVA, the sums of squares are matrices—called the **sums of squares and cross-products (SSCP)** matrices. These are square matrices (equal number of rows and columns) for which the diagonal elements are the sums of squares for each variable and for which the off-diagonal elements are the sums of cross-products for each pair of variables. SSCP matrices are directly analogous to sample variance-covariance matrices (see Equation 12.3), but we need three of them.

The among-groups SSCP matrix is the matrix **H**:

$$\mathbf{H} = q\sum_{k=1}^{g}(\overline{\mathbf{Y}}_{k.} - \overline{\mathbf{Y}}_{..})(\overline{\mathbf{Y}}_{k.} - \overline{\mathbf{Y}}_{..})^{T} \quad (12.12)$$

In Equation 12.12, $\overline{\mathbf{Y}}_{k.}$ is the vector of sample means in group k of the q observations in that group:

$$\overline{\mathbf{Y}}_{k.} = \frac{1}{q}\sum_{l=1}^{q}\mathbf{y}_{kl}$$

and $\overline{\mathbf{Y}}_{..}$ is the vector of sample means over all treatment groups

$$\overline{\mathbf{Y}}_{..} = \frac{1}{kq}\sum_{k=1}^{g}\sum_{l=1}^{q}\mathbf{y}_{kl}$$

The within-group SSCP matrix is the matrix **E**:

$$\mathbf{E} = \sum_{k=1}^{g}\sum_{l=1}^{q}(\mathbf{y}_{kl} - \overline{\mathbf{Y}}_{k.})(\mathbf{y}_{kl} - \overline{\mathbf{Y}}_{k.})^{T} \quad (12.13)$$

with $\overline{\mathbf{Y}}_{k.}$ again being the vector of sample means of treatment group k. In a univariate analysis, the analog of **H** in Equation 12.12 is the among-group sum of squares (see Equation 10.2, Chapter 10), and the analog of **E** in Equation 12.13 is the within-group sum of squares (Equation 10.3, Chapter 10). Lastly, we compute the total SSCP matrix:

$$\mathbf{T} = \sum_{k=1}^{g}\sum_{l=1}^{q}(\mathbf{y}_{kl} - \overline{\mathbf{Y}}_{..})(\mathbf{y}_{kl} - \overline{\mathbf{Y}}_{..})^{T} \quad (12.14)$$

TEST STATISTICS Four test-statistics are generated from the **H** and **E** matrices: Wilk's lambda, Pillai's trace, Hotelling-Lawley's trace, and Roy's greatest root (see Scheiner 2001). These are calculated as follows:

$$\text{Wilk's lambda} = \Lambda = \frac{|\mathbf{E}|}{|\mathbf{E}+\mathbf{H}|}$$

where | | denotes the determinant of a matrix (see Appendix).

$$\text{Pillai's trace} = \sum_{i=1}^{s}\left(\frac{\lambda_i}{\lambda_i+1}\right) = trace\left[(\mathbf{E}+\mathbf{H})^{-1}\mathbf{H}\right]$$

where s is the smaller of either the degrees of freedom (number of groups $g-1$) or the number of variables n; λ_i is the ith **eigenvalue** (see Appendix Equation A.17) of $\mathbf{E}^{-1}\mathbf{H}$; and *trace* is the trace of a matrix (see Appendix).

$$\text{Hotelling-Lawley's trace} = \sum_{i=1}^{n}\lambda_i = trace\left(\mathbf{E}^{-1}\mathbf{H}\right)$$

and

$$\text{Roy's greatest root} = \theta = \frac{\lambda_1}{\lambda_1+1}$$

where λ_1 is the largest (first) eigenvalue of $\mathbf{E}^{-1}\mathbf{H}$.

For large sample sizes, Wilk's lambda, Hotelling-Lawley's trace, and Pillai's trace all converge to the same *P*-value, although Pillai's trace is the most forgiving of violations of assumptions such as multivariate normality. Most software packages will give all of these test statistics.

Each of these four test statistics can be transformed into an F-ratio and tested as in ANOVA. However, their degrees of freedom are not the same. Wilk's lambda, Pillai's trace, and Hotelling-Lawley's trace are computed from all of the eigenvalues and hence have a sample size of nm (number of variables $n \times$ number of samples m) with $n(g-1)$ numerator degrees of freedom (g is the number of groups). Roy's largest root uses only the first eigenvalue, so its numerator degrees of freedom are only n. The degrees of freedom for Wilk's lambda often are fractional, and the associated F-statistic is only an approximation (Harris 1985). The choice among these MANOVA test statistics is not critical. They usually give very similar results, as in the analysis of the *Darlingtonia* data (Table 12.2).

COMPARISONS AMONG GROUPS As with ANOVA, if the MANOVA yields significant results, you may be interested in determining which particular groups dif-

TABLE 12.2 **Results of a one-way MANOVA**

I. H matrix

	Height	**Mouth**	**Tube**	**Keel**	**Wing1**	**Wing2**	**Wspread**
Height	35118.38	5584.16	–1219.21	–2111.58	17739.19	11322.59	18502.78
Mouth	5584.16	1161.14	–406.04	–395.59	2756.84	1483.22	1172.86
Tube	–1219.21	–406.04	229.25	127.10	–842.68	–432.92	659.62
Keel	–2111.58	–395.59	127.10	142.49	–1144.14	–706.07	–748.92
Wing1	17739.19	2756.84	–842.68	–1144.14	12426.23	9365.82	10659.12
Wing2	11322.59	1483.22	–432.92	–706.07	9365.82	7716.96	8950.57
Wspread	18502.78	1172.86	659.62	–748.92	10659.12	8950.57	21440.44

II. E matrix

	Height	**Mouth**	**Tube**	**Keel**	**Wing1**	**Wing2**	**Wspread**
Height	834836.33	27526.88	8071.43	1617.17	37125.87	46657.82	18360.69
Mouth	27526.88	2200.17	324.05	–21.91	4255.27	4102.43	3635.34
Tube	8071.43	324.05	671.82	196.12	305.16	486.06	375.09
Keel	1617.17	–21.91	196.12	265.49	–219.06	–417.36	–632.27
Wing1	37125.87	4255.27	305.16	–219.06	31605.96	28737.10	33487.39
Wing2	46657.82	4102.43	486.06	–417.36	28737.10	39064.26	41713.77
Wspread	18360.69	3635.34	375.09	–632.27	33487.39	41713.77	86181.61

III. Test statistics

Statistic	**Value**	**F**	**Numerator df**	**Denominator df**	***P*-value**
Pillai's trace	1.11	6.45	21	231	3×10^{-14}
Wilk's lambda	0.23	6.95	21	215.91	3×10^{-15}
Hotelling-Lawley's trace	2.09	7.34	21	221	3×10^{-16}
Roy's greatest root	1.33	14.65	7	77	5×10^{-12}

The original data are measurements of 7 morphological variables on 89 individuals of the cobra lily *Darlingtonia californica* collected at 4 sites (see Table 12.1). The **H** matrix is the among-groups sums of squares and cross-products (SSCP) matrix. The diagonal elements are the sums of squares for each variable, and the off-diagonal elements are the sums of cross-products for each pair of variables (see Equation 12.12). The **H** matrix is analogous to the univariate calculation of the among-groups sum of squares (see Equation 10.2). The **E** matrix is the within-groups SSCP matrix. The diagonal elements are the within-group deviations for each variable, and the off-diagonal elements are the residual cross-products for each pair of variables (see Equation 12.13). The **E** matrix is analogous to the univariate calculation of the residual sum of squares (see Equation 10.3). In these matrices, we depart from the good practice of reporting only significant digits in order to avoid round-off error in our calculations. If we were to report these data in a scientific publication, they would not have more significant digits that we had in our measurements (see Table 12.1). The four test statistics (Pillai's trace, Wilk's lambda, Hotelling-Lawley's trace, and Roy's greatest root) represent different ways to test for differences among groups in a multivariate analysis of variance (MANOVA). All of these measures generated very small *P*-values, suggesting that the 7-element vector of morphological measurements for *Darlingtonia* differed significantly among sites.

fer from one another. For post-hoc comparisons, Hotelling's T^2 test with the critical value adjusted using the Bonferroni correction (see Chapter 10) can be used for each pair-wise comparison. Discriminant analysis (described later in this chapter) also can be used to determine how well the groups can be separated.

All of the ANOVA designs described in Chapter 10 have direct analogues in MANOVA when the response variables are multivariate. The only difference is that the SSCP matrices are used in place of the among- and within-group sums of squares. Computing SSCP matrices can be difficult for complex experimental designs (Harris 1985; Hand and Taylor 1987). Scheiner (2001) summarizes MANOVA techniques for ecologists and environmental scientists.

ASSUMPTIONS OF MANOVA In addition to the usual assumptions of ANOVA (observations are independent and randomly sampled, and within-group errors are equal among groups and normally distributed), MANOVA has two additional assumptions. First, similar to the requirements of Hotelling's T^2 test, the covariances are equal among groups (the assumption of **sphericity**). Second, the multivariate variables used in the analysis and the error terms $\mathbf{e}_{kl}$ in Equation 12.11 must conform to a multivariate normal distribution. In the next section, we describe the multivariate normal distribution and how to test for departures from it.

Because Pillai's trace is robust to modest violations of these assumptions, in practice MANOVA is valid if the data do not depart substantially from sphericity or multivariate normality. Analysis of Similarity (ANOSIM) is an alternative method to MANOVA (Clarke and Green 1988; Clarke 1993) that does not depend on multivariate normality, but ANOSIM can be used only for one-way and fully crossed or nested two-way designs. ANOSIM has lower power (high probability of Type II statistical error) if strong gradients are present in the data (Somerfield et al. 2002).

The Multivariate Normal Distribution

Most multivariate methods for testing hypotheses require that the data being analyzed conform to the **multivariate normal** (or **multinormal**) **distribution.** The multivariate normal distribution is the analog of the normal (or Gaussian) distribution for the multidimensional variable $\mathbf{Y} = [Y_1, Y_2, Y_3, \ldots, Y_n]$. As we saw in Chapter 2, the normal distribution has two parameters: μ (the mean) and σ^2 (the variance). In contrast, the multivariate normal distribution[3] is defined by the mean vector $\boldsymbol{\mu}$ and the covariance matrix $\boldsymbol{\Sigma}$; the number of parameters in the multivariate normal distribution depends on the number of random variables Y_i in $\mathbf{Y}$. As you know, a one-dimensional normal distribution has a bell-shaped curve (see Figure 2.6). For a two-dimensional variable $\mathbf{Y} = [Y_1, Y_2]$, the bivariate normal distribution looks like a bell or hat (Figure 12.1). Most of the probability mass is concentrated near the center of the hat, closest to the means of the two variables. As we move in any direction toward the rim of the hat, the probability density becomes thinner and thinner. Although it is difficult to draw a multivariate distribution for variables of more than two dimensions, we can calculate and interpret them the same way.

For most purposes, it is convenient to use the standardized multivariate normal distribution, which has a mean vector $\boldsymbol{\mu} = [0]$. The bivariate normal distribution illustrated in Figure 12.1 is actually a standardized bivariate normal distribution. We also can illustrate this distribution as a contour plot (Figure 12.1E) of concentric ellipses. The central ellipse corresponds to a slice of the peak of the distribution, and as we move down the hat, the elliptical slices get larger. When all the variables Y_n in $\mathbf{Y}$ are independent and uncorrelated, all the off-diagonal correlation coefficients $r_{mn'}$ in the correlation matrix $\mathbf{P} = 0$, and the elliptical slices are perfectly circular. However, variables in most multivariate data sets are correlated, so the slices usually are elliptical in shape, as they are in Figure 12.1E. The stronger the correlation between a pair of variables, the more drawn out are the contours of the ellipse. The idea can be extended to more than two variables in $\mathbf{Y}$, in which case the slices would be ellipsoids or hyperellipsoids in multi-dimensional space.

[3] The probability density function for a multivariate normal distribution assumes an n-dimensional random vector $\mathbf{Y}$ of random variables $[Y_1, Y_2, \ldots, Y_n]$ with mean vector $\boldsymbol{\mu} = [\mu_1, \mu_2, \ldots, \mu_n]$ and variance-covariance matrix

$$\Sigma = \begin{bmatrix} \sigma_1^2 & \gamma_{12} & \cdots & \gamma_{1,n} \\ \gamma_{2,1} & \sigma_2^2 & \cdots & \gamma_{2,n} \\ \vdots & \vdots & \ddots & \vdots \\ \gamma_{n,1} & \gamma_{n,2} & \cdots & \sigma_n^2 \end{bmatrix}$$

in which σ_i^2 is the variance of Y_i and γ_{ij} is the covariance of Y_i and Y_j. For this vector $\mathbf{Y}$, the probability density function for the multivariate normal distribution is

$$f(\mathrm{Y}) = \frac{1}{\sqrt{(\pi)^n |\Sigma|}} e^{\left[-\frac{1}{2}(\mathrm{Y}-\mu)^{\mathrm{T}} \Sigma^{-1} (Y-\mu)\right]}$$

See the Appendix for details on matrix determinants (| |), inverses (matrices raised to the (–1) power) and transposition (T). The number of parameters needed to define the multivariate normal distribution is $2n + n(n - 1)/2$, where n is the number of variables in $\mathbf{Y}$. Notice how this equation looks like the probability density function for the univariate normal distribution (see Footnote 10, Chapter 2, and Table 2.4).

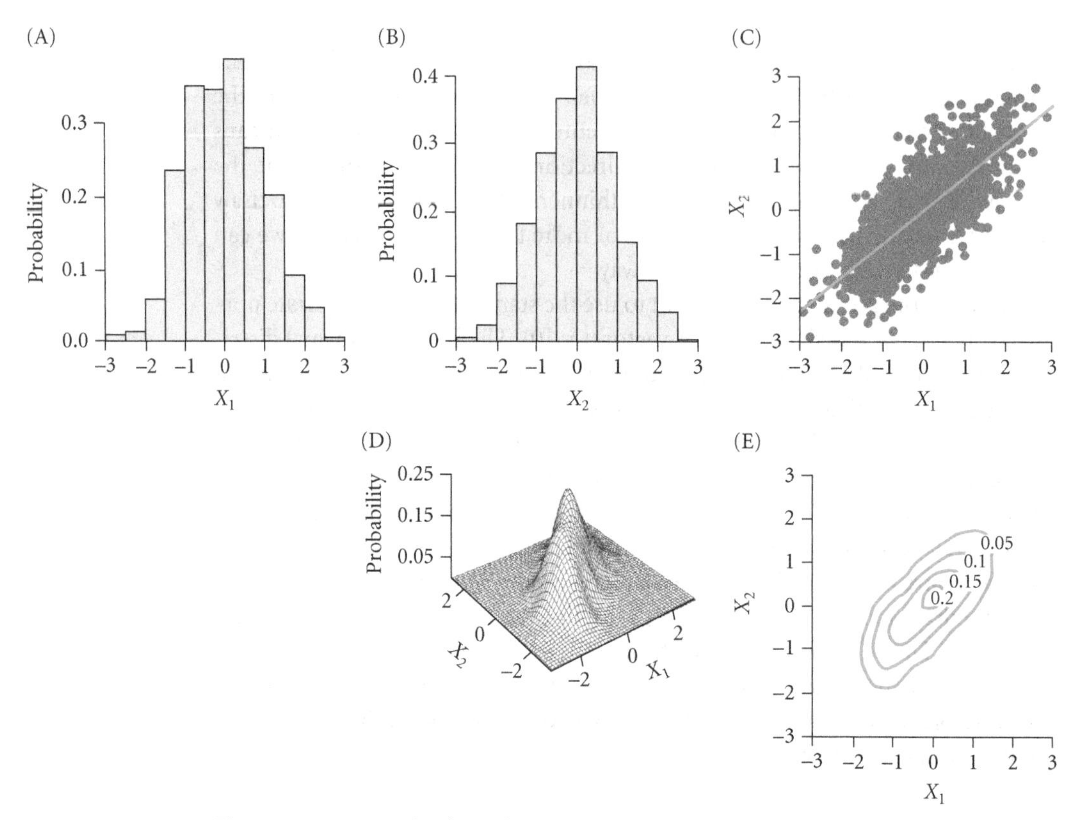

Figure 12.1 Example of a multivariate normal distribution. The multivariate random variable **X** is a vector of two univariate variables, X_1 and X_2, each with 5000 observations: $\mathbf{X} = [X_1, X_2]$. The mean vector $\boldsymbol{\mu} = [0,0]$, as both X_1 and X_2 have means = 0. The standard deviations of both X_1 and $X_2 = 1$, hence the matrix $\mathbf{D}(\mathbf{s})$ of sample standard deviations = $\begin{bmatrix} 1 & 0 \\ 0 & 1 \end{bmatrix}$. (A) Histogram that illustrates the frequency distribution of X_1; (B) histogram that illustrates the frequency distribution of X_2. The correlation between X_1 and $X_2 = 0.75$ and is illustrated in the scatterplot (C). Thus, the matrix **P** of sample correlations = $\begin{bmatrix} 1 & 0.75 \\ 0.75 & 1 \end{bmatrix}$. The joint multivariate probability distribution of **X** is shown as a mesh plot in (D). This distribution can also be represented as a contour plot (E), where each contour represents a slice through the mesh plot. Thus, for both X_1 and X_2 close to 0, the contour label = 0.2 means that the joint probability of obtaining these two values is approximately 0.2. The direction and eccentricity of the contour ellipses (or slices) reflect the correlation between the variables X_1 and X_2 (C). If X_1 and X_2 were completely uncorrelated, the contours would be circular. If X_1 and X_2 were perfectly correlated, the contours would collapse to a single straight line in bivariate space.

Testing for Multivariate Normality

Many multivariate tests assume that the multivariate data or their residuals are multivariate normally distributed. Tests for univariate normality (described in Chapter 11) are used routinely with regression, ANOVA, and other univariate statistical analyses, but tests for multivariate normality are performed rarely. This is not for the lack of availability of such tests. Indeed, more than fifty such tests for multivariate normality have been proposed (reviews in Koizol 1986, Gnanadesikan 1997, and Mecklin and Mundfrom 2003). However, these tests are not included in statistical software packages, and no single test yet has been found to account for the large number of ways that multivariate data can depart from normality (Mecklin and Mundfrom 2003). Because these tests are complex, and often yield conflicting results, they have been called "academic curiosities, seldom used by practicing statisticians" (Horswell 1990, quoted in Mecklin and Mundfrom 2003).

A common shortcut for testing for multivariate normality is simply to test whether each individual variable within the multivariate dataset is normally distributed (see Chapter 11). If any of the individual variables is not normally distributed, then it is not possible for the multivariate dataset to be multivariate normally distributed. However, the converse is not true. Each univariate measurement can be normally distributed but the entire dataset still may not be multivariate normally distributed (Looney 1995). Additional tests for multivariate normality should be performed even if each individual variable is found to be univariate normally distributed.

One test for multivariate normality is based on measures of multivariate skewness and kurtosis (see Chapter 3). This test was developed by Mardia (1970), and extended by Doornik and Hansen (2008). The computations involved in Doornik and Hansen's test are comparatively simple and the algorithm is provided completely in their 2008 paper.[4]

We tested the *Darlingtonia* data (see Table 12.1) for multivariate normality using Doornik and Hansen's test. Even though all the individual measurements were normally distributed ($P > 0.5$, all variables, using the Kolmogorov-Smirnov goodness-of-fit test), the multivariate data departed modestly from multivari-

---

[4] Doornik and Hansen's (2008) extension is more accurate (it yields the expected Type I error in simulations) and has better statistical power than Mardia's test both for small ($50 > N > 7$) and large ($N > 50$) sample sizes (Doornik and Hansen 2008; Mecklin and Mundfrom 2003). It is also easier to understand and program than the alternative procedure described by Royston (1983).

For a multivariate dataset with m observations and n measurements, Doornik and Hansen's test statistic E_n is calculated as $E_n = \mathbf{Z}_1^T\mathbf{Z}_1 + \mathbf{Z}_2^T\mathbf{Z}_2$, where $\mathbf{Z}_1$ is an n-dimensional column vector of transformed skewnesses of the multivariate data and $\mathbf{Z}_2$ is an n-dimensional column vector of transformed kurtoses of the multivariate data. Transformations are given in Doornik and Hansen (2008). The test-statistic E_n is asymptotically distributed as a χ^2 random variable with $2n$ degrees of freedom. (AME has written an R function to carry out Doornik and Hansen's test. The code can be downloaded from harvardforest.fas.harvard.edu/ellison/publications/primer/datafiles.)

ate normality ($P = 0.006$). This lack of fit was due entirely to the measurement of pitcher keel-width. With this variable removed, the remaining data passed the test for multivariate normality ($P = 0.07$).

Measurements of Multivariate Distance

Many multivariate methods quantify the difference among individual observations, samples, treatment groups, or populations. These differences most frequently are expressed as distances between observations in multivariate space. Before we plunge into the analytical techniques, we describe how distances are calculated between individual observations or between centroids that represent means for entire groups. Using the sample data in Table 12.1, we could ask: How far apart (or different) are two individual plants from one another in the multivariate space created using the morphological variables?

Measuring Distances between Two Individuals

Let's start by considering only two individuals, Plant 1 and Plant 2 of Table 12.1, and two of the variables, pitcher height (Variable 1) and spread of its fishtail appendage (Variable 2). For these two plants, we could measure the morphological distance between them by plotting the points in two dimensions and measuring the shortest distance between them (Figure 12.2). This distance, obtained by applying the Pythagorean theorem,[5] is called the **Euclidean distance** and is calculated as

$$d_{i,j} = \sqrt{(y_{i,1} - y_{j,1})^2 + (y_{i,2} - y_{j,2})^2} \quad (12.15)$$

In Equation 12.15, the plant is indicated by the subscript letter i or j, and the variable by subscript 1 or 2. To compute the distance between the measurements, we square the difference in heights, add to this the square of the difference in spreads, and take the square root of the sum. The result for these

Pythagoras

[5] Pythagoras of Samos (ca. 569–475 B.C.) may have been the first "pure" mathematician. He is remembered best for being the first to offer a formal proof of what is now known as the Pythagorean Theorem: the sum of the squares of the lengths of the two legs of a right triangle equals the square of the length of its hypotenuse: $a^2 + b^2 = c^2$. Note that the Euclidean distance we are using is the length of the hypotenuse of an implicit right triangle (see Figure 12.2). Pythagoras also knew that the Earth was spherical, although by placing it at the center of the universe he misplaced it a few billion light-years away from its actual position.

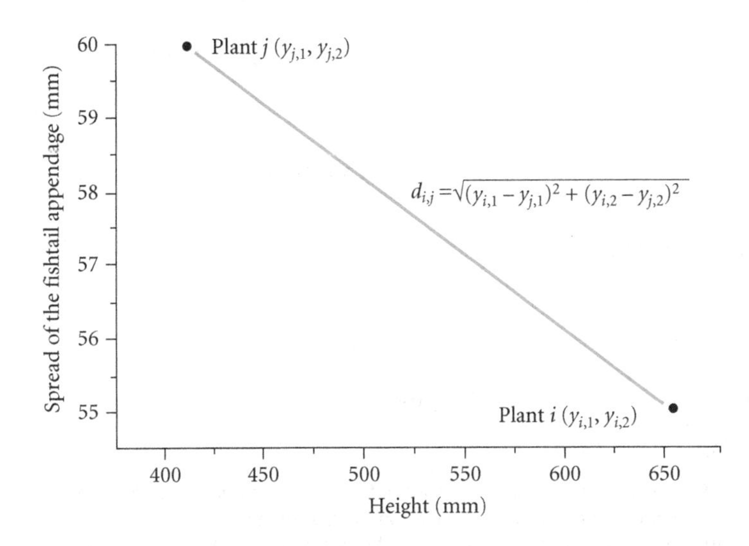

Figure 12.2 In two dimensions, the Euclidean distance $d_{i,j}$ is the straight-line distance between the points. In this example, the two morphological variables that we measured are plant height and spread of the fishtail appendage of the carnivorous cobra lily, *Darlingtonia californica* (see Table 12.1). Here the measurements for two individual plants are plotted in bivariate space. Equation 12.15 is used to calculate the Euclidean distance between them.

two plants is a distance $d_{i,j}$ = 241.05 mm. Because the original variables were measured in millimeters, when we subtract one measure from another, our result is still in millimeters. Applying subsequent operations of squaring, summing and taking the square root brings us back to a distance measurement that is still in units of millimeters. However, the units of the distance measurements will not stay the same unless all of the original variables were measured in the same units.

Similarly, we can calculate the Euclidean distance between these two plants if they are described by three variables: height, spread, and mouth diameter (Figure 12.3). We simply add another squared difference—here the difference between mouth diameters—to Equation 12.15:

$$d_{i,j} = \sqrt{(y_{i,1} - y_{j,1})^2 + (y_{i,2} - y_{j,2})^2 + (y_{i,3} - y_{j,3})^2} \quad \textbf{(12.16)}$$

Pythagoras founded a scientific-*cum*-religious society in Croton, in what is now southern Italy. The inner circle of this society (which included both men and women) were the *mathematikoi*, communitarians who renounced personal possessions and were strict vegetarians. Among their core beliefs were that reality is mathematical in nature; that philosophy is the basis of spiritual purification; that the soul can obtain union with the divine; and that some symbols have mystical significance. All members of the order were sworn to the utmost loyalty and secrecy, so there is scant biographical data about Pythagoras.

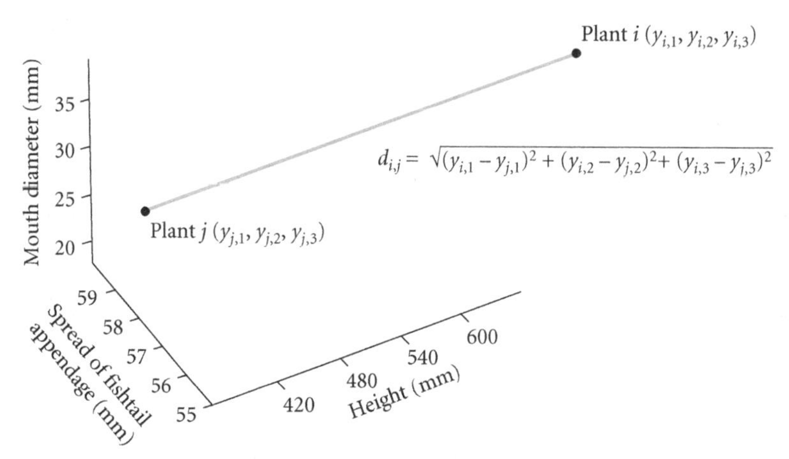

Figure 12.3 Measuring Euclidean distance in three-dimensional space. Mouth diameter is added to the two morphological variables shown in Figure 12.2, and the three variables are plotted in three-dimensional space. Equation 12.16 is used to calculate Euclidean distance, which is 241.58. Notice that this distance is virtually identical to the Euclidean distance measured in two dimensions (241.05; see Figure 12.2). The reason is that the third variable (mouth diameter) has a much smaller mean and variance than the first two variables and so it does not affect the distance measurement very much. For this reason, variables should be standardized (using Equation 12.17) before calculating distances between individuals.

The result of applying Equation 12.16 to the two plants is a morphological distance of 241.58 mm—a mere 0.2% change from the distance obtained using only the two variables height and spread.

Why are these two Euclidean distances not very different? Because the magnitude of plant height (hundreds of millimeters) is much greater than the magnitude of either spread or mouth diameter (tens of millimeters), the measurement of plant height dominates the distance calculations. In practice, therefore, it is essential to standardize the variables prior to computing distances. A convenient standardization is to transform each variable by subtracting its sample mean from the value of each observation for that variable, and then dividing this difference by the sample standard deviation:

$$Z = \frac{(Y_i - \overline{Y})}{s} \qquad \textbf{(12.17)}$$

The result of this transformation is called a ***Z*-score**. The Z-score controls for differences in the variance of each of the measured variables, and can also be

TABLE 12.3 **Standardization of multivariate data**

Site	Plant	Height	Mouth	Tube	Keel	Wingl	Wing2	Wspread	Hoodmass	Tubemass	Wingmass
TJH	1	0.381	1.202	−1.062	0.001	0.516	0.156	−1.035	1.568	0.602	0.159
TJH	2	−2.014	−1.388	−0.878	−0.228	−0.802	−1.984	−0.893	−0.881	−1.281	−0.276
⋮	⋮	⋮	⋮	⋮	⋮	⋮	⋮	⋮	⋮	⋮	⋮

The original data are measurements of 7 morphological variables and 3 biomass variables on 89 individuals of *Darlingtonia californica* collected at four sites (see Table 12.1). The first two rows of the data are illustrated after standardization. Standardized values are calculated by subtracting the sample mean of each variable from each observation and dividing this difference by the sample standard deviation (see Equation 12.17).

used to compare measurements that are not in the same units. The standardized values (Z-scores) for Table 12.1 are given in Table 12.3. The distance between these two plants for the two standardized variables height and spread is 2.527, and for the three standardized variables height, spread, and mouth diameter is 2.533. The absolute difference between these two distances is a modest 1%. However, this difference is five times greater than the difference between the distances calculated for the non-standardized data.

When the data have been transformed with Z-scores, what are the units of the distances? If we were working with our original variables, the units would be in millimeters. But standardized variables are always dimensionless; Equation 12.17 gives us units of mm × mm^{-1}, which cancel each other out. Normally, we think of Z-scores as being in units of "standard deviations"—how many standard deviations a measurement is from the mean. For example, a plant with a standardized height measurement of 0.381 is 0.381 standard deviations larger than the average plant.

Although we can't draw four or more axes on a flat piece of paper, we can still calculate Euclidean distances between two individuals if they are described by n variables. The general formula for the Euclidean distance based on a set of n variables is:

$$d_{i,j} = \sqrt{\sum_{k=1}^{n}(y_{i,k} - y_{j,k})^2} \qquad \textbf{(12.18)}$$

Based on all seven standardized morphological variables, the Euclidean distance between the two plants in Table 12.3 is 4.346.

Measuring Distances between Two Groups

Typically, we want to measure distances between samples or experimental treatment groups, not just distances between individuals. We can extend our formula for Euclidean distance to measure the Euclidean distance between means of any arbitrary number of groups g:

$$d_{i,j} = \sqrt{\sum_{k=1}^{g} (\overline{Y}_{i,k} - \overline{Y}_{j,k})^2} \quad \textbf{(12.19)}$$

where $\overline{Y}_{i,k}$ is the mean of variable i in group k. The Y_i in Equation 12.19 can be univariate or multivariate. Again, so that no one variable dominates the distance calculation, we usually standardize the data (see Equation 12.17) before calculating means and the distances between groups. From Equation 12.19, the Euclidean distance between the populations at DG and TJH based on all seven morphological variables is 1.492 standard deviations.

Other Measurements of Distance

Euclidean distance is the most commonly used distance measure. However, it is not always the best choice for measuring the distance between multivariate objects. For example, a common question in community ecology is how different two sites are based on the occurrence or abundance of species found at the two sites. A curious, counterintuitive paradox is that two sites with no species in common may have a smaller Euclidean distance than two sites that share at least some species! This paradox is illustrated using hypothetical data (Orlóci 1978) for three sites x_1, x_2, and x_3, and three species y_1, y_2, y_3. Table 12.4 gives the site × species matrix, and Table 12.5 gives all the pairwise Euclidean

TABLE 12.4 **Site × species matrix**

Site	Species		
	y_1	y_2	y_3
x_1	0	1	1
x_2	1	0	0
x_3	0	4	4

This matrix illustrates the paradox of using Euclidean distances for measuring similarity between sites based on species abundances. Each row represents a site, and each column a species. The values are the number of individuals of each species y_i at site x_j. The paradox is that the Euclidean distance between two sites with no species in common (such as sites x_1 and x_2) may be smaller than the Euclidean distance between two sites that share at least some species (such as sites x_1 and x_3).

TABLE 12.5 **Pair-wise Euclidean distances between sites**

	Site		
Site	x_1	x_2	x_3
x_1	0	1.732	4.243
x_2	1.732	0	5.745
x_3	4.243	5.745	0

This matrix is based on the species abundances given in Table 12.4. Each entry of the matrix is the Euclidean distance between site x_j and x_k. Note that the distance matrix is symmetric (see Appendix); for example, the distance between site x_1 and x_2 is the same as the distance between site x_2 and x_1. The diagonal elements are all equal to 0, because the distance between a site and itself = 0. This matrix of distance calculations demonstrates that Euclidean distances may give counterintuitive results if the data include many zeros. Sites x_1 and x_2 share no species in common, but their Euclidean distance is smaller than the Euclidean distance between sites x_1 and x_3, which share the same set of species.

distances between the sites. In this simple example, sites x_1 and x_2 have no species in common, and the Euclidean distance between them is 1.732 species (see Equation 12.14):

$$d_{x_1,x_2} = \sqrt{(1-0)^2 + (1-0)^2 + (1-0)^2} = 1.732$$

In contrast, sites x_1 and x_3 have all their species in common (both $\mathbf{y}_2$ and $\mathbf{y}_3$ occur at these two sites), but the distance between them is 4.243 species:

$$d_{x_1,x_3} = \sqrt{(0-0)^2 + (1-4)^2 + (1-4)^2} = 4.243$$

Ecologists, environmental scientists, and statisticians have developed many other measures to quantify the distance between two multivariate samples or populations. A subset of these are given in Table 12.6; Legendre and Legendre (1998) and Podani and Miklós (2002) discuss many more. These distance measurements fall into two categories: **metric distances** and **semi-metric distances**.

Metric distances have four properties:

1. The minimum distance = 0, and if two objects (or samples) $\mathbf{x}_1$ and $\mathbf{x}_2$ are identical, then the distance d between them also equals 0: $\mathbf{x}_1 = \mathbf{x}_2 \Rightarrow d(\mathbf{x}_1,\mathbf{x}_2) = 0$.
2. The distance measurement d is always positive if two objects $\mathbf{x}_1$ and $\mathbf{x}_2$ are not identical: $\mathbf{x}_1 \neq \mathbf{x}_2 \Rightarrow d(\mathbf{x}_1,\mathbf{x}_2) > 0$.
3. The distance measurement d is symmetric: $d(\mathbf{x}_1,\mathbf{x}_2) = d(\mathbf{x}_2,\mathbf{x}_1)$.
4. The distance measurement d satisfies the **triangle inequality**: for three objects $\mathbf{x}_1$, $\mathbf{x}_2$ and $\mathbf{x}_3$, $d(\mathbf{x}_1,\mathbf{x}_2) + d(\mathbf{x}_2,\mathbf{x}_3) \geq d(\mathbf{x}_1,\mathbf{x}_3)$.

TABLE 12.6 **Some common measures of distance or dissimilarity used by ecologists**

Name	Formula	Property
Euclidean	$d_{i,j} = \sqrt{\sum_{k=1}^{n} (y_{i,k} - y_{j,k})^2}$	Metric
Manhattan (*aka* City Block)	$d_{i,j} = \sum_{k=1}^{n} \lvert y_{i,k} - y_{j,k} \rvert$	Metric
Chord	$d_{i,j} = \sqrt{2 \times \left(1 - \frac{\sum_{k=1}^{n} y_{i,k} y_{j,k}}{\sqrt{\sum_{k=1}^{n} y_{i,k}^2 \sum_{k=1}^{n} y_{j,k}^2}} \right)}$	Metric
Mahalanobis	$d_{\mathbf{y}_i,\mathbf{y}_j} = \mathbf{d}_{i,j} \mathbf{V}^{-1} \mathbf{d}_{i,j}^{\mathrm{T}}$ $\mathbf{V} = \frac{1}{m_i + m_j - 2} [(m_i - 1)\mathbf{C}_i + (m_j - 1)\mathbf{C}_j]$	Metric
Chi-square	$d_{i,j} = \sqrt{\sum_{i=1}^{m} \sum_{j=1}^{m} y_{ij}} \times \sqrt{\sum_{k=1}^{n} \left[\frac{1}{\sum_{k=1}^{n} y_{jk}} \times \left(\frac{y_{ik}}{\sum_{k=1}^{n} y_{ik}} - \frac{y_{jk}}{\sum_{k=1}^{n} y_{jk}} \right)^2 \right]}$	Metric
Bray-Curtis	$d_{i,j} = \frac{\sum_{k=1}^{n} \lvert y_{i,k} - y_{j,k} \rvert}{\sum_{k=1}^{n} (y_{i,k} + y_{j,k})}$	Semi-metric
Jaccard	$d_{i,j} = \frac{a+b}{a+b+c}$	Metric
Sørensen's	$d_{i,j} = \frac{a+b}{a+b+2c}$	Semi-metric

The Euclidean, Manhattan (or City Block), Chord, and Bray-Curtis distances are used principally for continuous numerical data. Jaccard and Sørensen's distances are used for measuring distances between two samples that are described by presence/absence data. In Jaccard and Sørensen's distances, a is the number of objects (e.g., species) that occur only in y_i, b is the number of objects that occur only in y_j, and c is the number of objects that occur in both y_i and y_j. The Mahalanobis distance applies only to *groups* of samples $\mathbf{y}_i$ and $\mathbf{y}_j$, each of which contains respectively m_i and m_j samples. In the equation for Mahalanobis distance, $\mathbf{d}$ is the vector of differences between the means of the m samples in each group, and $\mathbf{V}$ is the pooled within-group sample variance-covariance matrix, calculated as shown, where $\mathbf{C}_i$ is the sample variance-covariance matrix for $\mathbf{y}_i$ (see Equation 12.3).

Euclidean, Manhattan, Chord, Mahalanobis, chi-square, and Jaccard distances[6] are all examples of metric distances.

[6] If you've used the Jaccard index of similarity before, you may wonder why we refer to it in Table 12.6 as a distance or dissimilarity measure. Jaccard's (1901) coefficient was developed to describe how similar two communities are in terms of shared species $s_{i,j}$

$$s_{i,j} = \frac{c}{a+b+c}$$

where *a* is the number of species that occur only in Community *i*, *b* is the number of species that occur only in Community *j*, and *c* is the number of species they have in common. Because measures of similarity take on their maximum value when objects are most similar, and measures of dissimilarity (or distance) take on their maximum value when objects are most different, any measure of similarity can be transformed into a measure of dissimilarity or distance. If a measure of similarity *s* ranges from 0 to 1 (as does Jaccard's coefficient), it can be transformed into a measure of distance *d* by using one of the following three equations:

$$d = 1 - s,\ d = \sqrt{1-s},\ \text{or}\ d = \sqrt{1-s^2}$$

Thus, in Table 12.6, the Jaccard distance equals 1 – the Jaccard coefficient of similarity. The reverse transformation (e.g., $s = 1 - d$) can be used to transform measures of distance into measures of similarity. If the distance measurement is not bounded (i.e., it ranges from 0 to ∞), it must be normalized to range between 0 and 1:

$$d_{norm} = \frac{d}{d_{\max}} \quad \text{or} \quad d_{norm} = \frac{d - d_{\min}}{d_{\max} - d_{\min}}$$

Although these algebraic properties of similarity indices are straightforward, their statistical properties are not. Similarity indices as used in biogeography and community ecology are very sensitive to variation in sample size (Wolda 1981; Jackson et al. 1989). In particular, small communities may have relatively high similarity coefficients even in the absence of any unusual biotic forces because their faunas are dominated by a handful of common, widespread species. Rare species are found mostly in larger faunas, and they will tend to reduce the similarity index for two communities even if the rare species also occur at random. Similarity indices and other biotic indices should be compared to an appropriate null model that controls for variation in sample size or the number of species present (Gotelli and Graves 1996). Either Monte Carlo simulations or probability calculations can be used to determine the expected value of a biodiversity or similarity index at small sample sizes (Colwell and Coddington 1994; Gotelli and Colwell 2001). The measured similarity in species composition of two assemblages depends not only on the number of species they share, but the dispersal potential of the species (most simple models assume species occurrences are equiprobable), and the composition and size of the source pool (Connor and Simberloff 1978; Rice and Belland 1982). Chao et al. (2005) introduce an explicit sampling model for the Jaccard index that accounts not only for rare species, but also for species that are shared but were undetected in any of the samples.

Semi-metric distances satisfy only the first three of these properties, but may violate the triangle inequality. Bray-Curtis and Sørensen's measures are semi-metric distances. A third type of distance measure, not used by ecologists, is **non-metric**, which violates the second property of metric distances and may take on negative values.

Ordination

Ordination techniques are used to order (or ordinate) multivariate data. Ordination creates new variables (called **principal axes**) along which samples are scored or ordered. This ordering may represent a useful simplification of patterns in complex multivariate data sets. Used in this way, ordination is a data-reduction technique: beginning with a set of *n* variables, the ordination generates a smaller number of variables that still illustrate the important patterns in the data. Ordination also can be used to discriminate or separate samples along the axis.

Ecologists and environmental scientists routinely use five different types of ordination—principal component analysis, factor analysis, correspondence analysis, principal coordinates analysis, and non-metric multidimensional scaling—each of which we discuss in turn. We discuss in depth the details of principal component analysis because the concepts and methods are similar to those used in other ordination techniques. Our discussion of the other four ordination methods is somewhat briefer. Legendre and Legendre (1998) is a good ecological guide to the details of basic multivariate analysis.

Principal Component Analysis

Principal component analysis (PCA) is the most straightforward way to ordinate data. The idea of PCA is generally credited to Karl Pearson (1901),[7] of Pearson chi-square and correlation fame, but Harold Hotelling[8] (of the Hotelling T^2 test) developed the computational methods in 1933. The primary use of PCA is to reduce the dimensionality of multivariate data. In other words, we use PCA to create a few key variables (each of which is a composite of many of our original variables) that characterize as fully as possible the variation in a multivariate dataset. The most important attribute of PCA is that the new variables are not correlated with one another. These uncorrelated variables can be used in multiple regression (see Chap-

[7] Pearson's goal in his 1901 paper was to assign individuals to racial categories based on multiple biometric measurements. See Footnote 3 in Chapter 11 for a biographical sketch of Karl Pearson.

ter 9) or ANOVA (see Chapter 10) without fear of multicollinearity. If the original variables are normally distributed, or have been transformed (see Chapter 8) or standardized (see Equation 12.17) prior to analysis, the new variables resulting from the PCA also will be normally distributed, satisfying one of the key requirements of parametric tests that are used for hypothesis testing.

PCA: THE CONCEPT Figure 12.4 illustrates the basic concept of a principal component analysis. Imagine you have counted the number of individuals of two species of grassland sparrows in each of ten prairie plots. These data can be plotted in a graph in which the abundance of Species A is on the x-axis and the abundance of Species B is on the y-axis. Each point in the graph represents the data from a different plot. We now create a new variable, or **first principal axis**,[9] that passes through the center of the cloud of the data points. Next, we calculate the value for each of the ten plots along this new axis, and then graph them on the axis. For each point, this calculation uses the numbers of both Species A and Species B to generate a single new value, which is the **principal component score** on the first principal axis. Although the original multivariate data had two observations for each plot, the principal component score reduces these two observations to a single number. We have effectively reduced the dimensionality of the data from two dimensions (Species A abundance and Species B abundance) to one dimension (first principal axis).

In general, if you have measured $j = 1$ to n variables Y_j for each replicate, you can generate n new variables Z_j that are all uncorrelated with one another (all the off-diagonal elements in $\mathbf{P} = 0$). We do this for two reasons. First, because the Z_j's are uncorrelated, they each can be thought of as measuring a different

Harold Hotelling

[8] Harold Hotelling (1895–1973) made significant contributions to statistics and economics, two fields from which ecologists have pilfered many ideas. His 1931 paper extending the *t*-test to the multivariate case also introduced confidence intervals to statisticians, and his 1933 paper developed the mathematics of principal components. As an economist, he was best known for championing approaches using marginal costs and benefits—the neoclassical tradition in economics based on Vilfredo Pareto's *Manual of Political Economy*. This tradition also forms the basis for many optimization models of ecological and evolutionary trade-offs.

[9] The term "principal axis" is derived from optics, where it is used to refer to the line (also known as the optical axis) that passes through the center of curvature of a lens so that a ray of light passing along that line is neither reflected nor refracted. It is also used in physics to refer to an axis of symmetry, such that an object will rotate at a constant angular velocity about its principal axis without applying additional torque.

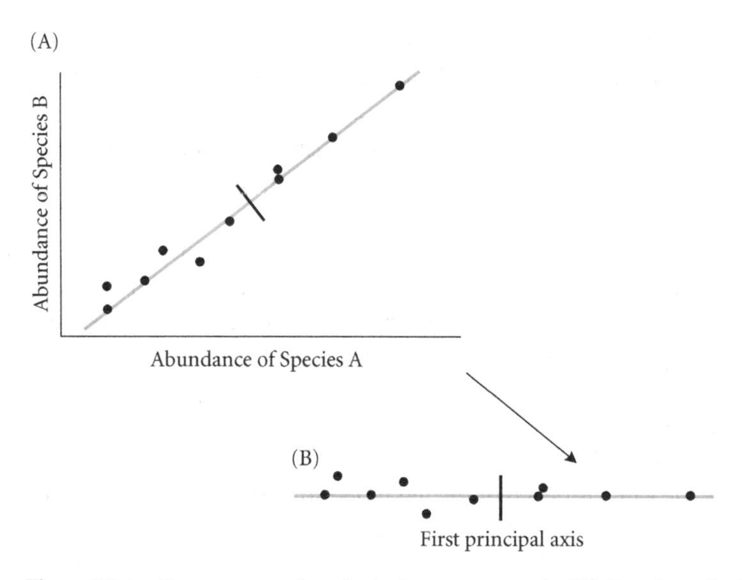

Figure 12.4 Construction of a principal component axis. (A) For a hypothetical set of 10 prairie plots, the number of individuals of two species of grassland sparrows are measured in each plot. Each plot can be represented as a point in a bivariate scatterplot with the abundance of each species graphed on the two axes. The first principal axis (blue line) passes through the major axis of variation in the data. The second principal axis (short black line) is orthogonal (perpendicular) to the first, and accounts for the small amount of residual variation not already incorporated in the first principal axis. (B) The first principal axis is then rotated, becoming the new axis along which the plot scores are ordinated (ordered). The vertical black line indicates the crossing of the second principal axis. Points to the right of the vertical line have positive scores on the first principal axis, and points to the left of the vertical line have negative scores. Although the original multivariate data had two measurements for each plot, most of the variation in the data is captured by a single score on the first principal component, allowing the plots to be ordinated along this axis.

and independent "dimension" of the multivariate data. Note that these new variables are not the same as the Z-scores described earlier (see Equation 12.17).

Second, the new variables can be ordered based on the amount of variation they explain in the original data. Z_1 has the largest amount of variation in the data (it is called the **major axis**), Z_2 has the next-largest amount of variation in the data, and so on ($\text{var}(Z_1) \geq \text{var}(Z_2) \geq \ldots \geq \text{var}(Z_n)$). In a typical ecological data set, most of the variation in the data is captured in the first few Z_j's, and we can discard the subsequent Z_j's that account for the residual variation. Ordered this way, the Z_j's are called **principal components**. If PCA is informative, it reduces a large number of original, correlated variables to a small number of new, uncorrelated variables.

Principal component analysis is successful to the extent that there are strong intercorrelations in the original data. If all of the variables are uncorrelated to begin with, we don't gain anything from the PCA because we cannot capture the variation in a handful of new, transformed variables; we might as well base the analysis on the untransformed variables themselves. Moreover, PCA may not be successful if it is used on unstandardized variables. Variables need to be standardized to the same relative scale, using transformations such as Equation 12.17, so that the axes of the PCA are not dominated by one or two variables that have large units of measurement.

AN EXAMPLE OF A PCA So how does it work? By way of example, let's examine some of the results of a PCA applied to the *Darlingtonia* data in Table 12.1. For this analysis, we used all ten variables—height, mouth diameter, tube and keel diameters, lengths and spread of the wings of the fishtail appendage, and masses of the hood, tube, and appendage—which we will call Y_1 through Y_{10}. The first principal component, Z_1, is a new variable (don't panic, details will follow), whose value for each observation is

$$Z_1 = 0.31Y_1 + 0.40Y_2 - 0.002Y_3 - 0.18Y_4 + 0.39Y_5 + 0.37Y_6 + 0.26Y_7 + 0.40Y_8 + 0.38Y_9 + 0.23Y_{10} \quad \textbf{(12.20)}$$

We generate the first principal component score (Z_1) for each individual plant by multiplying each measured variable times the corresponding coefficient (the **loading**) and summing the results. Thus, we take 0.31 times plant height, plus 0.40 times plant mouth diameter, and so on. For the first replicate in Table 12.1, $Z_1 = 1.43$. Because the coefficients for some of the Y_j's may be negative, the principal component score for a particular replicate may be positive or negative. The variance of $Z_1 = 4.49$ and it accounts for 46% of the total variance in the data. The second and third principal components have variances of 1.63 and 1.46, and all others are < 0.7.

Where did Equation 12.20 come from? Notice first that Z_1 is a **linear combination** of the ten Y variables. As we described above, we have multiplied each variable Y_j by a coefficient a_{ij}, also known as the loading, and added up all these products to give us Z_1. Each additional principal component is another linear combination of all of the Y_j's:

$$Z_j = a_{i1}Y_1 + a_{i2}Y_2 + \ldots + a_{in}Y_n \quad \textbf{(12.21)}$$

The a_{ij}'s are the coefficients for factor i that are multiplied by the measured value for variable j. Each principal component accounts for as much variance in the data as possible, subject to the condition that all the Z_j's are uncorrelated. This condition ensures that the Z_j's are independent and orthogonal. Therefore, when we plot the Z_j's, we are seeing relationships (if there are any) between independent variables.

CALCULATING PRINCIPAL COMPONENTS The calculation of both the coefficients a_{ij} and their associated variance is relatively simple. We start by standardizing our data by applying Equation 12.17, and then computing the sample variance-covariance matrix **C** (see Equation 12.3) of the standardized data. Note that **C** computed using standardized data is the same as the correlation matrix **P** (see Equation 12.5) computed using the raw data.

We then calculate the **eigenvalues** $\lambda_1 \ldots \lambda_n$ of the sample variance-covariance matrix and their associated **eigenvectors** $\mathbf{a}_j$ (see the Appendix for methods). The jth eigenvalue is the variance of Z_j, and the loadings a_{ij} are the elements of the eigenvectors. The sum of all the eigenvalues is the total variance explained:

$$\mathrm{var}_{total} = \sum_{j=1}^{n} \lambda_j$$

and the proportion of variance explained by each component Z_j is

$$\mathrm{var}_j = \frac{\lambda_j}{\sum_{j=1}^{n} \lambda_j}$$

If we multiply var_j by 100, we get the percent of variance explained. Because there are as many principal components as there are original variables, the total amount of variance explained by all the principal components = 100%. The results of these calculations for the *Darlingtonia* data in Table 12.1 are summarized in Tables 12.7 and 12.8.

HOW MANY COMPONENTS TO USE? Because PCA is a method for simplifying multivariate data, we are interested in retaining only those components that explain the bulk of the variation in the data. There is no absolute cutoff point at which we discard principal components, but a useful graphical tool to examine the contribution of each principal component to the overall PCA is a **scree plot** (Figure 12.5). This plot illustrates the percent of variance (derived from the eigenvalue) explained by each component in decreasing order. The scree plot resembles the profile of a mountainside down which a lot of rubble has fallen (known as a scree slope). The data used to construct the scree plot are tabulated in Table 12.7. In a scree plot, we look for a sharp bend or change in slope of the ordered eigenvalues. We retain those components that contribute to the mountainside, and ignore the rubble at the bottom. In this example, Components 1–3 appear

TABLE 12.7 **Eigenvalues from principal component analysis**

Principal component	Eigenvalue λ_i	Proportion of variance explained	Cumulative proportion of variance explained
Z_1	4.51	0.458	0.458
Z_2	1.65	0.167	0.625
Z_3	1.45	0.148	0.773
Z_4	0.73	0.074	0.847
Z_5	0.48	0.049	0.896
Z_6	0.35	0.036	0.932
Z_7	0.25	0.024	0.958
Z_8	0.22	0.023	0.981
Z_9	0.14	0.014	0.995
Z_{10}	0.05	0.005	1.000

These eigenvalues were obtained from analysis of the standardized measurements of 7 morphological variables and 3 biomass variables from 89 individuals of the cobra lily *Darlingtonia californica* collected at 4 sites (see Table 12.3). In a principal component analysis, the eigenvalue measures the proportion of variance in the original data explained by each principal component. The proportion of variance explained and the cumulative proportion explained are calculated from the sum of the eigenvalues ($\Sigma\lambda_j = 9.83$). The proportion of variance explained is used to select a small number of principal components that capture most of the variation in the data. In this data set, the first 3 principal components account for 77% of the variance in the original 10 variables.

TABLE 12.8 **Eigenvectors from a principal component analysis**

$$\mathbf{a}_1 = \begin{bmatrix} 0.31 \\ 0.40 \\ -0.002 \\ -0.18 \\ 0.39 \\ 0.37 \\ 0.26 \\ 0.40 \\ 0.38 \\ 0.23 \end{bmatrix} \quad \mathbf{a}_2 = \begin{bmatrix} -0.42 \\ -.025 \\ -0.10 \\ -0.07 \\ 0.28 \\ 0.37 \\ 0.43 \\ -0.18 \\ -0.41 \\ 0.35 \end{bmatrix} \quad \mathbf{a}_3 = \begin{bmatrix} 0.17 \\ -0.11 \\ -.74 \\ -.58 \\ -0.004 \\ 0.09 \\ 0.19 \\ -0.08 \\ 0.04 \\ 0.11 \end{bmatrix}$$

The first three eigenvectors from a principal component analysis of the measurements in Table 12.1. Each element in the eigenvector is the coefficient or loading that is multiplied by the value of the corresponding standardized variable (see Table 12.3). The products of the loadings and the standardized measurements are summed to give the principal component score. Thus, using the first eigenvector, 0.31 is multiplied by the first measurement (plant height), and this is added to 0.40 multiplied by the second measurement (plant mouth diameter), and so on. In matrix notation, we say that the principal component scores Z_j for each observation $\mathbf{y}$, which consists of ten measurements y_1 through y_{10}, are obtained by multiplying $\mathbf{a}_i$ by $\mathbf{y}$ (see Equations 12.20–12.23).

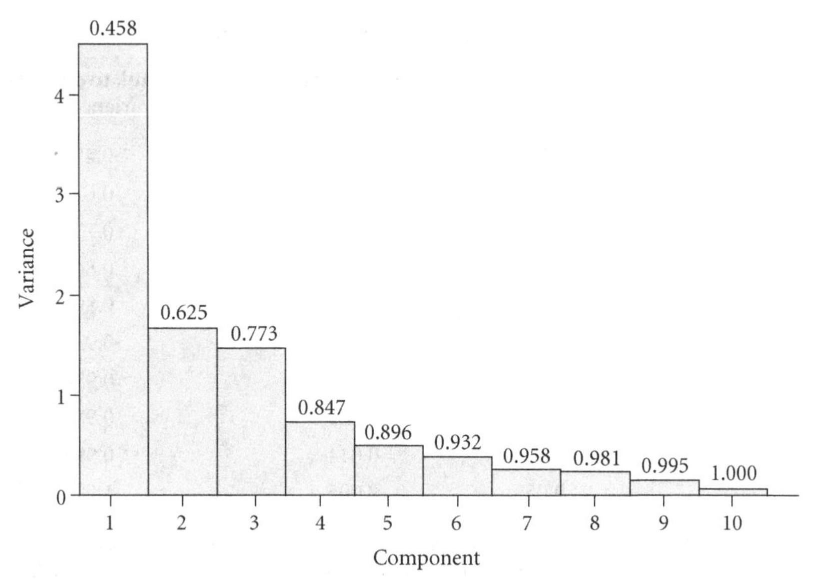

Figure 12.5 A scree plot for principal component analysis of the standardized data in Table 12.3. In a scree plot, the percent of variance (the eigenvalue) explained by each component (see Table 12.7) is shown in decreasing order. The scree plot resembles the profile of a mountainside down which a lot of rubble has fallen (known as a scree slope). In this scree plot, Component 1 clearly dominates the mountainside, although components 2 and 3 are also informative. The remainder is rubble.

useful, in total explaining 77% of the variance, whereas Components 4–10 look like rubble, none of them explaining more than 7% of the remaining variance. Jackson (1993) discusses a wide variety of heuristic methods (such as the scree plot) and statistical methods for choosing how many principal components to use. His results suggest that scree plots tend to overestimate by one the number of statistically meaningful components.

WHAT DO THE COMPONENTS MEAN? Now that we have selected a small number of components, we want to examine them in more detail. Component 1 was described already, in Equation 12.20:

$$Z_1 = 0.31Y_1 + 0.40Y_2 - 0.002Y_3 - 0.18Y_4 + 0.39Y_5 + 0.37Y_6 + 0.26Y_7 + 0.40Y_8 + 0.38Y_9 + 0.23Y_{10}$$

This component is the matrix product of the first eigenvector $\mathbf{a}_1$ (see Table 12.8) and the matrix of multivariate observations $\mathbf{Y}$ that had been standardized using Equation 12.17. Most of the loadings (or coefficients) are positive, and all are of approximately the same magnitude except the two loadings for variables related to the leaf tube (tube diameter Y_3 and keel diameter Y_4). Thus, this first component Z_1 seems to be a good measure of pitcher "size"—plants with tall pitchers, large mouths, and large appendages will all have larger Z_1 values.

Similarly, Component 2 is the product of the second eigenvector $\mathbf{a}_2$ (see Table 12.8) and the matrix of standardized observations $\mathbf{Y}$:

$$Z_2 = -0.42Y_1 - 0.25Y_2 - 0.10Y_3 - 0.07Y_4 + 0.28Y_5 + 0.37Y_6 + 0.43Y_7 - 0.18Y_8 - 0.41Y_9 + 0.35Y_{10} \quad \textbf{(12.22)}$$

For this component, loadings on the six variables related to pitcher height and diameter are negative, whereas loadings on the four variables related to the fishtail appendage are positive. Because all our variables have been standardized, relatively short, skinny pitchers will have negative height and diameter values, whereas relatively tall, stout pitchers will have positive height and diameter values. The value of Z_2 consequently reflects trade-offs in shape. Short plants with large appendages will have large values of Z_2, whereas tall plants with small appendages will have small values of Z_2. We can think of Z_2 as a variable that describes pitcher "shape." We generate the second principal component score for each individual plant by again substituting its observed values y_1 through y_{10} into Equation 12.22.

Finally, Component 3 is the product of the third eigenvector $\mathbf{a}_3$ (see Table 12.8) and the matrix of standardized observations $\mathbf{Y}$:

$$Z_3 = 0.17Y_1 - 0.11Y_2 + 0.74Y_3 + 0.58Y_4 - 0.004Y_5 + 0.09Y_6 + 0.19Y_7 - 0.08Y_8 + 0.04Y_9 + 0.11Y_{10} \quad \textbf{(12.23)}$$

This component is overwhelmingly dominated by large positive coefficients for tube (Y_3) and keel (Y_4) diameters—measurements of structures that support the pitcher. Z_3 will be large for "fat" plants and small for "skinny" plants. Because insects are trapped within the tube, plants with large Z_3 scores may be able to trap larger prey than plants with smaller Z_3 scores.

ARE ALL THE LOADINGS MEANINGFUL? In creating the principal component scores, we used the loadings from all of the original variables. But some loadings are large and others are close to zero. Should we use them all, or only those that are above some cutoff value? There is little agreement on this point (Peres-Neto et al. 2003), and if PCA is used only for exploratory data analysis, it probably doesn't matter much whether or not small loadings are retained. But if you use principal component scores for hypothesis testing, you must explicitly state which loadings were used, and how you decided to include or exclude them. If you are testing for interactions among principal axes, it is important to retain all of them.

USING THE COMPONENTS TO TEST HYPOTHESES Finally, we may want to examine differences among the four sites in their principal component scores. We can illustrate these differences by plotting the principal component scores for each plant on two axes, and coding the points (= replicates) for each group using different colors or symbols (Figure 12.6). In this plot, we illustrate the scores for the first two principal components, Z_1 and Z_2. We could construct similar plots for the other meaningful principal components (e.g., Z_1 versus Z_3; Z_2 versus Z_3). Figure 12.6 was generated using all the loadings in Equations 12.20 and 12.22.

An important feature of PCA is that the positions of the principal component scores for each replicate (such as the points in Figure 12.6) have the same Euclidean distance between them as do the original data in multivariate space. This property holds only when the Euclidean distances are calculated using *all* the principal components (Z_j's). If you calculate Euclidean distances using only the first few principal components, they will not be equal to the distances between the original data points. If the data originally were collected along with

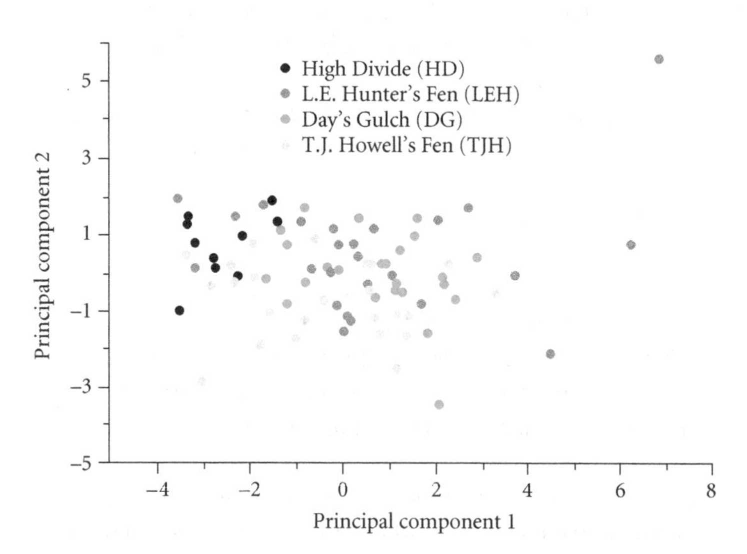

Figure 12.6 Plot illustrating the first two principal component scores from a PCA on the data in Table 12.3. Each point represents an individual plant, and the colors represent the different sites in the Siskiyou Mountains of southern Oregon. Although there is considerable overlap among the groups, there is also some distinct separation: the TJH samples (light blue) have lower scores on principal component 2, and the HD samples (black) have lower scores on the first principal component. Univariate or multivariate tests can be used to compare principal component scores among the populations (see Figure 12.7).

spatial information (such as x,y or latitude-longitude coordinates where the data were collected), the principal component scores could be plotted against their spatial coordinates. Such plots can provide information on multivariate spatial relationships.

We can treat the principal component scores as a simple univariate response variable and use ANOVA (see Chapter 10) to test for differences among treatment groups or sample populations. For example, an ANOVA on the first principal component score of the *Darlingtonia* data gives the result that there are significant differences among sites ($F_{3,85} = 9.26$, $P = 2 \times 10^{-5}$), and that all pairwise differences among sites, except for the DG–LEH comparison, also are significantly different (see Chapter 10 for details on calculating a posteriori comparisons). We conclude that plant size varies systematically across sites (Figure 12.7).

We obtain similar results for plant shape ($F_{3,85} = 5.36$, $P = 0.002$), but only three of the pairwise comparisons—DG versus TJH; HD versus TJH; and LEH versus TJH—are significantly different. This result suggests that plants at TJH have a different shape from plants at all the other sites. The concentration of light blue points (the principal component scores for the plants at TJH) in Figure 12.6 reveals that these plants have lower scores for the second principal component —that is, they are relatively tall plants with small fishtail appendages.

Factor Analysis

Factor analysis[10] and PCA have a similar goal: the reduction of many variables to few variables. Whereas PCA creates new variables as linear combinations of the original variables (see Equation 12.21), factor analysis considers each of

Charles Spearman

[10] Factor analysis was developed by Charles Spearman (1863–1945). In his 1904 paper, Spearman used factor analysis to measure general intelligence from multiple test scores. The factor model (see Equation 12.24) was used to partition the results from a battery of "intelligence tests" (the Y_j's) into factors (the F_j's) that measured general intelligence from factors specific to each test (the e_j's). This dubious discovery provided the theoretical underpinnings for Binet's general tests of intelligence (Intelligence Quotient, or I.Q., tests). Spearman's theories were used to develop the British "Plus-Elevens" examinations, which were administered to schoolboys at age 11 in order to determine whether they would be able to attend a University or a technical/vocational school. Standardized testing at all levels (e.g., MCAT, SAT, and GRE in the United States) is the legacy of Spearman's and Binet's work. Stephen Jay Gould's *The Mismeasure of Man* (1981) describes the unsavory history of IQ testing and the use of biological measurements in the service of social oppression.

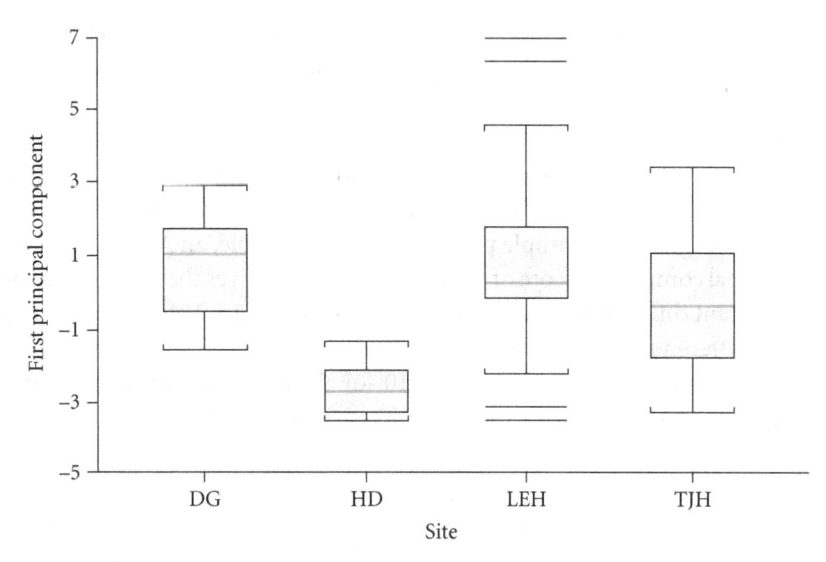

Figure 12.7 Box-plots of the scores of the first principal component from the PCA on the data in Table 12.3. Each box plot (see Figure 3.6) represents one of the four sites specified in Figure 12.6. The horizontal line indicates the sample median, and the box encompasses 50% of the data, from the 25th to the 75th percentile. The "whiskers" encompass 90% of the data, from the 10th to the 90th percentile. Four extreme data points (beyond the 5th and 95th percentiles) are shown as horizontal bars for the LEH site. Scores on Principal Component 1 varied significantly among sites ($F_{3,79} = 9.26$, $P = 2 \times 10^{-5}$), and were lowest (and least variable) at HD (see Figure 12.6).

the original variables to be a linear combination of some underlying "factors." You can think of this as PCA in reverse:

$$Y_j = a_{i1}F_1 + a_{i2}F_2 + \ldots + a_{in}F_n + e_j \qquad \textbf{(12.24)}$$

To use this equation, the variables Y_j must be standardized (mean = 0, variance = 1) using Equation 12.17. Each a_{ij} is a **factor loading**, the F's are called **common factors** (which each have mean = 0 and variance = 1), and the e_j's are factors specific to the jth variable, also uncorrelated with any F_j, and each with mean = 0.

Being a "PCA-in-reverse," factor analysis usually begins with a PCA and uses the meaningful components as the initial factors. Equation 12.21 gave us the formula for generating principal components:

$$Z_j = a_{i1}Y_1 + a_{i2}Y_2 + \ldots + a_{in}Y_n$$

Because the transformation from Y to Z is orthogonal, there is a set of coefficients a^*_{ij} such that

$$Y_j = a^*_{i1}Z_1 + a^*_{i2}Z_2 + \ldots + a^*_{in}Z_n \quad (12.25)$$

For a factor analysis, we keep only the first m components, such as those determined to be important in the PCA using a scree plot:

$$Y_j = a^*_{i1}Z_1 + a^*_{i2}Z_2 + \ldots + a^*_{im}Z_m + e_j \quad (12.26)$$

To transform the Z_j's into factors (which are standardized with variance = 1), we divide each of them by its standard deviation, $\sqrt{\lambda_i}$, the square root of its corresponding eigenvalue. This gives us a **factor model**:

$$Y_j = b_{i1}F_1 + b_{i2}F_2 + \ldots + b_{im}F_m + e_j \quad (12.27)$$

where $F_j = Z_j / \sqrt{\lambda_j}$ and $b_{ij} = a^*_{ij}\sqrt{\lambda_j}$.

ROTATING THE FACTORS Unlike the Z_j's generated by PCA, the factors F_j in Equation 12.27 are not unique; more than one set of coefficients will solve Equation 12.27. We can generate new factors F^*_j that are linear combinations of the original factors:

$$F^*_j = d_{i1}F_1 + d_{i2}F_2 + \ldots + d_{im}F_m \quad (12.28)$$

These new factors also explain as much of the variance in the data as the original factors. After generating factors using Equations 12.25–12.27, we identify the values for the coefficients d_{ij} in Equation 12.28 that give us the factors that are the easiest to interpret. This identification process is called **rotating** the factors.

There are two kinds of factor rotations: **orthogonal rotation** results in new factors that are uncorrelated with one another, whereas **oblique rotation** results in new factors that may be correlated with one another. The best rotation is one in which the factor coefficients d_{ij} are either very small (in which case the associated factor is unimportant) or very large (in which case the associated factor is very important). The most common type of orthogonal factor rotation is called **varimax rotation**, which maximizes the sum of the variance of the factor coefficients in Equation 12.28:

$$\text{var}_{\max} = \text{maximum of var}\left(\sum_{j=1}^{n} d_{ij}^2\right)$$

COMPUTING AND USING FACTOR SCORES Factor scores for each observation are computed as the values of the F_j's for each Y_i. Because the original factor scores F_j are linear combinations of the data (working backward from Equation 12.27 to Equation 12.25), we can re-write Equation 12.28 in matrix notation as:

$$\mathbf{F}^* = (\mathbf{DTD})^{-1}\mathbf{DTY} \tag{12.29}$$

where $\mathbf{D}$ is the $n \times n$ matrix of factor coefficients d_{ij}, $\mathbf{Y}$ is the data matrix, and $\mathbf{F}^*$ is the matrix of rotated factor scores. Solving this equation directly gives us the factor scores for each replicate.

Be careful when carrying out hypotheses tests on factor scores. Although they can be useful for describing multivariate data, they are not unique (as there are infinitely many choices of F_j), and the type of rotation used is arbitrary. Factor analysis is recommended only for exploratory data analysis.

A BRIEF EXAMPLE We conducted a factor analysis on the same data used for the PCA: 10 standardized variables measured on 89 *Darlingtonia* plants at four sites in Oregon and California (see Table 12.3). The analysis used four factors, which respectively accounted for 28%, 26%, 13%, and 8%, or a total of 75%, of the variance in the data.[11] After using varimax rotation and solving Equation 12.29, we plotted the first two factor scores for each plant in a scatterplot (Figure 12.8). Like PCA, the factor analysis suggests some discrimination among groups: plants from site HD (black points) have low Factor 1 scores, whereas plants from site TJH (light blue points) have low Factor 2 scores.

Principal Coordinates Analysis

Principal coordinates analysis (PCoA) is a method used to ordinate data using any measure of distance. Principal component analysis and factor analysis are used when we analyze quantitative multivariate data and wish to preserve Euclidean distances between observations. In many cases, however, Euclidean distances between observations make little sense. For example, a binary presence-absence matrix is a very common data structure in ecological and environmental studies: each row of the matrix is a site or sample, and each col-

[11] Some factor analysis software use a goodness-of-fit test (see Chapter 11) to test the hypothesis that the number of factors used in the analysis is sufficient to capture the bulk of the variance versus the alternative that more factors are needed. For the *Darlingtonia* data, a chi-square goodness-of-fit analysis failed to reject the null hypothesis that four factors were sufficient ($\chi^2 = 17.91$, 11 degrees of freedom, $P = 0.084$).

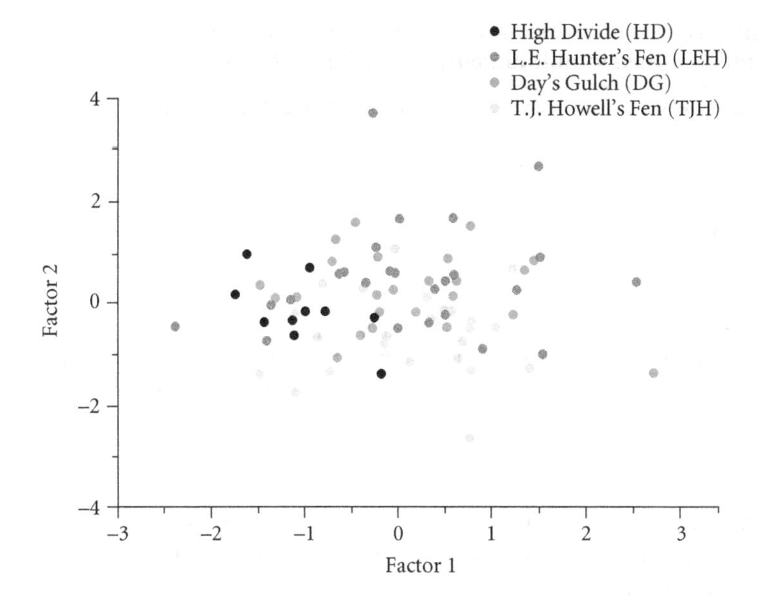

Figure 12.8 Plot illustrating the first two factors resulting from a factor analysis of the data in Table 12.3. Each point represents an individual plant, and the colors represent the different sites as in Figure 12.6. Results are qualitatively similar to the principal component analysis shown in Figure 12.6, with lower scores for HD on Factor 1 and lower scores for TJH on Factor 2.

umn is a species or taxon. The matrix entries indicate the absence (0) or presence (1) of a species in a site. As we saw earlier (see Tables 12.4 and 12.5), Euclidean distances measured for such data may give counterintuitive results. Another example for which Euclidean distances are not appropriate is a genetic distance matrix of the relative differences in electrophoretic bands. In both of these examples, PCoA is more appropriate than PCA. Principal component analysis itself is a special case of PCoA: the eigenvalues and associated eigenvectors of a PCA that we calculated from the variance-covariance matrix also can be recovered by applying PCoA to a Euclidean distance matrix.

BASIC CALCULATIONS FOR A PRINCIPAL COORDINATES ANALYSIS Principal coordinates analysis follows five steps:

1. Generate a distance or dissimilarity matrix from the data. This matrix, **D**, has elements d_{ij} that correspond to the distance between two samples i and j. Any of the distance measurements given in Table 12.6 can be used to calculate the d_{ij}'s. **D** is a square matrix, with the number of rows i = number of columns j = number of observations m.
2. Transform **D** into a new matrix **D***, with elements

$$d^*_{i,j} = -\frac{1}{2}d^2_{i,j}$$

TABLE 12.9 **Presence-absence matrix used for principal coordinates analysis (PCoA), correspondence analysis (CA), and non-metric multidimensional scaling (NMDS)**

	Aphaenogaster	*Brachymyrmex*	*Camponotus*	*Crematogaster*	*Dolichoderus*	*Formica*	*Lasius*
CT	1	0	1	0	0	0	1
MA mainland	1	1	1	0	1	1	1
MA islands	0	0	0	1	1	1	1
Vermont	1	0	1	0	0	1	1

Each row represents samples from different locations: Connecticut (CT), the Massachusetts (MA) mainland, islands off the coast of Massachusetts, and Vermont (VT). Each column represents a different ant genus collected during a survey of forest uplands surrounding small bogs. (Data compiled from Gotelli and Ellison 2002a,b and unpublished.)

This transformation converts the distance matrix into a coordinate matrix that preserves the distance relationship between the transformed variables and the original data.

3. Center the matrix **D*** to create the matrix **Δ** with elements $\delta_{i,j}$ by applying the following transformation to all the elements of **D***:

$$\delta_{ij} = d^*_{ij} - \bar{d}^*_i - \bar{d}^*_j + \bar{d}^*$$

where $\bar{d}^*_i$ is the mean of row i of **D***, $\bar{d}^*_j$ is the mean of column j of **D***, and $\bar{d}^*$ is the mean of all the elements of **D***.

4. Compute the eigenvalues and eigenvectors of **Δ**. The eigenvectors $\mathbf{a}_k$ must be scaled to the square root of their corresponding eigenvalues:

$$\sqrt{\mathbf{a}_k^T \mathbf{a}_k} = \sqrt{\lambda_k}$$

5. Write the eigenvectors $\mathbf{a}_k$ as columns, with each row corresponding to an observation. The entries are the new coordinates of the objects in principal coordinates space, analogous to the principal component scores.

A BRIEF EXAMPLE We use PCoA to analyze a binary presence-absence matrix of four sites (Connecticut, Vermont, Massachusetts mainland, and Massachusetts Islands) and 16 genera of ants found in forested uplands surrounding small bogs (Table 12.9). We first generated a distance matrix for the four sites using Sørensen's measure of dissimilarity (Table 12.10). The 4 × 4 distance matrix contains the values of Sørensen's measure of dissimilarity calculated for all pair-wise combinations of sites. We then computed the eigenvectors, which are used to

Leptothorax	*Myrmecina*	*Myrmica*	*Nylanderia*	*Prenolepis*	*Ponera*	*Stenamma*	*Stigmatomma*	*Tapinoma*
1	0	1	0	0	0	1	0	1
1	1	1	0	1	0	1	1	0
1	1	1	1	1	0	1	1	1
1	0	1	0	0	1	1	0	1

compute the principal coordinate scores for each site. Principal coordinate scores for each site are plotted for the first two principal coordinate axes (Figure 12.9). Sites differ based on dissimilarities in their species compositions, and can be ordinated on this basis. The sites are ordered CT, VT, MA mainland, MA islands from left to right along the first principal axis, which accounts for 80.3% of the variance in the distance matrix. The second principal axis mostly separates the mainland MA sites from the island MA sites and accounts for an additional 14.5% of the variance in the distance matrix.

Correspondence Analysis

Correspondence analysis (CA), also known as reciprocal averaging (RA) (Hill 1973b) or indirect gradient analysis, is used to examine the relationship of species assemblages to site charcteristics. The sites usually are selected to span an environmental gradient, and the underlying hypothesis or model is that species abundance distributions are unimodal and approximately normal (or Gaussian) across the environmental gradient (Whittaker 1956). The axes resulting from a

TABLE 12.10 **Measures of dissimilarity used in principal coordinates analysis (PCoA) and non-metric multidimensional scaling (NMDS)**

	CT	MA mainland	MA islands	VT
CT	0	0.37	0.47	0.13
MA mainland	0.37	0	0.25	0.33
MA islands	0.47	0.25	0	0.43
VT	0.13	0.33	0.43	0

The original data consist of a presence-absence matrix of forest ant genera in four New England sites (see Table 12.9). Each entry is the Sørensen's dissimilarity between each pair of sites (see Table 12.6 for formula). The more two sites differ in the genera they contain, the higher the dissimilarity. By definition, the dissimilarity matrix is symmetric and the diagonal elements equal 0.

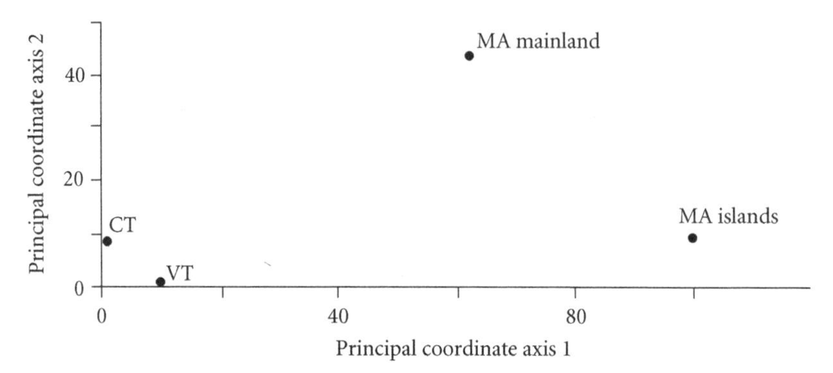

Figure 12.9 Ordination of four upland sites based on their ant species assemblages (see Table 12.9) using Principal Coordinates Analysis (PCoA). A pairwise distance matrix for the four sites is calculated from the data in Table 12.10 and used in the PCoA. The sites are ordered CT, VT, MA mainland, and MA islands from left to right along the first principal axis, which accounts for 80.3% of the variance in the distance matrix. The second principal axis mostly separates the mainland MA sites from the island MA sites, and accounts for an additional 14.5% of the variance.

CA maximize the separation of species abundances along each axis of peaks. Correspondence analysis takes as input a site × species matrix that is manipulated as if it were a contingency table (see Chapter 11). Correspondence analysis also can be considered a special case of PCoA; a PCoA using a chi-square distance matrix results in a CA.

BASIC CALCULATIONS FOR A CORRESPONDENCE ANALYSIS Correspondence analysis is not much more complex to interpret than PCA or factor analysis, but it requires more matrix manipulations.

1. Begin with a row × column table (such as a contingency table or a site [rows] × species [columns] matrix) with m rows and n columns, and for which $m \geq n$. This is often a restriction of CA software. Note that if $m \leq n$ in the original data matrix, transposing it will meet the condition that $m \geq n$, and the interpretation of the results will be identical.
2. Create a matrix **Q** whose elements q_{ij} are proportional to chi-square values:

$$q_{i,j} = \frac{\left(\dfrac{\text{observed} - \text{expected}}{\sqrt{\text{expected}}}\right)}{\sqrt{\text{grand total}}}$$

The expected values are calculated as we described in Chapter 11 for chi-square tests for independence (see Equation 11.4).

3. Apply singular-value decomposition to **Q**: $\mathbf{Q} = \mathbf{UWV}^T$, where **U** is an $m \times n$ matrix, **V** is an $n \times n$ matrix, **U** and **V** are **orthonormal matrices**, and **W** is a diagonal matrix (see Appendix for details).
4. Observe that the product $\mathbf{Q}^T\mathbf{Q} = \mathbf{VW}^T\mathbf{WV}^T$.
5. The new diagonal matrix $\mathbf{W}^T\mathbf{W}$, written as $\boldsymbol{\Lambda}$, has elements λ_i that are the eigenvalues of $\mathbf{Q}^T\mathbf{Q}$. The matrix **V** has columns that are the eigenvectors in which elements are the loadings for the columns of the original data matrix. The matrix **U** has columns that are the eigenvectors for which the elements are the loadings for the rows of the original data matrix.
6. Use matrices **U** and **V** to separately plot the locations of the rows and columns in ordination space. The locations also can be plotted together in an ordination **bi-plot** so that you can see relationships between, say, sites and species. However, redundancy analysis (RDA; see below) is a more direct method for analyzing joint relationships between site (environmental) and species (compositional) data.

Like PCA and factor analysis, CA results in principal axes and scores, although in this case we get scores for both the rows (e.g., sites) and columns (e.g., species). As with PCA and factor analysis, the first axis from a CA has the largest eigenvalue (explains the most variance in the data); it also maximizes the association between rows and columns. Subsequent axes account for the residual variation and have successively smaller eigenvalues. Rarely do we use more than two or three axes from CA, because each axis is thought to represent a particular environmental gradient, and a system with more than three unrelated gradients is difficult to interpret.

A BRIEF EXAMPLE We use CA to examine the joint relationship of ant genera composition at the four sites in Connecticut, Massachusetts, and Vermont (see Table 12.9). These sites differ slightly in latitude (≈3°) and whether they are on the mainland or on islands.

The first two axes of the CA account for 58% and 28% (total = 86%) of the variance in the data, respectively. The graph of the ordination of sites (Figure 12.10A) shows discrimination among the sites similar to that observed with the PCoA: CT and VT group together, and are separated from MA along the first principal axis. The second principal axis separates the MA mainland from the MA island sites. Separation of genera with respect to sites is also apparent (Figure 12.10B). Unique genera (such as *Ponera* in the VT sample, *Brachymyrmex* in the MA mainland sample, and *Crematogaster* and *Paratrechina* in the MA

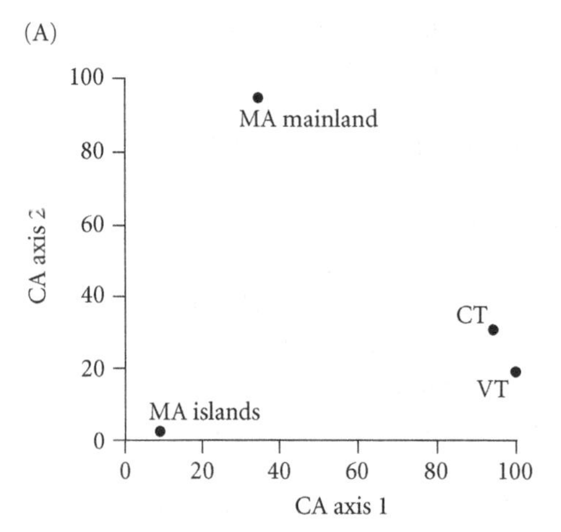

Figure 12.10 Results of a correspondence analysis (CA) on the site × ant genus matrix in Table 12.9. Plot A shows the ordination of sites, plot B the ordination of genera, and plot C is a bi-plot showing the relationship between the two ordinations. The results of the site ordination are similar to those of the principal coordinates analysis (see Figure 12.9), whereas the ordination of genera separates unique genera such as *Ponera* in the VT sample, *Brachymyrmex* in the MA mainland sample, and *Crematogaster* and *Paratrechina* in the MA island sample. The ordination results have been scaled so that both the site and species ordinations can be shown on the same plot. In (C), sites are indicated by black dots and ant genera by blue dots.

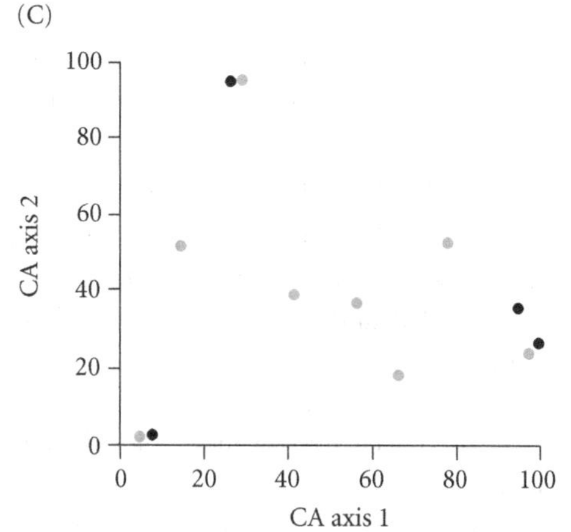

island sample) separate out in the same way as the sites. The remaining genera are arrayed more toward the center of the ordination space (Figure 12.10B). The overlay of the two ordinations captures the relationships of these genera and the sites (compare the site × genus matrix in Table 12.9 with Figure 12.10C). Using more information about the sites (Gotelli and Ellison 2002a), we could attempt to infer additional information about the relationships between individual genera and site characteristics.

You should be cautious about drawing such inferences, however. The difficulty is that correspondence analysis carries out a simultaneous ordination of the rows and columns of the data matrix, and asks how well the two ordinations are associated with each other. Thus, we expect a fair degree of overlap between the results of the two ordinations (as shown in Figure 12.10C). Because data used for CA essentially are the same as those used for contingency table analysis, hypothesis testing and predictive modeling is better done using the methods described in Chapter 11.

THE HORSESHOE EFFECT AND DETRENDING Correspondence analysis has been used extensively to explore relationships among species along environmental gradients. However, like other ordination methods, it has the unfortunate mathematical property of compressing the ends of an environmental gradient and accentuating the middle. This can result in ordination plots that are curved into an arch or a horseshoe shape (Legendre and Legendre 1998; Podani and Miklós 2002), even when the samples are evenly spaced along the environmental gradient. An example of this is seen in Figure 12.10A. The first axis would array the four sites linearly, and the second axis pulls the MA mainland site up. These four points form a relatively smooth arch. In many instances, the second CA axis simply may be a quadratic distortion of the first axis (Hill 1973b; Gauch 1982).

A reliable and interpretable ordination technique should preserve the distance relationships between points; their original values and the values created by the ordination should be equally distant (to the extent possible given the number of axes that are selected) when measured with the same distance measure. For many common distance measures (including the chi-square measure used in CA), the **horseshoe effect** distorts the distance relationships among the new variables created by the ordination. **Detrended correspondence analysis** (DCA; Gauch 1982) is used to remove the horseshoe effect of correspondence analysis and presumably illustrate more accurately the relationship of interest. However, Podani and Miklós (2002) have shown that the horseshoe is a mathematical consequence of applying most distance measures to species that have unimodal responses to underlying environmental gradients. Although DCA is implemented widely in ordination software, it is no longer recommended (Jackson and Somers 1991). The use of alternative distance measures (Podani and Miklós 2002) is a better solution to the horseshoe problem.

Non-Metric Multidimensional Scaling

The four previous ordination methods are similar in that the distances between observations in multivariate space are preserved to the extent possible after the multivariate data have been reduced to a smaller number of composite variables. In contrast, the goal of **non-metric multidimensional scaling** (NMDS) is to end up with a plot in which different objects are placed far apart in the ordination space, while similar objects are placed close together. Only the rank ordering of the original distances or dissimilarities is preserved.

BASIC COMPUTATIONS OF A NON-METRIC MULTIDIMENSIONAL SCALING Carrying out a non-metric multidimensional scaling requires nine steps, some of which are repeated.

1. Generate a distance or dissimilarity matrix **D** from the data. Any distance or dissimilarity measure can be used (see Table 12.6). The elements of d_{ij}, **D** are distances or dissimilarities between observations.
2. Choose the number of dimensions (axes) n to be used to draw the ordination. Use two or three axes because most graphs are two- (x- and y-axes) or three- (x-, y-, and z-axes) dimensional.
3. Start the ordination by placing the m observations in the n-dimensional space. Subsequent analyses depend strongly on this initialization, because NMDS finds its solution by local minimization (similar to non-linear regression; see Chapter 9). If some geographic information is available (such as latitude and longitude), that may be used as a good starting point. Alternatively, the output from another ordination (such as PCoA) can be used to determine the initial positions of observations in an NMDS.
4. Compute new distances δ_{ij} between the observations in the initial configuration. Normally, Euclidean distances are used to calculate δ_{ij}.
5. Regress δ_{ij} on d_{ij}. The result of this regression is a set of predicted values $\hat{\delta}_{ij}$. For example, if we use a linear regression, $\delta_{ij} = \beta_0 + \beta_1 d_{ij} + \varepsilon_{ij}$, then the predicted values are

$$\hat{\delta}_{ij} = \hat{\beta}_0 + \hat{\beta}_1 d_{ij}$$

6. Compute a goodness of fit between δ_{ij} and $\hat{\delta}_{ij}$. Most NMDS programs compute this goodness of fit as a **stress**:

$$\text{Stress} = \sqrt{\frac{\sum_{i=1}^{m}\sum_{j=1}^{n}(\delta_{ij} - \hat{\delta}_{ij})^2}{\sum_{i=1}^{m}\sum_{j=1}^{n}\hat{\delta}_{ij}^2}}$$

   Stress is computed on the lower triangle of the square **D** matrix. The numerator is the grand sum of the square of the difference between observed and expected values, and the denominator is the grand sum of the squared expected values. Note that the stress looks a lot like a chi-square test-statistic (see Chapter 11).

7. Change the position of the m observations in n-dimensional space slightly in order to reduce the stress.
8. Repeat steps 4–7 until the stress can no longer be reduced any further.
9. Plot the position of the m observations in n-dimensional space for which stress is minimal. This plot illustrates "relatedness" among observations.

A BRIEF EXAMPLE We use NMDS to analyze the ant data in Table 12.8. We used Sørensen's measure of dissimilarity for our distance measurement. The scree plot of stress scores indicates that two dimensions are sufficient for the analysis (Figure 12.11). As with the CA and PCoA, we have good discrimination among the sites (Figure 12.12A), and good separation among unique assemblages of genera (Figure 12.12B). Because scaling of axes in NMDS is arbitrary, we cannot overlay these two plots in an ordination bi-plot.

Advantages and Disadvantages of Ordination

Ecologists and environmental scientists use ordination to reduce complex multivariate data to a smaller, more manageable set of data; to sort sites on the basis of environmental variables of species assemblages; and to identify species responses to environmental gradients or perturbations. Because some ordination methods (such as PCA and factor analysis) are commonly available in most commercial statistical packages, we have easy, often menu-driven access to these tools. Used prudently, ordination can be a powerful tool for exploring data, illustrating patterns, and generating hypotheses that can be tested with subsequent sampling or experiments.

The disadvantages of ordination are not apparent because these techniques have been automated in many software packages, the mechanics are hidden, and the assumptions are rarely spelled out. Ordination is based on matrix algebra,

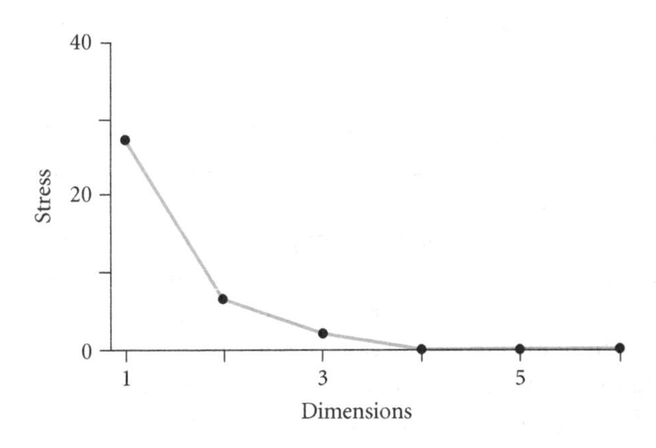

Figure 12.11 Scree plot for dimensions in a NMDS analysis of the site × ant genus data (see Table 12.9). A pairwise distance matrix for the four sites is calculated from these data (see Table 12.10) and used in the NMDS. In an NMDS, the stress in the data is a deviation measure similar to a chi-square goodness-of-fit statistic. This scree plot illustrates the successive reduction in stress with increasing dimension in the NMDS of the ant data. No further significant reduction in stress occurred after three dimensions, which account for most of the goodness-of-fit.

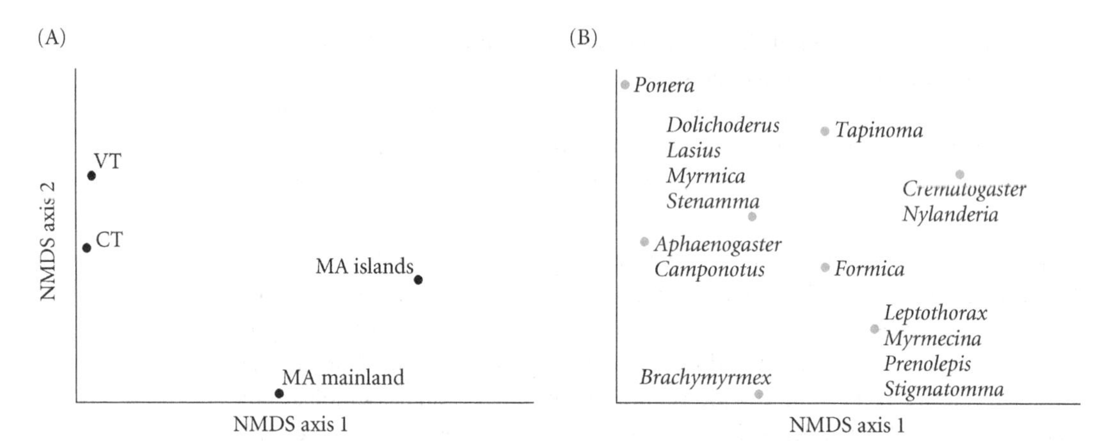

Figure 12.12 Plot of the first two dimensions of the NMDS analysis of the site × ant genus data in Table 12.9. A pairwise distance matrix for the four sites is calculated from the data in Table 12.10 and is used in the NMDS. (A) Ordination of sites. (B) Ordination of genera.

with which many ecologists and environmental scientists are unfamiliar. Because principal axes normally are rescaled prior to plotting principal scores, the scores are interpretable only relative to each other and it is difficult to relate them back to the original measurements.

RECOMMENDATIONS It is rarely obvious which ordination technique you should choose to best order observations, samples, or populations in multivariate space. Simple data reduction is best accomplished by principal component analysis (PCA) when Euclidean distances are appropriate and outliers or highly skewed data are not present. Principal coordinates analysis (PCoA) is appropriate when other distance measures are needed; PCA is equivalent to a PCoA when Euclidean distances are used. We recommend PCoA for most ordination applications in which the goal is to preserve the original multivariate distances between observations in the reduced (ordination) space. Non-metric multidimensional scaling (NMDS) only preserves the rank ordering of the distances, but it can be used with any distance measure. Correspondence analysis (CA) is a reasonable method of ordination for examining species distributions along environmental gradients, but its cousin, detrended correspondence analysis (DCA), should no longer be used. Likewise, the results of factor analysis are difficult to interpret and too dependent on subjective rotation methods. In all cases, it is important to remember the underlying ecological or environmental question of interest. For example, if the original question was to see how species or characteristics are distributed along an environmental gradient, then simply plotting the response variable against the appropriate measurement of the environment (or the first or second principal axes) may be more informative than any ordination plot.

Whichever method is employed, ordination should be used cautiously for testing hypotheses; it is best used for data exploration and pattern generation. However, classical hypothesis testing can be carried out on scores resulting from any ordination, as long as the assumptions of the tests are met.

Classification

Classification is the process by which we place objects into groups. Whereas the goal of ordination is to separate observations or samples along environmental gradients or biological axes, the goal of classification is to group similar objects into identifiable and interpretable classes that can be distinguished from neighboring classes. Taxonomy is a familiar application of classification—a taxonomist begins with a collection of specimens and must decide how to assign them to species, genera, families, and higher taxonomic groups. Regardless of whether taxonomy is based on morphology, gene sequences, or evolutionary history, the goal is the same: an inclusive classification system based on hierarchical clusters of groups.

Ecologists and environmental scientists may be suspicious of classifications because the classes are assumed to represent discrete entities, whereas we are more used to dealing with continuous variation in community structure or morphology that maps onto continuous environmental gradients. Ordination identifies patterns occurring within gradients, whereas classification identifies the endpoints or extremes of the gradients while ignoring the in-between.

Cluster Analysis

The most familiar type of classification analysis is **cluster analysis**. Cluster analysis takes m observations, each of which has associated with it n continuous numerical variables, and segregates the observations into groups. Table 12.11 illustrates the kind of data used in cluster analysis. Each of the nine rows of the table represents a different country in which samples were collected. Each of the three columns of the table represents a different morphological measurement taken on specimens of the marine snail *Littoraria angulifera*. The goal of this analysis is to form clusters of sites on the basis of similarity in shell shape. Cluster analysis also can be used to group sites on the basis of species abundances or presences-absences, or to group organisms on the basis of similarity in measured characteristics such as morphology or DNA sequences.

TABLE 12.11 **Snail shell measurements used for cluster analysis**

Country	Continent	Proportionality	Circularity	Spire Height
Angola	Africa	1.36	0.76	1.69
Bahamas	N. America	1.51	0.76	1.86
Belize	N. America	1.42	0.76	1.85
Brazil	S. America	1.43	0.74	1.71
Florida	N. America	1.45	0.74	1.86
Haiti	N. America	1.49	0.76	1.89
Liberia	Africa	1.36	0.75	1.69
Nicaragua	N. America	1.48	0.74	1.69
Sierra Leone	Africa	1.35	0.73	1.72

Shell shape in the snail *Littoraria angulifera* was measured for samples from 9 countries. Shell proportionality = shell height/shell width; shell circularity = aperture width/aperture height; and spire height = shell height/aperture length. Values in the table are averages based on 2 to 100 samples per site (data from Merkt and Ellison 1998). Cluster analysis groups the different countries based on similarity in shell morphology (see Figure 12.13).

Choosing a Clustering Method

There are several methods available for clustering data (Sneath and Sokal 1973), but we will focus on only two that are commonly used by ecologists and environmental scientists.

AGGLOMERATIVE VERSUS DIVISIVE CLUSTERING **Agglomerative clustering** proceeds by taking many separate observations and grouping them into successively larger clusters until one cluster is obtained. **Divisive clustering**, on the other hand, proceeds by placing all the observations in one group and then splitting them into successively smaller clusters until each observation is in its own cluster. For both methods, a decision must be made, often based on some statistical rule, as to how many clusters to use in describing the data.

We illustrate these two methods using the data in Table 12.11. Agglomerative methods begin with a square $m \times m$ distance matrix, in which the entries are the pairwise morphological distances measured for each pair of sites; Euclidean distances are the most straightforward to use (Table 12.12), but any distance measure (see Table 12.6) can be used in cluster analysis. An agglomerative cluster analysis starts with all the objects—sites in this analysis—separately and successively groups them. We see from the Euclidean distance matrix (see Table 12.12) that Brazil is nearest (in distance units) to Nicaragua, Angola is nearest to Liberia, Belize is nearest to Florida, and The Bahamas are nearest to Haiti.

TABLE 12.12 **Euclidean distance matrix for morphological measurements of snail shells**

	Angola	Bahamas	Belize	Brazil	Florida	Haiti	Liberia	Nicaragua	Sierra Leone
Angola	0								
Bahamas	0.23	0							
Belize	0.17	0.09	0						
Brazil	0.08	0.17	0.14	0					
Florida	0.19	0.06	0.04	0.15	0				
Haiti	0.24	0.04	0.08	0.19	0.05	0			
Liberia	0.01	0.23	0.17	0.07	0.19	0.24	0		
Nicaragua	0.12	0.17	0.17	0.05	0.17	0.20	0.12	0	
Sierra Leone	0.04	0.21	0.15	0.08	0.17	0.22	0.04	0.13	0

The original data are detailed in Table 12.11. Because distance matrices are symmetrical, only the lower half of the matrix is shown.

These clusters form the bottom row of the **dendrogram** (tree-diagram) shown in Figure 12.13. In this example, Sierra Leone is the odd site out (agglomerative clustering works upwards in pairs), and its addition to the Angola-Liberia cluster to form a new African cluster. The resulting four new clusters are clustered further in such a way that successively larger clusters contain smaller ones. Thus, the Bahamas–Haiti cluster is joined with the Belize–Florida cluster to form a new Caribbean–North Atlantic cluster; the other two clusters form a South Atlantic–Africa cluster.

A divisive cluster analysis begins with all the objects in one group, and then divides them up based on dissimilarity. In this case, the resulting clusters are the same ones that were produced with agglomerative clustering. More typically, divisive clustering results in fewer clusters, each with more objects, whereas agglomerative clustering results in more clusters, each with fewer objects. Divisive clustering algorithms also may proceed to a predetermined number of clusters that is specified by the investigator in advance of performing the analysis. A now rarely used method for community classification, TWINSPAN (for "two-way indicator-species analysis"; see Gauch 1982), uses divisive clustering algorithms.

HIERARCHICAL VERSUS NON-HIERARCHICAL METHODS Intuitively, clusters with few observations should be embedded within higher-order clusters with more observations. In other words, if observations *a* and *b* are in one cluster, and observations *c* and *d* are in another cluster, a larger cluster that includes *a* and *c* should also include *b* and *d*. This classification arrangement should be familiar from taxonomy: if two species are in one genus, and two other species are in anoth-

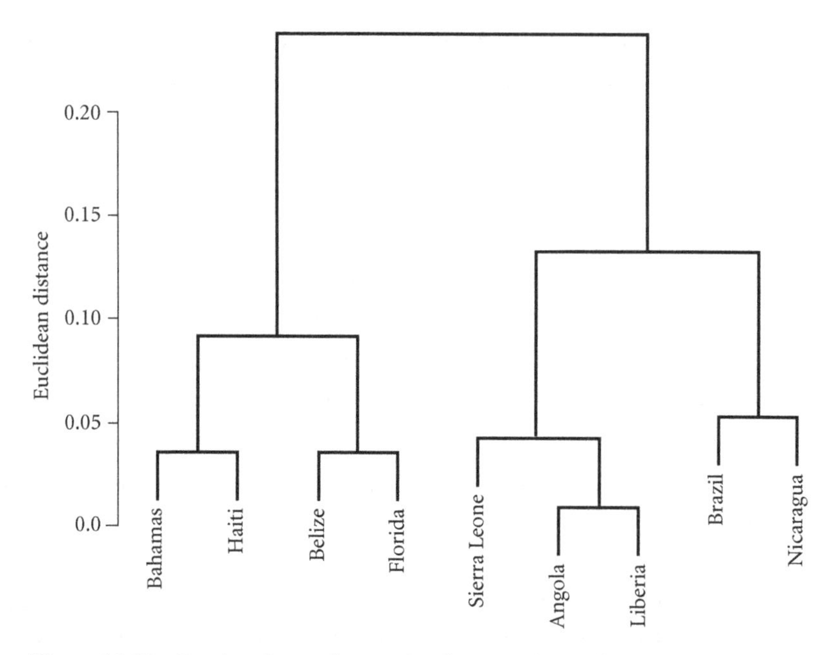

Figure 12.13 Results of an agglomerative cluster analysis. The original data (Merkt and Ellison 1998) consist of morphological ratios of snail (*Littoraria angulifera*) shell measurements from several countries (see Table 12.11). A dendrogram (branching figure) groups the sites into clusters based on similarity in shell morphology. The y-axis indicates the Euclidean distance for which sites (or lower-level clusters) are joined into a higher-level cluster.

er genus, and both genera are in the same family, then all four species are also in that same family. Clustering methods that follow this rule are called **hierarchical clustering methods.** Both agglomerative and divisive clustering methods can be hierarchical.

In contrast, **non-hierarchical clustering** methods group observations independently of an externally imposed reference system. Ordination can be considered analogous to non-hierarchical clustering—the results of an ordination often are independent of externally imposed orderings of the system. For example, two species in the same genus may wind up in different clusters produced by ordination of a set of morphological or habitat characteristics. Ecologists and environmental scientists generally use non-hierarchical methods to search for patterns in the data and to generate hypotheses that should then be tested with additional observations or experiments.

***K*-means clustering** is a non-hierarchical method that requires you to first specify how many clusters you want. The algorithm creates clusters in such a way that all the objects within one cluster are closer to each other (based on a distance measure) than they are to objects within the other clusters. *K*-means clustering minimizes the sums of squares of distances from each object to the centroid of its cluster.

We first applied *K*-means clustering to the snail data by specifying two clusters. One cluster, consisting of samples from Angola, Brazil, Liberia, Nicaragua, and Sierra Leone, had relatively small shells (centroid: proportionality = 1.39, circularity = 0.74, spire height = 1.70), and the other cluster (Bahamas, Belize, Florida, and Haiti) had relatively large shells (centroid = 1.47, 0.76, 1.87). The within-cluster sum of squares for the first (African–South Atlantic) cluster equals 0.14, and the within-cluster sum of squares for the second (Caribbean–North Atlantic) cluster equals 0.06. A *K*-means clustering of these data specifying three clusters split the African–South Atlantic cluster into two clusters, one for the African sites and one for Nicaragua and Brazil.

How do you choose how many clusters to use? There is no hard-and-fast rule, but the more clusters used, the fewer members in each cluster, and the more similar the clusters will be to one another. Hartigan (1975) suggests you first perform a *K*-means clustering with k and $k + 1$ clusters on a sample of m observations. If

$$\left[\frac{\sum SS_{(\text{within } k \text{ clusters})}}{\sum SS_{(\text{within } k+1 \text{ clusters})}} \times (m-k-1)\right] > 10 \qquad (12.30)$$

then it makes sense to add the $(k + 1)$th cluster. If the inequality in Equation 12.30 is less than 10, it is better to stop with k clusters. Equation 12.30 will generate a large number whenever adding a cluster substantially increases the within-group sum of squares. However, in simulation studies, Equation 12.30 tends to overestimate the true number of clusters (Sugar and James 2003). For the snail data in Table 12.11, Equation 12.30 = 8.1, suggesting that two clusters is better than three. Sugar and James (2003) review a wide range of methods for identifying the number of clusters in a dataset, an active area of research in multivariate analysis.

Discriminant Analysis

Discriminant analysis is used to assign samples to groups or classes that are defined in advance; it behaves like a cluster analysis in reverse. Continuing with our snail example, we can use the geographic locations to define groups in advance. The snail samples come from three continents: Africa (Angola, Liberia, and Sierra Leone), South America (Brazil), and North America (all the rest). If we were given only the shell shape data (three variables), could we assign the observations to the correct continent? Discriminant analysis is used to generate linear combinations of the variables (as in PCA; see Equation 12.21) that are then used to separate the groups as best as possible. As with MANOVA, discriminant analysis requires that the data conform to a multivariate normal distribution.

We can generate a linear combination of the original variables using Equation 12.21

$$Z_i = a_{i1}Y_1 + a_{i2}Y_2 + \ldots + a_{in}Y_n$$

If the multivariate means $\overline{\mathbf{Y}}$ vary among groups, we want to find the coefficients **a** in Equation 12.20 that maximize the differences among groups. Specifically, the coefficients a_{in} are those that would *maximize* the F-ratio in an ANOVA—in other words, we want the among-group sums of squares to be as *large* as possible for a given set of Z_i's.

Once again, we turn to our sums of squares and cross-products (SSCP) matrices **H** (among-groups) and **E** (within-groups, or error; see the discussion of MANOVA earlier in the chapter). The eigenvalues and eigenvectors are then determined for the matrix $\mathbf{E}^{-1}\mathbf{H}$. If we order the eigenvalues from largest to smallest (as in PCA, for example), their corresponding eigenvectors $\mathbf{a}_i$ are the coefficients for Equation 12.21.

The results of this discriminant analysis are given in Table 12.13. This small subset of the full dataset departs from multivariate normality ($E_n = 31.32$, $P = 2 \times 10^{-5}$, with 6 degrees of freedom), but the full dataset (of 1042 observations) passed this test of multivariate normality ($P = 0.13$). Although the multivariate means (Table 12.13A) do not differ significantly by continent (Wilk's lambda = 0.108, F = 2.7115, $P = 0.09$ with 6 and 8 degrees of freedom), we can still illustrate the discriminant analysis.[12] The first two eigenvectors (Table 12.13B) accounted for 97% of the variance in the data, and we used these eigenvectors in Equation 12.21 to calculate principal component scores for each of the original samples. These scores for the nine observations in Table 12.11 are plotted in Figure 12.14, with different colors corresponding to shells from different con-

[12] The full dataset analyzed by Merkt and Ellison (1998) had 1042 samples from 19 countries. Discriminant analysis was used to predict from which oceanographic current system (Caribbean, Gulf of Mexico, Gulf Stream, North or South Equatorial) a shell came from. The discriminant analysis successfully assigned more than 80% of the samples to the correct current system.

TABLE 12.13 **Discriminant analysis by continent of snail shell morphology data from 9 sites**

A. Variable means

	Africa	N. America	S. America
Proportionality	1.357	1.470	1.430
Circularity	0.747	0.752	0.740
Spire height	1.700	1.830	1.710

The first step is to calculate the variable means for the samples grouped by each continent.

B. Eigenvectors

	a_1	a_2
Proportionality	0.921	0.382
Circularity	–0.257	–0.529
Spire height	0.585	–0.620

Discriminant analysis generates linear combinations of these variables (similar to PCA) that maximally separate the samples by continent. The eigenvectors contain the coefficients (loadings) that are used to create a discriminant score for each sample; only the first two, which explain 97% of the variance, are shown.

C. Classification matrix

		Predicted (classified)			
		Africa	**N. America**	**S. America**	**% Correct**
Observed	**Africa**	3	0	0	100
	N. America	0	4	1	80
	S. America	0	0	1	100

The samples are assigned to one of the three continents according to their discriminant scores. In this classification matrix, all but one of the samples was correctly assigned to its original continent. This result is not too surprising because the same data were used to create the discriminant score and to classify the samples.

D. Jackknifed classification matrix

		Predicted (classified)			
		Africa	**N. America**	**S. America**	**% Correct**
Observed	**Africa**	2	0	1	67
	N. America	0	3	2	60
	S. America	0	1	0	0

A more unbiased approach is to use a jackknifed classification matrix in which the discriminant score is created from $m - 1$ samples and then used to classify the excluded sample. In this way, the discriminant function is independent of the sample being classified. The jackknifed classification did not perform nearly as well as the non-jackknifed matrix, with only 5 of the 9 samples correctly classified by continent.

The full data from which these ratios are drawn is detailed in Table 12.11. Discriminant analyses and other multivariate techniques should be based on much larger samples than we have used in this example. (Data from Merkt and Ellison 1998.)

tinents. We see an obvious cluster of the African samples, and another obvious cluster of the North American samples. However, one of the North American samples is closer to the South American sample than it is to the other North American samples.

The predicted classification is based on minimizing distances (usually the Mahalanobis or Euclidean distance; see Table 12.6) among the observations within groups. Applying this method to the data results in accurate predictions for the South American and African shells, but only 80% (4 of 5) of the North American shells are correctly classified. Table 12.13C summarizes these classifications as a **classification matrix**. The rows of the square classification matrix represent the observed groups and the columns represent the predicted (classified) groups. The entries represent the number of observations that were classified into a particular group. If the classification algorithm has no errors, then all of the observations will fall on the diagonal. Off-diagonal values represent errors in classification in which an observation was assigned incorrectly.

It should come as no surprise that the predicted classification in a discriminant analysis closely matches the observed data; after all, the observed data were used to generate the eigenvectors that we then used to classify the data! Thus, we expect the results to be biased in favor of correctly assigning our observations to the groups from which they came. One solution to overcoming this bias is to jackknife the classification matrix in Table 12.13C. Jack-

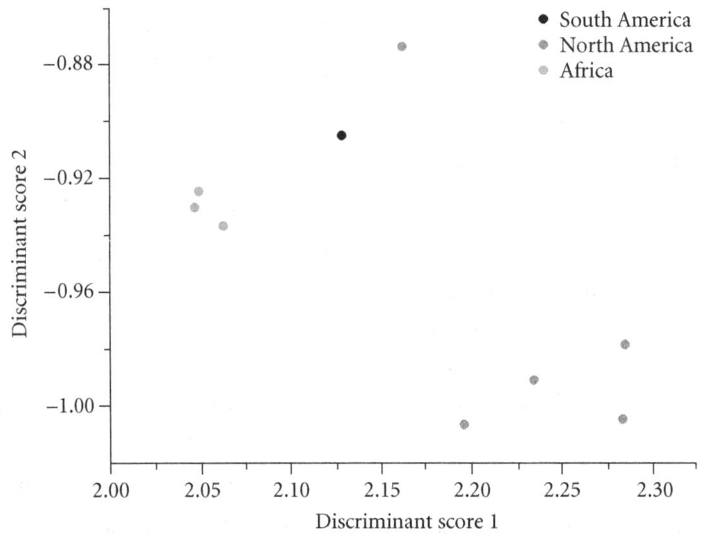

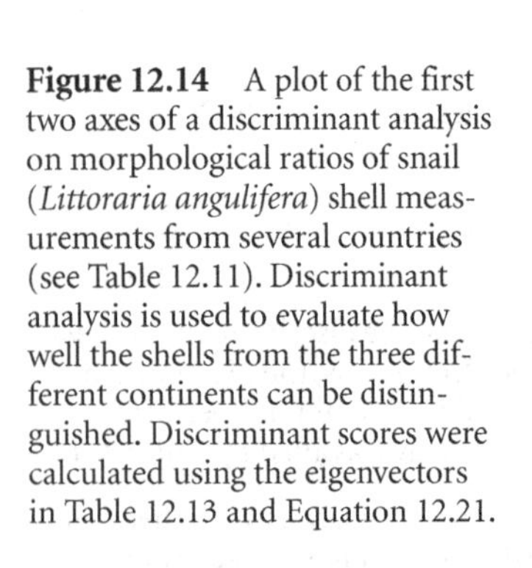
Figure 12.14 A plot of the first two axes of a discriminant analysis on morphological ratios of snail (*Littoraria angulifera*) shell measurements from several countries (see Table 12.11). Discriminant analysis is used to evaluate how well the shells from the three different continents can be distinguished. Discriminant scores were calculated using the eigenvectors in Table 12.13 and Equation 12.21.

knifing proceeds by dropping one individual observation from the dataset, re-analyzing the data without that observation, and then using the results to classify the dropped observation (see also "The Influence Function" in Chapter 9, and Footnote 2 in Chapter 5). The jackknifed classification matrix is given in Table 12.13D. This classification is much poorer, because the multivariate means for each continent were not very different. On the other hand, we normally would not conduct a discriminant analysis on such a small dataset. Another solution, if the original dataset is large, is to partition it randomly into two groups. Use one group to construct the classification and use the other to test it. New data can also be collected and classified with the discriminant function, although there needs to be some way to independently verify that the group assignments are correct.

Advantages and Disadvantages of Classification

Classification techniques are relatively easy to use and interpret, and are implemented widely in statistical software. They allow us either to group samples, or to assign samples to groups identified a priori. Dendrograms and classification matrices clearly summarize and communicate the results of cluster analysis and discriminant analysis.

Both cluster analysis and discriminant analysis must be used cautiously, however. There are many ways to carry out a cluster analysis, and different methods usually yield different results. It is most important to decide on a method beforehand instead of trying all the different methods and picking the one that looks the nicest or conforms to preconceived notions. Discriminant analysis is much more straightforward to perform, but it requires both ample data and assignment of groups a priori. Both cluster analysis and discriminant analysis are descriptive and exploratory methods. The MANOVA statistics described earlier in this chapter can be used on the results of a discriminant analysis to test hypotheses about differences among groups. Conversely, discriminent analysis can also be used as an a posteriori test to compare groups once the null hypothesis in a MANOVA has been rejected.

Indeed, cluster analysis and discriminant analysis begin with the assumption that distinct groups or clusters exist, whereas a proper null hypothesis would be that the samples are drawn from a single group and exhibit only random differences among one another. Strauss (1982) describes Monte Carlo methods for testing the statistical significance of clusters. Bootstrapping and other computer-intensive methods for testing the statistical significance of clusters are also widely used in phylogenetic analysis (Felsenstein 1985; Emerson et al. 2001;

Huelsenbeck et al. 2002; Sanderson and Shaffer 2002; Miller 2003; Sanderson and Driskell 2003).

Multivariate Multiple Regression

The methods for multivariate analysis described thus far have been primarily descriptive (ordination, classification) or limited to categorical predictor variables (Hotelling's T^2 test and MANOVA). The final topic for this chapter is the extension of regression (one continuous response variable and one or more continuous predictor variables) to multivariate response data. Two methods are used commonly by ecologists: **redundancy analysis (RDA)** and **canonical correspondence analysis (CCA)**. We only describe the mechanics of RDA, which is the direct extension of multiple regression to the multivariate case. RDA assumes a causal relationship between the independent and dependent variables, whereas CCA focuses on generating a unimodal axis with respect to the response variables (e.g., species occurrences or abundances) and a linear axis with respect to the predictor variables (e.g., habitat or environmental characteristics). A third technique, canonical correlation analysis (Hotelling 1936; Manly 1991) is based on symmetrical relationships between the two variables. In other words, canonical correlation analysis assumes error in both the predictor and response (see discussion of the assumptions of regression in Chapter 9).

Redundancy Analysis

One of the most common uses of RDA is to examine relationships between species composition, measured as the abundances of each of n species, and environmental characteristics, measured as a set of m environmental variables (Legendre and Legendre 1998). The species composition data represent the multivariate response variable, and the environmental variables represent the multivariate predictor variable. But RDA is not restricted to that kind of analysis—it can be used for any multiple regression involving multiple response and multiple predictor variables, as we demonstrate in our example.

THE BASICS OF A REDUNDANCY ANALYSIS As developed in Chapter 9, the standard multiple linear regression is

$$Y_j = \beta_0 + \beta_1 X_1 + \beta_2 X_2 + \ldots + \beta_m X_m + \varepsilon_j \quad (12.31)$$

The β_i values are the parameters to be estimated, and ε_j represents the random error. Standard least squares methods are used to fit the model and provide unbiased estimates $\hat{\beta}$ and $\hat{\varepsilon}$ of the β_i's and ε_j's. In the multivariate case, the single response variable Y is replaced by a matrix of n variables **Y**. Redundancy analysis regresses **Y** on a matrix of independent variables **X**, all measured for each observation. The steps of an RDA are:

1. Regress each of the individual response variables Y_j in **Y** on all the variables in **X** (using Equation 12.31). This results in a matrix of fitted values $\hat{\mathbf{Y}}=\left[\hat{Y}_j\right]$. This matrix is calculated as $\hat{\mathbf{Y}} = \mathbf{XB}$, where **B** is the matrix of regression coefficients:

$$\mathbf{B} = (\mathbf{X}^T\mathbf{X})^{-1}\mathbf{X}^T\mathbf{Y}$$

2. Use the $\hat{\mathbf{Y}}$ matrix in a principal component analysis of its standardized sample variance-covariance matrix, yielding a matrix **A** of eigenvectors. The eigenvectors have the same interpretation as they do in a PCA. If you are performing an RDA on a matrix of environmental characteristics × species, the elements of **A** are called the **species scores.**
3. Generate two sets of scores from this PCA.
   a. The first set of scores, **F**, are computed by multiplying the matrix of eigenvectors **A** by the matrix of response variables **Y**. The result, **F** = **YA**, is a matrix in which the columns are called **site scores.** The site scores **F** are an ordination relative to **Y**. If you are using an environmental characteristics × species matrix, the site scores are based on the original, *observed* species distributions.
   b. The second set of scores, **Z**, are computed by multiplying the matrix of eigenvectors **A** by the matrix of fitted values $\hat{\mathbf{Y}}$. The result, $\mathbf{Z} = \hat{\mathbf{Y}}\mathbf{A}$, is a matrix in which the columns are called **fitted site scores.** Because $\hat{\mathbf{Y}}$ is also equal to the matrix **XB**, the matrix of fitted site scores can also be written as **Z** = **XBA**. The fitted site scores **Z** are an ordination relative to **X**. If you are using an environmental characteristics × species matrix, the fitted site scores are based on the distribution of species that are *predicted* by the environment.
4. Next we want to know how the predictor variables in **X** contribute to each of the ordination axes. The most straightforward way to measure the contribution of the predictor variables is to observe that the **Z** matrix is composed of two parts: **X**, the predictor values, and **BA**, the matrix product of the regression coefficients and the eigenvectors of the $\hat{\mathbf{Y}}$ matrix. The matrix **C** = **BA** tells us the contribution of the **X** variables to the matrix of fitted site scores **Z**. This decomposition is equivalent to saying that the values in each of the columns of **C** are equal to standardized regression coefficients on the matrix **X**. An alternative method is to determine the correlation between the environmental (predictor) vari-

ables **X** and the site scores **F**, which is the product of the response variables and their eigenvectors.

5. Finally, we construct a bi-plot, as we did for corresponding analysis, in which we plot the principal axes of either **F** (the observed) or **Z** (the fitted) scores for the predictor variables. On this plot, we can place the locations of the response variables. In this way, we can visualize the multivariate relationship of the response variables to the predictor variables.

A BRIEF EXAMPLE To illustrate RDA, we return to the snail data (see Table 12.11). For each of these sites we also have measures of four environmental variables: annual rainfall (mm), number of dry months per year, mean monthly temperature, and mean height of the mangrove canopy in which the snails foraged (Table 12.14). The results of the four steps of the RDA are presented in Table 12.15. Because the response variables are measures of shell-shape, the **A** matrix is composed of shell-shape scores. The **F** and **Z** matrices are site scores and fitted site scores, respectively.

Figure 12.15 illustrates how sites co-vary with environmental variables (blue points and directional arrows) and how shell shapes co-vary with the sites and their environmental variables (black points and directional arrows). For example, the Brazilian, Nicaraguan, and African sites are characterized by greater canopy height and greater rainfall, whereas the Florida and Caribbean sites are characterized by higher temperatures and longer dry months. Shells from the Caribbean and Florida have larger proportionality and higher spire heights, whereas shells from Africa, Brazil, and Nicaragua are more circular.

TABLE 12.14 **Environmental variables used in the redundancy analysis (RDA)**

Country	Annual rainfall (mm)	No. dry months	Mean monthly temp. (°C)	Mean height of forest canopy (m)
Angola	363	9	26.4	30
Bahamas	1181	2	25.1	3
Belize	1500	2	29.5	8
Brazil	2150	4	26.4	30
Florida	1004	1	25.3	10
Haiti	1242	6	27.5	10
Liberia	3874	3	27.0	30
Nicaragua	3293	0	26.0	15
Sierra Leone	4349	4	26.6	35

Data from Merkt and Ellison (1998).

TABLE 12.15 **Redundancy analysis of small shell morphology, and environmental data**

A. Compute the matrix of fitted values $\hat{\mathbf{Y}} = \mathbf{X}(\mathbf{X}^T\mathbf{X})^{-1}\mathbf{X}^T\mathbf{Y}$.

$$\hat{\mathbf{Y}} = \begin{bmatrix} -0.05 & 0.01 & -0.04 \\ 0.09 & 0.00 & 0.09 \\ 0.00 & 0.01 & 0.08 \\ -0.05 & -0.01 & -0.08 \\ 0.05 & 0.00 & 0.03 \\ 0.04 & 0.02 & 0.09 \\ -0.05 & -0.01 & -0.07 \\ 0.03 & -0.01 & 0.00 \\ -0.06 & -0.01 & -0.10 \end{bmatrix}$$

This matrix is the result of regressing each of the three shape variables on the four environmental variables. Each row of $\hat{\mathbf{Y}}$ is one observation (from one site; see Table 12.11), and each column is one of the variables (proportionality, circularity, and spire height, respectively).

B. Use $\hat{\mathbf{Y}}$ in a PCA to compute the eigenvectors A.

$$\mathbf{A} = \begin{bmatrix} 0.55 & -0.81 & -0.19 \\ 0.08 & 0.27 & -0.96 \\ 0.83 & 0.52 & 0.21 \end{bmatrix}$$

These are the shell-shape scores, and could be used to calculate principal component values (Z_i) for shell shape. Columns represent the first three eigenvectors, and rows represent the three morphological variables. Component 1 accounts for 94% of the variance in shell shape. Shell circularity, which is roughly constant across sites (see Table 12.11), has little weight ($a_{21} = 0.08$) on Principal Component 1.

C. Compute the matrix of site scores F = YA and the matrix of fitted site scores $\mathbf{Z} = \hat{\mathbf{Y}}\mathbf{A}$.

$$\mathbf{F} = \begin{bmatrix} 2.21 & -0.03 & -0.63 \\ 2.44 & -0.06 & -0.62 \\ 2.38 & 0.00 & -0.61 \\ 2.26 & -0.08 & -0.62 \\ 2.40 & -0.02 & -0.59 \\ 2.45 & -0.03 & -0.61 \\ 2.21 & -0.03 & -0.62 \\ 2.28 & -0.13 & -0.63 \\ 2.23 & -0.01 & -0.59 \end{bmatrix} \qquad \mathbf{Z} = \begin{bmatrix} -0.06 & 0.02 & -0.01 \\ 0.13 & -0.03 & 0.00 \\ 0.07 & 0.04 & 0.01 \\ -0.09 & 0.00 & 0.00 \\ 0.05 & -0.02 & 0.00 \\ 0.10 & 0.02 & 0.00 \\ -0.09 & 0.00 & 0.00 \\ 0.01 & -0.03 & 0.00 \\ -0.12 & 0.00 & 0.00 \end{bmatrix}$$

Matrix **F** contains the principal component scores for shell-shape values obtained by multiplying the original values (**Y**) by the eigenvectors in matrix **A**. Matrix **Z** contains the principal component scores for shell-shape values obtained by multiplying the fitted values ($\hat{\mathbf{Y}}$), which are derived from the habitat matrix **X** (see Table 12.14), by the eigenvectors in **A**.

(continued)

TABLE 12.15 *(continued)*

D. Compute the matrix C = BA to yield the regression coefficients on X.

$$\mathbf{C} = \begin{bmatrix} 0.01 & 0.00 & 0.00 \\ 0.03 & 0.01 & 0.00 \\ 0.00 & 0.02 & 0.00 \\ -0.11 & 0.00 & 0.00 \end{bmatrix}$$

Rows are the four environmental variables, and columns are the first three components. Coefficients of 0.00 are rounded values of small numbers. Alternatively, we can compute the correlations of the environmental variables **X** with the site scores **F**. The cell entries are the correlation coefficients between **X** and the columns of **F**:

	Component 1	Component 2	Component 3
Rainfall	–0.55	–0.19	0.11
Dry months	–0.28	0.34	–0.18
Mean temperature	0.02	0.49	0.11
Canopy height	–0.91	0.04	–0.04

In this example, we have regressed snail shell-shape characteristics (see Table 12.11) on environmental data (see Table 12.13). Prior to analysis, the environmental data were standardized using Equation 12.17. (Data from Merkt and Ellison 1998.)

TESTING SIGNIFICANCE OF THE RESULTS The results of RDA (and CCA) can be tested for statistical significance using Monte Carlo simulation methods (see Chapter 5). The null hypothesis is that there is no relationship between the predictor variables and the response variables. Randomizations of the raw data—interchanging the rows of the **Y** matrix—are followed by repetition of the RDA. We then compare an F-ratio computed for the original data with the distribution of F-ratios for the randomized data. The F-ratio for this test is:

$$F = \frac{\left(\dfrac{\sum_{j=1}^{n} \lambda_j}{m}\right)}{\left(\dfrac{RSS}{(p-m-1)}\right)} \tag{12.32}$$

In this equation, $\sum \lambda_j$ is the sum of the eigenvalues; RSS is the residual sums of squares (computed as the sum of the eigenvalues of a PCA using $(\mathbf{Y} - \hat{\mathbf{Y}})$ in place of $\hat{\mathbf{Y}}$ (Table 12.15B); m is the number of predictor variables; and p is the number of observations. Although the sample size is too small for a meaning-

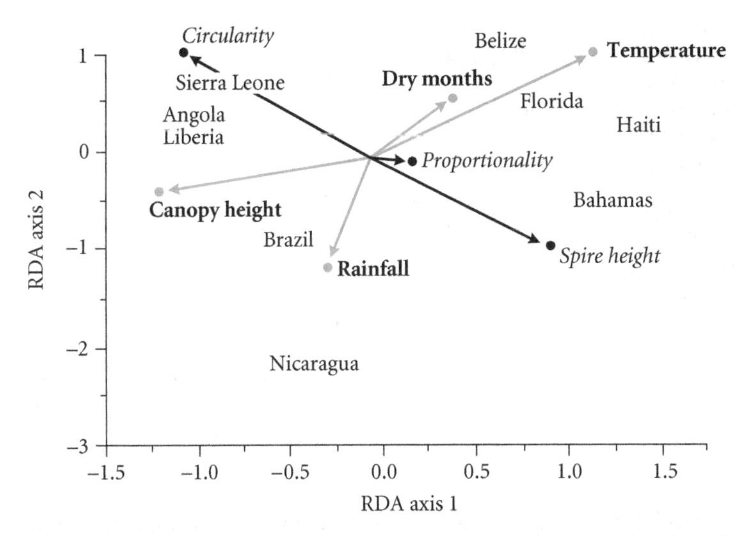

Figure 12.15 Bi-plot of the first two axes of a Redundancy Analysis (RDA) regressing snail shell data (see Table 12.11) on environmental data (see Table 12.14), all of which were measured for nine countries (Merkt and Ellison 1998). The country labels indicate the placement of each site in the ordination space (the **F** matrix of Table 12.15). The black symbols indicate the placement of the morphological variables in site space. Thus, shells from Belize, Florida, Haiti, and the Bahamas have larger proportionality and higher spire heights than shells in the other sites, whereas African, Brazilian, and Nicaraguan shells are more circular. Black arrows indicate the direction of increase of the morphological variable. Similarly, the blue points and arrows indicate how the sites were ordinated. All values were standardized to enable plotting of all matrices on similar scales. Thus, the lengths of the arrows do not indicate magnitude of effects. However, they do indicate their directionality or loading on the two axes.

ful analysis, we can nevertheless use Equation 12.32 to see if our results are statistically significant. The F-ratio for our original data = 3.27. A histogram of F-ratios from 1000 randomizations of our data is shown in Figure 12.16. Comparison of the observed F-ratio with the randomizations gives an exact *P*-value of 0.095, which is marginally larger than the traditional $P = 0.05$ cutoff. The narrow conclusion (based on a limited sample size) is that there is no significant relationship between snail morphology and environmental variation among these nine sites.[13] See Legendre and Legendre (1998) for further details and examples of hypothesis testing in RDA and CCA.

[13] In contrast, the analysis of the full dataset of 1042 observations from 19 countries yielded a significant association between snail shell shape and environmental conditions (Merkt and Ellison 1998).

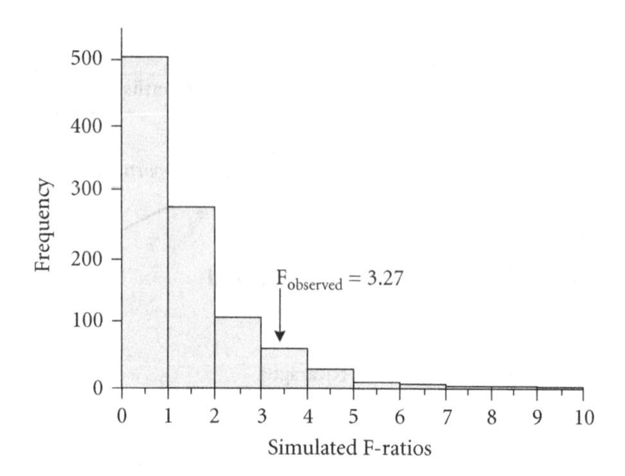

Figure 12.16 Frequency distribution of F-ratios resulting from 1000 randomizations and subsequent redundancy analysis (RDA) of the snail data (see Table 12.11), in which the country labels have been randomly re-shuffled among the samples. The F-ratio for the observed data = 3.27. Ninety-five of the randomizations had a larger F-ratio, so the *P*-value for the RDA in Figure 12.15 = 0.095. This result is marginally nonsignificant using an α-level equal to 0.05, reflecting the limitation of using RDA with such a small dataset.

An extension of RDA, distance-based RDA (or db-RDA) is an alternative method of testing complex multivariate models (Legendre and Anderson 1999; McArdle and Anderson 2001). Unlike a classical MANOVA, db-RDA does not require that data be multivariate normally distributed, that distance measurements between observations be Euclidean, or that there be more observations than there are measured response variables. Distance-based RDA can be used with any metric or semi-metric distance measure, and it allows for partitioning components of variation in complex MANOVA models. Like RDA, it uses randomization tests to determine significance of the results. Distance-based RDA is implemented in the vegan library in R.

Summary

Multivariate data contain multiple non-independent response variables measured for each replicate. Multivariate analysis of variance (MANOVA) and redundancy analysis (RDA) are the multivariate analogues of ANOVA and regression methods used in univariate analyses. Statistical tests based on MANOVA and RDA assume random, independent sampling, just as in univariate analyses, but only MANOVA requires the data conform to a multivariate normal distribution.

Ordination is a set of multivariate methods designed to order samples or observations of multivariate data, and to reduce multivariate data to a smaller number of variables for additional analysis. Euclidean distance is the most natural measure of the distance between samples in multivariate space, but this distance measure can yield counterintuitive results with datasets that have many zeros, such as species abundance or presence–absence data. Other distance measures, including Manhattan and Jaccard's distances, have better algebraic properties, although indices of biogeographic similarity based on these measures are very sensitive to sample size effects.

Principal components analysis (PCA) is one of the simplest methods for ordinating multivariate data, but it is not useful unless there are underlying correlations in the original data. PCA extracts new orthogonal variables that capture the important variation in the data and preserves the distances of the original samples in multivariate space. PCA is a special case of principal coordinates analysis (PCoA). Whereas PCA is based on Euclidean distances, PCoA can be used with any distance measure. Factor analysis is a type of reverse PCA in which measured variables are decomposed into linear combinations of underlying factors. Because the results of factor analysis are sensitive to rotation methods and because the same dataset can generate more than one set of factors, factor analysis is less useful than PCA.

Correspondence analysis (CA) is an ordination tool that reveals associations between species assemblages and site characteristics. It simultaneously ordinates species and sites and may reveal groupings along both axes. However, many problems have been identified in the related detrended correspondence analysis (DCA), which was developed to deal with the horsehoe effect common in CA (and other ordination techniques). An alternative to DCA is to conduct a CA using other distance measures. Non-metric multidimensional scaling (NMDS) preserves the rank ordering of the distances, but not the distances themselves, among variables in multivariate space. Ordination methods are useful for data reduction and for revealing patterns in data. Ordination scores can be used with standard methods for hypothesis testing, as long as the assumptions of the tests are met.

Classification methods are used to group objects into classes that can be identified and interpreted. Cluster analysis uses multivariate data to generate clusters of samples. Hierarchical clustering algorithms are either agglomerative (separate observations are sequentially grouped into clusters) or divisive (the entire dataset is sequentially divided into clusters). Non-hierarchical methods, such as K-means clustering, require the user to specify the number of clusters that will be created, although statistical stopping rules can be used to decide on the correct number of clusters. Discriminant analysis is a type of cluster analysis in reverse: the groups are defined a priori, and the analysis produces sample scores that provide the best classification. Jackknifing and testing of independent datasets can be used to evaluate the reliability of the discriminant function. Classification methods assume a priori that the groups or clusters are biologically relevant, whereas classic hypothesis testing would begin with a null hypothesis of random variation and no underlying groups or clusters.

PART IV
Estimation

CHAPTER 13

The Measurement of Biodiversity

In both applied and basic ecology, much effort has gone into quantifying patterns of biodiversity and understanding the mechanisms that maintain it. Ecologists and biogeographers have been interested in species diversity at least since the writings of Humboldt (1815) and Darwin (1859), and there is a new interest from the emerging "omics" disciplines in the biodiversity of genes and proteins (Gotelli et al. 2012).

Biodiversity itself encompasses a diversity of meanings (Magurran and McGill 2011), but in this chapter we focus on **alpha diversity**: the number of species (i.e., **species richness**) within a local assemblage and their relative abundances (**species evenness**).[1] As a definition, alpha diversity may seem restrictive, but it can be applied to any collection of "objects" (e.g., individuals, DNA sequences, amino acid sequences, etc.) that can be uniquely classified into a set of exclusive "categories" (e.g., species, genes, proteins, etc.).[2] By taking a small, random sample of such objects, we can apply methods from the probability calculus outlined in Chapters 1 and 2 to make inferences about the diversity of the entire assemblage.

---

[1] Total diversity at the regional scale is γ diversity (**gamma diversity**), and the turnover or change in diversity among sites within a region is β diversity (**beta diversity**). The quantification of β diversity, and the contribution of α and β diversity to γ diversity have been contentious issues in ecology for over a decade. See Jost (2007), Tuomisto (2010), Anderson et al. (2011), and Chao et al. (2012) for recent reviews.

[2] This simple definition can be modified to accommodate other traits of organisms. For example, all other things being equal, a community comprised of ten distantly related species, each in a different genus, would be more diverse than a community comprised of ten closely related species, all in a single genus. See Weiher (2011) and Velland et al. (2011) for recent reviews of methods of analysis for phylogenetic, functional, and trait diversity.

Although counting the number of species in a sample might seem a simple task,[3] estimating the number of species in an entire assemblage is tricky, and there are some common statistical pitfalls that should be avoided. In spite of decades of interest in the quantification of biodiversity, this is still a very active area of research, and many of the important methods described in this chapter have been developed only in the last 10 years. We first describe techniques for estimating and comparing species richness, and then consider measures of biodiversity that also incorporate species evenness. To illustrate the challenges associated with estimating biodiversity and the statistical solutions, let's begin with an empirical example.

Estimating Species Richness

A growing ecological problem is that more and more species are being transplanted (both deliberately and accidentally) beyond the limits of their historical geographic ranges and into novel habitats and assemblages. In eastern North America, the hem-

[3] Actually, even counting (or sampling) species is not a simple task at all. Conducting a quantitative biodiversity survey is very labor-intensive and time-consuming (Lawton et al. 1998). For any particular group of organisms (e.g., ground-foraging ants, desert rodents, marine diatoms) there are specific trapping and collecting techniques—each of which has its own particular biases and quirks (e.g., bait stations, Sherman live traps, plankton tows). And once a sample has been collected, geo-referenced, preserved, cleaned up, labeled, and catalogued, each specimen has to be identified to species based on its particular morphological or genetic characters.

For many tropical taxa, the majority of species encountered in a biodiversity survey may not have even been described before. Even for "well-studied" taxa in the temperate latitudes, many of the important taxonomic keys for making species identifications are buried in obscure out-of-print literature (birds are a conspicuous exception). And such keys can be difficult or impossible to use if you haven't already been taught the basics of species identification by someone who knows how. It can take many years of practice and study to become competent and confident in the identification of even one taxonomic group. You might think that anyone calling himself or herself an ecologist would have developed this skill, but many ecologists are painfully illiterate when it comes to "reading" the biodiversity of nature and being able to identify even common species in a garden or on a campus walk. Taxonomists and museum specialists are a disappearing breed, but the great museums and natural history collections of the world still contain a treasure-trove of data and information on biodiversity—*if* you know how to access it. See Gotelli (2004) for an ecologist's entrée into the world of taxonomy, and Ellison et al. (2012) for an example of how museum specimens can be compiled and analyzed to answer basic questions in ecology and biogeography.

lock woolly adelgid (*Adelges tsugae*)[4] is an introduced pest that is selectively killing eastern hemlock (*Tsuga canadensis*), a foundation species in eastern U.S. forests (Ellison et al. 2005). To study the long-term consequences of the adelgid invasion, researchers at the Harvard Forest have created large experimental plots to mimic the effects of adelgid invasion and the subsequent replacement of hemlock forests by hardwood forests (Ellison et al. 2010). In 2003, four experimental treatments were established in large (0.81 hectare) forest plots: (1) Hemlock Control (sites that had not yet been invaded by the hemlock wooly adelgid); (2) Logged (hemlock removal); (3) Girdled (the bark and cambium of each hemlock tree in the plot were cut, which slowly kills the standing hemlock tree, just as the adelgid does); and (4) Hardwood Control (the forest type expected to replace hemlock stands after the adelgid invasion). In 2008, invertebrates were censused in the four plot types by using an equal number of pitfall traps in each treatment (Sackett et al. 2011).

Table 13.1 summarizes the data for spiders collected in the four treatments. Although 58 species were collected in total, the number of species in each treatment ranged from 23 (Hemlock Control) to 37 (Logged). Does the number of spider species differ significantly among the four treatments? Although it would seem that more spider species were collected in the Logged treatment, we note that the Logged treatment also had the greatest number of individuals (252). The Hemlock Control had only 23 species, but perhaps that is not surprising, considering that only 106 individual spiders were collected. In fact, the rank order of species richness among the four treatments follows exactly the rank order of spider abundance (compare the last two rows of Table 13.1)!

If you are surprised by the fact that more individuals and species of spiders were collected in the Logged versus the Hemlock Control plots, think again. Logging generates a lot of coarse and fine woody debris, which is ideal microhabitat for many spiders. Moreover, the elimination of the cool shade provided by the hemlock trees boosts the air and soil temperatures in the plot (Lustenhouwer et al. 2012). This

[4] The hemlock woolly adelgid is native to East Asia. This small phloem-feeding insect was introduced to the United States in 1951 on nursery stock—Asian hemlock saplings destined for landscaping and garden stores—imported into Richmond, Virginia. It was next collected in the mid-1960s in the Philadelphia area, but did not attract any attention until the early 1980s, when its populations irrupted in Pennsylvania and Connecticut. The adelgid subsequently spread both north and south, leaving vast stands of dead and dying hemlocks in its wake (Fitzpatrick et al. 2012). In Northeast North America, the spread of the adelgid is limited by cold temperatures, but as regional climates warm, the adelgid continues to move northward. Although individual trees can be treated with insecticide to kill the adelgid, chemical control is impractical on a large scale. Biological control is being explored (Onken and Reardon 2011), but with limited success so far.

TABLE 13.1 Spider diversity data

Species	Hemlock Control	Girdled	Hardwood Control	Logged
Agelenopsis utahana	1	0	0	1
Agroeca ornata	2	15	15	10
Amaurobius borealis	27	46	59	22
Callobius bennetti	4	2	2	3
Castianeira longipalpa	0	0	0	3
Centromerus cornupalpis	0	0	1	0
Centromerus persolutus	1	0	0	0
Ceraticelus minutus	1	1	0	1
Ceratinella brunnea	3	6	0	4
Cicurina arcuata	0	1	5	1
.	.	.	.	.
.	.	.	.	.
.	.	.	.	.
Pardosa moesta	0	0	0	13
Zelotes fratris	0	0	0	7
Total number of individuals	106	168	250	252
Total number of species	23	26	28	37

Counts of the number of individuals of 58 spider species collected from four experimental treatment groups in the Harvard Forest Hemlock Removal Experiment. Data from Sackett et al. (2011); see the online resources for the complete data table.

matters because spiders (and all invertebrates) are ectotherms—that is, their body temperature is not internally regulated, as in so-called "warm-blooded" birds and mammals. Thus, the population sizes of ectotherms often will increase after logging or other disturbances because the extra warmth afforded by the elimination of shade gives individuals more hours in the day and more days in the year during which they can actively forage. Finally, the increase in temperature may increase the number of spiders collected even if their population sizes have not increased: spiders are more active in warmer temperatures, and their extra movement increases the chances they will be sampled by a pitfall trap. For all these reasons, we must make some kind of adjustment to our estimate of spider species richness to account for differences in the number of individuals collected.

You also should recognize that the sample plots are large, the number of invertebrates in them is vast, and the counts in this biodiversity sample represent only a tiny fraction of what is out there. From the perspective of a 5-mm long spider, a 0.81-ha plot is substantially larger than all five boroughs of New York City are to an individual human! In each of the plots, if we had collected more individual spiders, we surely would have accumulated more species, including species that have so far not been detected in any of the plots. This **sampling effect** is very strong, and it is pervasive in virtually all biodiversity surveys (whether it is recognized or not). Even when standard sampling procedures are used, as in this case, the number of individuals in different collections almost never will be the same, for both biological and statistical reasons (Gotelli and Colwell 2001).

One tempting solution to adjust for the sampling effect would be to simply divide the number of species observed by the number of individuals sampled. For the Logged treatment, this diversity estimator would be 37/252 = 0.15 species/individual. For the Hemlock Control treatment, this diversity estimator would be 23/106 = 0.22 species/individual. Although this procedure might seem reasonable, the sampling curve of species richness (and other diversity measures) is distinctly non-linear in shape. As a consequence, a simple algebraic rescaling (such as dividing species richness by sampling effort or area) will substantially over-estimate the actual diversity in large plots and should always be avoided.

Standardizing Diversity Comparisons through Random Subsampling

How can we validly compare species richness in the Logged plots versus the Hemlock Control plots, when there are 252 individual spiders sampled from the Logged plots, but only 106 sampled from the Hemlock Control plots? An intuitively appealing approach is to use a random subsample of the individuals from the larger sample, drawing out the same number of individuals as were found in the smaller sample. As an analogy (following Longino et al. 2002), imagine that each individual spider in the Logged plot is a single jelly bean. The jelly beans are of 37 colors, each one corresponding to a particular species of spider in the sample from the Logged treatment. Put all 252 jelly beans in a jar, mix them thoroughly, and then draw out exactly 106 jelly beans. Count the number of colors (i.e., species) in this random subsample, and then compare it to the observed number of species in the sample from the Hemlock Control treatment.

Such "candy-jar sampling"—known as **rarefaction**—allows for a direct comparison of the number of species in the Hemlock Control and Logged plots based on an equivalent number of individuals in both samples (106). This method of drawing random subsamples can be applied to any number of individuals less than or equal to the original sample size. To rarefy a sample in this manner is to make it less dense, and a rarefaction curve gives the expected number of species and its variance for random samples of differing sizes.[5]

Table 13.2 (see next page) shows the results of a computer simulation of this exercise. This particular subsample of 106 individuals from the Logged sample contains 24 species, which is pretty close to the 23 species observed in the Hemlock Control sample. Of course, if we were to replace all the jelly beans and repeat this exercise, we would get a slightly different answer, depending on the mix of individuals that are randomly drawn and the species they represent.[6] Figure 13.1 shows a histogram of the counts of species richness from 1000 such random draws. For this particular set of random draws, the simulated species richness ranges from 19 to 32, with a mean of 26.3. The observed richness of

---

[5] Rarefaction has an interesting history in ecology. In the 1920s and then again in the 1940s, several European biogeographers independently developed the method to compare taxonomic diversity ratios, such as the number of species per genus (S/G; Järvinen 1982). In his influential 1964 book, *Patterns in the balance of nature*, C. B. Williams argued that most of the observed variation in S/G ratios reflected sampling variation (Williams 1964). The early controversy over the analysis of S/G ratios foreshadowed a much larger conflict in ecology over the use of null models in the 1970s and 1980s (Gotelli and Graves 1996). In 1968, the marine ecologist Howard Sanders proposed rarefaction as a method for standardizing samples and comparing species diversity among habitats (Sanders 1968). Sanders' mathematical formula was incorrect, but his notion that sample size adjustments must be made to compare diversity was sound. Since then, the original individual-based rarefaction equation has been derived many times (Hurlbert 1971), as has the equation for sample-based rarefaction (Chiarucchi et al. 2008), and there continue to be important statistical developments in this literature (Colwell et al. 2012).

[6] Note that for a single simulation, the 106 individual jelly beans are randomly sampled *without replacement*. Drawing cards from a deck is an example of sampling without replacement. For the first card drawn, the probability of obtaining a particular suit (clubs, diamonds, hearts, or spades), is 13/52 = 1/4. However, these probabilities change once the first card is drawn, and they are conditional on the result of the first draw. For example, if the first card drawn is a spade, the probability that the second card drawn will also be a spade is 12/51. If the first card drawn is not a spade, the probability that the second card drawn is a spade is now 13/51. However, in the rarefaction example, all of the 106 jelly beans would be replaced before the start of the next simulation, sampling with replacement. A familiar example of sampling with replacement is the repeated tossing of a single coin. Sampling with or without replacement correspond to different statistical distributions and different formulae for the expectation and the variance. Most probability models (as in Chapters 1 and 2) assume sampling with replacement, which simplifies the calculations a bit. In practice, as the size of the sampling universe increases relative to the size of the sample, the expected values converge for the two models: if the jelly bean jar is large enough, the chances of getting a particular color are nearly unchanged whether or not you replace each jelly bean before you draw another one.

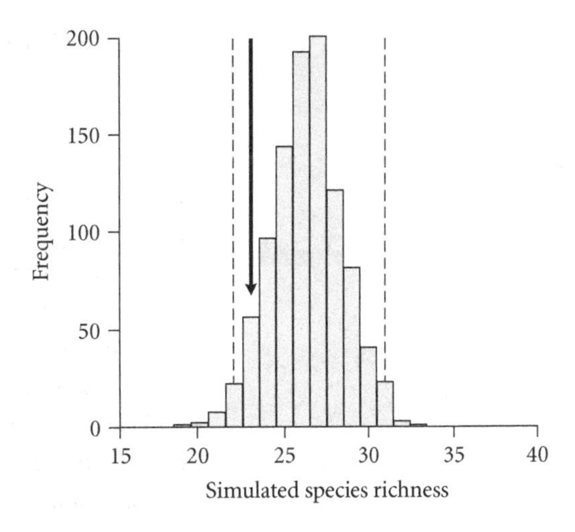

Figure 13.1 Histogram of species richness counts for 1000 random subsamples of 106 individuals from the Logged plots of the Harvard Forest Hemlock Removal Experiment (see data in Table 13.1). The blue line is the observed species richness (23) from the 106 individuals in the Hemlock Control plots, and the dashed lines bracket the 95% confidence interval for the simulated distribution.

23 species for the Hemlock Control treatment is on the low side ($P = 0.092$ for the lower tail of the distribution), but still falls within the 95% confidence interval of the simulated distribution (22 to 31 species). We would conclude from this analysis that the Hemlock Control and Logged treatments do not differ significantly in the number of species they support.

Rarefaction Curves: Interpolating Species Richness

The full rarefaction curve for the spiders of the Logged treatment, as shown in Figure 13.2, illustrates the general features characteristic of any **individual-based rarefaction** curve. We call this kind of rarefaction curve "individual-based" because the individual organism is the unit of sampling. The x-axis for the graph of an individual-based rarefaction curve is the number of individuals, and the

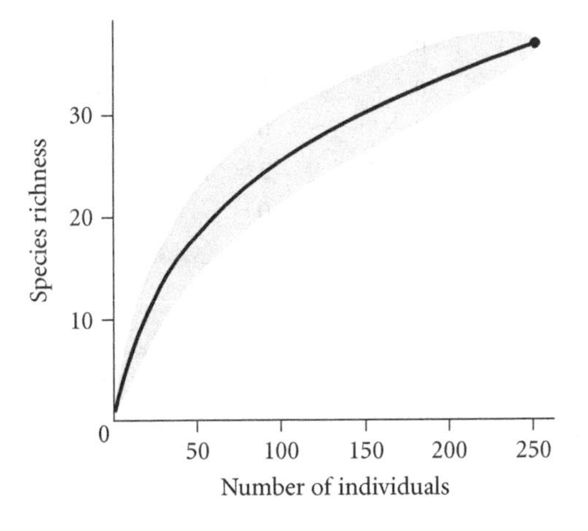

Figure 13.2 Individual-based rarefaction curve for the spider data from the Logged plots in the Harvard Forest Hemlock Removal Experiment (see data in Table 13.1). The simulation is based on 1000 random draws at every sampling level from 1 to 252. The black line is the average of the simulated values, and the blue envelope is the parametric 95% confidence interval. The point represents the original reference sample (252 individuals, 37 species).

TABLE 13.2 **Observed and randomly subsampled counts of spiders**

Species	Logged	Logged (random subsample)	Hemlock Control
Agelenopsis utahana	1	0	1
Agroeca ornata	10	3	2
Amaurobius borealis	22	10	27
Callobius bennetti	3	0	4
Castianeira longipalpa	3	2	0
Centromerus cornupalpis	0	0	0
Centromerus persolutus	0	0	1
Ceraticelus minutus	1	0	1
Ceratinella brunnea	4	2	3
Cicurina arcuata	1	0	0
Cicurina brevis	0	0	0
Cicurina pallida	0	0	0
Cicurina robusta	1	0	0
Collinsia oxypaederotipus	2	1	0
Coras juvenilis	0	0	1
Cryphoeca montana	2	1	2
Dictyna minuta	0	0	0
Emblyna sublata	0	0	0
Eperigone brevidentata	0	0	1
Eperigone maculata	0	0	1
Habronattus viridipes	4	3	0
Helophora insignis	0	0	0
Hogna frondicola	2	1	0
Linyphiid sp. 1	0	0	0
Linyphiid sp. 2	0	0	1
Linyphiid sp. 5	0	0	1
Meioneta simplex	8	3	2
Microneta viaria	1	0	0
Naphrys pulex	1	0	0
Neoantistea magna	15	7	11
Neon nelli	1	1	0
Ozyptila distans	0	0	0
Pardosa distincta	1	0	0

TABLE 13.2 **Continued**

Species	Logged	Logged (random subsample)	Hemlock Control
Pardosa moesta	13	6	0
Pardosa xerampelina	88	40	0
Pelegrina proterva	1	1	0
Phidippus whitmani	1	1	0
Phrurotimpus alarius	8	3	0
Phrurotimpus borealis	5	2	0
Pirata montanus	16	5	20
Pocadicnemis americana	0	0	0
Robertus riparius	1	1	0
Scylaceus pallidus	1	0	0
Tapinocyba minuta	4	1	2
Tapinocyba simplex	1	0	0
Tenuiphantes sabulosus	0	0	1
Tenuiphantes zebra	3	0	2
Trochosa terricola	2	0	0
Unknown morphospecies 1	0	0	0
Wadotes calcaratus	0	0	4
Wadotes hybridus	7	2	12
Walckenaeria digitata	0	0	0
Walckenaeria directa	3	1	4
Walckenaeria pallida	1	0	2
Xysticus elegans	0	0	0
Xysticus fraternus	0	0	0
Zelotes duplex	7	4	0
Zelotes fratris	7	5	0
Total number of individuals	252	106	106
Total number of species	37	24	23

The first and third columns give the counts of spiders from the Logged and Hemlock Control treatments. The second column gives the counts of spiders from a computer-generated random subset of 106 individuals from the Logged sample. By rarefying the data from the Logged treatment, direct comparisons can be made to observed species richness in the Hemlock Control treatment based on a standardized number of individuals (106).

y-axis is the number of species. At the low end, the individual-based rarefaction begins at the point (1,1) because a random draw of exactly 1 individual always yields exactly 1 species. With two individuals drawn, the observed species richness will be either 1 or 2, and the expected value for two individuals is thus an average that is between 1.0 and 2.0. The smoothed curve of expected values continues to rise as more individuals are sampled, and it does so with a characteristic shape: it is always steepest at the low end, with the slope becoming progressively shallower as sampling increases. Such a shape results because the most common species in the assemblage are usually picked up in the first few draws. With additional sampling, the curve continues to rise, but more slowly, because the remaining unsampled species are, on average, progressively less common. In other words, most of the new individuals sampled at that stage represent species that have already been added.

Because most of the individuals in the collection are included in larger random draws, the expected species number eventually rises to its highest point, indicated by the dot in Figure 13.2. The high point represents a sample of the same number of individuals as the original collection. Note that this end-point of species richness does *not* represent the true asymptote of richness for the entire assemblage—it represents only the number of species that were present in the original sample. If the sample had been larger to begin with, it probably would have contained more species. Of course, if empirical sampling is intensive enough, all of the species in an assemblage will have been encountered, and the sampling curve will reach a flat asymptote. In practice, the asymptote is almost never reached, because a huge amount of additional sampling is usually needed to find all of the rare species. Rarefaction is thus a form of interpolation (see Chapter 9): we begin with the observed data and rarefy down to progressively smaller sample sizes.

The variance in Figure 13.2 also has a characteristic shape: it is zero at the low end because all random samples of 1 individual will always contain only 1 species. It is also zero at the high end because all random samples of all individuals will always yield exactly the observed species richness of the original sample. Between these extremes, the variance reflects the uncertainty in species number associated with a particular number of individuals (see Figure 13.1). When estimated by random subsampling, the variance is conditional on the empirical sample and will always equal zero when the size of the subsample equals that of the original data. We will return to this variance calculation later in the next section.

Figure 13.3 again illustrates the rarefaction curve for the Logged treatment (and its 95% confidence interval), but with the added rarefaction curves for the other three treatments. It is easy to see that the Hemlock Control rarefaction

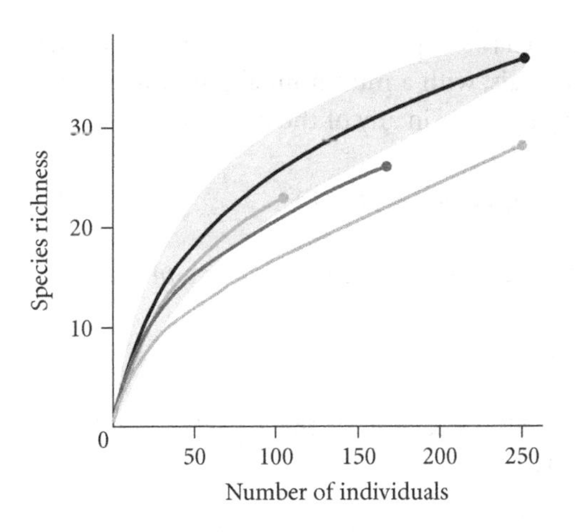

Figure 13.3 Individual-based rarefaction curves for the four experimental treatments in the Harvard Forest Hemlock Removal Experiment. The lines are the expected species richness based on 1000 random draws from the data for each plot (see Table 13.1). Black: Logged plots; light gray: Hardwood Control plots; dark gray: Girdled plots; blue: Hemlock Control plots. The blue area is the parametric 95% confidence interval for the Logged plots.

curve lies within the 95% confidence interval of the Logged rarefaction curve. The Hardwood Control rarefaction curve has the lowest species richness for a given sampling level, whereas the Girdled rarefaction curve lies very close to the Hemlock Control curve. Notice how all the curves converge and even cross at the lower end. When rarefaction curves cross, the ranking of species richness will depend on what sampling level is chosen.

The Expectation of the Individual-Based Rarefaction Curve

Formally, suppose an assemblage consists of N total individuals, each belonging to one of S different species. Species i has N_i individuals, which represents a proportion of the total $p_i = N_i/N$. By definition, these p_i values add up to 1.0 when summed across all species: $\sum_{i=1}^{S} p_i = 1.0$.

We stress that N, S, and p_i represent the complete or "true" values for an assemblage: if we only knew what these numbers were, we would not have to worry about sampling effects, or even about sampling! We could just compare these numbers directly. Instead, we have only a small representative sample of data from this assemblage. Specifically, we have a **reference sample** of only n individuals (where n is a number usually much smaller than N), with S_{obs} species counted, and X_i individuals of each species i. Because the sampling is incomplete, some species in the assemblage will be missing from the reference collection. Thus, $X_i = 0$ for some values of i. The estimated proportion of abundance represented by each species i is $\hat{p}_i = X_i / n$.

With individual-based rarefaction, we would like to estimate the expected number of species in a small subset of m individuals drawn from our reference sample of n. If we knew the underlying p_i values, we could derive the expected number of species using some simple laws of probability (see Chapter 2). First, if we randomly draw out m individuals, with a multinomial sampling model[7] the probability of *not* encountering species i in any of the m trials is:

$$p(\text{not sampling species } i) = (1 - p_i)^m \tag{13.1}$$

Therefore, the probability of capturing species i at least once with m trials is:

$$p(\text{sampling species } i) = 1 - (1 - p_i)^m \tag{13.2}$$

Summing these probabilities across all i species gives the expected number of species in a sample of m individuals:

$$S_{\text{ind}}(m) = \sum_{i=1}^{S}\left[1-(1-p_i)^m\right] = S - \sum_{i=1}^{S}(1-p_i)^m \tag{13.3}$$

Note that the larger m is, the smaller the term within the summation sign, and therefore the closer $S_{\text{ind}}(m)$ is to the complete S. For the empirical reference sample of n individuals, an unbiased estimator of the expected number of species from a small subsample of m individuals comes from the hypergeometric distribution:[8]

[7] The multinomial probability distribution generalizes the binomial distribution that was explained in detail in Chapter 2. Recall that a binomial distribution describes the numbers of successes of a set of n independent Bernoulli trials: experiments that have only two possible outcomes (present or absent, reproduce or not, etc.). If there are many possible (k) discrete outcomes (red, green, blue; male, female, hermaphrodite; 1, 2, 3, …, many species) and each trial results in only one of these possible outcomes, each with probability p_i ($i = 1, 2, \ldots, k$), and there are n trials, then the random variable X_i has a multinomial distribution with parameters n and p. The probability distribution function for a **multinomial random variable** is

$$P(X) = \frac{n!}{x_1! \cdot \ldots \cdot x_k!} p_1^{x_1} \cdot \ldots \cdot p_k^{x_k}$$

the expected value of any outcome $E\{X_i\} = np_i$, the variance of an outcome $V(X_i) = np_i(1 - p_i)$, and the covariance between two distinct outcomes ($i \neq j$) is $\text{Cov}(X_i, X_j) = -np_ip_j$.

[8] Like the binomial distribution, the hypergeometric distribution is a discrete probability distribution that models the probability of k successes in n draws from a population of size N with m successes. The difference between the two is that draws or trials in a

$$S_{\text{ind}}(m) = S_{\text{obs}} - \sum_{X_i > 0} \left[\binom{n - X_i}{m} \Big/ \binom{n}{m} \right] \quad (13.4)$$

Equation 13.4 gives the exact solution for the expected number of species, sampling m individuals without replacement from the original collection of n individuals. In practice, a very similar answer usually results from Equation 13.3, sampling with replacement and using X_i/n as an estimator for p_i. Both equations also can be estimated by the simple random sampling protocol that was used to generate Table 13.2 and Figures 13.1 and 13.2.

Heck et al. (1975) give an equation for the variance of Equation 13.4, but we don't need to use it here. However, we should note that the variance for standard rarefaction is conditional on the collection of n individuals: as the subsample m approaches the reference sample n in size, this variance and the width of the associated 95% confidence interval approaches zero (see Figure 13.2).

Sample-Based Rarefaction Curves: Massachusetts Ants

One difficulty with the "jelly bean" sampling model is that it assumes that the individual organism is the unit that is randomly sampled. However, most ecological sampling schemes use some larger collecting unit, such as a quadrat, trap, plot, transect, bait, or other device that samples multiple individuals. For example, the spider data compiled in Table 13.1 actually were pooled from two plots per treatment, with four pitfall traps per plot. It is these sampling units that represent statistically independent replicates, not the individuals themselves (see Chapters 6 and 7).

---

hypergeometric distribution are sampled *without* replacement, whereas draws in a binomial distribution are sampled *with* replacement. The probability distribution function of a **hypergeometric random variable** is

$$P(X) = \frac{\binom{m}{k}\binom{N-m}{n-k}}{\binom{N}{n}}$$

where $\binom{a}{b}$ is the binomial coefficient (see Chapter 2). The expected value of a hypergeometric random variable is

$$E(X) = n\frac{m}{N}$$

and its variance is

$$Var(X) = \left(n\frac{m}{N}\right)\left(\frac{N-m}{N}\right)\left(\frac{N-n}{N-1}\right)$$

In some kinds of studies, it may not even be possible to count individual organisms within a sample. For example, many marine invertebrates, such as corals and sponges, as well as many perennial plants, grow as asexual clones, making it difficult to define an "individual." In such cases, we can only score the presence or incidence of a species in a sample; we cannot count its abundance. But whether the individuals are counted or not, there are multiple sampling units for each treatment or habitat that is surveyed. We therefore apply **sample-based rarefaction** to such data sets.

For example, Table 13.3 gives the number of ant nests of different species that were detected within 12 standardized sample plots (Allen's Pond through Peaked Mountain) in Massachusetts cultural grasslands.[9] Each row is a different ant species, and each column is a different sample plot in a grassland habitat. In this data set, the entry is the number of ant nests of a particular species detected in a particular sample plot.[10] The data are organized as an **incidence matrix**, in which each entry is simply the presence (1) or absence (0) of a species in a sample. Incidence matrices are most often used for sample-based rarefaction. We have comparable data for 11 other plots from oak-hickory-white pine forests, and five plots from successional shrublands.

In the spider example, we used individual-based rarefaction to compare species richness among four experimental treatments. In this analysis, we use

[9] A few words on the origin of the phrase "cultural grasslands." What are called "natural" grasslands in New England are fragments of a coastal habitat that was more widespread during the relatively warm Holocene climatic optimum, which occurred 9000–5000 years ago. However, most contemporary grasslands in New England are rather different: they are the result of clearing, agriculture, and fire, and have been maintained for many centuries by the activities of Native Americans and European settlers. These cultural grasslands currently support unique assemblages of plants and animals, many of which do not occur in more typical forested habitats. Motzkin and Foster (2002) discuss the unique history of cultural grasslands and their conservation value in New England.

[10] The ant nests were discovered by standardized hand searching at each site for one person-hour within a plot of 5625 m^2. Hand-searching for nests is a bit of an unusual survey method for ants, which are usually counted at baits or pitfall traps. Although we can count individual worker, queen, or male ants, a quirky feature of ant biology complicates the use of these simple counts for biodiversity estimation. Ant workers originate from a nest that is under the control of one or more queens. The nest is in effect a "super-organism," and it is the nest, not the individual ant worker, that is the proper sampling unit for ants. When dozens of ant workers of the same species show up in a pitfall trap, it is often because they all came from a single nest nearby. Thus, counting individual ant workers in lieu of counting ant nests would be like counting individual leaves from the forest floor in lieu of counting individual trees. Therefore, pitfall data for ants are best analyzed with sample-based rarefaction. Gotelli et al. (2011) discuss some of the other challenges in the sampling and statistical analysis of ant diversity. The unique biological features of different plants and animals often dictate the way that we sample them, and constrain the kinds of statistical analyses we can use.

TABLE 13.3 **Data for sample-based rarefaction**

	Sites											
Species	**Allen's Pond**	**Boston Nature Center**	**Brooks Woodland**	**Daniel Webster**	**Doyle Center**	**Drumlin Farm**	**Elm Hill**	**Graves Farm**	**Moose Hill**	**Nashoba Brook**	**Old Town Hill**	**Peaked Mountain**
Aphaenogaster rudis complex	0	0	1	0	1	3	0	0	0	2	1	4
Brachymyrmex depilis	0	0	0	0	0	1	0	0	0	0	1	0
Formica incerta	0	0	0	0	0	0	1	0	1	0	3	2
Formica lasiodes	0	0	0	0	0	0	0	0	0	0	2	0
Formica neogagates	0	0	1	0	0	4	0	0	0	2	0	0
Formica neorufibarbis	0	0	0	0	1	0	0	0	0	0	0	0
Formica pergandei	0	0	0	0	0	0	0	0	0	0	0	2
Formica subsericea	0	0	0	0	0	2	0	0	0	0	0	1
Lasius alienus	0	0	0	0	0	0	0	0	0	0	3	0
Lasius flavus	0	0	1	0	1	0	0	0	0	0	0	0
Lasius neoniger	9	0	4	1	3	0	15	1	12	0	3	11
Lasius umbratus	0	0	0	0	0	2	0	0	1	0	0	1
Myrmica americana	0	0	0	0	0	0	0	0	5	0	2	0
Myrmica detritinodis	0	2	1	0	1	2	4	0	12	0	1	0
Myrmica nearctica	0	0	0	0	0	0	2	5	1	0	0	0
Myrmica punctiventris	0	1	2	0	0	0	0	0	0	0	0	0
Myrmica rubra	0	4	0	8	0	0	0	0	0	0	0	0
Ponera pennsylvanica	0	0	0	0	0	0	0	0	0	0	1	1
Prenolepis imparis	0	0	0	4	1	1	0	0	0	0	0	0
Solenopsis molesta	0	0	0	0	0	1	0	0	0	1	0	0
Stenamma brevicorne	0	0	0	7	0	1	0	0	3	0	0	0
Stenamma impar	0	0	0	0	0	1	0	0	0	0	0	0
Tapinoma sessile	1	4	2	0	0	0	0	3	0	2	1	4
Temnothorax ambiguus	0	0	0	0	0	2	0	1	0	2	0	1
Temnothorax curvispinosus	0	0	0	0	0	0	0	0	0	0	0	1
Tetramorium caespitum	0	0	0	0	4	0	1	0	0	13	1	0

Each row is a species, each column is a sampled cultural grassland site, and each cell entry is the number of ant nests recorded for a particular species in a particular site. Although this data set contains information on abundance, the same analysis could be carried out with an incidence matrix, in which each cell entry is either a 0 (species is absent) or a 1 (species is present). The data for the other two habitats—the oak-hickory-white pine forest and the successional shrublands—are available in the book's online resources.

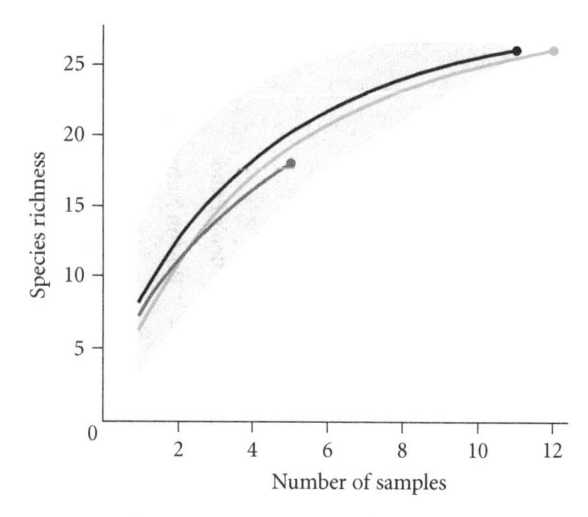

Figure 13.4 Sample-based rarefaction of number of ant species as a function of the number of sample plots surveyed in Massachusetts. Black: oak-hickory-white pine forests; blue: cultural grasslands (see data from Table 13.3); gray: successional shrublands. The blue band is the 95% confidence interval for the oak-hickory-white pine forest.

sample-based rarefaction to compare ant species richness among three habitat types. The principle is exactly the same, but instead of randomly sampling individuals, we randomly sample entire plots. Figure 13.4 illustrates the sample-based rarefaction curves for the three habitat types. As with individual-based rarefaction, the sample-based rarefaction curve rises steeply at first, as common species are encountered in the first few plots, and then more slowly, as rare species are encountered as more plots are included. The endpoint represents the number of species and samples in the original reference sample. In this example, the reference samples had plot numbers of 5 (successional shrublands), 11 (oak-hickory-white pine forests) and 12 (cultural grasslands).

The variance (and a 95% confidence interval) also can be constructed from the random samples. As before, the variance is zero at the (maximum) reference sample size because, when all of the plots are used, the number of species is exactly the same as in the original data set. However, the variance at the low end does not look like that of the individual-based rarefaction curve. In the individual-based rarefaction curve, a subsample of one individual always yields exactly one species. But in the sample-based rarefaction curve, a subsample of one plot may yield more than one species, and there is uncertainty in this number depending on which plot is randomly selected. Therefore, the sample-based rarefaction curve usually has more than one species at its minimum of one sample, and the variance associated with that estimate is greater than zero. As in the previous discussion of individual-based rarefaction, this variance estimator for sample-based rarefaction is conditional on the particular samples in hand.

Sample-based rarefaction curves effectively control for differences in the number of samples collected in the different habitats being compared. However, there may still be differences in the number of individuals per sample that can affect the species richness estimate. Because this data set has counts of individual ant nests in

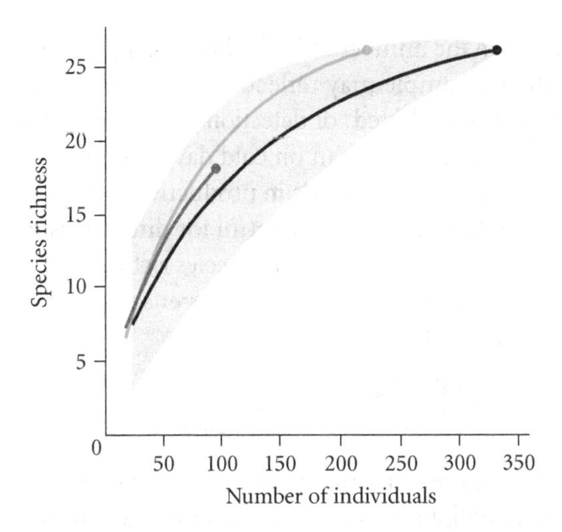

Figure 13.5 Sample-based rarefaction, rescaled to the average number of individual ant nests per sample on the *x*-axis. These are rarefaction curves of the same ant data used to generate Figure 13.4, but the *x*-axis now illustrates the average number of individuals per sample, not the number of samples. Note that the rank ordering of the rarefaction curves is different in the two graphs. Black: oak-hickory-white pine forests; blue: cultural grasslands (see data from Table 13.3); gray: successional shrublands.

each plot, we can take one further step, which is to re-plot the sample-based rarefaction curves (and their variances) against the average number of individuals (or incidences) per plot.[11] This potentially shifts and stretches each rarefaction curve depending on the abundance (or incidence) per sample (Gotelli and Colwell 2001).

For the Massachusetts ant data, the rank order of the sample-based rarefaction curves (see Figure 13.4) changes when they are plotted against abundance, as in Figure 13.5. Specifically, the cultural grassland curve now lies above the other two (it is barely within the confidence interval for the oak-hickory-white pine forest curve). This example illustrates that the estimate of species richness depends not only on the shape of the rarefaction curve, but also on the number of samples and the number of individuals or incidences per sample being compared.

Species Richness versus Species Density

Although hundreds of ecological papers analyze species richness, it is actually better to refer to the measured number of species as **species density**, the number of species per sample unit (James and Wamer 1982). As illustrated by individual-based rarefaction, species density depends on two components:

$$\frac{\text{Species}}{\text{Sample}} = \frac{\text{Individuals}}{\text{Sample}} \times \frac{\text{Species}}{\text{Individual}} \tag{13.5}$$

Two communities might differ in the number of species / sample because of either differences in the number of species / individual (which is quantified with

[11] With an incidence matrix, expected species richness can be re-plotted against the average number of incidences (species occurrences) per sample. However, incidences sometimes can be difficult to interpret because they may not correspond to a discrete "individual" or other sampling unit that contains biodiversity information.

the rarefaction curve) or differences in the number of individuals / sample. Variation in the number of individuals / sample may reflect differences in sampling effort (how many individuals were collected) or detection probability (e.g., pitfall traps catch more samples on warm days than on cold days; see Chapter 14), or differences in biological factors (e.g., gradients in productivity or available energy). Rarefaction is a straightforward way to control for differences in the number of individuals per sample and their effect on species richness.

For a sample-based data set, in which there are several plots per sample, and abundance is measured within each plot, the relationship is:

$$\frac{\text{Species}}{\text{Sample}} = \frac{\text{Plots}}{\text{Sample}} \times \frac{\text{Individuals}}{\text{Plot}} \times \frac{\text{Species}}{\text{Individual}} \quad (13.6)$$

As we noted earlier, it is never appropriate to estimate diversity by computing these values as simple ratios, but Equations 13.5 and 13.6 do illustrate how the sampling properties of the data contribute to the observed species density.

The Statistical Comparison of Rarefaction Curves

This chapter (along with Chapter 14) emphasizes the methods for the proper estimation of diversity. Once rarefaction curves have been calculated, species richness estimated for any level of sampling can be compared using conventional statistical methods, including regression (see Chapter 9) and analysis of variance (see Chapter 10).

Hypothesis testing with rarefaction curves is possible, although there has not been much published work in this area. To test whether species richness differs among rarefaction curves for a particular level of sampling, a simple, conservative test is to examine whether there is any overlap in the 95% confidence intervals calculated for each curve. Payton et al. (2003) recommend a less conservative approach with approximate 84% confidence intervals to control for Type I error. However, both methods assume equal variances in the two groups. The results will depend on the sampling level that is chosen for comparison (particularly if the rarefaction curves cross), and on the way in which the variance of the rarefaction curve is calculated (see next section).[12]

[12] What is needed is a general test for differences among rarefaction curves. In many ecological studies, the implicit null hypothesis is that the two samples differ no more in species richness than would be expected if they were both drawn randomly from the same assemblage. However, in many biogeographic studies, the null hypothesis is that the samples differ no more in species richness than would be expected if the underlying shape of the species abundance distribution were the same in both assemblages. For many biogeographic comparisons, there may be no shared species among the regions, but it is still of

Assumptions of Rarefaction

Whether using sample-based or individual-based rarefaction, the following assumptions should be considered:

SUFFICIENT SAMPLE SIZE As illustrated in Figures 13.3 and 13.4, rarefaction curves converge at small sample sizes, and the differences among curves can become highly compressed. Because all rarefaction curves have to be compared at the sampling intensity of the smallest reference sample, it is important that this sample be of sufficient size for comparison to the others.[13] There is no set amount, but in practice, rarefaction curves based on reference samples of fewer than 20 individuals or five samples usually are inadequate.

COMPARABLE SAMPLING METHODS As discussed earlier in this chapter (see Footnote 3), all sampling methods have their biases, such that some species will be overrepresented and others will be underrepresented or even missing. There is no such thing as a truly "random" sampling method that is completely unbiased. For this reason, it is important that identical sampling methods are used for all comparisons. If quadrat surveys are used in grasslands, but point counts are used

interest to ask whether they differ in species richness. We are starting to explore some bootstrapping analyses in which we measure the expected overlap among rarefaction confidence intervals calculated for different curves. Stay tuned!

[13] Alroy (2010), Jost (2010), and Chao and Jost (in press) recently have suggested a new approach to the standardization of diversity curves. Rather than comparing them at a constant number of samples or individuals, rarefaction curves can be compared at a common level of **coverage** or completeness. Coverage is the percentage of the total abundance of the assemblage that is represented by the species in the reference sample. Thus, a coverage of 80% means that the species represented in the reference sample constitute 80% of the abundance in the assemblage. The abundance of the undetected species in the reference sample constitutes the remaining 20%. A good estimator of coverage comes from the work of Alan Turing, a British computer scientist who made fundamental contributions to statistics (see Footnote 21 in this chapter for more details on Turing). Turing's simple formula is Coverage $\approx 1.0 - f_1/n$, where f_1 is the number of species represented by exactly 1 individual (i.e., the number of singletons), and n is the number of individuals in the reference sample (see the section Asymptotic Estimators: Extrapolating Species Richness for more on singletons). Coverage-based rarefaction standardizes the data to a constant level of completeness, which is not exactly the same as standardizing to a common number of individuals or samples. However, rarefaction curves based on coverage have the same rank order as rarefaction curves based on the number of individuals or the number of samples, so we still advocate the use of traditional rarefaction for ease of interpretation. See Chao and Jost (in press) for more details on coverage-based rarefaction.

in wetlands, the comparison of diversity in the two habitats will be confounded by differences in the sampling methods.[14]

TAXONOMIC SIMILARITIES OF SAMPLES The implicit null hypothesis in many ecological applications of rarefaction is that the samples for different habitats or treatments have been drawn from a single underlying assemblage. In such a case, the samples will often have many shared species, as in the spider data of Table 13.1. Thus, some degree of taxonomic similarity often is listed as an assumption of rarefaction (Tipper 1979). However, in biogeographic analyses (such as comparisons of Asian versus North American tree species richness), there may be few or no species shared among the comparison groups. For either biogeographic or ecological comparisons, rarefaction curves reflect only the level of sampling and the underlying relative abundance distribution of the component species. Rarefaction curves often do not reflect differences among assemblages in species composition,[15] and there are more powerful, explicit tests for differences in species composition among samples (Chao et al. 2005).

CLOSED COMMUNITIES OF DISCRETE INDIVIDUALS The jelly bean analogy for sampling assumes that the assemblage meets the **closure assumption**, or that the assemblage at least is circumscribed enough so that individuals sampled from

[14] Researchers sometimes combine the data from different survey methods in order to maximize the number of species found. Such a "structured inventory" is valid as long as the same suite of methods is used in all the habitats that are being compared (Longino and Colwell 1997). It is true that some species may show up only in one particular kind of trap. However, such species are often rare, and it is hard to tell whether they are only detectable with one kind of trapping, or whether they might show up in other kinds of traps if the sampling intensity were increased.

[15] The spider data in Table 13.2 illustrate this principle nicely. Although the rarefaction curves for the Logged and Hemlock Control treatments are very similar (see Figure 13.3), there are nevertheless differences in species composition that cannot be explained by random sampling. These differences in species composition are best illustrated with columns 2 and 3 of Table 13.2, which show the complete data for the Hemlock Control treatment (column 3, $n = 106$ individuals, 23 species) and a single rarefied sample of the data for the Logged treatment (column 2, $n = 106$ individuals, 24 species). The rarefied sample for the Logged treatment contained 40 individuals of *Pardosa xerampelina*, whereas zero individuals of this species were observed in the Hemlock Control treatment. Conversely, the Hemlock Control treatment contained 20 individuals of *Pirata montanus* whereas the rarefied sample from the Logged treatment contained only five individuals of this species. Even though total species richness was similar in the two habitats, the composition and species identity clearly were different.

this assemblage can be counted and identified. If individuals cannot be readily counted, then a sample-based rarefaction design is needed.[16]

CONSTANT DETECTION PROBABILITY PER INDIVIDUAL Rarefaction directly addresses the issue of undersampling, and the fact that there may be many undetected species in an assemblage. However, the model assumes that the probability of detection or capture *per individual* is the same among species and among comparison groups (habitats or treatments; see Chapter 14 for formal statistical models of detection probability). Consequently, any differences in the detection or capture probability *per species* reflect underlying differences in commonness and rarity of different species.[17]

SPATIAL RANDOMNESS OF INDIVIDUALS Individual-based rarefaction assumes that individuals of the different species are well mixed and occur randomly in space. However, this is often not true, and the more typical pattern is that individuals of a single species are clumped or spatially aggregated (see the discussion of the coefficient of dispersion in Chapter 3). In such cases, the spatial scale of sampling needs to be increased to avoid small-scale patchiness (see Chapter 6). Alternatively, use sample-based rarefaction.[18]

INDEPENDENT, RANDOM SAMPLING As with all of the statistical methods described in this book, the individuals or samples should be collected randomly and independently. Both rarefaction methods assume that the sampling itself does not

[16] The assumption of a closed community can be difficult to justify if the assemblage is heavily affected by migration from adjacent habitats (Coddington et al. 2009), or if there is temporal variation in community structure from ongoing habitat change (Magurran 2011). In many real communities, the jelly bean jar probably "leaks" in space and time.

[17] Hierarchical sampling models (Royle and Dorazio 2008) can be used to explicitly model the distinct components of the probability of occupancy and the probability of detection, given that a site is occupied. These models use either maximum likelihood or Bayesian approaches to estimate both probabilities and the effects of measured covariates. Chapter 14 introduces these models, which can be applied to many problems, including the estimation of species richness (Kéry and Royle 2008).

[18] You may be a tempted to pool the individuals from multiple samples and use individual-based rarefaction. However, if there is spatial clumping in species occurrences, the individual-based rarefaction curve derived from the pooled sample will consistently over-estimate species richness compared to the sample-based rarefaction curve. This is yet another reason to use sample-based rarefaction: it preserves the small-scale heterogeneity and spatial clumping that is inherent in biodiversity samples (Colwell et al. 2004).

affect the relative abundance of species, which is statistically equivalent to sampling with replacement (see Footnote 6 in this chapter). For most assemblages, the size of an ecological reference sample is very small compared to the size of the entire assemblage, so this assumption is easy to meet even when the individuals are not replaced during sampling.

Asymptotic Estimators: Extrapolating Species Richness

Rarefaction is an effective method of interpolating and estimating species richness within the range of observed sampling effort. However, with continued sampling of a closed assemblage, the **species accumulation curve** would eventually level out at an asymptote.[19] Once the asymptote is reached, no new species would be added with additional sampling. The asymptote of S species with N individuals thus represents the total diversity of the assemblage. In contrast, the reference sample consists of only S observed species (S_{obs}) with n individuals. Unless the reference sample is very large or there are very few species in the assemblage, $S_{obs} \ll S$, and $n \ll N$. This section provides useful methods for extrapolating estimates of N and S of the assemblage from n and S_{obs} of the reference sample.

Although there are several statistical strategies[20] for estimating S, we prefer non-parametric estimators, which do not assume any particular species abundance dis-

[19] There is a subtle but important distinction between a species accumulation curve and a rarefaction curve. A species accumulation curve is created by accumulating individuals or samples from an assemblage until species richness reaches an asymptote and rises no further. Because the diversity is so great in most assemblages (which may not be truly closed), the species accumulation curve is a mostly hypothetical construct that would be created by progressively adding more data, building the curve from the point (1,1) (i.e., 1 individual, 1 species) and moving towards the right. In contrast, a rarefaction curve is an empirical construct that is created by taking a reference sample of biodiversity (a collection of either individuals or samples) and randomly sampling with progressively smaller sample sizes, moving to the left to the minimum point of (1,1). The rarefaction curve is thus an estimator of the species accumulation curve in its lower reaches.

[20] Beginning with a seminal paper by R.A. Fisher (Fisher et al. 1943; see Footnote 5 in Chapter 5), a common strategy has been to estimate species richness by fitting a curve to a species abundance distribution. Imagine a histogram in which the *x*-axis is the number of individuals, and the *y*-axis is the number of species. Each bar in the histogram is the number of species represented by a particular number of individuals, and the area under the curve (equaling the sum of the bars) is the total number of species in the assemblage. Biodiversity data in this form can sometimes be approximated by mathematical distributions such as the log-normal or the geometric series (see Footnote 12 in Chapter 2), and these distributions have formed the basis for estimating

tribution, although they do incorporate the observation that some species in the assemblage are relatively common and some species are relatively rare. An important sampling principle for diversity data is that the greater the frequency of rare species in a reference sample, the greater the number of undetected species that are present in the assemblage, but missing from the reference sample.[21]

---

species richness by fitting the curve to a reference sample (McGill 2011). However, these methods yield good estimates only when the "true" form of the species abundance distribution is used in the model (O'Hara 2005). But the true form is rarely known, and some empirical data sets cannot be well-fit by any single parametric curve or may be equally well-fit by several (Connolly and Dornelas 2011).

A second, and simpler, approach to estimating total species richness is to extrapolate a rarefaction curve out to an asymptote (Soberón and Llorente 1993). This method also requires the user to specify a mathematical function (such as the Michaelis-Menton equation; see Figure 4.2) for a curve that rises to an asymptote. Perhaps because the extrapolation method does not use any information on species abundances or frequencies, it rarely provides a good fit to empirical rarefaction curves. Our preference is to use the family of non-parametric estimators—e.g., Chao1 (Equations 13.7, 13.8), Chao2 (Equations 13.10, 13.11), and others—because these have a solid foundation in statistical sampling theory and because they usually perform better than alternative methods in comparisons with simulated and real data sets (Gotelli and Colwell 2011).

[21] This theorem derives directly from the work of the mathematician Alan M. Turing (1912–1954), who is widely regarded as the father of computer science (see also Footnote 13 in this chapter). Turing built one of the first modern computers, and developed many fundamental theorems in computer science. He provocatively proposed an operational test for artificial intelligence based on whether an observer can have a conversation by teletype with a computer program and be fooled into thinking the responses are coming from another human. This Turing Test continues to be widely discussed and debated in the artificial intelligence literature. During World War II, Turing worked for Britain's Government Code and Cypher School at Bletchley Park. He and I. J. Good developed theorems in cryptographic analysis that were successfully used to crack the German Wehrmacht coding machine, The Enigma. It is these theorems that are used in the development of the non-parametric species richness estimators (Chao 1984). After the war, Turing worked on the mathematical biology of pattern formation in morphogenesis. In 1952, Turing was prosecuted for homosexual acts, which were illegal in Britain at that time. As an alternative to prison, he underwent hormone treatments (chemical castration). Turing died from cyanide poisoning in 1954. Although an inquest determined that his death was a suicide, the facts are still widely disputed. In 2009, following an Internet campaign, the British government officially apologized for "the appalling way he was treated."

Alan M. Turing

A simple but powerful minimum estimator of S for individual-based data is the Chao1 index:[22]

$$\text{Chao1} = S_{\text{obs}} + \frac{f_1^2}{2f_2} \text{ if } f_2 > 0 \tag{13.7}$$

$$\text{Chao1} = S_{\text{obs}} + \frac{f_1(f_1 - 1)}{2(f_2 + 1)} \text{ if } f_2 = 0 \tag{13.8}$$

where S_{obs} is the observed number of species in the reference sample, f_1 is the number of **singletons** (species in the reference sample represented by exactly 1 individual), and f_2 is the number of **doubletons** (species in the reference sample represented by exactly 2 individuals). Equation 13.7 is the standard form, and Equation 13.8 is a bias-corrected form that can be used when there are no doubletons ($f_2 = 0$). The Chao1 estimator is a *minimum* asymptotic estimator, so the true species richness will probably be at least as large as predicted by Equation 13.7. For example, in the Hemlock Control treatment (see Table 13.1), there were 23 spider species observed, with 9 species each represented by exactly one individual, and 6 species each represented by exactly two individuals. Thus, $S_{\text{obs}} = 23$, $f_1 = 9$, $f_2 = 6$, so Chao1 = 29.75. Additional sampling in the Hemlock Control treatment should yield a minimum of 6 or 7 previously undetected spider species.

[22] Many of the equations and methods presented in this chapter were developed by two contemporary scientists, Anne Chao and Robert K. Colwell. Anne Chao is the Tsing Hua Distinguished Chair Professor, Institute of Statistics, National Tsing Hua University, Taiwan. She received a PhD in 1977 from the University of Wisconsin. Her research interests include species abundance estimation, capture-recapture experiments, and the epidemiology of suicide. Robert K. Colwell is a Board of Trustees Distinguished Professor of Ecology and Evolutionary Biology at the University of Connecticut. He received a PhD in 1969 from the University of Michigan. In addition to biodiversity statistics, his research interests include biogeography and biodiversity, the ecology and evolution of tropical species interactions, and the development of database tools for biodiversity inventories. Colwell's popular software program EstimateS has been widely used in biodiversity analysis. Colwell and Coddington (1994) named the Chao1 and Chao2 indices and introduced them to ecologists.

A parametric 95% confidence interval (see Chapter 3) can be constructed from the variance of Chao1 (for $f_1 > 0$ and $f_2 > 0$):[23]

$$\sigma^2_{\text{Chao1}} = f_2\left[\frac{1}{2}\left(\frac{f_1}{f_2}\right)^2 + \left(\frac{f_1}{f_2}\right)^3 + \frac{1}{4}\left(\frac{f_1}{f_2}\right)^4\right] \quad (13.9)$$

For the Hemlock Control treatment, the estimated variance is 34.6, with a corresponding 95% confidence interval of 18.2 to 41.3 species.

For sample-based rarefaction, the formulas are similar, but instead of using f_1 and f_2 for the number of singletons and doubletons, we use q_1, the number of **uniques** (= species occurring in exactly 1 sample) and q_2, the number of **duplicates** (species occurring in exactly 2 samples). The formula for Chao2, the expected number of species with sample-based data, also includes a small bias correction for R, the number of samples in the data set:

$$\text{Chao2} = S_{\text{obs}} + \left(\frac{R-1}{R}\right)\frac{q_1^2}{2q_2} \text{ if } q_2 > 0 \quad (13.10)$$

$$\text{Chao2} = S_{\text{obs}} + \left(\frac{R-1}{R}\right)\frac{q_1(q_1-1)}{2(q_2+1)} \text{ if } q_2 = 0 \quad (13.11)$$

The corresponding variance estimate for sample-based incidence data (with $q_1 > 0$ and $q_2 > 0$) is:

$$\sigma^2_{\text{Chao2}} = q_2\left[\frac{A}{2}\left(\frac{q_1}{q_2}\right)^2 + A^2\left(\frac{q_1}{q_2}\right)^3 + \frac{A^2}{4}\left(\frac{q_1}{q_2}\right)^4\right] \quad (13.12)$$

where $A = (R\text{-}1)/R$. For example, in the cultural grasslands data matrix (see Table 13.3), there were 26 ant species observed in 12 plots, with six species each occurring in exactly one plot, and 8 species each occurring in exactly two plots. Thus, $S_{\text{obs}} = 26$, $q_1 = 6$, $q_2 = 8$, and $R = 12$, so Chao2 = 28.45. Sampling additional plots of cultural grasslands should yield a minimum of two or three additional species. The 95% confidence interval (calculated from the variance in Equation 13.12)

[23] See Colwell (2011), Appendix B of the EstimateS User's Guide for other cases and for the calculation of asymmetrical confidence intervals, which are also used in Figure 14.3.

is from 23.2 to 33.7 species. Table 13.4 summarizes the calculations for the individual-based spider data in each of the four treatments of the hemlock experiment, and Table 13.5 summarizes the calculations for the sample-based Massachusetts ant data in each of the three habitats.

The minimum estimated species richness and its confidence interval vary widely among samples. In general, the greater the difference between the observed species richness (S_{obs}) and the asymptotic estimator (Chao1 or Chao2), the larger the uncertainty and the greater the size of the resulting confidence interval. The most extreme example is from the Hardwood Control treatment in the spider data set. Although 28 species were recorded in a sample of 250 individuals, there were 18 singletons and only one doubleton. The resulting Chao1 = 190 species, with a (parametric) confidence interval from –162 to 542 species! The large number of singletons in these data signals a large number of undetected species, but the sample size is not sufficient to generate an extrapolated estimate with any reasonable degree of certainty.

How many additional individuals (or samples) would need to be collected in order to achieve these asymptotic estimators? Equation 13.7 provides a natural "stopping rule": sampling can stop when there are no singletons in the data set, so that all species are represented by at least two individuals. That amount of sampling often turns out to be very extensive: by the time enough individuals have been sampled to find a second representative for each singleton species, new singletons will have turned up in the data set.

TABLE 13.4 Asymptotic estimator summary statistics for individual-based sampling of spiders

Treatment	n	S_{obs}	f_1	f_2	Chao1	σ^2_{Chao1}	Confidence interval	n^* (g = 1.0)	n^* (g = 0.90)
Hemlock Control	106	23	9	6	29.8	34.6	(18, 41)	345	65
Girdled	168	26	12	4	44.0	207	(6, 72)	1357	355
Hardwood Control	250	28	18	1	190.0	32,238	(–162, 542)	17,676	4822
Logged	252	37	14	4	61.5	346.1	(25, 98)	2528	609

n = number of individuals collected in each treatment; S_{obs} = observed number of species; f_1= number of singleton species; f_2 = number of doubleton species; Chao1 = estimated asymptotic species richness; σ^2_{Chao1}= variance of Chao1; confidence interval = parametric 95% confidence interval; n^* (g = 1.0) = estimated number of additional individuals that would need to be sampled to reach Chao1; n^* (g = 0.90) = estimated number of additional individuals that would need to be sampled to reach 90% of Chao1.

TABLE 13.5 **Asymptotic estimator summary statistics for sample-based collections of ants**

Habitat	R	S_{obs}	q_1	q_2	Chao2	σ^2_{Chao2}	Confidence interval	R* (g = 1.0)	R* (g = 0.90)
Cultural grasslands	12	26	6	8	28.06	5.43	(23, 33)	17	2
Oak-hickory-white pine forest	11	26	6	8	28.05	5.36	(24, 33)	16	2
Successional shrubland	5	18	9	4	26.1	37.12	(12, 40)	26	6

R = number of sampled plots in each habitat; S_{obs} = observed number of species; q_1= number of unique species; q_2 = number of duplicate species; Chao2 = estimated asymptotic species richness; σ^2_{Chao2} = variance of Chao2; confidence interval = parametric 95% confidence interval; R^* (g = 1.0) = estimated number of additional samples that would need to be sampled to reach Chao2; R^* (g = 0.90) = estimated number of additional samples that would need to be sampled to reach 90% of Chao2.

But what is the number of individuals (or samples) that would be needed to eliminate all of the singletons (or uniques)? Chao et al. (2009) derive formulas (and provide an Excel-sheet calculator) for estimating the number of additional individuals (n^*) or samples (R^*) that would need to be collected in order to achieve Chao1 or Chao2. These estimates are provided in the last two columns of Tables 13.4 and 13.5. The column for n^* when $g = 1$ gives the sample size needed to achieve the asymptotic species richness estimator, and the column for n^* when $g = 0.9$ gives the sample size needed to achieve 90% of the asymptotic species richness.

The additional sampling effort varies widely among the different samples, depending on how close S_{obs} is to the asymptotic estimator. For example, in the oak-hickory-white pine forest, 26 ant species were collected in 11 sample plots. The asymptotic estimator (Chao2) is 28.1 species, and the estimated number of *additional* sample plots needed is 16, a 54% increase over the original sampling effort. At the other extreme, 28 spider species were recorded from a collection of 250 individuals in the Hardwood Control treatment of the hemlock experiment. To achieve the estimated 190 species (Chao1) at the asymptote, an additional 17,676 spiders would need to be sampled, a 70-fold increase over the original sampling effort. Other biodiversity samples typically require from three to ten times the original sampling effort in order to reach the estimated asymptotic species richness (Chao et al. 2009). The effort is usually considerable because a great deal of sampling in the thin right-hand tail of the species-abundance distribution is necessary to capture the undetected rare species. The sampling requirements are less burdensome if we can be satisfied with capturing some lesser fraction, say 90%, of the asymptotic species richness. For example, in the Hemlock Control treatment, 23 spider species were observed, with an estimated asymptote of 29.8. An additional 345 individual spiders would need to be censused to reach this asymptote, but if we would be satisfied with capturing 90% of asymptotic richness, only an additional 65 spiders would be needed.

Rarefaction Curves Redux: Extrapolation and Interpolation

Biodiversity sampling begins with a reference sample—a standardized collection of individuals (or samples) that has a measured number of species. With rarefaction, the data are interpolated to progressively smaller sample sizes to estimate the expected species richness. With asymptotic estimators, the same data are extrapolated to a minimum asymptotic estimator of species richness, with an associated sampling effort that would be needed to achieve that level of richness. Recently, Colwell et al. (2012) unified the theoretical frameworks of rarefaction and asymptotic richness estimation. They derived equations that link the interpolated part of the rarefaction curve with the extrapolated region out to the asymptotic estimator. In standard rarefaction, the variance is conditional on the observed data, and the confidence interval converges to zero at the observed sample size (see Figures 13.2 and 13.4). In the Colwell et al. (2012) framework, the rarefaction variance is unconditional. The reference sample is viewed more properly as a sample from a larger assemblage, and the unconditional variance can be derived based on the expectation and variance of the asymptotic estimator. We will forego the equations here, but Figure 13.6 illustrates these extended rarefaction/extrapolation curves for the Massachusetts ant data. These curves graphically confirm the results of the rarefaction analysis (see Figure 13.4), which is that ant species richness is fairly similar in all three habitats. However, the extrapolation of the successional shrublands data out to an asymptotic estimator is highly uncertain and generates a very broad confidence interval because it was based on a reference sample of only five plots.

Estimating Species Diversity and Evenness

Up until now, this chapter has addressed the estimation of species richness, which is of primary concern in many applied and theoretical questions. Focusing on species richness seems to ignore differences in the relative abundance of species, although both the asymptotic estimators and the shape of the rarefaction curve depend very much on the commonness versus rarity of species.

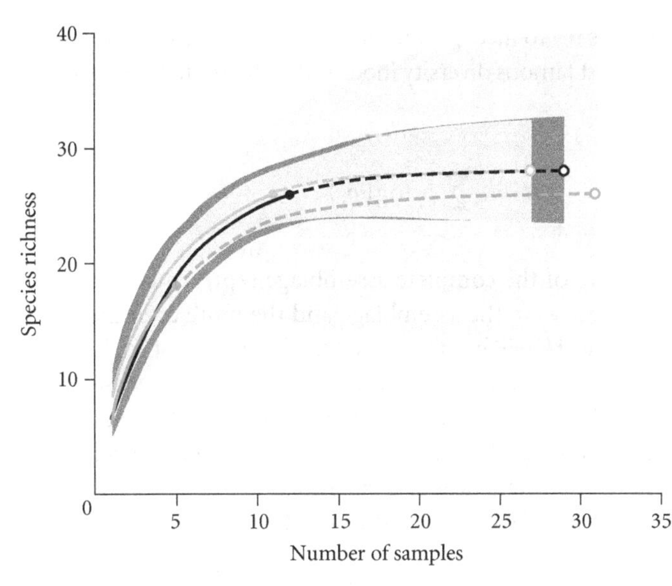

Figure 13.6 Interpolated and extrapolated sample-based rarefaction curves of the ant data. Light gray: oak-hickory-white pine forests; dark gray = cultural grasslands; blue = successional shrublands. The filled points are the actual samples and the open points are the extrapolated species richness (Chao2). The solid lines are the interpolated regions of each curve and the dashed lines are the extrapolated regions of each curve. The shaded regions indicate the approximate 95% confidence interval for each curve.

Ecologists have tried to expand the measure of species diversity to include components of both species richness and species evenness. Consider two forests that each consist of five species and 100 individual trees. In Forest A, the 100 trees are apportioned equally among the five species (maximum evenness), whereas, in Forest B, the first species is represented by 96 trees, and the remaining four species are represented by one tree each. Most researchers would say that Forest A is more diverse than Forest B, even though both have the same number of species and the same number of individuals. If you were to walk through Forest A, you would frequently encounter all five species, whereas, in Forest B, you would mostly encounter the first species, and rarely encounter the other four species.[24]

[24] The great explorer and co-discoverer of evolution, Alfred Russel Wallace (1823–1913) was especially impressed by the high diversity and extreme rarity of tree species in Old World tropical rainforests: "If the traveler notices a particular species and wishes to find more like it, he may often turn his eyes in vain in every direction. Trees of varied forms, dimensions and colours are around him, but he rarely sees any one of them repeated. Time after time he goes towards a tree which looks like the one he seeks, but a closer examination proves it to be distinct. He may at length, perhaps, meet with a second specimen half a mile off, or may fail altogether, till on another occasion he stumbles on one by accident" (Wallace 1878). Wallace's description is a nice summary of the aspect of diversity that is measured by *PIE*, the *p*robability of an *i*nterspecific *e*ncounter (see Equation 13.15; Hurlbert 1971).

Alfred Russel Wallace

Many diversity indices try to incorporate effects of both species richness and species evenness. The most famous diversity index is the Shannon diversity index, calculated as:

$$H' = -\sum_{i=1}^{s} p_i \log(p_i) \tag{13.13}$$

where p_i is the proportion of the complete assemblage represented by species $i(p_i = N_i/N)$. The more species in the assemblage and the more even their relative abundances, the larger H' will be. There are literally dozens of such indices based on algebraic transformations and summations of p_i. In most cases, these diversity indices do not have units that are easy to interpret. Like species richness, they are sensitive to the number of individuals and samples collected, and they do not always have good statistical performance.

An exception is Simpson's (1949) index of concentration:

$$D = \sum_{i=1}^{s} p_i^2 \tag{13.14}$$

This index measures the probability that two randomly chosen individuals represent the same species. The lower this index is, the greater the diversity. Rearranging and including an adjustment for small sample sizes, we have:

$$PIE = \frac{n}{(n-1)}\left(1.0 - \sum_{i=1}^{s} p_i^2\right) \tag{13.15}$$

PIE is the *p*robability of an *i*nterspecific *e*ncounter (Hurlbert 1971), the probability that two randomly chosen individuals from an assemblage will represent two different species.[25] There are 3 advantages of using *PIE* as a simple diversity

[25] In economics, the *PIE* index (without the correction factor $n/(n-1)$) is known as the Gini coefficient (see Equation 11.26; Morgan 1962). Imagine a graph in which the cumulative proportions of species (or incomes) are plotted on the *y*-axis, and the rank order of the species (in increasing order) is plotted on the *x*-axis. If the distribution is perfectly even, the graph will form a straight line. But if there is any deviation from perfect evenness, a concave curve, known as the Lorenz (1905) curve, is formed with the same starting and ending point as the straight line. The Gini coefficient quantifies inequality in income as the relative area between the straight line and the Lorenz curve. Of course, for income distributions, the absolute difference between the richest and the poorest classes may be more important than the relative evenness of the different classes. Ecologists have used the Gini coefficient to quantify the magnitude of competitive interactions in dense populations of plants (Weiner and Solbrig 1984).

index. First, it has easily interpretable units of probability and corresponds intuitively to a diversity measure that is based on the encounter of novel species while sampling (see Footnote 24 in this chapter). Second, unlike species richness, *PIE* is insensitive to sample size; a rarefaction curve of the *PIE* index (using the estimator $\hat{p}_i = X_i / n$ generates a flat line. Finally, *PIE* measures the slope of the individual-based rarefaction curve measured at its base (Olsweski 2004).

Returning to the example of two hypothetical forests, each with five species and 100 individuals, the maximally even Forest A (20, 20, 20, 20, 20) has $PIE = 0.81$, whereas the maximally uneven Forest B (96, 1, 1, 1, 1) has $PIE = 0.08$. In spite of the advantages of the *PIE* index as an intuitive measure of species diversity based on relative abundance, the index is bounded between 0 and 1, so that the index becomes "compressed" near its extremes and does not obey a basic **doubling property**: if two assemblages with the same relative abundance distribution but no species in common are combined with equal weight, diversity should double. Certainly species richness obeys this doubling property. However, if we double up Forest A (20, 20, 20, 20, 20) to form Forest AA (20, 20, 20, 20, 20, 20, 20, 20, 20, 20), diversity changes from $PIE_A = 0.81$ to $PIE_{AA} = 0.90$. If we double up Forest B (96, 1, 1, 1, 1) to form Forest BB (96, 96, 1, 1, 1, 1, 1, 1, 1, 1), diversity changes from $PIE_B = 0.08$ to $PIE_{BB} = 0.54$.

Hill Numbers

Hill numbers (Hill 1973a) are a family of diversity indices that overcome the problems of many of the diversity indices most commonly used by ecologists. Hill numbers preserve the doubling property, they quantify diversity in units of modified species counts, and they are equivalent to algebraic transformations of most other indices. Hill numbers were first proposed as diversity measures by the ecologist Robert MacArthur (MacArthur 1965; see Footnote 6 in Chapter 4), but their use did not gain much traction until 40 years later, when they were reintroduced to ecologists and evolutionary biologists in a series of important papers by Jost (2006, 2007, 2010).

The general formula for the calculation of a Hill number is:

$$^{q}D = \left(\sum_{i=1}^{s} p_i^q\right)^{1/(1-q)} \qquad (13.16)$$

As before, p_i is the "true" relative frequency of each of each species ($p_i = N_i / N$, for $i = 1$ to S) in the complete assemblage. The exponent q is a non-negative integer that defines the particular Hill number. Changing the exponent q yields a family of diversity indices. As q increases, the index puts increasing weight on the most common species in the assemblage, with rare species making less and less of a contribution to the summation. Once $q \gtrsim 5$, Hill numbers rapidly converge to the inverse of the relative abundance of the most common species. Negative values for

q are theoretically possible, but they are never used as diversity indices because they place too much weight on the frequencies of rare species, which are dominated by sampling noise. As q increases, the diversity index decreases, unless all species are equally abundant (maximum evenness). In this special case, the Hill number is the same for all values of q, and is equivalent to simple species richness.

Regardless of the exponent q, the resulting Hill numbers are always expressed in units of **effective numbers of species**: the equivalent number of equally abundant species. For example, if the observed species richness in a sample is ten, but the effective number of species is five, the diversity is equivalent to that of a hypothetical assemblage with five equally abundant species.[26]

The first three exponents in the family of Hill numbers are especially important. When $q = 0$,

$$^{0}D = \left(\sum_{i=1}^{s} p_i^0\right)^1 = S \tag{13.17}$$

because $p_i^0 = 1$, $^{0}D = \left(\sum_{i=1}^{S} 1\right)^1$, and $S^1 = S$. Thus, ^{0}D corresponds to ordinary species richness. Because ^{0}D is not affected by species frequencies, it actually puts more weight on rare species than any of the other Hill numbers.

For $q = 1$, Equation 13.16 cannot be solved directly (because the exponent of the summation, $1/(1 - q)$ is undefined), but in the limit it approaches:

$$^{1}D = e^{H'} = e^{\left(-\sum_{i=1}^{s} p_i \log p_i\right)} \tag{13.18}$$

Thus ^{1}D is equivalent to the exponential of the familiar Shannon measure of diversity (H') (see Equation 13.13). ^{1}D weights each species by its relative frequency.

Finally, for $q = 2$, Equation 13.16 gives:

$$^{2}D = \left(\sum_{i=1}^{s} p_i^2\right)^{-1} = \frac{1}{\left(\sum_{i=1}^{s} p_i^2\right)} \tag{13.19}$$

[26] The effective number of species is very similar in concept to the effective population size in population genetics, which is the equivalent size of a population with entirely random mating, with no diminution caused by factors such as bottlenecks or unequal sex ratios (see Footnote 3 in Chapter 3). The concept of effective number of species also has analogues in physics and economics (Jost 2006).

Thus ^2D is equivalent to the inverse of Simpson's index (see Equation 13.14). ^2D and ^qD for values of $q > 2$ give heavier weight to the more common species.

Hill numbers provide a useful family of diversity indices that consistently incorporate relative abundances while at the same time express diversity in units of effective numbers of species. Two caveats are important when using Hill numbers, however.

First, no diversity index can completely disentangle species richness from species evenness (Jost 2010). The two concepts are intimately related: the shape of the rarefaction curve is affected by the relative abundance of species, and any index of evenness is affected by the number of species present in the assemblage.

Second, Hill numbers are not immune to sampling effects. Our treatment of Hill numbers is based on the true parameters (p_i and S) from the complete assemblage, although calculations of Hill numbers are routinely based on estimates of those parameters ($\hat{p}_i$ and S_{obs}) from a reference sample.[27] Species richness is the Hill number with $q = 0$, and this chapter has emphasized that both the number of individuals and the number of samples have strong influences on S_{obs}. Sample size effects are important for all the other Hill numbers, although their effect diminishes as q is increased. Figure 13.7 depicts rarefaction curves for the spider data of the Hardwood Control, and illustrates the sampling effect with the Hill numbers ^0D, ^1D, ^2D, and ^3D.

Software for Estimation of Species Diversity

Software for calculating rarefaction curves and extrapolation curves for species richness and other diversity indices is available in Anne Chao's program SPADE (Species Prediction and Diversity Estimation: chao.stat.nthu.edu.tw/softwareCE.html), and in Rob Colwell's program EstimateS (purl.oclc.org/estimates). The vegetarian library in R, written by Noah Charney and Sydne Record, contains many useful functions for calculating Hill numbers (cran.r-project.org/web/packages/vegetarian/index.html). R code used to analyze and graph the data in this chapter is available in the data section of this book's website (harvardforest.fas.harvard.edu/ellison/publications/primer/datafiles) and as part of

[27] The problem with using the standard "plug-in" estimator $\hat{p}_i = X_i / n$ is that, on average, it over-estimates p_i for the species that are present in the reference sample. Chao and Shen (2003) derive a low-bias estimator for the Shannon diversity index (see Equation 13.13) that adjusts for the both biases in $\hat{p}_i$ and for the missing species in the reference sample. Anne Chao and her colleagues are currently developing asymptotic estimators and variances for other Hill numbers.

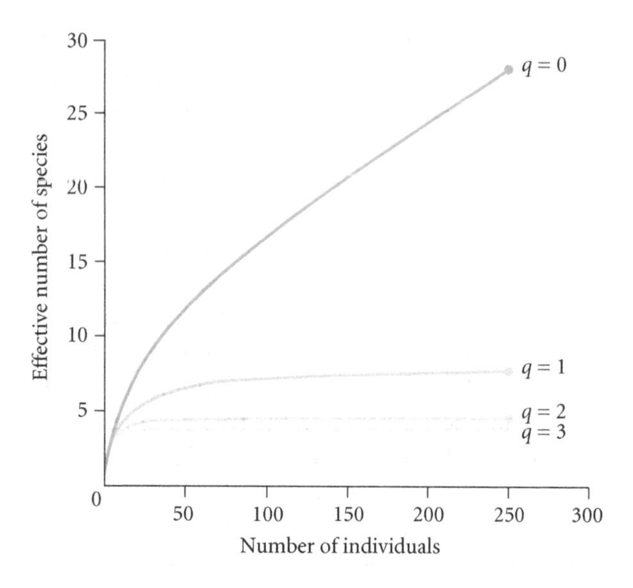

Figure 13.7 Individual-based rarefaction curves for the family of Hill numbers (see Equation 13.16) with exponents $q = 0$ to 3. Each curve is based on 1000 randomization of the Hardwood Control reference sample of 250 spiders and 28 species from the Harvard Forest Hemlock Removal Experiment (see Table 13.1). Note that $q = 0$ is simple species richness, so the curve is identical to the light gray rarefaction curve shown in Figure 13.3.

EcoSimR, a set of R functions and scripts for null model analysis (www.uvm.edu/~ngotelli/EcoSim/EcoSim.html).

Summary

The measurement of biodiversity is a central activity in ecology. However, biodiversity data (samples of individuals, identified to species, or records of species incidences) are labor-intensive to collect, and in addition usually represent only a tiny fraction of the actual biodiversity present. Species richness and most other diversity measures are highly sensitive to sampling effects, and undetected species are a common problem. Comparisons of diversity among treatments or habitats must properly control for sampling effects. Rarefaction is an effective method for interpolating diversity data to a common sampling effort, so as to facilitate comparison. Extrapolation to asymptotic estimators of species richness is also possible, although the statistical uncertainty associated with these extrapolations is often large, and enormous sample sizes may be needed to achieve asymptotic richness in real situations. Hill numbers provide the most useful diversity measures that incorporate relative abundance differences, but they too are sensitive to sampling effects. Once species diversity has been properly estimated and sampling effects controlled for, the resulting estimators can be used as response or predictor variables in many kinds of statistical analyses.

CHAPTER 14

Detecting Populations and Estimating their Size

We began this book with an introduction to probability and sampling and the central tasks of measurement and quantification. Questions such as "Are two populations different in size?", "What is the density of *Myrmica* ants in the forest?", or "Do these two sites have different numbers of species?" are fundamental to basic ecological studies and many applied problems in natural resource management and conservation. The different inferential approaches (see Chapter 5) and methods for statistical estimation and hypothesis testing (see Chapters 9–13) that we have discussed up until now uniformly assume not only that the samples are randomized, replicated, and independent (see Chapter 6), but also that the investigator has complete knowledge of the sample. In other words, there are no errors or uncertainty in the measurements, and the resulting data actually quantify the parameters in which we are interested. Are these reasonable assumptions?

Consider the seemingly simple problems of comparing the nest density (see Chapter 5) or species richness (see Chapter 13) of ground-foraging ants in fields and forests. The measurements for these examples include habitat descriptors and counts of the number of ant nests per sampled quadrat (see Table 5.1) or the number of species encountered in different sites (see Table 13.5). We used these data to make inferences about the size of the population or the species richness of the assemblage, which are the parameters we care about. We can be confident of our identifications of habitats as forests or fields, but are we really certain that there were only six ant nests in the second forest quadrat or 26 species in the average oak-hickory-white pine forest? What about that *Stenamma brevicorne* nesting in between two decomposing leaves that we missed, or the nocturnal *Camponotus castaneus* that was out foraging while we were home sleeping? How easy would it be to count all of the individuals accurately, or even to find all of the species that are present?

This chapter introduces methods to address the problem that the probability of detecting all of the individuals in a sample or all of the species in a habitat is always less than 1. Because sampling is imperfect and incomplete, the observed count C is always less than the true number of individuals (or species) N. If we assume that there is a number p equal to the fraction of individuals (or species) of the total population that we sampled, then we can relate the count C to the total number N as:

$$C = Np \tag{14.1}$$

If we had some way to estimate p, we could then estimate the total population size (or number of species) N, given our observed count C:

$$\hat{N} = C / \hat{p} \tag{14.2}$$

As in earlier chapters, the "hats" (ˆ) atop N and p in Equation 14.2 indicate that these values are statistical estimates of the unknown, true values of these parameters. As we develop models in this chapter based on Equations 14.1 and 14.2, we will usually add one or more subscripts to each of C, N, and p to indicate a particular sample location or sample time.

Isolating $\hat{N}$ on the left-hand side of Equation 14.2 indicates that we are explicitly interested in estimating the total number of objects (see Chapter 13)—e.g., individuals in a population, species in an assemblage, etc. But $\hat{N}$ is not the only unknown quantity in Equation 14.2. We first need to estimate $\hat{p}$, the proportion of the total number of objects that were sampled. In Chapter 13, we faced a similar problem in estimating the proportional abundance of a given species as $\hat{p}_i = X_i / n$, where X_i was the number of individuals of species i and n was the number of individuals in the reference sample, which was always less than the total number of individuals in the assemblage (N). The rearrangement of Equation 14.2 highlights the necessity of estimating both $\hat{p}$ *and* $\hat{N}$.

In summary, a simple idea—that we are unable to have seen (or counted) everything we intended to see or count (as in Figure 14.1)—and its encapsulation in a simple equation (Equation 14.2), brings together many of the ideas of probability, sampling, estimation, and testing that we have presented in the preceding chapters. The apparent simplicity of Equation 14.2 belies a complexity of issues and statistical methods. We begin by estimating the probability that a species is present or absent at a site (occupancy), and then move on to estimate

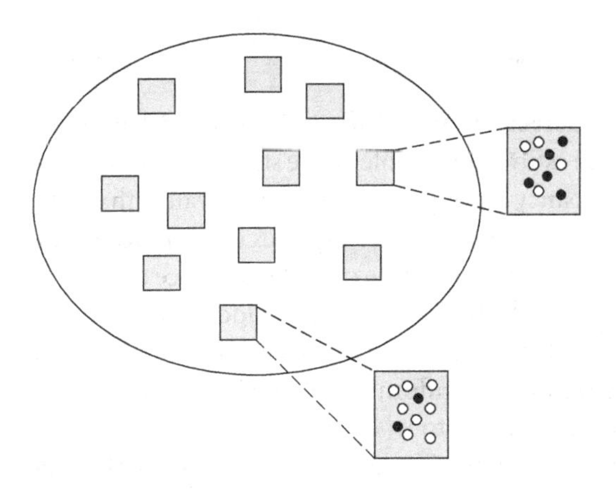

Figure 14.1 Sample replication and detection probability. Within a spatially defined sample space (large circle), only a subset of plots can be sampled (blue squares). Within each plot, individuals of interest (circles) will either be found (detected, indicated by black circles) or not found (undetected, indicated by white circles). (After MacKenzie et al. 2006.)

how many individuals of that species occur at the site (abundance), given that (i.e., conditional on) it is present.[1]

Occupancy

Before we can estimate the size of a local population, we need to determine whether the species is present at all; we refer to a site where a species is present as an occupied site, and one where a species is absent as an unoccupied site. We define **occupancy** as the probability that a sampling unit or an area of interest (henceforth, a site) that is selected at random contains at least one individual of the species of interest. In brief, occupancy is the probability that a species is present at a site. If we sample x sites out of a total of s possible sites,[2] we can estimate occupancy as:

[1] Although it is logical to first ask whether a species is present or not, and then ask how many individuals of a given species there are, methods for estimating the latter were actually developed first. We suspect that estimates of population size were derived before estimates of occupancy because mark-recapture methods were initially applied to common species of commercially important animals, such as fish or deer. There was no question these animals were present, but their population sizes needed to be estimated. Later, biologists became interested in occupancy models as a tool for studying species that are too rare to find and count easily. Unfortunately, many of the species that are currently extinct or very rare were formerly common, but have been overexploited by hunting or harvesting.

[2] Be aware that the notation among the chapters of this book is not consistent. In the species richness literature (see Chapter 13), s refers to species, and n refers to sites or samples. But in the occupancy literature, s refers to samples, and n refers to species or individuals. In both of these chapters, we have stuck with the variable notation that is most commonly used in cited papers in these subdisciplines rather than imposing our own scheme to keep the notation consistent among chapters. *Caveat lector*!

$$\hat{\psi} = \frac{x}{s} \tag{14.3}$$

If there is no uncertainty in **detection probability** (the value of $\hat{p}$ in Equation 14.2), then we have a simple estimation problem. We start by assuming that there are no differences in occupancy among all of the sampled sites (that is $\psi_1 = \psi_2 = \ldots = \psi_s \equiv \psi$) and that we can always detect a species if it is indeed present (the detection probability $p = 1$). In this simple case, the presence or absence of an individual species at a sampled site is a Bernoulli random variable (it is either there or it is not; see Chapter 2), and the number of sites in which the species is present, x, will be a binomial random variable with expected value given by Equation 14.3 and variance equal to $\psi(1 - \psi)/s$.

If the detection probability $p < 1$, but is still a known, fixed quantity, then the probability of detecting the species in at least one of t repeated surveys is

$$p' = 1 - \underbrace{(1-p)\times(1-p)\times\ldots\times(1-p)}_{t \text{ times}} = 1-(1-p)^t \tag{14.4}$$

Equation 14.4 is simply 1 minus the probability of not having detected the species in all t of the surveys. With the detection probability p' now adjusted for multiple surveys, we can move on to estimating occupancy, again as a binomial random variable. Equation 14.3 is modified to adjust occupancy by the detection probability given by Equation 14.4, resulting in:

$$\hat{\psi} = \frac{x}{sp'} \tag{14.5}$$

Note that the numerator x in Equation 14.5 is the number of sites in which the species is detected, which is less than or equal to the number of sites that the species actually occupies.

The variance of this estimate of occupancy is larger than the variance of the estimate of occupancy when the detection probability equals one:

$$\text{Var}(\hat{\psi}) = \frac{1-\psi p'}{sp'} = \frac{\psi(1-\psi)}{s} + \frac{\psi(1-p')}{sp'} \tag{14.6}$$

The first term on the right-hand side of Equation 14.6 is the variance when $p = 1$, and the second term is the inflation in the variance caused by imperfect detection.

The real challenge, however, arises when detection probability p is unknown, so now the number of sampled sites in which the species is actually present also has to be estimated:

$$\hat{\psi} = \frac{\hat{x}}{s} \tag{14.7}$$

Because there is uncertainty in both detection probability and occupancy,[3] Equation 14.7 is the most general formula for estimating occupancy. Two approaches to this have been developed to estimating occupancy and detection probabilities. In one approach, we first estimate detection probability and then use Equation 14.5 to estimate occupancy. In the second approach, both occupancy and detection are estimated simultaneously with either a maximum likelihood or a Bayesian model (see Chapter 5). Although the first approach is computationally easier (and usually can be done with pencil and paper), it assumes that occupancy and detection probabilities are simple constants that do not vary in time or space. The second approach can accommodate covariates (such as habitat type or time of day) that are likely to affect occupancy or detection probabilities. With the second approach, we can also compare the relative fit of models for the same data set that incorporate different covariates. Both approaches are described in detail by MacKenzie et al. (2006); we present only the second approach here.

The Basic Model: One Species, One Season, Two Samples at a Range of Sites

The basic model for jointly estimating detection probabilities and occupancy was developed by MacKenzie et al. (2002). In our description of occupancy models, we continue to use the notation introduced in Equations 14.3–14.7: occupancy (ψ) is the probability that a species is present at site i; p_{it} is the probability that the species is detected at site i at sample time t; T is the total number of sampling times; s is the total number of sites surveyed; x_t is the number of sites

[3] For charismatic megafauna such as pandas, mountain lions, and Ivory-billed Woodpeckers that are the focus of high-profile hunting, conservation, or recovery efforts, occupancy probabilities may be effectively 0.0, but uncertainty in detection is very high and unconfirmed sightings are unfortunately the norm. The Eastern Mountain Lion (*Puma concolor couguar*) was placed on the U.S. Fish & Wildlife Service's Endangered Species List (www.fws.gov/endangered/) in 1973 and is now considered extinct; the last known individual was captured in 1938 (McCollough 2011). Similarly, the last specimen of the Ivory-billed Woodpecker (*Campephilus principalis*) was collected in the southeastern United States in 1932, and the last verified sighting was in 1944. But both species are regularly "rediscovered." For the Ivory-billed Woodpecker (Gotelli et al. 2012), the probability of these rediscoveries being true positives ($< 6 \times 10^{-5}$) is much smaller than the probability that physicists have falsely discovered the Higgs boson after 50 years of searching. Hope springs eternal, however, and signs and sightings of both Eastern Mountain Lions in New England and Ivory-billed Woodpeckers in the southeastern U.S. are regularly reported—but never with any verifiable physical or forensic evidence (McKelvey et al. 2003). There is also even a fraction of the tourist economy that caters to the searchers. Bigfoot in the Pacific Northwest and "The King" in Graceland now have lots of company!

for which the species was detected at time t; and x is the total number of sites at which the species was detected on at least one of the sampling dates.

The simplest model for estimating occupancy requires at least two samples from each of one or more sites. In addition, we assume that, for the duration of all of our samples, the population is closed to immigration, emigration, and death: that is, individuals neither enter nor leave a site between samples (we relax this closure assumption in a subsequent section). We also assume that we can identify accurately the species of interest; false positives are not allowed. However, false negatives are possible, with probability equal to [1 – occupancy]. Finally, we assume that all sites are independent of one another, so that the probability of detecting the species at one site is not affected by detections at other sites.

In Chapter 9, we illustrated how the number of New England forest ant species changes as a function of latitude and elevation. In addition to sampling ants in forests, we also sampled ants in adjacent bogs, which harbor a number of ant species specialized for living in sunny, water-logged habitats. In the summer of 1999, we sampled each of 22 bogs twice, with samples separated by approximately six weeks. The data for a single species, *Dolichoderus pustulatus*, which builds carton nests inside of old pitcher-plant leaves, are shown in Table 14.1 (from Gotelli and Ellison 2002a). We collected *D. pustulatus* at $x = 16$ of the $s = 22$ bogs ($x_1 = 11$ detections in the first sample, $x_2 = 13$ detections in the second sample). We are interested in estimating ψ, the probability that a randomly chosen bog was indeed occupied by *D. pustulatus*.

For each site at one of the two sampling dates, *D. pustulatus* was either detected (1) or not (0). If it was detected on only one of the two dates, we assumed it was present but not detected at the other date (because of the closure assumption of this model). The **detection history** for a single site can be written as a 2-element string of ones and zeros, in which a 1 indicates that the individual was detected (collected) and a 0 indicates that it was not. Each site in this example has one of four possible detection histories, each with 2 elements: (1,1)—*D. pustulatus* was collected in both samples; (1,0)—*D. pustulatus* was collected only in the first sample; (0,1)—*D. pustulatus* was collected only in the second sample; and (0,0)—*D. pustulatus* was collected in neither sample. Figure 14.2 illustrates the sampling histories for all 22 sites.

If we collected *D. pustulatus* at site i in the first sample, but not in the second (sampling history [1,0]), the probability that we would detect it at that site is:

$$P(x_i) = \psi_i p_{i1}(1 - p_{i2}) \tag{14.8}$$

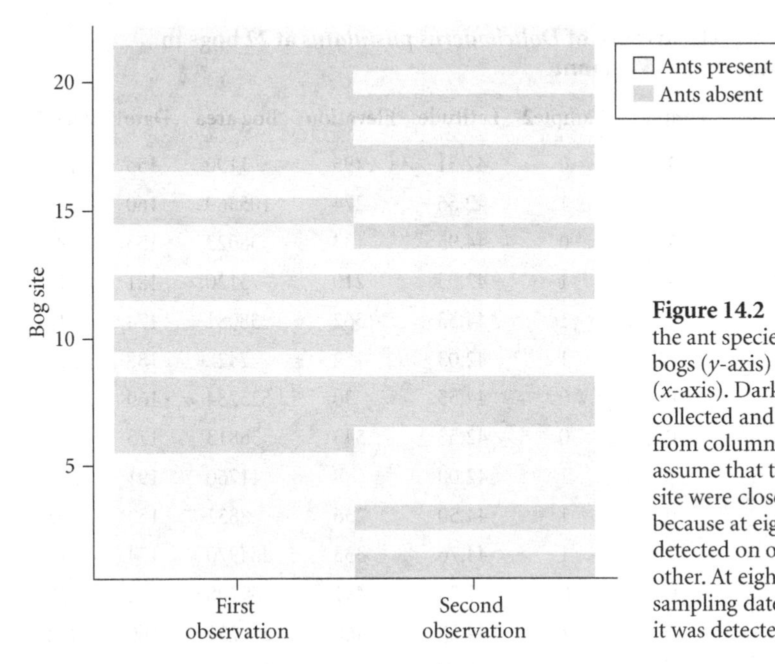

Figure 14.2 The collection histories of the ant species *Dolichoderus pustulatus* at 22 bogs (y-axis) on each of two sampling dates (x-axis). Dark blue indicates this species was collected and light blue that it was not (data from columns 2 and 3 of Table 14.1). If we assume that the populations of ants at each site were closed, detection was imperfect, because at eight of the 22 sites the species was detected on one sampling date but not the other. At eight sites it was detected on both sampling dates, and at the remaining six sites it was detected on neither sampling date.

In words, this probability is equal to *D. pustulatus* occupying the site (ψ_i) *and* being detected at time 1 (p_{i1}) *and* not being detected at time 2 ($1 - p_{i2}$). On the other hand, if we never collected *D. pustulatus* at site j, the collection history would be (0,0), and the probability that we did not detect it would be the sum of the probabilities that it did not occupy the site ($1 - \psi_j$), *or* it did occupy the site (ψ_j) *and* was not collected at either time ($\prod_{t=1}^{2}(1-p_{jt})$):

$$P(x_j) = (1-\psi_j) + \psi_j \prod_{t=1}^{2}(1-p_{jt}) \tag{14.9}$$

In general, if we assume that both occupancy and detection probabilities are the same at every site, we can combine the general forms of Equations 14.8 and 14.9 to give the full likelihood:

$$L(\psi, p_1, \ldots, p_x) = \left[\psi^x \prod_{t=1}^{T} p_t^{x_t}(1-p_t)^{x-x_t}\right] \times \left[(1-\psi) + \psi\prod_{t=1}^{T}(1-p_t)\right]^{s-x} \tag{14.10}$$

For the data in Table 14.1, $s = 22$, $x_1 = 11$, $x_2 = 13$, $x = 16$, and $T = 2$. In this context, the likelihood is the *value* of ψ that would be most probable (= likely), given the data on detection histories ($p_1, \ldots, p_x$).

TABLE 14.1 Occurrence of *Dolichoderus pustulatus* at 22 bogs in Massachusetts and Vermont

Site	Sample1	Sample2	Latitude	Elevation	Bog area	Date1	Date2
ARC	1	0	42.31	95	1190	153	195
BH	1	1	42.56	274	105369	160	202
CAR	1	0	44.95	133	38023	153	295
CB	1	1	42.05	210	73120	181	216
CHI	1	1	44.33	362	38081	174	216
CKB	0	1	42.03	152	7422	188	223
COL	0	0	44.55	30	623284	160	202
HAW	0	0	42.58	543	36813	175	217
HBC	1	1	42.00	8	11760	191	241
MOL	0	1	44.50	236	8852	153	195
MOO	1	1	44.76	353	864970	174	216
OB	0	0	42.23	491	89208	174	209
PEA	1	1	44.29	468	576732	160	202
PKB	1	0	42.19	47	491189	188	223
QP	0	1	42.57	335	40447	160	202
RP	1	1	42.17	78	10511	174	209
SKP	0	0	42.05	1	55152	191	241
SNA	0	1	44.06	313	248	167	209
SPR	0	0	43.33	158	435	167	209
SWR	0	1	42.27	121	19699	153	195
TPB	0	0	41.98	389	2877	181	216
WIN	1	1	42.69	323	84235	167	202
Total occurrences	11	13					

Values shown are the name of each bog (site); whether *D. pustulatus* was present (1) or absent (0) on the first (sample1) or second (sample2) sample date; the latitude (decimal degrees), elevation (meters above sea level) and area (m^2) of the bog mat; and the days of the year (January 1 1999 = 1) on which each bog was sampled (date1, date2).

In addition, we can model both occupancy and detection probability as a function of different covariates. Occupancy could be a function of site-based covariates such as latitude, elevation, bog area, or average annual temperature. In contrast, detection probability is more likely to be a function of time-based covariates that are associated with each census. Such time-based covariates might include the day of the year on which a site was sampled, the time of day we arrived at the site, or weather conditions during sampling. In general, both occupancy and detection probability can be modeled as functions of either site-based or time-based covariates. Once the model structure is selected, the covariates can be incorporated into the model using logistic regression (see Chapter 9):

$$\psi = \frac{\exp(\boldsymbol{\beta}\mathbf{X}_{\text{site}})}{1+\exp(\boldsymbol{\beta}\mathbf{X}_{\text{site}})} \quad \textbf{(14.11a)}$$

$$p = \frac{\exp(\boldsymbol{\beta}\mathbf{X}_{\text{time}})}{1+\exp(\boldsymbol{\beta}\mathbf{X}_{\text{time}})} \quad \textbf{(14.11b)}$$

In these two equations, the parameters and covariates are bold-faced because they represent vectors, and the regression model can be specified using matrix operations (see Appendix Equations A.13 and A.14). If we include covariates in our estimation of occupancy, then we estimate the average occupancy as:

$$\bar{\hat{\psi}} = \frac{\sum_{i=1}^{s}\hat{\psi}_i}{s} \quad \textbf{(14.12)}$$

We used Equations 14.10–14.12 to estimate the occupancy of *Dolichoderus* in Massachusetts and Vermont bogs.[4] We first estimated occupancy without assuming the influence of any covariates, then fit three alternative models with different sets of covariates: (1) a model in which detection probability varied with time of sample (sampling date); (2) a model in which occupancy varied with three site-specific covariates (latitude, elevation, and area); and (3) a model in which occupancy varied with the three site-specific geographic covariates and detection probability varied with time of sample. We compared the fit of the four different models to the data using Akaike's Information Criteria (AIC; see "Model Selection Criteria" in Chapter 9).

The results, shown in Table 14.2, suggest that the best model is the simplest one—constant detection probability at the two sample dates and occupancy that does not co-vary geographically—but that the model that includes a temporally varying detection probability fits almost as well (the difference in the AIC

[4] All of these models were fit using the occu function in the R library unmarked (Fiske and Chandler 2011). Spatial and temporal covariates were first rescaled as *Z* scores (see Chapter 12).

TABLE 14.2 **Estimates of occupancy and detection probability of *Dolichoderus pustulatus* in 22 New England bogs**

	Model 1	Model 2	Model 3	Model 4
Description	No covariates	Detection probability varies with time of sample	Occupancy varies with bog geographic characteristics	Detection varies with time and occupancy varies geographically
Number of parameters to estimate	2	3	5	6
Estimated occupancy ($\hat{\psi}$)	0.82 (0.13)	0.83 (0.13)	0.80 (0.10)	0.88 (0.11)
Estimated detection probability ($\hat{p}$)	0.67 (0.11)	0.66 (0.12)	0.76 (0.09)	0.67 (0.10)
AIC	63.05	64.53	66.45	68.03

Dark blue shading indicates the best-fitting models as determined by AIC; numbers in parentheses are the standard error of the given parameter.

between these two models is less than 2). All four of the models estimated occupancy at between 80% and 90%, which suggests that we missed finding *D. pustulatus* in at least two of the bogs that were sampled.

Note that even the simplest model without any covariates requires that we estimate two parameters—occupancy and detection (see Equation 14.7). Occupancy is of primary interest, whereas detection probability is considered a **nuisance parameter**—a parameter that is not of central concern, but that we nonetheless must estimate to get at what we really want to know. Our estimates of detection probability ranged from 67% in the simple models to 76% in the model in which occupancy was a function of three geographic covariates. Even though we have been studying pitcher plants and their prey, and collecting ants in bogs for nearly 15 years, we can still overlook common ants that build their nests inside of unique and very apparent plants.

The standard errors in Table 14.2 of all of the occupancy estimates of *D. pustulatus* were relatively large because only two samples (the absolute minimum necessary) were used to fit the model. With a greater number of sampling times, the uncertainty in the estimate of occupancy would decrease. In simulated data sets, for a sample size s of 40 sites with detection probabilities (p) of 0.5, at least five samples are needed to get accurate estimates of occupancy (MacKenzie et al. 2002). In our example, the standard error of occupancy could have been reduced by as much as 50% if we had had visited each bog five times instead of two. We return to a discussion of general issues of sampling at the end of this chapter.

Occupancy of More than One Species

In Chapter 13, we described asymptotic species richness estimators (Chao1 and Chao2) which are used to estimate the number of undetected species in a reference sample. These estimators are based on the numbers of singletons, or uniques, and doubletons, or duplicates (species in the reference sample represented by exactly one or two individuals, respectively; see Equations 13.7 and 13.8). We also estimated the number of additional samples that would be needed to find all of the undetected species (see Tables 13.4 and 13.5; Chao et al. 2009). As with rarefaction, these estimators assume that the probability of detection per individual is constant and does not differ among individuals of different species. Therefore, the detection probability of different species is proportional to their relative frequency in the assemblage (see Equations 13.1–13.4). However, it would be more realistic to assume that occupancy and detection probabilities vary among species, through time, and as a function of site-based variables, and to incorporate these factors into species richness estimators (or other diversity measures). The simple occupancy model for a single species (see Equations 14.7–14.12) can be expanded to incorporate multiple species and used to estimate the number of undetected species. Our ant data from New England bogs, of which the *D. pustulatus* example (see Tables 14.1 and 14.2) are a subset, provide an illustrative example (Dorazio et al. 2011).

At each of the 22 sites in Table 14.1, on each of the two sampling dates, we collected ant occurrence data from a grid of 25 pitfall traps. For the $s = 22$ sites, the occurrence data can be summarized as an n (species) × s (sites) matrix of observations y_{ik}, where $i = \{1, \ldots, n\}$, $k = \{1, \ldots, s\}$, and y_{ik} is the number of pitfall traps in which each ant species was captured at the kth of the 22 sites, summed over the two sampling times (Table 14.3). Thus, y_{ik} is an integer between 0 (the species was never collected in a site) and 50 (in both sample periods, the species was found at site k in every one of the 25 pitfall traps in the grid). We observed n species among all the s sites, but we are interested in the total number N species present at all sites that could have been captured. N is unknown, but probably includes some species that

TABLE 14.3 **Observed collection data of ants in New England bogs**

Species i	Abundance at site k			
1	y_{11}	y_{12}	...	y_{1s}
2	y_{21}	y_{22}	...	y_{2s}
⋮	⋮	⋮	⋮	⋮
n	y_{n1}	y_{n2}	...	y_{ns}

Each of n species was observed in y_{ik} 25 pitfall traps set at site k (one of 22 bogs). Because 25 pitfall traps were set at each bog on each of two dates, $0 \leq y_{ik} \leq 50$.

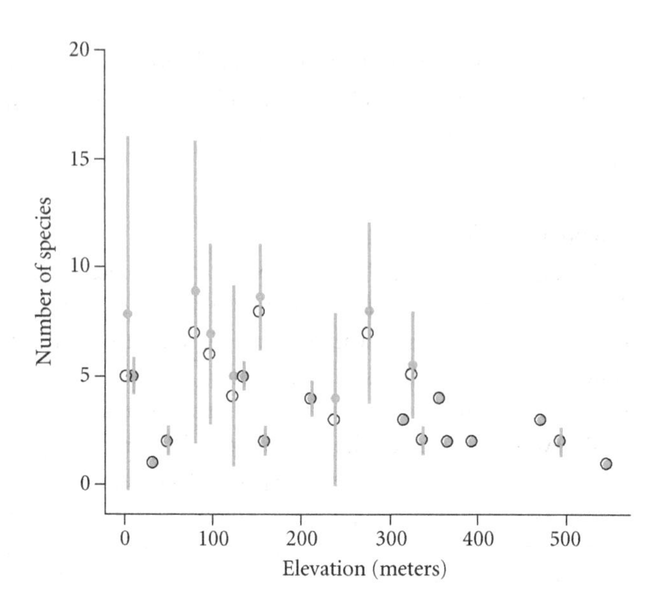

Figure 14.3 Observed (black open circles) and Chao2 estimates (blue filled circles) of species richness of ants in each of 22 New England bogs as a function of their elevation. Vertical lines are 95% confidence intervals (from Equation 13.12). The width of confidence intervals equals zero when doubletons are present in the collection but singletons are not present in the collection.

were present but undetected at all of the sites. Although N is at least equal to n, it probably exceeds it ($n \leq N$).

In this data set, $n = 19$ species were detected among 22 bogs. Treating captures in pitfall traps as incidence data (Gotelli et al. 2011), the Chao2 estimate (see Equations 13.10–13.12) for regional diversity is 32 species, with a confidence interval of 8.5–54.9. Chao2 estimates of species richness at each bog are shown in Figure 14.3. As discussed in Chapter 13, if there are no uniques in the data set (species recorded at only a single site), the Chao2 estimator equals the observed species richness: there are estimated to be no undetected species, and both the variance and confidence interval equal zero (see Footnote 23 in Chapter 13). However, this index (and others; see Chapter 13) does not incorporate variation among species, sites, or times in occupancy or detection probabilities.

Including occupancy, detection probability, and time- and site-specific covariates into species richness estimation is a formidable problem that is just beginning to be addressed (e.g., Dorazio and Royle 2005; MacKenzie et al. 2006; Royle and Dorazio 2008; Dorazio et al. 2011; Dorazio and Rodríguez 2012; Royle and Dorazio 2012). It is important, however, to keep in mind the common goal of both the methods presented in Chapter 13, and the methods presented here: estimation of the total number of N species present at all sites that could be found.

Estimation of N is a multi-step process. First, we assume that there are some number of species, $N - n$, which were present in the bog but that we did not collect. Thus, we can expand the $n \times s$ sample matrix (see Table 14.3) to an $N \times s$ matrix, in which the unobserved species are designated by additional rows with matrix entries assigned to equal zero, as shown in Table 14.4. Second, we assume that the N species in each bog (our alpha diversity; see Chapter 13) are part of a regional species pool of size M (the gamma diversity $M \gg N$), and expand the matrix accordingly to an $M \times s$ matrix; see Table 14.5 for this further step. M puts a finite upper bound on regional species richness and constrains subsequent analyses.[5]

Third, we identify from the regional pool of M species those that can or cannot occur at a site (perhaps due to physiological tolerances, special habitat requirements, etc.) and provide a dummy variable for each species ($w_i = 1$ if it can occur, 0 if it cannot). Finally, we complete the matrix in Table 14.5 with a $N \times s$ set of site-specific and species-specific occurrences z_{ik} that is equivalent to a species × site incidence matrix in which rows are species, columns are sites, and cell entries are either 1 (present at a site) or 0 (absent at a site) (cf. Table 13.3). In Chapter 13, incidence (or abundance) matrices were fixed observations, but in the framework of occupancy models, the incidence matrix is itself a random variable.

TABLE 14.4 **Expanded matrix to include both the observed collection data of ants in New England bogs and the ($N - n$) species that were present but not observed**

Species i	Abundance at site k			
1	y_{11}	y_{12}	...	y_{1s}
2	y_{21}	y_{22}	...	y_{2s}
⋮	⋮	⋮	⋮	⋮
n	y_{n1}	y_{n2}	...	y_{ns}
$n + 1$	0	0	...	0
$n + 2$	0	0	...	0
⋮	⋮	⋮	⋮	⋮
N	0	0	...	0

The dark blue portion of the matrix is the observed collection data (see Table 14.3), and the light blue portion represents species that could have been present but were not collected.

A Hierarchical Model for Parameter Estimation and Modeling

Putting all of these ideas together yields a type of nested or **hierarchical model** (see Chapter 11) described fully in Equations 14.13–14.18 and summarized in Equation 14.19. To begin with, whether a species can be present or not (w_i in Table 14.5) is assumed to be an independent random variable, and not dependent on the presence or absence of other species.[6] Because it can take on values of either 1 or 0, w_i can be modeled as a Bernoulli random variable:

$$w_i \sim \text{Bernoulli}(\Omega) \tag{14.13}$$

[5] This upper bound on species richness must be identified independently of the sample data, from sources such as regional checklists of species or collated museum records (Ellison et al. 2012).

[6] Null models and randomization tests of unreplicated incidence matrices have traditionally been used for the analysis of species co-occurrence (Gotelli and Graves 1996). However, with replicated hierarchical sampling, these problems are beginning to be studied with occupancy models (see Chapter 8 in MacKenzie et al. 2006).

TABLE 14.5 **The complete matrix for estimating species richness**

Species i	Abundance				Incidence				w
1	y_{11}	y_{12}	...	y_{1s}	z_{11}	z_{12}	...	z_{1s}	w_1
2	y_{21}	y_{22}	...	y_{2s}	z_{21}	z_{22}	...	z_{2s}	w_2
⋮	⋮	⋮	⋮	⋮	⋮	⋮	⋮	⋮	⋮
n	y_{n1}	y_{n2}	...	y_{ns}	z_{n1}	z_{n2}	...	z_{ns}	w_s
$n+1$	0	0	...	0	$z_{n+1,1}$	$z_{n+1,2}$	...	$z_{n+1,s}$	w_{n+1}
$n+2$	0	0	...	0	$z_{n+2,1}$	$z_{n+2,2}$	...	$z_{n+2,s}$	w_{n+2}
⋮	⋮	⋮	⋮	⋮	⋮	⋮	⋮	⋮	⋮
N	0	0	...	0	z_{N1}	z_{N2}	...	z_{Ns}	w_N
$N+1$	0	0	...	0	$z_{N+1,1}$	$z_{N+1,2}$	...	$z_{N+1,s}$	w_{N+1}
$N+2$	0	0	...	0	$z_{N+2,1}$	$z_{N+2,2}$	...	$z_{N+2,s}$	w_{N+2}
⋮	⋮	⋮	⋮	⋮	⋮	⋮	⋮	⋮	⋮
M	0	0	...	0	z_M	z_M	...	z_M	w_M

This "augmented" matrix builds on the observed abundances of species (number of pitfall traps in which ants were collected) at each site (medium blue portion of the matrix, using data from Table 14.3) and the additional rows of zeros for abundances of species uncollected but present at each site (dark blue, from Table 14.4). Additional rows and columns augment this matrix (lightest blue): species potentially available for collecting from the regional species pool (rows $N + 1$ to M); an incidence matrix of z_{ik}; and a column vector of parameters w indicating whether or not a species in the regional species pool could have been collected. See Dorazio et al. (2011) for additional details.

where Ω is the probability that a species in the full $N \times s$ matrix is present and could be captured. The parameter of interest N, the total number of species in the community, is derived as:

$$N = \sum_{i=1}^{M} w_i \tag{14.14}$$

If we can estimate Ω and w_i, we can solve for N.

Getting there takes several steps. First, examine the $N \times s$ incidence matrix of z_{ik}s. Each z_{ik} is either a 1 (species present) or a 0 (species absent), but note that any species i that is not a possible member of the community ($w_i = 0$) must have

$z_{ik} = 0$ as well. In other words, z_{ik} is conditional on w_i (see Chapter 1 for a refresher on conditional probabilities):

$$z_{ik} \mid w_i \sim \text{Bernoulli}(w_i \psi_{ik}) \tag{14.15}$$

As above, ψ is occupancy, here of species i at site k. Occupancy ψ_{ik} is multiplied by community membership w_i because z_{ik} will be 1 with occupancy probability ψ_{ik} if species i can be present in the community; if not, then z_{ik} will always be 0. Thus, if $w_i = 1$, then $P(z_{ik} = 1 \mid w_i = 1) = \psi_{ik}$ (the probability of occupancy). Otherwise, if $w_i = 0$, then $P(z_{ik} = 0 \mid w_i = 0) = 1$.

Second, to estimate ψ, which is a function of some set of site- or time-specific covariates, we use Equation 14.11a. Recall from Equation 14.5, however, that the estimates of ψ_{ik} are also dependent on detection probability p_{ik}. If we assume that each of the ith species of ant present at the kth site has equal probability of being captured in any of the J_k pitfall traps at the two sampling dates ($J = 1, \ldots, 50$), then their probability of captures (= detection) in a pitfall trap can be modeled as:

$$y_{ik} \mid z_{ik} \sim \text{Binomial}(J_k, z_{ik} p_{ik}) \tag{14.16}$$

Once again, note that p_{ik} is a conditional probability of capture, given that the species is there. If it is absent from the site, it won't ever be captured.

Finally, we assume that capture probabilities are not dependent on site- or time-based covariates. So instead of using Equation 14.11b, we model (on the **logit scale**) the probability of capture of each species as a constant:

$$\text{logit}(p_{ik}) = a_{0i} \tag{14.17}$$

Equation 14.17 specifies that each species has a different (but constant) detection probability, but how different? As noted above, we assume for this analysis that ant species occurrences are independent of one another and not strongly affected by species interactions. We also assume that all ant species have "similar" behaviors so we can take advantage of information about the "average" ant to make assumptions about capture probability of rare ants.[7]

[7] Such borrowing of information is a compromise between estimating a separate parameter for each species and estimating a single parameter for all species. We don't have enough data for estimating all the parameters for the very rarest species, but we have enough data for many of them. Thus, we "shrink" the estimates of each species' parameters toward their average value as a function of the amount of information that we actually have (Gelman et al. 2004).

To complete the hierarchical model, there is variation in detection probabilities (a_{0i}) and variation in occurrence probabilities (b_{0i}), each of which we can treat as a normal random variable:[8]

$$\begin{bmatrix} b_{0i} \\ a_{0i} \end{bmatrix} \sim \text{Normal}\left(\begin{bmatrix} \beta_0 \\ \alpha_0 \end{bmatrix}, \begin{bmatrix} \sigma_{b0}^2 & \rho\sigma_{b0}\sigma_{a0} \\ \rho\sigma_{b0}\sigma_{a0} & \sigma_{a0}^2 \end{bmatrix} \right) \quad \textbf{(14.18a)}$$

$$b_{li} \sim \text{Normal}\left(\beta_1 \sigma_{b_l}^2\right) \quad \textbf{(14.18b)}$$

In Equation 14.18a, the σ's are the true magnitudes of the variation in the detection probability (σ_{a0}^2) and occupancy (σ_{b0}^2) of each species at the average level of the covariates, and ρ is the covariance between species occurrence and detection probability. From Equation 14.18b, we assume no covariance among the $l = 1 \ldots p$ covariates.

To summarize this hierarchical model, we are interested in estimating the **marginal probability** of the actual observations y_{ik}, given all the unknowns:

$$p\left(y_k \mid w_i, \mu_\psi, \mu_p, \sigma_a^2, \sigma_b^2, \rho\right) \quad \textbf{(14.19)}$$

where the μ's are the average occupancy and detection probabilities across the sites. To estimate y_k, we need to integrate the likelihood for each of the i species, $L\left(y_{ik} \mid \psi_i, p_i\right)$ over the joint distribution of a and b (see Footnote 7 in Chapter 5 for additional discussion of the likelihood function $L(\cdot)$). In the end, we estimate the total number of species, $\hat{N}$, as:

$$\hat{N} = \frac{x}{1 - p\left(0 \mid \hat{\mu}_\psi, \hat{\mu}_p, \hat{\sigma}_a^2, \hat{\sigma}_b^2, \hat{\rho}\right)} \quad \textbf{(14.20)}$$

[8] Where did that b_{0i} term come from? We estimate ψ as function of site- and time-specific covariates (see Equation 14.11a). If we expand that equation, we have logit(ψ_{ik}) = $b_{0i} + b_{1i}x_{1k} + \ldots + b_{pi}x_{pk}$, where b_{0i} is the intercept for species i, and b_{pk} is the parameter for the effect of covariate x_i on the probability of occurrence of species i. If we center and scale the data with a Z-score (see Chapter 12), then b_{0i} can be interpreted as the probability of occurrence of species i at the average value of the covariates.

[9] Equation 14.20 is solved by using the two parts of the matrix in Table 14.5: the actual observations (the $n \times s$ matrix of y_{ik}s) and the estimated parameters for the unknown community of size M. We then estimate N from Equation 14.14.

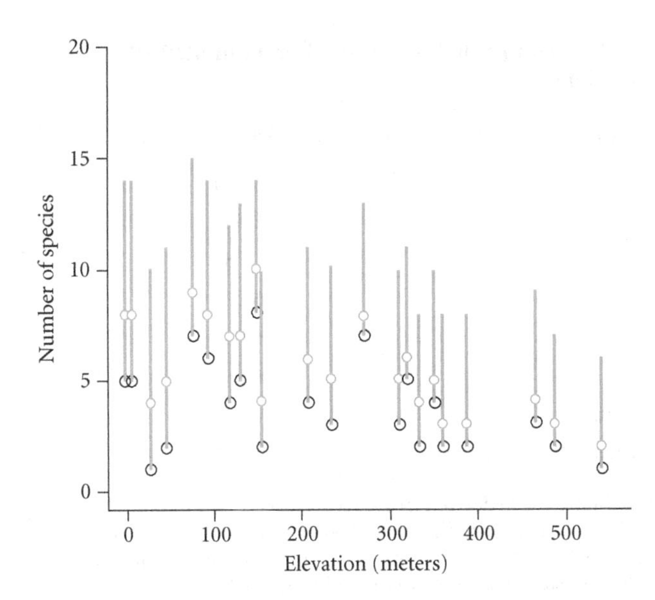

Figure 14.4 Observed (black open circles) and estimated (blue open circles) species richness of ants in each of 22 New England bogs as a function of their elevation. Vertical lines are 95% credible intervals (see Chapter 5) from the full hierarchical model (see Equation 14.20). (After Dorazio et al. 2011.)

The denominator of Equation 14.20 is the probability of detecting any average species in the community.[9]

All of this is accomplished using numerical integration in a Bayesian framework (see Chapter 5).[10] Our analysis identified two models that fit the data equally well: a simple model with no covariates affecting detection probability or occupancy, or the model with only bog elevation as a covariate of occupancy (Table 14.6). This model provides an estimate of ant species richness in New England bogs ($\hat{N} = 25$), as well as estimates of species richness for ants at each bog (Figure 14.4) as a function of the key covariate (elevation). It is interesting to compare these estimates to the much simpler Chao2 asymptotic estimators for the same data. Both estimators give similar results, but the estimates from the hierarchical model are usually greater than those of the Chao2 estimator (Figure 14.5). This difference probably reflects the fact that the hierarchical model takes into account site- and species-specific differences in occupancy and detection probability.

The hierarchical model also yields estimates of occupancy and detection probability for each species. For *Dolichoderus pustulatus*, the estimated capture probability was 0.09, and the estimated occupancy was 0.70. These estimates are both somewhat lower than those from the basic model with only a single species

[10] Model code for solving this is given in the Data and Code section of the book's website, as well as in Dorazio et al. (2011).

TABLE 14.6 Posterior probabilities of different models fit to the bog ant data

Covariates included in model	Posterior probability of the model
Elevation	0.424
None	0.342
Latitude	0.082
Area + Elevation	0.060
Latitude + Elevation	0.045
Area	0.038
Latitude + Area	0.006
Latitude + Area + Elevation	0.004

Posterior probabilities were derived from five independent Markov chains, each run for 250,000 iterations. The first 50,000 iterations were discarded, and the remaining 200,000 observations were thinned by taking only every 50th point. These posterior samples were used to estimate model parameters and 95% credible intervals. Uninformative priors were used for all of the parameters. Counts and covariances had uniform priors: $\Omega \sim$ Uniform(0,1), $N \sim$ Uniform to each integer $\{0, 1, \ldots, M\}$, $\rho \sim$ Uniform(–1,1). Variances were assigned prior probabilities from a half-Cauchy distribution, $f(\sigma) = 2/[\pi(1 + \sigma^2])$ (Gelman 2006). Logit-scale covariates (α_0, β_0, β_1) were assigned prior probabilities from a t-distribution with $\sigma = 1.566$ and $\upsilon = 7.763$, which approximates a Uniform(0,1) distribution for p within the interval (–5,5) (Gelman et al. 2008). Alternative uninformative priors (e.g., Jeffrey's prior) gave similar results (Dorazio et al. 2011).

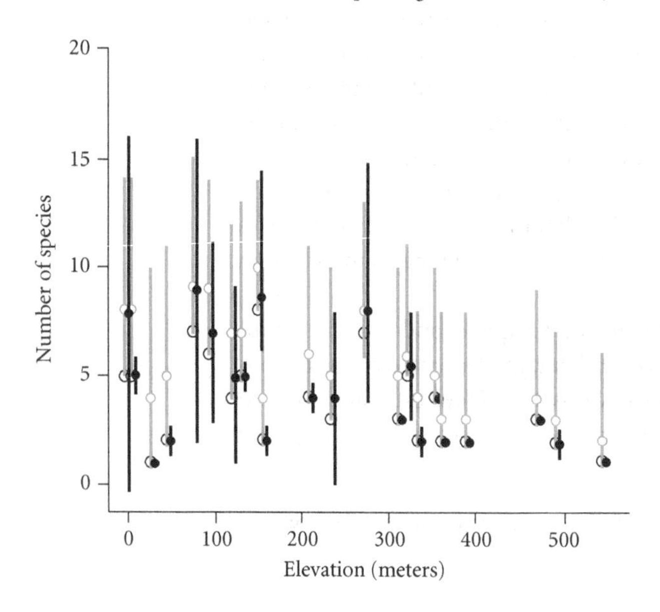

Figure 14.5 Comparison of observed number of species in each bog (black open circles) with Chao2 estimates of species richness (black filled circles), and estimates from the hierarchical model that account for variation in occupancy and detection probability (blue open circles). Black and blue vertical lines indicate 95% confidence intervals for Chao2 and for the hierarchical model estimator of species richness, respectively.

and no hierarchical structure (see Table 14.2). The full results are reported in Dorazio et al. (2011).

Occupancy Models for Open Populations

One of the most important assumptions of the occupancy models discussed above, and for the rarefaction and extrapolation estimates of species richness discussed in Chapter 13, is that the population is closed to immigration, emigration, and extinction. In contrast, important community and population models, such as MacArthur and Wilson's (1967) theory of island biogeography and Levins's (1969) **metapopulation** model, describe open systems with migration. However, occupancy in classic mathematical models in community and population ecology was either not specified or else ambiguously defined, and, until recently, empirical tests of these models did not account for detection errors.

As we have seen above, incorporating variation in detection probability can alter estimates of occupancy for a single species or species richness for a community. Relaxing the closure assumption of the basic, single species model and at the same time accounting for local colonization and extinction (Figure 14.6) can extend the reach of site occupancy models. Occupancy models of open populations are called **dynamic occupancy models** because they allow for changes in occupancy as a function of time. Dynamic occupancy models are analogous to metapopulation models (Hanski 1991), because, at each site, the measured occurrence of a population is a function of occupancy, local colonization, and local extinction. The addition of detection probability can yield new insights, although the models are complex and require lots of data to estimate their parameters (e.g., Harrison et al. 2011).

At a minimum, a dynamic occupancy model requires samples at replicate sites. Each site must then be sampled at two or more time periods (**seasons**) between which local colonization and extinction could occur. As in the simple occupancy model, we are considering only presence/absence (occurrence) data, such as annual surveys for insect infestations. Here we illustrate dynamic occupancy models with results from a long-term survey of forest stands for

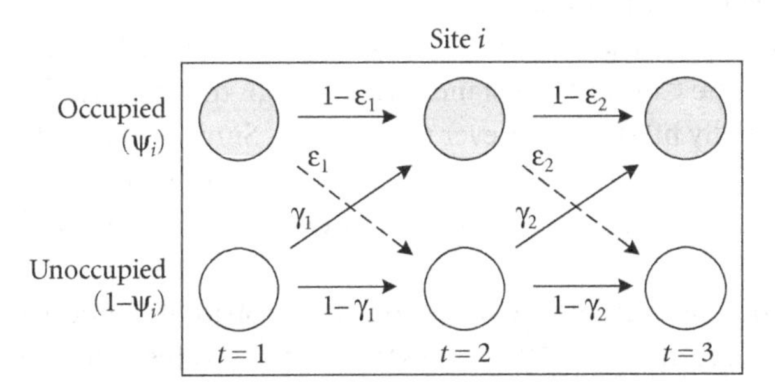

Figure 14.6 Effects of colonization (γ) and extinction (ε) on occupancy of a given site. An occupied site can remain occupied if a species does not go extinct locally or if it is colonized or recolonized between time intervals. An occupied site can become unoccupied if is a species dies out or emigrates and the site is not recolonized between time intervals.

TABLE 14.7 **Sample data (10 of 62 sites) of the occurrence of hemlock woolly adelgid surveyed every two years since 2003 in central and western Massachusetts**

Sampled stand	2003	2005	2007	2009	2011
Athol #1	0	0	0	1	1
Belchertown #1	1	1	1	1	1
Bernardston #4	0	0	1	1	1
Easthampton #4	0	1	1	0	1
Greenfield #4	1	0	1	1	1
Hampden #4	1	1	1	1	1
Orange #5	0	0	0	1	0
Shutesbury #3	1	0	1	1	1
Warren #1	0	0	1	1	1
Winchendon #3	0	0	0	1	0

Cell entries indicate presence (1) or absence (0) of the adelgid. Only ten lines of the data are shown here; the full data set is available from the Data and Code section of the book's website. Initial analyses of the adelgid data, along with data collected in Connecticut beginning in 1997, were published by Preisser et al. (2008, 2011).

the presence of the hemlock woolly adelgid (Table 14.7), the non-native insect species that motivated the hemlock canopy manipulation experiment described in Chapter 13.

Beginning in 2003, the occurrence of the adelgid was surveyed biennially in 140 hemlock stands across a 7500 km^2 transect in southern New England, an area extending from Long Island Sound in southern Connecticut to the Vermont border of north-central Massachusetts (Preisser et al. 2008, 2011). In each stand, multiple branches were sampled on each of 50 hemlocks, and the adelgid's presence and density (in logarithmic categories) were recorded. We illustrate dynamic occupancy models only for the Massachusetts portion of the data set, because densities of the adelgid in the Connecticut stands were so high throughout the sampling period that virtually no stand was ever unoccupied. Similarly, we only explore temporal variability in the model's parameters; effects of stand-level (site-level) covariates on adelgid density are explored by Orwig et al. (2002) and Preisser et al. (2008, 2011).

The parameters to estimate from these data include the probability of occupancy at time t (ψ_t), the colonization (γ_t) and extinction (ε_t) rates, and the changes in these rates through time (e.g., $\gamma_t = (\psi_{t+1} / \psi_t)$. The basic occupancy model (see Equation 14.10) is expanded to a dynamic one with two key steps (MacKenzie et al. 2003). We observe that if we know occupancy at time (season) $t = 1$ (ψ_1), we can estimate it at season $t = 2$ recursively as:

$$\underbrace{\psi_2}_{\text{occupied in season 2}} = \underbrace{\psi_1(1-\varepsilon_1)}_{\substack{\text{occupied in season 1}\\ \text{and did not go extinct}\\ \text{between season 1 and season 2}}} + \underbrace{(1-\psi_1)}_{\substack{\text{not occupied}\\ \text{in season 1}}} \times \underbrace{\gamma_1}_{\substack{\text{colonized between}\\ \text{season 1 and season 2}}} \tag{14.21}$$

The likelihood for ψ_1 is:

$$L(\psi_1,\varepsilon,\gamma,\mathbf{p}|\mathbf{X}_1,\ldots,\mathbf{X}_s) = \prod_{i=1}^{s} \Pr(\mathbf{X}_i) \tag{14.22}$$

where

$$\Pr\left(X_i = \phi_0 \prod_{t-1}^{T-1} D(\mathbf{p}_{X,t})\phi_t \mathbf{p}_{X,T}\right) \tag{14.23a}$$

$$\phi_0 = [\psi_1 \;\; 1-\psi_1] \tag{14.23b}$$

$$\phi_t = \begin{bmatrix} 1-\varepsilon_1 & \varepsilon_1 \\ \gamma_1 & 1-\gamma_1 \end{bmatrix} \tag{14.23c}$$

and

$$\mathbf{p}_{x=1,t} = \begin{bmatrix} p_t \\ 0 \end{bmatrix} \text{ (if site is occupied)} \tag{14.23d}$$

$$\mathbf{p}_{x=0,t} = \begin{bmatrix} (1-p_t) \\ 1 \end{bmatrix} \text{ (if site is unoccupied)} \tag{14.23e}$$

In this set of equations, $\mathbf{X}_i$ is the detection history (see Equations 14.8 and 14.9) at site i at each of the t seasons; γ and ε are the vectors of probabilities of local colonization, extinction, and detection at each of the seasons (Figure 14.7); and $\mathbf{p}_{X,t}$ is a column vector of detection probability conditional on occupancy state.[11]

[11] In a more general model that we do not consider here, each site would be sampled multiple times within seasons (see Figure 14.9 and our discussion of mark-recapture models for open populations), and the first entry of the column vector in Equation 14.23d would be the probability of the detection history *within* a season (e.g., Equations 14.8 and 14.9). See MacKenzie et al. (2003, 2006) for additional discussion.

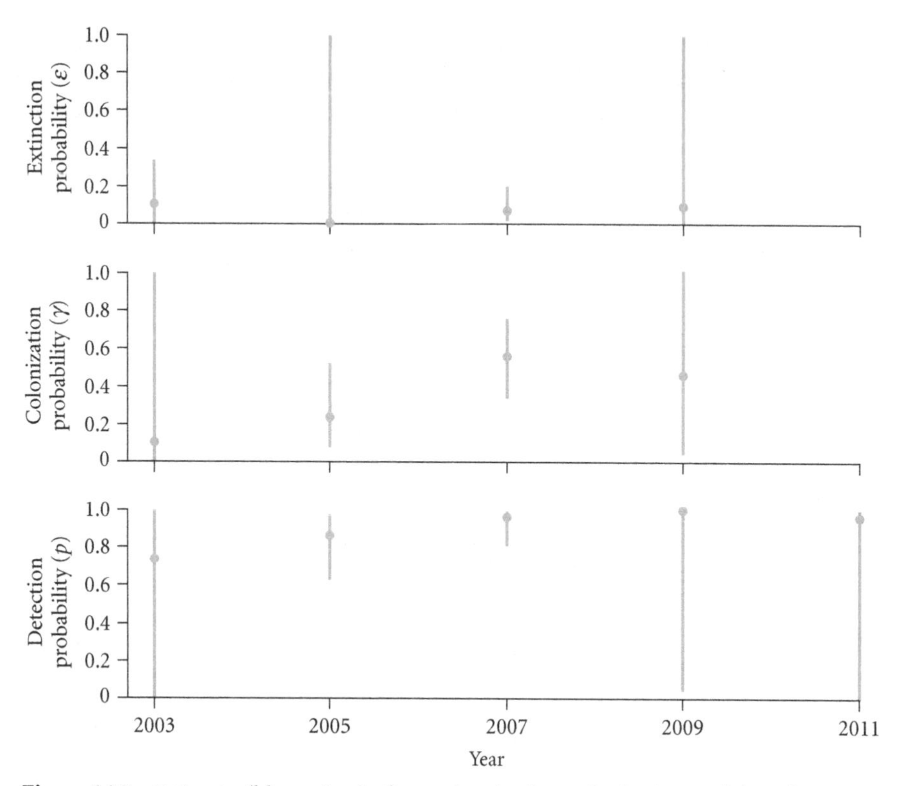

Figure 14.7 Estimates (blue points) of annual extinction, colonization, and detection probabilities of the hemlock woolly adelgid at 62 sites in Massachusetts. Vertical lines are asymmetric 95% confidence intervals. Note that estimates of extinction and colonization are only made through 2009, as 2011 estimates would only be available after the 2013 census (see Figure 14.6). Detection probability, however, can be estimated for every census.

Equations 14.22 and 14.23 assume that each parameter is the same across all sites at any given time, but variation in the parameters can be accounted for by including a matrix of covariates, as in the closed population models (see Equation 14.11):

$$\theta = \frac{\exp(\mathbf{Y}\boldsymbol{\beta})}{1+\exp(\mathbf{Y}\boldsymbol{\beta})} \tag{14.24}$$

In Equation 14.24, $\boldsymbol{\theta}$ represents one of the parameters of the model (ψ, γ, ε, or p), $\mathbf{Y}$ is the matrix of site-specific covariates, and $\boldsymbol{\beta}$ is the column vector of regression coefficients to be estimated (see Equation A.14 in the Appendix).

Dynamic Occupancy of the Adelgid in Massachusetts

We used Equations 14.22 and 14.23 to fit two dynamic occupancy models to the adelgid survey data in Massachusetts (see Table 14.7).[12] The first model assumed that all parameters were constant across sites and seasons, whereas the second model assumed that all parameters varied through time.

Point estimates for the model with constant parameters (Table 14.8) suggest that approximately one-half of the sites were occupied in 2003 ($\hat{\psi}_1 = 0.47$; the observed occupancy was 0.45), about 40% of the sites are colonized each year ($\hat{\gamma} = 0.40$); a site once colonized rarely loses the adelgid ($\hat{\varepsilon} = 0.06$); and that detection probability, conditional on a site being occupied, is reasonably high ($\hat{p} = 0.95$).[13] When we fit a model with parameters that varied among years, there was substantial year-to-year variability and large variance in some parameters, which is often the sign of a poorly fitted model. Comparison of the two models confirms that the simple model with constant parameters fit the data much better than the more complex model with time-varying parameters (AIC = 326.3 versus AIC = 332.8).

TABLE 14.8 **Results of an initial dynamic occupancy model for the hemlock woolly adelgid data**

Parameter	Estimate	95% confidence interval
Occupancy $\hat{\psi}$	0.47	0.34–0.60
Colonization $\hat{\gamma}$	0.38	0.28–0.49
Extinction $\hat{\varepsilon}$	0.06	0.02–0.18
Detection $\hat{p}$	0.95	0.80–0.99

Data used in these models can be found in Table 14.7. Estimated occupancy is shown only for the first season (2003); estimates of ψ for subsequent years would be determined from the recursive formula, Equation 14.21.

[12] We used the colext function in the R library unmarked (Fiske and Chandler 2011) to fit these models. See the Data and Code section of the book's website for the R code.

[13] In a study in which we actually modeled detection probability from multiple surveys at a single site in a single year, using Equation 14.10, we estimated detection probability for experienced observers, such as those who collected the data in Table 14.7, to range from 0.78 to 0.94 (Fitzpatrick et al. 2009).

Estimating Population Size

Now that we have a modeling framework to estimate occupancy and detection probability, we are ready to estimate population size. There are three basic options. First, we could count every individual in a single census of the population, assuming we could reliably detect them.[14] Alternatively, we could resample a population at different times, which allows for estimates of detection probability and population size. Multiple samples come in two basic forms. In mark-recapture data (and its relatives, such as mark-resight and band-return; see Pollock 1991), individuals are marked or tagged so that subsequent censuses can distinguish between marked individuals (which have been previously counted) and unmarked ones (which have not). In unmarked censuses, populations are repeatedly surveyed, but there is no way to distinguish new individuals from those that may have been previously captured. Mark-recapture data are far more widespread, and are widely used in studies of birds, mammals, and fish. Estimating abundance from unmarked populations requires far more restrictive assumptions, and provides less reliable estimates of population sizes, but it may be the only option for censuses of insects, plankton, microbes, and other organisms that cannot be easily tagged.

[14] Unfortunately, this is the path taken by many national census bureaus in regular population censuses. In the U.S., a census every 10 years is mandated by Congress. The results are inherently controversial because they are used to apportion seats in the House of Representatives, as well as to allocate dollars for many federally funded projects. The first such census was conducted in 1790, with an estimated population size $\hat{N} = 3{,}929{,}326$. However, now that the U.S. population is over 300 million, it is not realistic to try and count everyone. Direct counting will systematically underestimate the sizes of many large groups of people such as the homeless, migrant workers, and others who do not maintain a place of "usual residence." Stratified sampling is a better strategy for obtaining an unbiased estimate, but even this is difficult to properly implement because of the huge spatial variation in population density. It may take much more effort to accurately estimate the human population size within a square mile of Manhattan that to estimate it within an entire county of North Dakota. As we describe in this chapter, mark-recapture methods are powerful. However, they present significant ethical dilemmas for use in estimating human population size. Nonetheless, progress is being made, beginning with mark-recapture studies of at-risk populations of intravenous drug users, alcoholics, and homeless individuals (Bloor 2005).

Mark-Recapture: The Basic Model

One of the earliest applications of probability theory to an important ecological question was to estimate population sizes of animals destined for the dinner table (Petersen 1896).[15] Petersen was interested in the number of plaice (*Pleuronectes platesa*), a flounder-like flatfish caught in the Limfjørd waterway, an approximately 200-km long fjord that cuts across northern Jutland in Denmark.[16]

Petersen's estimator of population size was

$$\hat{N} = \frac{n_1 n_2}{m_2} \tag{14.25}$$

where n_1 is the number of individuals captured at time $t = 1$, all of which were marked when they were captured; n_2 is the number of individuals captured at time $t = 2$; and m_2 is the number of individuals captured at time $t = 2$ that had

[15] Carl Georg Johannes Petersen (1860–1928) was a Danish biologist who studied fish and fisheries. He invented the eponymous Petersen disc-tag for marking fish; the earliest versions were made of bone or brass and followed from his efforts to mark fish simply by punching one or two holes in their dorsal fins. The Petersen tag is actually two discs, one attached to each side of the body and connected to each other by a metal wire or pin pushed through the fish's dorsal fin or body. Modern versions are made of plastic.

Carl Georg Johannes Petersen

[16] This is also an interesting example of an early study of an accidental species introduction. In the same article in which he describes the Petersen disc for tagging fish and develops the Petersen mark-recapture estimator, Petersen wrote that "it must be remembered that this fish is a newcomer to the Fjord, as it was not found there before breaking through of the German Sea in the beginning of this century" (Petersen 1896: 5). In 1825, a flood caused the Limfjørd to break through a series of sand bars and meandering channels into the North Sea. This event also changed the western waters of the Limfjørd from fresh to brackish, altering its biota. By 1895, Limfjørd's plaice fishery alone was worth 300,000 Danish Kroner (approximately US $350,000 in 2012), but by 1900, it was already considered overfished (Petersen 1903).

been marked at time $t = 1$.[17] Frederick C. Lincoln (1930) independently came up with this same formula for estimating the population size of waterbirds.[18] Equation 14.25 is now known as the Lincoln-Petersen estimator.[19] Chapman

[17] This formula can be derived in several different ways. If the first sample yields n_1 individuals out of an unknown total population of size N, then the proportion of the total population sampled is simply n_1/N. If the second sample yields n_2 individuals, m_2 of which were captured and marked the first time, *and if* we assume that all individuals are equally likely to be sampled in either of the two censuses, then the proportion of marked individuals captured in the second census should equal the proportion of total individuals captured in the first census: $\frac{n_1}{N} = \frac{m_2}{n_2}$. Since n_1, n_2, and m_2 are all known, we can solve for N (technically, $\hat{N}$) using Equation 14.25.

Alternatively, we simply estimate the probability of capturing an individual $\hat{p}$ in Equation 14.1 as m_2/n_2 and use Equation 14.2 to estimate $\hat{N} = \frac{n_1}{\hat{p}} = \frac{n_1 n_2}{m_2}$.

Finally, we can use probability theory (see Chapters 1 and 2) to come up with the formal probability distribution for a mark-recapture study with two sampling periods:

$$P(n_1, n_2, m_2 \mid N, p_1, p_2) = \frac{N!}{m_2!(n_1 - m_1)!(n_2 - m_2)!(N - r)!} \times$$
$$(p_1 p_2)^{m_2} \left[p_1(1-p_2)\right]^{n_1 - m_2} \left[(1-p_1)p_2\right]^{n_2 - m_2} \left[(1-p_1)(1-p_2)\right]^{N-r}$$

The terms in this equation represent the number of individuals captured at time 1 (n_1) and time 2 (n_2), and the number of individuals captured at time 2 that had also been captured and marked at time 1 (m_2). All of these quantities are known, but they are conditional on the total population size (N) and the probability of capture at time 1 and time 2 (p_1 and p_2, respectively); r is the total number of individuals captured ($r = n_1 + n_2 - m_2 \leq N$). This equation assumes that a capture is a binomial random variable (see Equation 2.3 in Chapter 2). A unique probability is assigned to individuals captured (p) or not captured ($1-p$) at each sample (subscripts 1 or 2). The Lincoln-Petersen estimate of $\hat{N}$ (see Equation 14.25) is the maximum likelihood estimate of the probability distribution given above when $p_1 = p_2$ (see Footnote 13 in Chapter 5 for additional discussion of maximum likelihood estimators).

[18] In addition to reinventing the Petersen method for estimating population size, Frederick Charles Lincoln (1892–1960) developed the concept of the "flyway," which undergirds all regulations for conservation and management (including hunting) of migratory birds. As a long-time employee of the U.S. Bureau of Biological Survey, Lincoln organized the continental-scale bird-banding program, which he ran through 1946 (Gabrielson 1962; Tautin 2005).

Frederick Charles Lincoln

[19] The earliest known formulation of Equation 14.25 was attributed by Pollock (1991) to Laplace (1786). Laplace used this equation to estimate the population size of France in 1793 to be between 25 and 26 million people (see Footnote 14 in Chapter 2 for more information about Pierre Laplace). It was also derived independently by Jackson (1933) while he was studying tsetse flies in Tanganyika (what is now Tanzania).

(1951, after Schnabel 1938) proposed a slight modification of Equation 14.25 to minimize its bias:

$$\hat{N} = \frac{(n_1+1)(n_2+1)}{(m_2+1)} \tag{14.26}$$

Chapman (1951) also provided an estimator for its variance:

$$\text{var}(\hat{N}) = \frac{(n_1+1)(n_2+1)(n_1-m_2)(n_2-m_2)}{(m_2+1)^2(m_2+2)} \tag{14.27}$$

We illustrate the basic Lincoln-Petersen model with a simple example of estimating the number of pink lady's slipper orchids (*Cypripedium acaule*) in a small patch (approximately 0.5 hectare) of Massachusetts woodland.[20] The data were collected as two plant ecologists walked a few hundred meters along a circular trail on three successive days in May, 2012. Along this trail, every orchid that was encountered was marked and whether or not it was flowering was noted.[21] The data—spatial maps of recorded plants—are illustrated in Figure 14.8. It is immediately apparent that different numbers of plants were found each day, and that not every plant was "recaptured" on the second or third visit. Based on these data, what is the best estimate of the number of lady's slipper orchids present along the trail?

---

[20] Estimation of population sizes from mark-recapture data is almost always applied to animals, but there is no reason we can't also apply these models to plants. Although our intuition tells us that animals are harder to find or see than plants, and that a plant, once found, is unlikely to move or be lost from a sample plot, it turns out that many plants have different life stages, including dormant corms and underground tubers. Not only may these hidden life stages be difficult to track, but even very sharp-eyed botanists and naturalists can regularly overlook conspicuous flowering plants they saw only one day before! We therefore illustrate population estimation from mark-recapture data of plants to encourage plant ecologists and botanists to catch up with their zoologically inclined colleagues and use mark-recapture models (Alexander et al. 1997).

[21] Plants were "marked" by recording their location on a hand-held GPS. For individual plants separated by more than 1 m, the GPS coordinates were sufficiently accurate to determine whether an orchid was "recaptured" the following day. For plants in small clumps, the GPS coordinates were not accurate enough to clearly identify recaptures, so those plants have been permanently tagged for future censuses. The actual motivation for these orchid walks was to establish a long-term monitoring study, using a robust design to account for detection errors (E. Crone, personal communication). Perhaps we'll have the results in time for the third edition of this book!

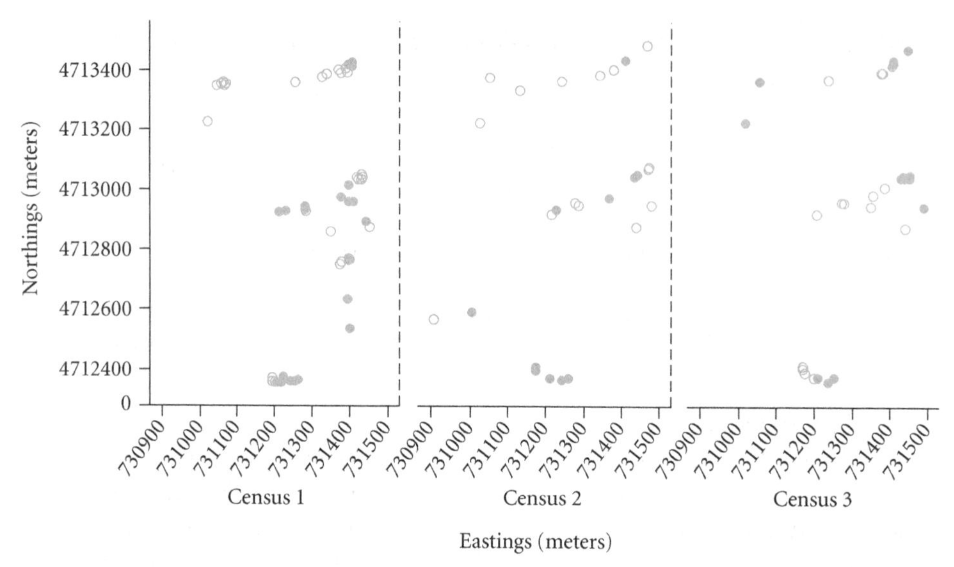

Figure 14.8 Maps of pink lady's slipper orchids marked and recaptured along a trail at Harvard Forest on three successive days in May, 2012. The x and y coordinates are in meters. Open circles are non-flowering plants and solid circles are flowering plants.

The Lincoln-Petersen model requires only two sampling events; thus we begin the analysis of this example by using only the first two days of orchid mark-recapture data. Before we can analyze the data, we have to organize them, which is done in a manner similar to organizing occupancy data. In files of mark-recapture data, each row portrays the detection history of a single individual: the row is a series of ones and zeros, where a 1 indicates that the individual was captured and a 0 indicates that it was not. Time proceeds from left to right, so the first number in the row corresponds to the first sampling time and the second number in the row corresponds to the second sampling time. Table 14.9 illustrates a subset of the capture/recapture history of the plants observed on the first two sampling dates (a total of 214 plants were located and marked over the full three-day period).

In total, 161 plants were found and marked on the first day and 133 plants were found on the second day. Of the latter, 97 of them had been marked on the first day and 36 were new finds on the second day. Solving Equation 14.25, with $n_1 = 161$, $n_2 = 133$, and $m_2 = 97$

$$\hat{N} = \frac{161 \times 133}{97} = 220.7$$

results in an estimate of approximately 221 plants (the less-biased version, Equation 14.26, gives an estimate of 221.5, with a variance, from Equation 14.27, of 52.6).

As with the application of any statistical model, several crucial assumptions must be met for the Lincoln-Petersen estimator to be valid and unbiased. The first assumption, and the one that distinguishes two broad classes of mark-recapture models for estimating population size, is that the population is closed to gains or losses of new individuals. Although most populations change in size through time, models such as the Lincoln-Petersen estimator assume that the time interval between consecutive censuses is short relative to the lifespan of the organism. During that interval, it is assumed that no individuals were born or immigrated into the population, and that no individuals died or emigrated from the population. Alternative models for open populations, such as the Jolly-Seber model, are discussed later in this chapter.

The second assumption is that the marks or tags are not lost in between the two censuses and that if a marked animal is recaptured its tag will indeed be seen again. Any ecologist who has tried to "permanently" mark even stationary organisms, such as by nailing numbered metal tags to trees, appreciates that this assumption may be reasonable only in the short term.

TABLE 14.9 **Initial capture histories of 11 (of 214) lady's slipper orchids**

Plant ID number	Capture history	Flowering?
1	01	Yes
2	01	No
3	11	Yes
4	11	Yes
5	11	Yes
6	10	Yes
7	10	Yes
8	11	Yes
9	11	Yes
10	10	Yes
...	...	...
214	10	Yes

Each orchid found was assigned a unique identification number. The capture history is a string of zeros and ones in which the first digit indicates whether it was found (1) or not found (0) on the first sampling date, and the second digit indicates whether it was found (1) or not found (0) on the second sampling date. There are four possible capture histories: (1,1)—a plant was found on both the first and second sampling dates; (1,0)—a plant was found on the first sampling date but not on the second sampling date; (0,1)—a plant was found on the second sampling date but not on the first sampling date; and (0,0)—a plant was found neither on the first sampling date nor on the second sampling date.

The third assumption is that capture probability is the same at each sample for every individual in the population. In other words, you are just as likely to find and capture an individual at time $t = 1$ as you are to find, capture, or recapture an individual at time $t = 2$. This assumption rarely can be met. The probability of capture may depend on characteristics of each individual, such as its size, age, or sex. Some animals change their behavior once they are captured and marked. They may become trap-happy, returning day after day to the same baited trap for a guaranteed reward of high-energy food, or they may become trap-shy, avoiding the area in which they were first trapped. The probability of capture also may depend on external factors at the time the population is sampled, such as how warm it is, whether it is raining, or the phase of the moon. Finally, the probability of capture may be a function of the state of the researcher. Are you well-rested or tired? Are you confident in, or unsure of, your abilities to catch animals, sight plants, identify their species and sex, and find their tags? Do you trust in, or doubt, the reliability and repeatability of your own numbers?

Incorporating any or all of these sources of heterogeneity can lead to a seemingly endless proliferation of different mark-recapture models.[22] We fit five different closed-population models to the full data set of orchids sampled on three successive days (Tables 14.10 and 14.11).[23] In the first model, we assumed that the probability of finding an individual was the same for all plants at all dates ($p_1 = p_2 = p_3 \equiv p$). In the second model, we assumed that the probability of finding a plant differed depending on whether it was flowering or not, but not on what day it was sampled ($p_{\text{flowering plant}} \neq p_{\text{non-flowering plant}}$). In the third model, we assumed that the probability of finding a plant differed depending on whether a plant had been seen before (i.e., the probability of initial detection [p] and relocation [c] were different). In the fourth model, we assumed that the probability of finding a plant was different each day ($p_1 \neq p_2 \neq p_3$) but the probabilities of initial detection and relocation were the same ($p_i = c_i$). Finally, in the fifth model, we assumed different probabilities of finding a plant each day *and* that the probabilities of initial detection and relocation differed each day, too

[22] For example, version 6.2 (2012) of the widely-used program MARK (White and Burnham 1999) includes 12 different models for mark-recapture data that assume closed populations. These range from simple likelihood-based estimation without heterogeneity to models that allow for misidentification of individuals or their tags and different types of heterogeneity in capture probabilities. Similarly, the native R package Rcapture (Baillargeon and Rivest 2007) can fit eight closed-population mark-recapture models that depend on variation in capture probabilities through time, among individuals, and as a function of change in the behavior of individuals (e.g., trap-happiness).

[23] Each of these models estimates population size by maximizing the likelihood in which the total population size N does *not* enter as a model parameter. Rather, the likelihood is conditioned on only the individuals actually encountered (the value of r in Footnote 17). In these models, originally developed by Huggins (1989), we first maximize the likelihood $P(\{x_{ij}\ldots\} \mid r, p_1, p_2, \ldots)$, where each x_{ij} is a capture history (as in Table 14.10), r is the total number of individuals captured, and the p_is are the probabilities of capture at each sample time t. We first find estimators $\hat{p}_i$ for the capture probabilities at each sample and then estimate the total number of individuals N as

$$\hat{N} = \sum_{i=1}^{r} \frac{1}{1-\left[1-\hat{p}_1(x_i)\right]\left[1-\hat{p}_2(x_i)\right]\cdots\left[1-\hat{p}_t(x_i)\right]}$$

The advantage of using this approach is that covariates describing individuals (e.g., their behavior, size, reproductive status) can be incorporated into the model. We cannot add in covariates for individuals that we never captured!

TABLE 14.10 **Complete capture histories of 11 (of 214) lady's slipper orchids**

Plant ID number	Capture history	Flowering?
1	010	Yes
2	010	No
3	111	Yes
4	111	Yes
5	111	Yes
6	100	Yes
7	100	Yes
8	111	Yes
9	111	Yes
10	100	Yes
...	...	...
214	100	Yes

Each orchid found was assigned a unique identification number. The capture history is a string of zeros and ones in which the first digit indicates whether it was found (1) or not found (0) on the first sampling date, the second digit indicates whether it was found (1) or not found (0) on the second sampling date, and the third digit indicates whether it was found (1) or not found (0) on the third sampling date. With three sampling dates there are eight possible capture histories: 111, 110, 101, 100, 011, 010, 001, 000.

TABLE 14.11 **Summary of the capture histories of the lady's slipper orchids**

Capture history	Number flowering	Number not flowering	Probability to be estimated
111	56	34	$p_1c_2c_3$
110	3	4	$p_1c_2(1-c_3)$
101	19	13	$p_1(1-c_2)c_3$
011	9	12	$(1-p_1)p_2c_3$
100	22	10	$p_1(1-c_2)(1-c_3)$
010	8	7	$(1-p_1)p_2(1-c_3)$
001	11	6	$(1-p_1)(1-p_2)p_3$
000	—	—	$(1-p_1)(1-p_2)(1-p_3)$

This is a summary of the seven different observable capture histories (111–001) of the lady's slipper orchids and the one unobservable one (000). The different probabilities that could be estimated for each sample date are p_i, the probability that an orchid is first found on sample date i ($i = \{1, 2, 3\}$) and c_i, the probability that an orchid found on a previous sample date i is relocated on a subsequent sample date ($i = \{2, 3\}$).

($p_1 \neq p_2 \neq p_3$, $p_i \neq c_i$ and $c_2 \neq c_3$). The results of these models and the estimates of the orchid population size are shown in Table 14.12.

The short summary is that the best-fitting model was Model 4, in which the probability of finding a plant differs each day ($0.60 < \hat{p}_i < 0.75$) with an esti-

TABLE 14.12 Estimates of orchid population size based on three-sample mark-recapture models with different assumptions

	Model 1	Model 2	Model 3	Model 4	Model 5
Description	Single probability of finding a plant	Different probabilities for finding flowering and nonflowering plants	Different probabilities for initially finding and then relocating plants	Different probabilities for finding plants each day	Different probabilities for finding plants each day, and different probabilities for initially finding plants and relocating them
Number of model parameters to estimate	1	2	2	3	5
Maximum likelihood parameter estimate (and 95% confidence interval)	$\hat{p} = 0.69$ (0.64, 0.72)	$\hat{p}_{flower} = 0.68$ (0.63, 0.73) $\hat{p}_{non\text{-}flower} = 0.69$ (0.62, 0.75)	$\hat{p} = 0.72$ (0.65, 0.78) $\hat{c} = 0.68$ (0.62, 0.72)	$\hat{p}_{day1} = 0.73$ (0.66, 0.79) $\hat{p}_{day2} = 0.60$ (0.54, 0.67) $\hat{p}_{day3} = 0.73$ (0.66, 0.78)	$\hat{p}_{day1} = 0.67$ (0. 67, 0. 67) $\hat{p}_{day2} = 0.45$ (0.45, 0.45) $\hat{p}_{day3} = 0.39$ (0.39, 0.39) $\hat{c}_{day1} = 0.60$ (0.53, 0.68) $\hat{c}_{day2} = 0.73$ (0.66, 0.78)
AIC	766.29	768.29	767.19	759.90	761.93
Estimated number of flowering plants	132	132	131	132	144
Estimated number of non-flowering plants	89	89	89	89	97
Estimated total population size	221	221	220	221	241

The best-fitting model identified by AIC is shaded in dark blue.

mated population size $\hat{N} = 221$ individuals. But let's work through the model results in detail, from top to bottom.

Model 1 is the simplest model. It is the extension of the Lincoln-Petersen model to more than two samples, and its solution is based on maximizing the likelihood conditional only on the plants actually found, then estimating the population size as described in Footnote 17. This relatively simple model requires estimating only a single parameter (the probability of finding a plant) and it assumes that every plant is equally likely to be found. It should not be surprising that this model yielded an estimate of the total population size (221) identical (within round-off error) to that of the Lincoln-Petersen estimate based on only the first two censuses (221): the Lincoln-Petersen estimate $m_2/n_2 = 0.6917$ and the Model 1 estimate = 0.6851. Both models also suggested the researchers didn't find all of the marked plants on subsequent visits. What were they missing?

Our first thought was that the flowering orchids, with their large, showy, pink petals, were more likely to be found than orchids without flowers; the latter have only two green leaves that lie nearly flat on the ground. The data in Table 14.11 might suggest that it was easier to find flowering orchids—the researchers found 128 flowering orchids but only 86 non-flowering ones—but the parameter estimates in Model 2 falsified this hypothesis: detection probabilities of flowering and non-flowering plants were nearly identical. Assessment of model fit using AIC also revealed that Model 2 fit the data more poorly than did Model 1. Therefore, in the remaining three models, we assumed a single capture probability for flowering and non-flowering plants.

The three remaining models included either different initial and subsequent capture probabilities (Model 3), different probabilities for finding plants on different days (Model 4), or a combination of these two (Model 5). The results of fitting Model 3 suggested that there was no difference in the probability of initially finding an orchid and finding it again; both probabilities were close to 0.70.

In contrast, the results of Model 4 implied that something very different happened on Day 2, when the probability of finding a plant declined to 0.60. The researchers suggested that after Day 1, they felt quite confident in their abilities to find orchids, and so they moved more rapidly on Day 2 than they had on Day 1. Fortunately, they looked at the data immediately after returning from the field on the second day (recall our discussion "Checking Data" in Chapter 8). They discovered that not only had they located only 74% of the plants they had found on the first day, but also that they had found 36 new ones. They therefore searched much more carefully on the third day. This extra effort appeared to pay off: not only were nearly as many plants (160) found on Day 3 as on Day 1 (161), but far fewer new plants (17) were found on Day 3 than on Day 2 (36). Model 4 fit the data better than any of the other models (lowest AIC) and so is considered the most appropriate model.

The final model was the most complex. Like Model 4, Model 5 had separate parameters for probabilities of finding plants each day. Like Model 3, Model 5 also had separate parameters for probabilities of relocating plants once they had been found the first time. Despite having the most parameters of any of the models, Model 5 fit the data most poorly (highest AIC in Table 14.12). A careful examination of the parameter estimates revealed other odd features of this model. The first three parameters we estimated, p_{day1}, p_{day2}, and p_{day3}, all had no variance, so their standard errors and confidence intervals equaled 0. These two results, along with the high AIC and the very different estimator of total population size, suggests that Model 5 is grossly over-fitting the data and is unreliable.[24]

Mark-Recapture Models for Open Populations

If mark-recapture studies extend beyond the lifespan of the organisms involved, or if individuals have the opportunity to enter or leave the plot between samples, then the key closure assumption of closed-population mark-recapture models no longer holds. In open-population models, we also need to estimate changes in population size due to births and deaths, immigration and emigration. These can be pooled into estimates of net population growth: [births + immigrants] – [death + emigrants].

The basic open-population model for mark-recapture data is the Jolly-Seber model, which refers to two papers published back-to-back in *Biometrika* (Jolly 1965; Seber 1965). The Jolly-Seber model makes a number of assumptions that parallel those of closed-population models: tags are not lost between samples, are not missed if a marked animal is recaptured, and are read accurately; each individual has the same probability of being caught at each sampling time, regardless of whether it is marked or not; and all individuals have the same probability of surviving from one sample to another. Finally, although the time between samples does not have to be fixed or constant, samples are assumed to be a "snapshot" that represents a very short time interval relative to the time between samples. Unlike the Lincoln-Petersen estimate or other closed-popu-

[24] As a general rule, when variances of fitted parameters are either very large or very small, or when the parameter estimates themselves seem unreasonable, you should suspect that the model has a poor fit, either because it is biologically unrealistic or because it is being fit with not enough data. In such cases, optimization and fitting algorithms in R and other software packages often perform poorly and generate unstable parameter estimates. This is another good reason to take the time to fit multiple models and make sure the results are well-behaved.

lation mark-recapture models that require only two samples, Jolly-Seber models require at least three samples to estimate population size.

As before (see Equation 14.2), population size at time t is estimated as the quotient of the number of marked individuals available for capture at time t and the proportion of the total population that has been sampled:

$$\hat{N}_t = \frac{\hat{M}_t}{\hat{p}_t} \tag{14.28}$$

Just as we did in the closed-population models, we estimate the denominator in Equation 14.28, p_t, as the ratio of marked animals to total animals caught in a sample (using the unbiased form from Equation 14.26):

$$\hat{p}_t = \frac{\hat{m}_t + 1}{n_t + 1} \tag{14.29}$$

Note that estimating p_t requires at least two samples, because the denominator is the sum of the marked and unmarked individuals captured at time t, and having marked individuals at time t implies a previous round of captures and marking at time $t - 1$.

Estimating the numerator in Equation 14.28 is a bit more complicated. We must include an estimate of survivorship because some of the marked animals may have died or emigrated between samples:

$$\hat{M}_t = \frac{(s_t + 1)Z_t}{R_t + 1} + m_t \tag{14.30}$$

In this equation, m_t is the total number of marked animals caught in the tth sample; R_t is the number of individuals captured at sample t (whether marked earlier or for the first time at this sample) that were then released but subsequently re-captured in a later sample (time $t + 1$ or later); s_t is the total number of individual captured and released at time t (some of these may not be captured ever again); and Z_t is the number of individuals marked before sample t, *not caught* in sample t, but caught in another, later sample ($t + 1$ or later). Note that determining Z_t requires at least three samples.

Naturally, all of the assumptions of the basic Jolly-Seber model rarely will apply. Individuals are likely to vary in their probability of capture or detection either intrinsically (based on age, sex, etc.) or behaviorally (trap-happiness, trap-shyness). Estimating survivorship accurately can be especially problematic

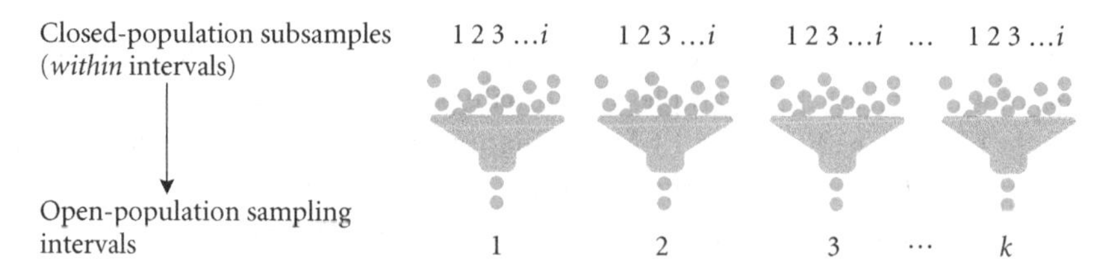

Figure 14.9 Schematic representation of a mark-recapture study design developed by Pollock (1982) that combines the best of open- and closed-population models. Data from closed-population sub-samples are modeled with methods that incorporate individual heterogeneity in capture probabilities to yield estimated numbers of marked animals in each sampling interval. The pooled results from each set of within-interval subsamples are then used in open-population Jolly-Seber-type models to estimate overall population size and population growth rate. This kind of sampling is also used to estimate within-season detection probabilities in hierarchical models of occupancy (see Equations 14.22 and 14.23d).

(Lebreton et al. 1992),[25] and there are dozens of models that can be applied. But if estimation of population size is the goal, a particularly elegant solution is the one proposed by Pollock (1982) that combines the best attributes of both open and closed population models.

In brief, Pollock's approach, illustrated in Figure 14.9, is to take repeated samples within a time interval in which it can be assumed reliably that the population is closed. These repeated samples are then pooled to determine whether an individual has been captured at least once in the larger time interval encompassing the closed-population samples. As we saw in the orchid example, it is straightforward to incorporate heterogeneity in capture probability in closed-population models, but several subsamples are needed for robust results (Pollock 1982 suggests at least five). If this is done over several (at least three) larger sampling intervals, we can then apply the Jolly-Seber model to the pooled subsamples to estimate overall population size and rates of change.

Occupancy Modeling and Mark-Recapture: Yet More Models

Ecologists and statisticians continue to develop occupancy and mark-recapture models for estimating changes in population size and population growth rates that relax assumptions of the classic models, incorporate individual heterogeneity in new ways, and are specific to particular situations (e.g., Schofield and Barker

[25] In terms of the variables used in Equations 14.28–14.30, survivorship from sample t to $t+1$ can be estimated as: $\hat{\phi}_t = \hat{M}_{t+1} / \left(\hat{M}_t - m_t + s_t\right)$. Given this estimate of survival, recruitment (births + immigrants) from sample t to $t+1$ can be estimated as: $\hat{B}_t = \hat{N}_{t+1} - \hat{\phi}_t\left(\hat{N}_t - n_t + s_t\right)$.

2008; McGarvey 2009; Thomas et al. 2009; King 2012). Some of these models use genetic markers (e.g., Watts and Paetkau 2005; Ebert et al. 2010), which are appropriate when it is impossible, or nearly so, to recapture a previously marked individual (e.g., Pollock 1991). Other models are used to compare parameter estimates for different populations. Still others focus on particular aspects of occupancy or mark-recapture processes, such as movement of individuals (e.g., Schick et al. 2008). Software also continues to evolve (e.g., White and Burnham 1999; Baillargeon and Rivest 2007; Thomas et al. 2010), and we can expect new methods for specialized data structures to appear on a regular basis. However, the basic models presented in this chapter for both occupancy analysis and the estimation of population size should match the structure of the majority of ecological data sets that ecologists routinely collect.

Sampling for Occupancy and Abundance

In both this chapter and Chapter 13, we have focused on the estimation of species richness, occupancy, and abundance. We have assumed throughout that the sampling was consonant with the questions of interest and that the data necessary for the analyses were collected appropriately. The standard principles of experimental design presented in Chapter 6 also apply: there should be a clear question being asked; samples should be independent and randomized; and replication should be adequate. As awareness of the importance of estimating occupancy and detection probabilities has increased among ecologists, there has been a parallel increase in attention to how to design studies to provide reliable estimates of parameters while minimizing uncertainty. Entrées to this burgeoning literature begin with Robson and Regier (1964), MacKenzie and Royle (2005), and Bailey et al. (2007); Lindberg (2012) provides a thorough review of this topic.

For studies of closed assemblages (no births, deaths, immigration, or emigration during the study period), MacKenzie and Royle (2005) identified analytically the optimum number of surveys to be done at each site for estimating occupancy and minimizing its variance subject to the constraints that all sites are surveyed an equal number of times and that the cost of the first survey is greater than the cost of all subsequent surveys.[26] Not surprisingly, the optimal

[26] MacKenzie and Royle (2005) also assessed the double sampling design, in which replicate surveys are conducted at only a subset of sites and the remaining sites are surveyed only once. They showed through simulation that a double sampling design is only more efficient than the standard design in which every site is sampled repeatedly when the cost of the first survey is less than the cost of all subsequent surveys. This seems unlikely to occur in reality, and so they conclude that there are few if any circumstances in which a double sampling design would be appropriate.

number of surveys per site declines as detection probability increases. If detection probability ($p > 0.5$), then in most cases three to four surveys per site are sufficient if the cost of the first survey is less than ten times as costly as the subsequent ones (Figure 14.10). But the number of surveys needed rapidly increases as detection probability drops below 0.4, regardless of occupancy. At the extreme, for a detection probability of 0.1, more than 30 surveys per site would be needed even when occupancy is > 0.8. This is especially important to keep in mind when surveying for rare species. MacKenzie and Royle (2005) and Pacifici et al. (2012) provide general guidance: for rare species it is better to sample more sites fewer times each; for common species it is better to sample fewer sites, each more frequently. Other specialized sampling designs can take advantage of information that may be available from other sources (MacKenzie et al. 2005; Thompson 2012).

Estimation of population sizes from closed mark-recapture studies requires a minimum of two samples (one to capture and mark, the other to recapture and resample). However, the number of individuals that must be sampled to obtain an accurate estimate of total population size can be quite large. Robson and Regier (1964) point out that in general, the product of the number of

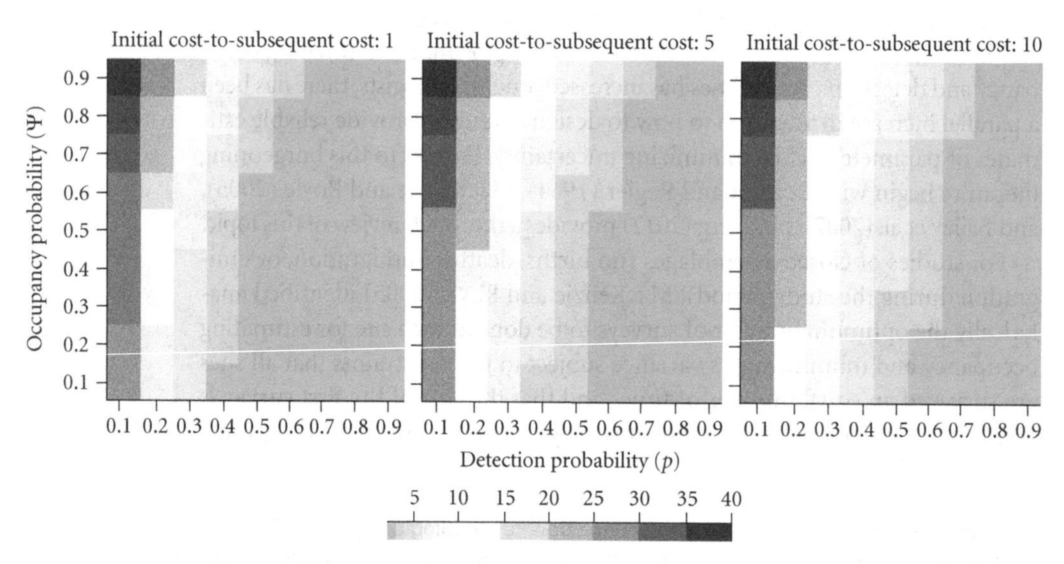

Figure 14.10 Optimal number of surveys per site as a function of occupancy, detection probability, and relative costs of first and subsequent surveys. Shading indicates the number of surveys necessary to minimize the variance in estimates of occupancy. In general, more surveys per site are required as detection probability decreases and occupancy probability increases. (Data from MacKenzie and Royle 2005.)

marked individuals (n_1 in Equation 14.25) and the number of individuals captured at the second sample (n_2 in Equation 14.25) should exceed four times one's initial estimate of the true population size. They provide look-up charts to estimate sample sizes for a desired degree of accuracy in estimating $\hat{N}$. Even for moderately sized populations (hundreds to thousands of individuals), if detection (or recapture) probabilities are $\ll 1.0$, many hundreds of individuals may need to be marked to yield estimates of the population size with reasonable accuracy (± 10%).

There are dozens of sampling designs available for mark-recapture studies of open populations (Lindberg 2012). Typically, a minimum of three samples is needed to estimate all of the necessary parameters (see discussion of Equation 14.30). Different designs are used when animals can be recaptured alive or when only dead individuals are recovered (as in the recovery of banded birds shot by hunters).

Finally, Pollock's (1982) sampling design, combining as it does both within-season (closed) population sampling and between-season (open) population sampling (see Figure 14.9), can be used both in occupancy studies with unmarked individuals and in population studies with marked individuals. More elaborate versions have been developed to study finer-scale demographic changes in populations (reviewed by Lindberg 2012).

Software for Estimating Occupancy and Abundance

For occupancy modeling, the most extensive software package is PRESENCE, maintained by Jim Hines of the US Geological Survey, Patuxent Wildlife Research Center. Hines also developed GENPRES, which simulates presence-absence data that can be used to estimate sample sizes needed and potential biases in experimental studies. The standard software for mark-recapture data is MARK. Originally developed by White and Burnham (1999), the version we used for the orchid example was 6.2, released in spring 2012. These three packages, along with a range of others for more specific occupancy, detection, and population estimation problems, can be accessed through the Patuxent Wildlife Research Center Software Archive (www.mbr-pwrc.usgs.gov/software/). Remi Choquet and his colleagues at the Centre d'Ecologie Fonctionnelle & Evolutive in Montpellier, France, have produced the E-SURGE, M-SURGE, and U-CARE packages for analysis of complex, multistate capture-recapture models (www.cefe.cnrs.fr/en/biostatistics-and-biology-of-populations/software). The R library unmarked can be used to estimate occupancy, detection, and abundance in closed and open populations using hierarchical models. The RMark library provides an R interface to MARK. Both are available through the R project for statistical computing (www.r-project.org). Finally, the web-based forum at www.phidot.org/ is an online community of people working on design, analysis, and software related to occupancy modeling, detection probability, and estimation of population size.

Summary

Detecting species and estimating the sizes of their populations are crucial problems for ecologists studying biological diversity, the distribution of species, population dynamics, and interspecific interactions. Modelers use these estimates to forecast changes in species distributions under different climate-change scenarios. Natural resource managers need these estimates to set realistic quotas for harvests of fish and game. Conservation biologists work with these estimates to determine whether to list species as threatened or endangered and to establish effective reserves to protect them. Three questions are of primary interest: is a species actually present in a location (i.e., is the location occupied); what is the probability of detecting it, given that it is present; and how many are there, given that it is present and that it can be detected. A wide range of methods has been developed to estimate these probabilities and quantities. These methods differ principally in whether they assume the population is closed or open to changes in species occurrence or population size (due to births, deaths, immigration, or extinction).

Closed-population models can be used in many practical situations, and results from them also are used routinely as inputs into open-population models. Closed-population models, including occupancy models developed by MacKenzie and his colleagues and Lincoln-Petersen mark-recapture models, require at least two successive samples in at least one site during one season, but reliable estimates generally require anywhere from three or four to more than 30 samples within a season, depending on the true (underlying) occupancy and detection probabilities. For open population models, estimates of survivorship are also required, i.e., [births + immigrants] – [deaths + emigrants]. In addition to replicate samples within a season, at least three seasons of sampling are required to estimate survivorship rates in mark-recapture models. Hierarchical models, in which parameters are estimated with either likelihood or Bayesian methods, allow for the inclusion of site- and time-specific covariates that potentially affect detection probabilities and occupancy, but these models usually demand large amounts of data. A large number of software solutions are available for fitting occupancy and mark-recapture models and estimating their parameters.

APPENDIX

Matrix Algebra for Ecologists

Many statistical techniques can be illustrated most concisely using matrix algebra. In particular, the multivariate techniques described in Chapter 12 rely on matrix algebra, and regression (see Chapter 9), ANOVA (see Chapter 10), and both basic and hierarchial occupancy models (see Chapter 14) often are presented in matrix notation. If you understand the basics of matrix algebra, you may be able to write simple programs to implement new statistical methods that are introduced in journal articles. In this appendix, we briefly explain the fundamentals of matrix algebra. If you want to learn more about using matrix algebra in statistics, Harville (1997) is an excellent and readable reference.

What Is a Matrix?

A basic definition of a matrix is a table of numbers in which the rows are replicates and the columns are the measured variables. More formally, a **matrix** is a rectangular array of numbers that has m rows and n columns. Matrices are written in expanded form with coefficients for each entry. In compact form, a matrix is symbolized by an uppercase, boldfaced letter:

$$\mathbf{A} = \begin{bmatrix} a_{11} & a_{12} & \cdots & a_{1n} \\ a_{21} & a_{22} & \cdots & a_{2n} \\ \vdots & \vdots & \ddots & \vdots \\ a_{m1} & a_{m2} & \cdots & a_{mn} \end{bmatrix}$$

Each **element** of the matrix is indexed with two subscripts: the first subscript indicates the row, and the second subscript indicates the column. Thus, a_{34} is the element of matrix **A** that is in the third row and the fourth column.

The **dimension** of a matrix is a pair of numbers specifying the number of rows and number of columns in the matrix:

$$\text{Dim}(\mathbf{A}) = (m,n)$$

Matrices can have any number of rows or columns, but three shapes of matrices are of particular importance. A **square matrix** is a matrix for which the number of rows equals the number of columns ($m = n$):

$$\mathbf{A} = \begin{bmatrix} a_{11} & a_{12} & \cdots & a_{1n} \\ a_{21} & a_{22} & \cdots & a_{2n} \\ \vdots & \vdots & \ddots & \vdots \\ a_{n1} & a_{n2} & \cdots & a_{nn} \end{bmatrix}$$

The sum of the elements of a square matrix that fall along the main diagonal, $a_{11} + a_{22} + \ldots + a_{nn}$, is called the **trace** of the matrix.

A **column vector** is a matrix with m rows and one column:

$$\mathbf{A} = \begin{bmatrix} a_{11} \\ a_{21} \\ \vdots \\ a_{m1} \end{bmatrix}$$

A **row vector** is a matrix with one row and n columns:

$$\mathbf{A} = \begin{bmatrix} a_{11} & a_{12} & \cdots & a_{1n} \end{bmatrix}$$

In matrix algebra, a single number k is called a **scalar**, in order to distinguish it from a matrix with one row and one column:

$$\mathbf{A} = [a_{11}]$$

Basic Mathematical Operations on Matrices

Addition and Subtraction

Adding and subtracting matrices is done element by element. For two matrices **A** and **B** with elements $\{a_{ij}\}$ and $\{b_{ij}\}$, respectively, their sum **A** + **B** is a new matrix **C** whose i,jth element equals $a_{ij} + b_{ij}$:

$$\begin{bmatrix} 3 & 4 \\ -1 & 5 \\ 0 & 2 \end{bmatrix} + \begin{bmatrix} 4 & 8 \\ 1 & -3 \\ -2 & 6 \end{bmatrix} = \begin{bmatrix} 3+4 & 4+8 \\ -1+1 & 5+(-3) \\ 0+(-2) & 2+6 \end{bmatrix} = \begin{bmatrix} 7 & 12 \\ 0 & 2 \\ -2 & 8 \end{bmatrix} \quad \textbf{(A.1)}$$

Similarly, the difference between two matrices **A** – **B** is a new matrix **C** whose i,jth element equals $a_{ij} - b_{ij}$. Because matrix addition and subtraction is done on an element-by-element basis, only matrices with the same dimension (i.e., matrices with the same number of rows and the same number of columns) can be added to or subtracted from each other.

Matrix addition is **commutative**, which means that for two matrices **A** and **B** of equal dimension, **A** + **B** = **B** + **A**. Matrix addition is also **associative**: for three matrices **A**, **B**, and **C** of equal dimension, **A** + (**B** + **C**) = (**A** + **B**) + **C**. You may have seen these commutative and associative properties before, because they also apply to ordinary arithmetic.

Multiplication

Two kinds of multiplication are possible with matrices. When we multiply a matrix **A** of any dimension by a single number, or scalar k, the product $k\mathbf{A}$ is a new matrix of the same dimension as **A** whose i,jth element equals ka_{ij}:

$$3\begin{bmatrix} 1 & 3 & 2 & -5 \\ 2 & 0 & 0 & 3 \end{bmatrix} = \begin{bmatrix} 3\times1 & 3\times3 & 3\times2 & 3\times(-5) \\ 3\times2 & 3\times0 & 3\times0 & 3\times3 \end{bmatrix} = \begin{bmatrix} 3 & 9 & 6 & -15 \\ 6 & 0 & 0 & 9 \end{bmatrix} \quad \text{(A.2)}$$

Like matrix addition and subtraction, scalar multiplication is commutative ($k\mathbf{A} = \mathbf{A}k$) and associative ($c(k\mathbf{A}) = (ck)\mathbf{A}$). Scalar multiplication also is **distributive** with respect to addition: $k(\mathbf{A} + \mathbf{B}) = k\mathbf{A} + k\mathbf{B}$. As with ordinary arithmetic, $1 \times \mathbf{A} = \mathbf{A}$.

It is also possible to multiply two matrices **A** and **B**, with dimensions (m,n) and (p,q), respectively, but only if the number of columns of the first matrix **A** equals the number of rows of the second matrix **B** (i.e., if $n = p$). The product **AB** of two such matrices is a new matrix whose i,jth element equals

$$\sum_{k=1}^{n} a_{ik}b_{kj} = a_{i1}b_{1j} + a_{i2}b_{2j} + \ldots + a_{in}b_{nj}$$

For example:

$$\begin{bmatrix} 1 & 3 & 3 \\ 2 & 0 & -2 \end{bmatrix} \begin{bmatrix} 1 & -2 & 2 \\ 3 & 2 & 2 \\ -1 & 3 & 3 \end{bmatrix}$$

$$= \begin{bmatrix} 1(1)+3(3)+3(-1) & 1(-2)+3(2)+3(3) & 1(2)+3(2)+3(3) \\ 2(1)+0(3)+(-2)(-1) & 2(-2)+0(2)+(-2)(3) & 2(2)+0(2)+(-2)(3) \end{bmatrix}$$

$$= \begin{bmatrix} 7 & 13 & 17 \\ 4 & -10 & -2 \end{bmatrix} \quad \text{(A.3)}$$

The dimension of **AB** = (m,q), where m is the number of rows of **A** and q is the number of columns of **B**.

For three matrices **A**, **B**, and **C**, matrix multiplication is associative [**A**(**BC**) = (**AB**)**C**], and is distributive with respect to addition [**A**(**B** + **C**) = **AB** + **AC** and (**A** + **B**)**C** = **AC** + **BC**]. Matrix multiplication generally is not commutative, however: **AB** usually does not equal **BA**. In fact, you cannot even multiply a pair of matrices in both directions unless $n = p$ and $m = q$. If both **A** and **B** are square matrices of identical dimensions ($m = n = p = q$), both **AB** and **BA** are defined (with Dim(**AB**) = (n,n)), but it is still almost never the case that **AB** = **BA**. For example:

$$\begin{bmatrix} 1 & 0 \\ 0 & 4 \end{bmatrix}\begin{bmatrix} 0 & 2 \\ 2 & 1 \end{bmatrix} = \begin{bmatrix} 0 & 2 \\ 8 & 4 \end{bmatrix} \neq \begin{bmatrix} 0 & 8 \\ 2 & 4 \end{bmatrix} = \begin{bmatrix} 0 & 2 \\ 2 & 1 \end{bmatrix}\begin{bmatrix} 1 & 0 \\ 0 & 4 \end{bmatrix} \tag{A.4}$$

We call **AB** the **preproduct** or the **premultiplication** of **B** by **A**. Equivalently, we say that **AB** is the **postproduct** or **postmultiplication** of **A** by **B**.

Notice that there is a difference between multiplying a matrix by a scalar k**A** and multiplying a matrix by a matrix with one row and one column $[k]$**A**. The former operation is possible for a matrix with any number of rows and columns (see Equation A.2), but the latter operation is possible only for a matrix with one row.

If we multiply a row vector [dimension = $(1,n)$] by a column vector [dimension = $(p,1)$], the result is a 1×1 matrix. This product is often called a **scalar product** (because there is only one element in the product of these two vectors), and many authors treat it as a scalar (e.g., Legendre and Legendre 1998), but most software packages will treat this result as a matrix.

Two vectors whose product equals 0 are called **orthogonal**. For example, a pair of orthogonal a priori ANOVA contrasts (see Chapter 10) are two vectors of numbers such that sum of their products equals 0. In matrix algebra terms, the pair of contrasts represents a row and column vector whose matrix product is 0, and they are therefore orthogonal.

Transposition

Interchanging the rows and columns of a matrix yields its **transpose**. For a matrix **A** with dimension (m,n), the transpose of **A**, written either as **A′** or $\mathbf{A}^T$ is a matrix with dimension (n,m) whose i,jth element equals the j,ith element of **A**:

$$\begin{bmatrix} 3 & 0 \\ -1 & -2 \\ 2 & -1 \end{bmatrix}^T = \begin{bmatrix} 3 & -1 & 2 \\ 0 & -2 & -1 \end{bmatrix} \tag{A.5}$$

Transposing a transposed matrix recovers the original matrix: $(\mathbf{A}^T)^T = \mathbf{A}$.

We call a matrix **symmetric** if it does not change following transposition: $\mathbf{A}^T = \mathbf{A}$. Only square matrices can be symmetric. The elements of a symmetric square matrix are the same on the corresponding sides of the diagonal:

$$\begin{bmatrix} 1 & 2 & 2 & -1 \\ 2 & 4 & 3 & 2 \\ 2 & 3 & -4 & 0 \\ -1 & 2 & 0 & 8 \end{bmatrix}$$

Another example of a symmetric square matrix is the variance-covariance matrix (see Footnote 5 in Chapter 9). For a set of n variables, the $n \times n$ variance-covariance matrix contains the covariance σ_{ij} between variables i and j, with the variances of variable i, σ_{ii}, as the diagonal elements. Because the covariance $\sigma_{ij} = \sigma_{ji}$, the elements on the two sides of the diagonal are identical and the matrix is symmetric.

There are two symmetric matrices that are especially useful. A **diagonal matrix D** is a square matrix for which all the elements equal 0 except for those along the main diagonal:

$$\mathbf{D} = \begin{bmatrix} a_{11} & 0 & 0 & 0 & 0 \\ 0 & a_{22} & 0 & 0 & 0 \\ 0 & 0 & a_{33} & 0 & 0 \\ \vdots & \vdots & \vdots & \ddots & \vdots \\ 0 & 0 & 0 & 0 & a_{nn} \end{bmatrix} \quad \textbf{(A.6)}$$

A diagonal matrix for which all the diagonal terms equal 1 is called the **identity matrix**, which is normally written as **I**:

$$\mathbf{I} = \begin{bmatrix} 1 & 0 & 0 & 0 & 0 \\ 0 & 1 & 0 & 0 & 0 \\ 0 & 0 & 1 & 0 & 0 \\ \vdots & \vdots & \vdots & \ddots & \vdots \\ 0 & 0 & 0 & 0 & 1 \end{bmatrix} \quad \textbf{(A.7)}$$

A matrix is called **triangular** if all of the elements on one side of the diagonal equal 0. If all of the elements above the diagonal equal zero, the matrix is called a **lower triangular matrix:**

$$\mathbf{A} = \begin{bmatrix} 1 & 0 & 0 & 0 \\ 13 & 2 & 0 & 0 \\ 2 & -1 & 2 & 0 \\ 4 & 2 & -2 & -1 \end{bmatrix} \quad \textbf{(A.8)}$$

In contrast, if all of the elements below the diagonal equal zero, the matrix is called an **upper triangular matrix:**

$$\mathbf{A} = \begin{bmatrix} 1 & 10 & 3 & -2 \\ 0 & 2 & 3 & 5 \\ 0 & 0 & 2 & -3 \\ 0 & 0 & 0 & -1 \end{bmatrix} \quad \textbf{(A.9)}$$

Note that transposing a lower triangular matrix results in an upper triangular matrix, and vice-versa.

INVERSION For a simple variable or a scalar k, we know that its inverse k^{-1} is defined such that $k \times k^{-1} = 1$. By analogy, for a matrix **A**, we define its inverse $\mathbf{A}^{-1}$ to be the matrix that satisfies the equation $\mathbf{A} \times \mathbf{A}^{-1} = \mathbf{I}$, the identity matrix. Inverse matrices are defined only for square matrices, but not all square matrices have inverses. If $\mathbf{A}^{-1}$ does exist, then $\mathbf{A}\mathbf{A}^{-1} = \mathbf{A}^{-1}\mathbf{A} = \mathbf{I}$. An example of an **inverse matrix** is:

$$\begin{bmatrix} 1 & 3 \\ 2 & 4 \end{bmatrix}^{-1} = \begin{bmatrix} -2 & 1.5 \\ 1 & -0.5 \end{bmatrix} \quad \textbf{(A.10)}$$

You can verify that these two matrices are inverses by multiplying them together. You should find that their product is the identity matrix **I**. For the special case where the inverse of a matrix equals its transpose, $\mathbf{A}^{-1} = \mathbf{A}^{T}$, the matrix **A** is called **orthogonal.** Calculating the inverse of a matrix is best done using mathematical or statistical software.

QUADRATIC FORMS If we combine transposition and matrix multiplication in a particular way, we obtain an expression called a **quadratic form**. Given a square matrix $\mathbf{A}$ of dimension (n,n), and column vector $\mathbf{x}$ of dimension $(n,1)$, the expression

$$\mathbf{Q} = \mathbf{x}^{\mathrm{T}}\mathbf{A}\mathbf{x} \tag{A.11}$$

is called a quadratic form. $\mathbf{Q}$ is a 1×1 matrix with its element equal to

$$\mathbf{Q} = \sum_{i=1}^{n}\sum_{j=1}^{n} x_i a_{ij} x_j \tag{A.12}$$

$\mathbf{Q}$ is usually considered to be a scalar, although some mathematical or statistical software packages may treat it as a 1×1 matrix.

SOLVING SYSTEMS OF LINEAR EQUATIONS One of the most important applications of matrix algebra is to solve systems of linear equations. In particular, estimating the parameters β_0 and β_1 for the basic linear regression model (see Chapter 9) is equivalent to solving a system of linear equations:

$$\begin{aligned} Y_1 &= \beta_0 + \beta_1 X_1 \\ Y_2 &= \beta_0 + \beta_1 X_2 \\ Y_3 &= \beta_0 + \beta_1 X_3 \\ &\vdots \\ Y_n &= \beta_0 + \beta_1 X_n \end{aligned} \tag{A.13}$$

If $\mathbf{Y}$ is the column vector

$$\begin{bmatrix} Y_1 \\ Y_2 \\ \vdots \\ Y_n \end{bmatrix}$$

representing each of the n responses; $\mathbf{X}$ is an $n \times 2$ matrix

$$\begin{bmatrix} 1 & X_1 \\ 1 & X_2 \\ 1 & \vdots \\ 1 & X_n \end{bmatrix}$$

representing each of the predictors; and **b** is the column vector

$$\begin{bmatrix} \beta_0 \\ \beta_1 \end{bmatrix}$$

of parameters for the linear regression model, then the system of linear equations (Equation A.13) can be represented as $\mathbf{Y} = \mathbf{Xb}$. The values of β_0 and β_1 that provide the best fit to the regression model (the least-squares model described in Chapter 9) can be calculated as

$$\mathbf{b} = [\mathbf{X}^T\mathbf{X}]^{-1}[\mathbf{X}^T\mathbf{Y}] \quad \text{(A.14)}$$

This equation works equally well for simple and multiple linear regression.

EIGENVALUES AND EIGENVECTORS A special set of linear equations

$$\begin{aligned} a_{11}x_1 + a_{12}x_2 + \ldots + a_{1n}x_n &= \lambda x_1 \\ a_{21}x_1 + a_{22}x_2 + \ldots + a_{2n}x_n &= \lambda x_2 \\ &\vdots \\ a_{n1}x_1 + a_{n2}x_2 + \ldots + a_{nn}x_n &= \lambda x_n \end{aligned} \quad \text{(A.15)}$$

can be written in matrix form as

$$\mathbf{Ax} = \lambda\mathbf{x} \quad \text{(A.16)}$$

where **A** is a square matrix with elements a_{ij}, **x** is an $n \times 1$ column vector, and λ is a scalar. Using the matrix algebra rules we have learned so far, Equation A.16 can be rewritten as

$$(\mathbf{A} - \lambda\mathbf{I})\mathbf{x} = \mathbf{0} \quad \text{(A.17)}$$

where **0** is an $n \times 1$ column vector whose elements all equal 0. If Equation A.17 (or A.16 or A.15) can be solved (which is not always the case), it can be solved only for certain values of λ. These values are called the **eigenvalues** of the matrix **A**. Because there are n equations in Equation A.15, there can be up to n eigenvalues, which may be complex numbers (always) or real numbers (sometimes). For each one of these eigenvalues λ_i, there is a corresponding **eigenvector**, which is the column vector **x** that satisfies Equation A.16 or A.17. Lastly, the sum of the eigenvalues of a matrix **A** equals its trace.

DETERMINANTS The equation for calculating eigenvalues (Equation A.17) has a trivial solution when $\mathbf{x} = \mathbf{0}$ (a column vector with all elements equal to zero). In addition, Equation A.17 has another solution:

$$|\mathbf{A} - \lambda\mathbf{I}| = 0 \quad \textbf{(A.18)}$$

The two vertical bars (| |) in Equation A.18 refer to the **determinant** of the difference between matrices **A** and λ**I**. The determinant of a square matrix **A** is a scalar that is calculated as the sum of all possible signed (positive or negative) products containing one row element and one column element from **A**. The sign of each product is determined by a rule that will at first seem arbitrary, but bear with us.

The square matrix **A** has elements a_{ij}. For any product of a pair of elements, a_{ij} and a_{km}, where $i \neq k$ and $j \neq m$, we define the product $a_{ij}a_{km}$ to be a *negative pair* if one of the elements is located above and to the right of the other element. This is the case if $k > i$ and $m < j$ or if $k < i$ and $m > j$. Otherwise, the product is a *positive pair*. The assignment of the sign can be visualized using the following table:

	$k > i$	$k < i$
$m > j$	+	−
$m < j$	−	+

For example, in an $n \times n$ matrix where $n > 4$, the pair a_{34} and a_{22} is a positive pair, whereas the pair a_{34} and a_{41} is a negative pair. The assignment of positive and negative pairings is a based on the relative position of the two elements within a matrix and has no relationship to the actual sign (positive or negative) of the elements or their product.

There are n elements of a square matrix **A** such that no two of them fall either in the same row or the same column. An example of these are the $i_1j_1, \ldots,$ i_nj_nth elements of **A**, where $i_1, \ldots, i_n$ and $j_1, \ldots, j_n$ are permutations of the first n integers $(1, \ldots, n)$. A total of $\binom{n}{2}$ pairs can be formed from these n elements. We use the symbol $\sigma_n(i_1,j_1; \ldots ; i_n,j_n)$ to represent the number of these pairs that are negative.

The determinant of a square matrix **A**, written as $|\mathbf{A}|$, is defined as

$$|\mathrm{A}| = \sum_{i=1}^{n} (-1)^{\sigma_n(i_1,j_1;\ldots;i_n,j_n)} a_{1j_1} \ldots a_{nj_n} \quad \textbf{(A.19)}$$

where $j_1,\ldots,j_n$ is a permuation of the first n integers and the summation is over all such permutations.

The recipe for calculating the determinant is

1. Compute all possible products of each of n factors that can be obtained by choosing one and only one element from each row and column of the square matrix **A**.
2. For each product, count the number of negative pairs among the $\binom{n}{2}$ pairs that can be found for the n elements in this product. If the number of negative pairs is an even number, the sign of the product is +. If the number of negative pairs is an odd number, the sign of the product is –.
3. Add up the signed products.

For a 2×2 matrix, the determinant is

$$\begin{vmatrix} a_{11} & a_{12} \\ a_{21} & a_{22} \end{vmatrix} = a_{11}a_{22} - a_{12}a_{21} \quad \textbf{(A.20)}$$

For a 3×3 matrix, the determinant is

$$\begin{vmatrix} a_{11} & a_{12} & a_{13} \\ a_{21} & a_{22} & a_{23} \\ a_{31} & a_{32} & a_{33} \end{vmatrix} = \begin{matrix} a_{11}a_{22}a_{33} + a_{12}a_{23}a_{31} + a_{13}a_{21}a_{32} \\ -a_{11}a_{23}a_{32} - a_{12}a_{21}a_{33} - a_{13}a_{22}a_{31} \end{matrix} \quad \textbf{(A.21)}$$

Because determinants can be computed using either rows or columns of a matrix, the determinant of a matrix **A** equals the determinant of its transpose: $|\mathbf{A}| = |\mathbf{A}^T|$. A useful property of a triangular matrix is that its determinant equals the product of the entries on the main diagonal (as all the other products in Equation A.19 = 0).

In addition to providing a mechanism for identifying eigenvalues (see Equation A.18), determinants have other useful properties:

1. Determinants transform the result of matrix multiplication into a scalar: $|\mathbf{AB}| = |\mathbf{A}||\mathbf{B}|$.
2. A square matrix whose determinant equals zero has no inverse, and is called a **singular matrix**.

3. The determinant of a matrix equals the product of its eigenvalues:

$$|\mathbf{A}| = \prod_{i-1}^{n} \lambda_i$$

4. For any scalar k and any $n \times n$ matrix **A**, $|k\mathbf{A}| = k^n|\mathbf{A}|$.

As noted in Item 2 (previous page), not all matrices have inverses. In particular, the inverse of a matrix, $\mathbf{A}^{-1}$, only exists if the determinant of **A**, $|\mathbf{A}|$, does not equal zero.

SINGULAR VALUE DECOMPOSITION A useful property of any $m \times n$ matrix **A** is that it can be rewritten, or **decomposed**, using the following formula:

$$\mathbf{A} = \mathbf{VWU}^{\mathrm{T}} \tag{A.22}$$

where **V** is a matrix of the same dimension ($m \times n$) as **A**; **U** is an $n \times n$ matrix; **V** and **U** are **column-orthonormal matrices:** orthogonal matrices whose columns have been normalized so that the Euclidean distance of each column is

$$\sqrt{\sum_{k=1}^{n} (y_{ik} - y_{jk})^2} = 1$$

and **W** is an $n \times n$ diagonal matrix whose diagonal values w_{ii} are called the **singular values** of **A**.

Equation A.22, known as **singular value decomposition**, can be used to determine if a square matrix **A** is a singular matrix. For example, solving the system of linear equations (see Equation A.14) requires evaluating the expression $\mathbf{Y} = \mathbf{Xb}$ where **X** is a square matrix. As we showed in Equation A.14, solving for **b** requires taking the inverse of $\mathbf{X}^{\mathrm{T}}\mathbf{X}$. To simplify the rest of the discussion, we'll write $\mathbf{X}^*$ to represent $\mathbf{X}^{\mathrm{T}}\mathbf{X}$.

Using Equation A.22, we can represent $\mathbf{X}^*$ as $\mathbf{VWU}^{\mathrm{T}}$. Because the inverse of the preproduct of two matrices $[\mathbf{AB}]^{-1}$ equals the postproduct of their inverses $\mathbf{B}^{-1}\mathbf{A}^{-1}$, we can calculate the inverse of $\mathbf{X}^*$ as

$$\mathbf{X}^{*-1} = [\mathbf{VWU}^{\mathrm{T}}]^{-1} = [\mathbf{U}^{\mathrm{T}}]^{-1}\mathbf{W}^{-1}\mathbf{V}^{-1} \tag{A.23}$$

Because **U** and **V** are orthogonal, their inverses equal their transposes (which are much easier to calculate), and

$$\mathbf{X}^{*-1} = \mathbf{UW}^{-1}\mathbf{V}^{\mathrm{T}} \tag{A.24}$$

Further, the inverse of a diagonal matrix is simply a new diagonal matrix whose elements are the reciprocals of the original. If any of the diagonal elements of $\mathbf{W} = 0$, then their reciprocals are infinite ($1/0 = \infty$) and the inverse of $\mathbf{X}^*$ would not be defined. Thus, we can determine more easily whether $\mathbf{X}^*$ is singular. It is still possible to solve the equation $\mathbf{Y} = \mathbf{Xb}$ when $\mathbf{X}^T\mathbf{X}$ is singular using computationally intensive methods (Press et al. 1986).

Glossary

Numerals in brackets indicate the chapter(s) in which the most complete discussion of that term occurs. The letter A indicates a term found in the Appendix.

abscissa The horizontal, or x-axis, on a plot. The opposite of **ordinate.** [7]

accuracy The degree to which a measurement on an object comes close to the true value of the object. Contrast with **precision.** [2]

additive design An experimental design used to study competition between a target or focal species of interest and its competitors. In an additive design, the density of the target species is kept constant and the different treatments represent different numbers of the competitors. Contrast with **substitutive** and **response surface designs.** [7]

adjusted mean The expected value of a treatment group if all groups had the same mean covariate. [10]

adjusted r^2 How well a regression model fits the data, discounted by the number of parameters in the model. It is used to avoid over-fitting a model simply by including additional predictor variables. See also **coefficient of determination.** [9]

agglomerative clustering The process of grouping (or clustering) objects by taking many smaller clusters and grouping them on the basis of their similarity into larger clusters. See also **cluster analysis**; contrast with **divisive clustering.** [12]

Akaike's information criterion (AIC) A measure of the explanatory power of a statistical model that accounts for the number of parameters in the model. When comparing among multiple models for the same phenomenon (the models must share at least some parameters), the model with the lowest value of Akaike's information criterion is considered to be the best model. Used commonly for model selection in regression and path analysis. [9, 14]

alpha The first letter (α) of the Greek alphabet. In statistics, it is used to denote the acceptable probability of committing a Type I error. [4]

alpha diversity Species diversity measured within a single reference sample. Contrast with **beta diversity** and **gamma diversity.** [13]

alternative hypothesis The hypothesis of interest. It is the statistical hypothesis that corresponds to the scientific question being examined with the data. Unlike the null hypothesis, the alternative hypothesis posits that "something is going on." Contrast with **null hypothesis** and **scientific hypothesis.** [4]

analysis of covariance (ANCOVA) A hybrid statistical model, somewhere between regression and ANOVA, that includes a continuous predictor variable (the covariate) along with categorical predictor variables of an ANOVA model. [10]

analysis of variance (ANOVA) A method of partitioning the sums of squares, due to Fisher. It is used primarily to test the statistical null hypotheses that distinct treatments have no effects on measured response variables. [5, 9, 10]

analysis of variance design The class of experimental layouts used to explore relationships between categorical predictor variables and continuous response variables. [7]

angular transformation The function $Y^* = \text{arcsine}\sqrt{Y}$, where arcsine($X$) is the function that returns the angle θ for which the sine of θ = X. Also called the **arcsine transformation** or the **arcsine square-root transformation.** [8]

ANCOVA See **analysis of covariance.** [10]

ANOVA See **analysis of variance.** [5, 9, 10]

arcsine transformation See **angular transformation.** [8]

arcsine square-root transformation See **angular transformation.** [8]

ARIMA See **autoregressive integrated moving average.** [7]

arithmetic mean The average of the set of observations of a given variable. It is calculated as the sum of all of the observations divided by the number of observations, and is symbolized by placing a horizontal line over the variable name, for example, $\bar{Y}$. [3]

assemblage A complete community consisting of all species represented in their true relative abundances. Most biodiversity data represent only a small sample of individuals from an assemblage. [9, 10, 12, 13]

associative In mathematics, the property that $(a + b) + c = a + (b + c)$. [A]

asymptotic statistics Statistical estimators that are calculated based on the assumption that if the experiment were repeated infinitely many times, the estimator would converge on the calculated value. See also **frequentist inference.** [3]

audit trail The set of files and associated documentation describing changes to an original data set. It describes all the changes made to the original data that resulted in the processed data set that was used for analysis. [8]

autoregressive A statistical model for which the response variable (Y) depends on one or more previous values. Many population growth models are autoregressive, as for example, models in which the size of a population at time $t + 1$ is dependent on the size of the population at time t. [6]

autoregressive integrated moving average (ARIMA) A statistical model used to analyze time-series data that incorporates both temporal dependency in the data (see **autoregressive**) and parameters that estimate changes due to experimental treatments. [7]

average The arithmetic mean of variable. Often used incorrectly to mean the most common value of a variable. Compare and contrast with **mean, median,** and **mode.** [3]

axiom An established (mathematical) principle that is universally accepted, self-evidently true, and has no need of a formal proof. [1]

BACI design An experimental method for evaluating the effects of a treatment (often an adverse environmental impact) relative to an untreated population. In a BACI (*b*efore-*a*fter-*c*ontrol-*i*mpact) design, there often is little spatial replication, and so the replication is in time. Effects on both control and treated populations are measured both before and after the treatment. [6, 7]

back-transformed The consequence of changing a variable or statistical estimator back into the original units of measurement. Back-transformation is only necessary if variables were transformed prior to statistical analysis. See also **transformation.** [3, 8]

backward elimination A method of stepwise regression, in which you begin with the model containing all possible parameters and eliminate them one at a time. See also **stepwise regression**; compare with **forward selection.** [9]

base The logarithmic function $\log_b a = X$ is described by two parameters, a and b, such that $b^X = a$ is equivalent to $\log_b a = X$. The parameter b is the base of the logarithm. [8]

Bayes' factor The relative posterior probability of the null hypothesis to the alternative hypothesis. If the prior probabilities of the two hypotheses are equal, it can be calculated as the ratio of the likelihoods of the two hypotheses. [9]

Bayes' Theorem The formula for calculating the probability of a hypothesis of interest, given observed data and any prior knowledge about the hypothesis.

$$P(\mathrm{H} \mid Y) = \frac{f(Y \mid \mathrm{H})P(\mathrm{H})}{P(Y)}$$

[1]

Bayesian inference One of three major frameworks for statistical analysis. It is often called inverse probability. Bayesian inference is used to estimate the probability of a statistical hypothesis given (or conditional on) the data at hand. The result of Bayesian inference is a new (or posterior) probability distribution. Contrast with **frequentist**, **parametric**, and **Monte Carlo** analysis. [5]

Bayesian information criterion (BIC) A method to compare the posterior probability distributions of two alternative models when the prior probability distributions are uninformative. It discounts the Baye's factor by the number of parameters in each of the models. As with Akaike's information criterion, the model with the lowest Bayesian information criterion is considered to be the best model. [9]

Bernoulli random variable The outcome of an experiment in which there are only two possible results, such as present/absent; dead/alive; heads/tails. See also **random variable.** [2]

Bernoulli trial An experiment that has only two possible results, such as present/absent, dead/alive, heads/tails. A Bernoulli trial results in a Bernoulli random variable. [2]

beta The second letter (β) of the Greek alphabet. In statistics, it is used to denote the acceptable probability of committing a Type II error. It also shows up, with subscripts, as the symbol for regression coefficients and parameters. [4]

beta diversity Differences in species diversity measured among a set of reference samples. The product of alpha diversity (within plot) and beta diversity (between plots) is the total gamma diversity. Contrast with **alpha diversity** and **gamma diversity.** [13]

between-subjects factor In ANOVA, the term that accounts for variance among treatment groups. In a repeated measures design, the between-subjects factor is equivalent to the whole-plot factor of a split-plot design. [7]

BIC See **Bayesian information criterion.** [9]

binomial coefficient The constant by which we multiply the probability of a Bernoulli random variable to obtain a binomial random variable. The binomial coefficient adjusts the probability to account for multiple, equivalent combinations of each of the two possible results. It is written as

$$\binom{n}{X}$$

and is read as "n choose X." It is calculated as

$$\frac{n!}{X!(n-X)!}$$

where "!" indicates factorial. [2]

binomial random variable The outcome of an experiment that consists of multiple Bernoulli trials. [2]

bi-plot A single graph of the results of two related ordinations. Used commonly to illustrate the results of correspondence analysis, canonical correspondence analysis, and redundancy analysis. [12]

bivariate data Data consisting of two variables for each observation. Often illustrated using scatterplots. [8]

block An area or time interval within which experimentally unmanipulated factors are considered to be homogeneous. [7]

Bonferroni method An adjustment of the critical value, α, at which the null hypothesis is to be rejected. It is used when multiple comparisons are made, and is calculated as the nominal α (usually equal to 0.05) divided by the number of comparisons. Contrast with **Dunn-Sidak method.** [10]

bootstrap Something with which to pull oneself up. In statistics, the bootstrap is a randomization (or Monte Carlo) procedure by which the observations in a dataset are re-sampled with replacement from the original dataset, and the re-sampled dataset is then used to re-calculate the test statistic of interest. This is repeated a large number of times (usually 1000–10,000 or more). The test statistic computed from the original dataset is then compared with the distribution of test statistics resulting from repeatedly pulling up the data by its own bootstraps in order to provide an exact *P*-value. See also **jackknife.** [5]

Box-Cox transformation A family of power transformations defined by the formula $Y^* = (Y^\lambda - 1)/\lambda$ (for $\lambda \neq 0$) and $Y^* = \ln(Y)$ (for $\lambda = 0$). The value of λ that results in the closest fit to a normal distribution is used in the final data transformation. This value must be found by computer-intensive methods. [8]

box plot A visual tool that illustrates the distribution of a univariate dataset. It illustrates the median, upper and lower quartiles, upper and lower deciles, and any outliers. [8]

canonical correspondence analysis (CCA) A method of ordination, typically used to relate species abundances or characteristics to environmental or habitat variables. This method linearly ordinates the predictors (environment or habitat variables) and ordinates the responses (species variables) to produce a unimodal, or humped, axis relative to the predictors. It is also a type of multivariate regression analysis. Compare with **redundancy analysis.** [12]

categorical variable Characteristics that are classified into two or more distinct groups. Contrast with **continuous variable.** [7, 11]

CCA See **canonical correspondence analysis.** [12]

CDF See **cumulative distribution function.** [2]

cell A box in a spreadsheet. It is identified by its row and column numbers, and contains a single entry, or datum. [8]

Central Limit Theorem The mathematical result that allows the use of parametric statistics on data that are not normally distributed. The Central Limit Theorem shows that any random variable can be transformed to a normal random variable. [2]

central moment The average (or arithmetic mean) of the differences between each observation of a variable and the arithmetic mean of that variable, raised to the *r*th power. The term central moment must always be qualified: e.g., first central moment, second central moment, etc. The first central moment equals 0, and the second central moment is the variance. [3]

centroid The mean vector of a multivariate variable. Think of it as the center of mass of a cloud of points in multidimensional space, or the point at which you could balance this cloud on the tip of a needle. [12]

change of scale The transformation of a normal random variable brought about by multiplying it by a constant. See also **shift operation.** [2]

circularity The assumption in analysis of variance that the variation among samples within subplots or blocks is the same across all subplots or blocks. [7]

classification The process of placing objects into groups. A general term for multivariate methods, such as cluster analysis or discriminant analysis, that are concerned with grouping observations. Contrast with **ordination.** [12]

classification matrix One outcome of a discriminant analysis. A table where the rows are the observed groups to which each observation belongs and the columns are the predicted groups to which each observation belongs. Entries in this table are the number of observations placed in each (observed, predicted) pair. In a perfect classification (a discriminant analysis with no error), only the diagonal elements of the classification matrix will have non-zero entries. [12]

classification tree A model or visual display of categorical data that is constructed by splitting the dataset into successively smaller groups. Splitting continue until no further improvement in model fit is obtained. A classification tree is analogous to a taxonomic key, except that divisions in a classification tree reflect probabilities, not identities. See also **cluster diagram**; **impurity**; **regression tree.** [11]

closed interval An interval that includes its endpoints. Thus, the closed interval [0, 1] is the set of all numbers between 0 and 1, including the numbers 0 and 1. Contrast with **open interval.** [2]

closure A mathematical property of sets. Sets are considered to be closed for a particular mathematical operation, for example addition or multiplication, if the result of applying the operation to a member or members of the set is also a member of the set. For example, the set of even integers $\{2, 4, \ldots\}$ is closed under multiplication because the product of any two (or more) even integers is also an even integer. In contrast, the set of whole numbers $\{0, 1, 2, \ldots\}$ is not closed under subtraction because it is possible to subtract one whole number from another and get a negative number (e.g., $3 - 5 = -2$), and negative numbers are not whole numbers. [1]

closure assumption In occupancy and mark-recapture models, the assumption that individuals (or species) do not enter (through births or immigration) or leave (through deaths or emigration) a site between samples. [13, 14]

cluster analysis A method for grouping objects based on the multivariate distances between them. [12]

coefficient of determination (r^2) The amount of variability in the response variable that is accounted for by a simple linear regression model. It equals the regression sum of squares divided by the total sums of squares. [9]

coefficient of dispersion The quantity obtained by dividing the sample variance by the sample mean. The coefficient of dispersion is used to determine whether or not individuals are arranged in clumped, regular, random, or hyperdispersed spatial patterns. [3]

coefficient of variation (CV) The quantity obtained by dividing the sample standard deviation by the sample mean. The coefficient of variation is useful for comparing variability among different populations, as it has been adjusted (standardized) for the population mean. [3]

collinearity The correlation between two predictor variables. See also **multicollinearity.** [7, 9]

column-orthonormal matrix An $m \times n$ matrix whose columns have been normalized so that the Euclidean distance of each column = 1. [A]

column total The marginal sum of the entries in a single column of a contingency table. See also **row total; grand total.** [11]

column vector A matrix with many rows but only one column. Contrast with **row vector.** [12]

combination An arrangement of n discrete objects taken X at a time. The number of combinations of a group of n objects is calculated using the binomial coefficient. Contrast with **permutation.** [2]

common factors The parameters generated by a factor analysis. Each original variable can be rewritten as a linear combination of the common factors, plus one additional factor that is unique to that variable. See also **factor analysis** and **principal component analysis.** [12]

commutative In mathematics, the property that $a + b = b + a$. [A]

complement In set theory, the complement A^c of a set A is everything that is not included in set A. [1]

complex event In set theory, complex events are collections of simple events. If A and B are two independent events, then $C = A$ *or* B, written as $C = A \cup B$ and read as "A union B," is a complex event. If A and B are independent events each, with associated probabilities P_A and P_B, then the probability of C (P_C) = $P_A + P_B$. Contrast with **shared events.** [1]

component of variation How much each factor in an analysis of variance of model accounts for the variability in the observed data. Initially, the total variation in the data can be partitioned into the variation among groups and the variation within groups. In some ANOVA designs, the variation can be partitioned into additional components. [10]

conditional probability The probability of a shared event when the associated simple events are not independent. If event B has already happened, and the probability of observing event A depends on the outcome of event B, then we say that the probability of A is *conditional* on B. We write this as $P(A|B)$, and calculate it as

$$P(A \mid B) = \frac{P(A \cap B)}{P(B)}$$

[1]

confidence interval An interval that includes the true population mean n% of the time. The term confidence interval must always be qualified with the n%—as in 95% confidence interval, 50% confidence interval, etc. Confidence intervals are calculated using frequentist statistics. They are commonly are misinterpreted to represent an n% chance that the mean is in the confidence interval. Compare with **credibility interval.** [3]

confounded The consequence of a measured variable being associated with another variable that may not have been measured or controlled for. [4, 6]

contingency table A way of organizing, displaying, and analyzing data to illustrate relationships between two or more categorical variables. Entries in contingency tables are the frequencies (counts) of observations in each category. [11]

contingency table analysis Statistical methods used to analyze datasets for which both response and predictor variables are categorical. [7, 11]

continuous monotonic function A function that is defined for all values in an interval, and which preserves the ordering of the operands. [8]

continuous random variable A result of an experiment that can take on any quantitative value. Contrast with **discrete random variable.** [2]

continuous variable Characteristics that are measured using integer or real numbers. Contrast with **categorical variable.** [7]

contrast A comparison between treatment groups in an ANOVA design. Contrasts refer to those comparisons decided on *before* the experiment is conducted. [10]

correlation The method of analysis used to explore relationships between two variables. In correlation analysis, no cause-and-effect relationship is hypothesized. Contrast with **regression.** [7, 9]

correlative A type of relationships that is observed, but has not yet been experimentally investigated in a controlled way. The existence of correlative data leads to the maxim "correlation does not imply causation." [4]

correspondence analysis An ordination method used to relate species distributions to environmental variables. Also called indirect gradient analysis. [12]

covariate A continuous variable that is measured for each replicate in an analysis of covariance design. The covariate is used to account for residual variation in the data and increase the power of the test to detect differences among treatments. [10]

covariance The sum of the product of the differences between each observation and its expected value, divided by the sample size. If the covariance is not divided by the sample size, it is called the **sum of cross products.** [9]

coverage The completeness or thoroughness of a biodiversity sample. Coverage is the proportion of the total abundance in the assemblage that is represented by the species detected in the reference sample. [13]

credibility interval An interval that represents the n% chance that the population mean lies therein. Like a confidence interval, a credibility interval always must be qualified: 95% credibility interval, 50% credibility interval, etc. Credibility intervals are determined using Bayesian statistics. Contrast with **confidence interval.** [3, 5]

critical value The value a test statistic must have in order for its probability to be statistically significant. Each test statistic is related to a probability distribution that has an associated critical value for a given sample size and degrees of freedom. [5]

cumulative distribution function (CDF) The area, up to a point X, under a curve that describes a random variable. Formally, if X is a random variable with probability distribution function $f(x)$, then the cumulative distribution function $F(X)$ equals

$$P(x < X) = \int_{-\infty}^{X} f(x)dx$$

The cumulative distribution function is used to calculate the tail probability of an hypothesis. [2]

CV See **coefficient of variation.** [3]

---

decile A tenth. Usually refers to the upper and lower deciles—the upper 10% and lower 10%—of a distribution. [3]

decomposition In matrix algebra, the re-expression of any $m \times n$ matrix $\mathbf{A}$ as the product of three matrices, $\mathbf{VWU}^T$, where $\mathbf{V}$ is an $m \times n$ matrix of the same dimension as $\mathbf{A}$; $\mathbf{U}$ and $\mathbf{W}$ are square matrices (dimension $n \times n$); $\mathbf{V}$ and $\mathbf{U}$ are column-orthonormal matrices; and $\mathbf{W}$ is a diagonal matrix with singular values on the diagonal. [A]

deduction The process of scientific reasoning that proceeds from the general (all swans are white) to the specific (this bird, being a swan, is white). Contrast with **induction.** [4]

degrees of freedom (df) How many independent observations we have of a given variable that can be used to estimate a statistical parameter. [3]

dendrogram A tree (*dendro*) diagram (*gram*). A method of graphing the results of a cluster analysis. Similar in spirit to a classification tree. Also used extensively to illustrate relatedness among species (phylogenies). [12]

dependent variable In a statement of cause-and-effect, the response variable, or the affected object for which one is trying to determine the cause. Contrast with **independent variable** and **predictor variable.** [7]

detection history A string of 1s and 0s that indicates the presence or absence of a species in a particular site over a sequence of repeated censuses. [14]

detection probability The probability of finding an individual at a specific location, given that it is occupying the site. As a first approximation, it can be estimated as the fraction of individuals (or species) of the total population (or species assemblage) that was sampled, but life is usually more complicated than that. See also **occupancy.** [14]

determinant In matrix algebra, the scalar computed as the sum of all possible signed products containing one row element and one column element of a given matrix. [A]

detrended correspondence analysis A variant of correspondence analysis used to ameliorate the horseshoe effect. [12]

deviation Difference between the model predictions and the observed values. [6]

diagonal matrix A square matrix in which all the entries except those along the main diagonal are equal to zero. [12]

dimension A 2-dimensional vector specifying the number of rows and number of columns in a matrix. [A]

discrete outcome A result or observation that can take on separate or integer values. [1]

discrete random variable A result of an experiment that takes on only integer values. Contrast with **continuous random variable.** [2]

discrete sets A set of discrete outcomes. [1]

discriminant analysis A method for classifying multivariate observations when groups have been assigned in advance. Discriminant analysis can also be used as a post hoc test for detecting differences among groups when a multivariate analysis of variance (MANOVA) has yielded significant results. [12]

dissimilarity How different two objects are. Normally quantified as a **distance.** [12]

distance How far apart two objects are. There are many ways to measure distances, the most familiar being the **Euclidean distance.** [12]

distributive In mathematics, the property that $c(a + b) = ca + cb$. [A]

divisive clustering The process of grouping (or clustering) objects by taking one large cluster and splitting it, based on dissimilarity, into successively smaller clusters. See also **cluster analysis**, and contrast with **agglomerative clustering.** [12]

doubletons The number of species represented by exactly two individuals in an individual-based reference sample. See also **duplicates**, **singletons**, and **uniques.** [13]

doubling property If two assemblages have an identical relative abundance distribution but share no species in common, the diversity index will double when the two assemblages are combined. Species richness and other members of the Hill family of diversity indices obey the doubling property, but most common diversity indices do not. [13]

Dunn-Sidak method An adjustment of the critical value, α, at which the null hypothesis is to be rejected. It is used when multiple comparisons are made, and is calculated as $\alpha_{adjusted}$ equals $1 - (1 - \alpha_{nominal})^{1/k}$ where $\alpha_{nominal}$ is the critical value of α (usually equal 0.05), and k is the number of comparisons. Contrast with **Bonferroni method**. [10]

duplicates The number of species represented in exactly two samples in a sample-based reference sample. See also **doubletons, singletons**, and **uniques**. [13]

dynamic occupancy models Models for estimating occupancy probability that violate the **closure assumption** and allow for changes in occupancy between samples. See also **seasons**. [14]

EDA See **graphical exploratory data analysis**. [8]

effect size The expected difference among means of groups that are subjected to different treatments. [6]

effective numbers of species The diversity equivalent to that of a hypothetical community in which all of the species are equally abundant. [13]

eigenvalue The solution or solutions to a set of linear equations of the form $\mathbf{Ax} = \lambda\mathbf{x}$. In this equation, **A** is a square matrix and **x** is a column vector. The eigenvalue is the scalar, or single number, λ. Compare with **eigenvector**. [12]

eigenvector The column vector **x** that solves the system of linear equations $\mathbf{Ax} = \lambda\mathbf{x}$. Compare with **eigenvalue**. [12]

element One entry in a matrix. It is usually indicated as a_{ij}, where i and j are the row and column numbers, respectively, of the matrix. [12]

empty set A set with no members or elements. [1]

error A value that does not represent the original measurement or observation. It can be caused by mistaken entry in the field, malfunctioning instruments, or typing errors made when transferring values from original notebooks or data sheets into spreadsheets. [8]

error variation What's left over (unexplained) by a regression or ANOVA model. Also called the residual variation or the residual sum of squares, it reflects measurement error and variation due to causes that are not specified by the statistical model. [9, 10]

Euclidean distance One measure of distance or dissimilarity between two objects. If each object can be described as a set of coordinates in n-dimensional space, the Euclidean distance is calculated as

$$d_{i,j} = \sqrt{\sum_{k=1}^{n}(y_{i,k} - y_{j,k})^2}$$

[12]

event A simple observation or process with a clearly defined beginning and a clearly defined end. [1]

exact probability The probability of obtaining only the result actually observed. Contrast with **tail probability**. [2, 5]

exhaustive A set of outcomes is said to be exhaustive if the set includes all possible values for the event. Contrast with **exclusive**. [1]

exclusive A set of outcomes is said to be exclusive if there is no overlap in their values. Contrast with **exhaustive**. [1]

expectation The average, mean, or expected value of a probability distribution. [1, 2]

expected value The most likely value of a random variable. Also called the expectation of a random variable. [2]

experiment A replicated set of observations or trials, usually carried out under controlled conditions. [1]

experiment-wide error rate The actual critical value of α at which point the null hypothesis would be rejected, after correction for multiple comparisons. It can be obtained using Bonferroni or Dunn-Sidak methods, among others. [10]

extent The total spatial area or temporal range that is covered by all of the sampling units in an experimental study. Compare with **grain**. [6]

extrapolation Estimation of unobserved values outside of the range of observed values. Generally less reliable than **interpolation**. [9, 13]

extrinsic hypothesis A statistical model for which the parameters are estimated from sources other than the data. Contrast with **intrinsic hypothesis.** [11]

---

factor In ANOVA designs, a set of levels of a single experimental treatment. [7]

factor analysis An ordination method that rescales each original observation as a linear combination of a small number of parameters, called common factors. It was developed in the early twentieth century as a means of calculating "general intelligence" from a battery of test scores. [12]

factor loading The linear coefficients by which the common factors in a factor analysis are multiplied in order to give an estimate of the original variable.[12]

factor model The result of a factor analysis after the common factors and their corresponding factor loadings have been standardized, and one that retains only the useful (or significant) common factors. [12]

factorial The mathematical operation indicated by an exclamation point (!). For a given value n, the value $n!$, or "n -factorial," is calculated as

$$n \times (n-1) \times (n-2) \times \ldots \times (3) \times (2) \times (1)$$

Hence, $5! = 5 \times 4 \times 3 \times 2 \times 1 = 120$. By definition, $0! = 1$, and factorials cannot be computed for negative numbers or non-integers. [2]

factorial design An analysis of variance design that includes all levels of the treatments (or factors) of interest. [7]

F-distribution The expected distribution of an F random variable, calculated as one variance (usually the sums of squares among groups, or $SS_{among\ groups}$) divided by another (usually the sums of squares within groups, or $SS_{within\ groups}$). [5]

first principal axis The new variable, or axis, resulting from a principal component analysis, that accounts for more of the variability in the data than any other axis. Also called the first principal component. [12]

Fisher's combined probability A method for determining if the results of multiple comparisons are significant, without recourse to adjusting the experiment-wide error rate. It is calculated as the sum of the logarithm of all the P-values, multiplied by –2. This test statistic is distributed as a chi-square random variable with degrees of freedom equal to 2 times the number of comparisons. [10]

Fisher's F-ratio The F-statistic used in analysis of variance. [5]

fit How consistent predictions relate to observations. Used to describe how well data are predicted by a given model, or how well models are predicted by available data. [9, 10, 11, 12]

fitted site scores One result of a redundancy analysis. The fitted site scores are the matrix **Z** resulting from an ordination of the original data relative to the environmental characteristics

$$Z = \hat{Y}A$$

where $\hat{Y}$ is the predicted values of the multivariate response variable and **A** is the matrix of eigenvectors resulting from a principal component analysis of the expected (fitted) values of the response variable. Contrast with **site scores** and **species scores.** [12]

fixed effects In an analysis of variance design, the set of treatment levels that represents all possible treatment levels. Also known as a fixed factor. Contrast with **random effects.** [7, 10]

flat files Computer files that contain data organized as a table with only two dimensions (rows and columns). Each datum is entered in a different cell in the table. [8]

forward selection A method of stepwise regression that begins with the model containing only one of several possible parameters. Additional parameters are tested and added sequentially one at a time. See also **stepwise regression**, and compare with **backward elimination.** [9]

frequency The number of observations in a particular category. The values on the *y*-axis (ordinate) of a histogram are the frequencies of observations in each group identified on the *x*-axis (abscissa). [1]

frequentist inference An approach to statistical inference in which probabilities of observations are estimated from functions that assume an infinite number of trials. Contrast with **Bayesian inference.** [1, 3]

F-statistic The result, in an analysis of variance (ANOVA) or regression, of dividing the sums of squares among groups,($SS_{among\ groups}$) by the sums of squares within groups ($SS_{within\ groups}$). It has two different degrees of freedom—the numerator degrees of freedom are those associated with $SS_{among\ groups}$, and the denominator degrees of freedom associated with $SS_{within\ groups}$. These two values also define the shape of the F-distribution. To determine its statistical significance, or *P*-value, the F-statistic is compared to the critical value of the F-distribution, given the sample size, probability of Type I error, and numerator and denominator degrees of freedom. [5]

fully crossed A multiple-factor analysis of variance design in which all levels of two or more treatments are tested at the same time in a single experiment. [7, 10]

function A mathematical rule for assigning a numerical value to another numerical value. Usually, functions involve applying mathematical operations (such as addition or multiplication) to the initial (input) value to obtain the result, or output of the function. Functions are usually written with italic letters: the function

$$f(x) = \beta_0 + \beta_1 x$$

means take x, multiply it by β_1, and add β_0 to the product. This gives a new value that is related to the original value x by the operation of the function $f(x)$. [2]

gamma diversity The total diversity measured in a region or among a set reference samples. Gamma diversity is the product of alpha diversity (within plot) and beta diversity (between plots). Contrast with **alpha diversity** and **beta diversity.** [13]

Gaussian random variable See **normal random variable.** [2]

geometric mean The *n*th root of the product of *n* observations. A useful measure of expectation for log-normal variables. The geometric mean is always less than or equal to the arithmetic mean. [3]

goodness-of-fit Degree to which a dataset is accurately predicted by a given distribution or model. [11]

goodness-of-fit test A statistical test used to determine goodness-of-fit. [11]

grain The spatial size or temporal extent of the smallest sampling unit in an experimental study. Compare with **extent.** [6]

grand total The sum of all the entries in a contingency table. It also equals the number of observations. See also **row total; column total.** [11]

graphical exploratory data analysis (EDA) The use of visual devices, such as graphs and plots, to detect patterns, outliers, and errors in data. [8]

group A set that possesses four properties. First, the set is closed under a mathematical operation $\oplus$ (such as addition or multiplication). Second, the operation is associative such that for any three elements a, b, c of the set, $a \oplus (b \oplus c) = (a \oplus b) \oplus c$ (where $\oplus$ indicates any mathematical operation). Third, the set has an identity element I such that for any element a of the set $a \oplus \mathrm{I} = a$. Fourth, the set has inverses such that for any element a of the set, there exists another element b for which $ab = ba = \mathrm{I}$. See also **closure.** [1]

haphazard Assignment of individuals or populations to treatment groups that is not truly random. Contrast with **randomization.** [6]

harmonic mean The reciprocal of the arithmetic mean of the reciprocals of the observations. The harmonic mean is a useful statistic to use to calculate the effective population size (the size of a population with completely random mating). The harmonic mean is always less than or equal to the geometric mean. [3]

heteroscedastic The property of a dataset such that the residual variances of all treatment groups are unequal. Contrast with **homoscedastic.** [8, 9]

hierarchical clustering A method of grouping objects that is based on an externally imposed reference system. In hierarchical clusters, if a and b are in one cluster, and observations c and d are in another cluster, a larger cluster that includes a and c should also include b and d. Linnaean taxonomy is an example of a hierarchical clustering system. [12]

hierarchical model A statistical model for which the parameters of simpler models are a subset of the parameters of more complex models. [11, 14]

Hill numbers A family of diversity indices with an exponent q that determines how much weight is given to rare species. Hill numbers obey the doubling property and express diversity in units of effective numbers of species. [13]

hinge The location of the upper or lower quartile in a stem-and-leaf plot. [8]

histogram A bar chart that represents a frequency distribution. The height of each of the bars equals the frequency of the observation identified on the x-axis. [1, 2]

homoscedastic The property of a dataset such that the residual variances of all treatment groups are equal. Contrast with **heteroscedastic.** [8]

horseshoe effect The undesirable property of many ordination methods that results from their exaggerating the mean and compressing the extremes of an environmental gradient. It leads to ordination plots that are curved into arches, horseshoes, or circles. [12]

hypergeometric random variable The outcome of an experiment that consists of multiple Bernoulli trials, sampled without replacement. Compare with **binomial random variable.** [13]

hypothesis A testable assertion of cause and effect. See also **alternative hypothesis** and **null hypothesis.** [4]

hypothetico-deductive method The scientific method championed by Karl Popper. Like induction, the hypothetico-deductive method begins with an individual observation. The observation could be predicted by many hypotheses, each of which provides additional predictions. The predictions are then tested one by one, and alternative hypotheses are eliminated until only one remains. Practitioners of the hypothetico-deductive method do not ever confirm hypotheses; they only reject them, or fail to reject them. When textbooks refer to "the scientific method," they are usually referring to the hypothetico-deductive method. Contrast with **induction, deduction,** and **Bayesian inference.** [4]

identity matrix A diagonal matrix in which all the diagonal terms = 1 and all other terms = 0. [A]

IID Stands for "independent and identically distributed." Good experimental design results in replicates and error distributions that are IID. See also **white noise, independent,** and **randomization.** [6]

impurity A measure of uncertainty in the endpoint of a classification tree. [11]

incidence matrix A site × species matrix in which each row represents a single species, each column represents an individual site, and cell entries are either 1 (indicating the species is present at the site) or 0 (indicating the species is absent at the site). [13, 14]

independent Two events are said to be independent if the probability of one event occurring is not related in any way to the probability of the occurrence of the other event. [1, 6]

independent variable In a statement of cause and effect, the predictor variable, or the object that is postulated to be the cause of the observed effect. Contrast with **dependent variable** and **response variable.** [7]

individual-based rarefaction Rarefaction by random subsampling of individual organisms from a reference sample, followed by calculation of a diversity index for the smaller subsample. The reference sample consists of a vector of abundances of different species. Individual-based rarefaction should be used only when individual organisms in a reference sample can be counted and identified to species (or a similar grouping). See also **sample-based rarefaction.** [13]

induction The process of scientific reasoning that proceeds from the specific (this pig is yellow) to the general (all pigs are yellow). Contrast with **deduction.** [4]

inference A conclusion that logically arises from observations or premises. [4]

influence function A diagnostic plot to assess the impact of each observation on the slope and intercept of a regression model. In this plot, the jackknifed intercepts (on the y-axis) are plotted against the jackknifed slopes (on the x-axis). The slope of this plot should always be negative, but outliers point at observations with undue effect (or influence) on the regression model. [9]

information criterion Any method for selecting among alternative statistical models that accounts for the number of parameters in the different models. Usually qualified (as in Akaike's information criterion or Bayesian information criterion). [9]

interaction The joint effects of two or more experimental factors. Specifically, the interaction represents a response that cannot be predicted simply knowing the main effect of each factor in isolation. Interactions result in non-additive effects. They are an important component of multi-factor experiments. [7, 10]

intercept The point at which a regression line crosses the y-axis; it is the expected value of Y when X equals zero. [9]

interpolation Estimation of unobserved values within the of range of observed values. Generally more reliable than **extrapolation.** [9, 13]

intersection In set theory, the elements that are common to two or more sets. In probability theory, the probability of two events occurring at the same time. In both cases, intersection is written with the symbol $\cap$. Contrast with **union.** [1]

interval A set of numbers with a defined minimum and a defined maximum. [2]

intrinsic hypothesis A statistical model for which the parameters are estimated from the data themselves. Contrast with **extrinsic hypothesis.** [11]

inverse matrix A matrix $\mathbf{A}^{-1}$ such that $\mathbf{AA}^{-1} = \mathbf{I}$, the identity matrix. [A]

inverse prediction interval Based on a regression model, the range of possible *X* (or predictor) values for a given *Y* (response) value. [9]

jackknife A general-purpose tool. In statistics, the jackknife is a randomization (or Monte Carlo) procedure by which one observation is deleted from the dataset, and then the test statistic is recomputed. This is repeated as many times as there are observations. The test statistic computed from the original dataset is then compared with the distribution of test statistics resulting from the repeated application of the jackknife to provide an exact *P*-value. See also **bootstrap** and **influence function.** [5, 12]

***K*-means clustering** A non-hierarchical method of grouping objects that minimizes the sums of squares of distances from each object to the centroid of its cluster. [12]

kurtosis A measure of how clumped or dispersed a distribution is relative to its center. Calculated using the fourth central moment, and symbolized as g_2. See also **platykurtic** and **leptokurtic.** [3]

Latin square A randomized block design in which n treatments are laid out in an $n \times n$ square, and in which each treatment appears exactly once in each row and in each column of the field. [7]

Law of Large Numbers A fundamental theorem of probability, the Law of Large Numbers says that the larger the sample, the closer the arithmetic mean comes to the probabilistic expectation of a given variable. [3]

least-trimmed squares A regression method for minimizing the effects of outliers. It computes regression slopes after removing some proportion (specified by the analyst) of the largest residual sums of squares. [9]

left-skewed A distribution that has most of its observations larger than the arithmetic mean but a long tail of observations smaller than the arithmetic mean. The skewness of a left-skewed distribution is less than 0. Compare with **right-skewed.** [3]

leptokurtic A probability distribution is said to be leptokurtic if it has relatively few values in the center of the distribution and relatively fat tails. The kurtosis of a leptokurtic distribution is greater than 0. Contrast with **platykurtic.** [3]

level The individual value of a given experimental treatment or factor in an analysis of variance or tabular design. [7, 11]

likelihood An empirical distribution that can allow for quantifying preferences among hypotheses. It is proportional to the probability of a specific dataset given a particular statistical hypothesis or set of parameters: $L(\text{hypothesis} \mid \text{observed data}) = cP(\text{observed data} \mid \text{hypothesis})$, but unlike a probability distribution, the likelihood is not constrained to fall between 0.0 and 1.0. It is one of two terms in the numerator of Bayes' Theorem. See also **maximum likelihood.** [5]

likelihood ratio test A method for testing among hypotheses based on the ratios of their relative probabilities. [11]

linear combination A re-expression of a variable as combination of other variables, in which the parameters of the new expression are linear. For example, the equation $Y = b_0X_0 + b_1X_1 + b_2X_2$ expresses the variable Y as a linear combination of X_0 to X_2 because the parameters b_0, b_1, and b_2 are linear. In contrast, the equation

$$Y = a_0X_0 + X_1^{a_1}$$

is not a linear combination, because the parameter a_1 is an exponential parameter, not a linear one. Note that the second equation can be transformed to a linear combination by taking logarithms of both sides of the equations. It is not always the case, however, that non-linear combinations can be transformed to linear combinations. [12]

linear regression model A statistical model for which the relationship between the predictor and response variables can be illustrated by a straight line. The model is said to be linear because its parameters are constants and affect the predictor variables only additively or multiplicatively. Contrast with **non-linear regression model**. [9]

loadings In a principal component analysis, the parameters that are multiplied by each variable of a multivariate observation to give a new variable, or principal component score for each observation. Each of these scores is a linear combination of the original variables. [12]

location The place in a probability distribution where most of the observations can be found. Compare with **spread**. [3]

log-likelihood function The logarithm of the likelihood function. Maximizing it is one way to determine which parameters of a formula or hypothesis best account for the observed data. See also **likelihood**. [8]

log-linear model A statistical model that is linear in the logarithms of the parameters. For example, the model $Y = aX^b$ is a log-linear model, because the logarithmic transform results in a linear model: $\log(Y) = \log(a) + b \times \log(X)$. [11]

log-normal random variable A random variable whose natural logarithm is a normal random variable. [2]

logarithmic transformation The function $Y^* = \log(Y)$, where log is the logarithm to any useful base. It often is used when means and variances are correlated positively with each other. [8, 9]

logic tree An example of the hypothetico-deductive method. An investigator works through a logic tree by choosing one of two alternatives at each decision point. The tree goes from the general (e.g., animal or vegetable?) to the specific (e.g., 4 toes or 3?). When there are no more choices left, the solution has been reached. A dichotomous key is an example of a logic tree. [4]

logistic regression A statistical model for estimating categorical response variables from continuous predictor variables. [9]

logit scale A rescaling of a probability from the range (0,1) to $(-\infty, +\infty)$:

$$\text{logit}(p_i) = \log\frac{p_i}{(1-p_i)}$$

where the logarithm is taken to the base e (i.e., the natural logarithm). Note that the fraction

$$\frac{p_i}{(1-p_i)}$$

is called the odds ratio, and represents the ratio of a "win" to a "loss." If $p_i = 0.5$, then the odds are even and the logit equals zero. Odds against winning (odds ratio < 0.5) yield a logit < 0, and odds in favor of winning (odds ratio > 0.5) yield a logit > 0.

The logit scale is often used in models where probabilities are a function of **covariates**, because probabilities are restricted to the range (0,1) but the covariates may have values outside of this range. See also **logistic regression** and **logit transformation.** [14]

logit transformation An algebraic transformation to convert the results of a logistic regression, which yields an S-shaped curve, to a straight line. [9]

lower triangular matrix A matrix in which all of the elements above the diagonal = 0. [A]

M-estimators A regression method for minimizing the effects of outliers. It computes regression slopes after down-weighting the largest sums of squares. [9]

main effect The individual effect of a single experimental factor. In the absence of any interaction, main effects are additive: the response of an experiment can be accurately predicted knowing the main effects of each individual factor. Contrast with **interaction.** [7, 10]

major axis In a principal component analysis, the first principal axis. Also, the longest axis of an ellipse. [12]

MANOVA See **multivariate analysis of variance.** [12]

manipulative experiment An experiment in which the investigator deliberately applies one or more treatments to a sample population or populations, and then observes the outcome of the treatment. Contrast with **natural experiment.** [6]

marginal probability When a model contains many random variables, the marginal probability is the probability of one of the random variables when the other random variables are fixed at their means. See also **hierarchical model.** [14]

marginal total The sum of the row or column frequencies in a contingency table. [7, 11]

Markov process A sequence of events for which the probability of a given event occurring depends only on the occurrence of the immediately preceding event. [3]

matched-pairs layout A form of a randomized block design in which replicates are selected to be similar in body size or other characteristic and then assigned randomly to treatments. [7]

matrix A mathematical object (often used as a variable) consisting of many observations (rows) and many variables (columns). [12]

maximum likelihood The most likely value of the likelihood distribution, computed by taking the derivative of the likelihood and determining when the derivative equals 0. For many parametric or frequentist statistical methods, the asymptotic test statistic is the maximum likelihood value of that statistic. [5]

mean The most likely value of a random variable or a set of observations. The term mean should be qualified: arithmetic mean, geometric mean, or harmonic mean. If unqualified, it is assumed to be the arithmetic mean. Compare with **median** and **mode.** [3]

mean square The average of the squared differences between the value of each observation and the arithmetic mean of the values. Also called the variance, or the second central moment. [3]

mean vector The vector of most likely values of multivariate random variables or multivariate data. [12]

median The value of a set of observations that is in the exact middle of all the values: one-half the values are above it and one-half are below it. The center of the fifth decile, or the 50th percentile. Compare with **mean** and **mode.** [3]

metadata Data about data. Prose descriptors that describe all the variables and features of a dataset. [8]

metapopulation A group of spatially distinct populations of a single species that are linked together by dispersal and migration. [14]

metric A type of a distance (or dissimilarity) measure that satisfies four properties: (1) the minimum distance equals 0 and the distance between identical objects equals 0; (2) the distance between two non-identical objects is positive; (3) distances are symmetric: the distance from object a to object b is the same as the distance from object b to object a; and (4) distance measurements satisfy the triangle inequality. Contrast with **semi-metric** and **non-metric.** [12]

mixed model An ANOVA model that includes both random effects and fixed effects. [10]

mode The most common value in a set of observations. Colloquially, the average (or population mean) often is mistaken for the mode. Compare with **mean** and **median.** [3]

model A mathematical function that relates one set of variables to another via a set of parameters. According to George E. P. Box, all models are wrong, but some are useful. [11]

Monte Carlo analysis One of three major frameworks for statistical analysis. It uses Monte Carlo methods to estimate P-values. See Monte Carlo methods. Contrast with **Bayesian inference** and **parametric analysis.** [5]

Monte Carlo methods Statistical methods that rely on randomizing or reshuffling the data, often using bootstrap or jackknife methods. The result of a Monte Carlo analysis is an exact probability value for the data, given the statistical null hypothesis. [5]

mosaic plot A graphical display designed for illustrating contingency tables. The size of the tiles in the mosaic is proportional to the relative frequency of each observation. [11]

multicollinearity The undesirable correlations between more than two predictor variables. See also **collinearity.** [7, 9]

multi-factor design An analysis of variance design that includes two or more treatments or factors. [7]

multinomial random variable The outcome of an experiment that consists of multiple categorical trials, each with more than two possible results. Compare with **binomial random variable.** [11, 13]

multinormal Abbreviation for multivariate normal. [12]

multiple regression A statistical model that predicts one response variable from more than one predictor variable. It may be linear or non-linear. [9]

multivariate analysis of variance (MANOVA) A method for testing the null hypothesis that the mean vectors of multiple groups are all equal to each other. An ANOVA for multivariate data. [12]

multivariate data Data consisting of more than two variables for each observation. Often illustrated using scatterplot matrices or other higher-dimensional visualization tools. [8, 12]

multivariate normal random variable The analog of a normal (Gaussian) random variable for multivariate data. Its probability distribution is symmetrical around its mean vector, and it is characterized by an n-dimensional mean vector and an $n \times n$ variance-covariance matrix. [12]

---

natural experiment A comparison among two or more groups that have not been manipulated in any way by the investigator; instead, the design relies on natural variation among groups. Natural experiments ideally compare groups that are identical in all ways, except for the single causative factor that is of interest. Although natural experiments can be analyzed with many of the same statistics as manipulative experiments, the resulting inferences may be weaker because of uncontrolled confounding variables. [1, 6]

nested design Any analysis of variance design for which there is subsampling within replicates.
[7, 10]

NMDS See **non-metric multidimensional scaling.** [12]

non-hierarchical clustering A method of grouping objects that is not based on an externally imposed reference system. Contrast with **hierarchical clustering.** [12]

non-linear regression model A statistical model for which the relationship between the predictor and response variables is not a straight line, and for which the predictor variables affect the response variables in non-additive or non-multiplicative ways. Contrast with **linear regression model.** [9]

non-metric A type of distance (or dissimilarity) measure that satisfies two properties: (1) the minimum distance equals 0 and the distance between identical objects equals 0; (2) distances are symmetric—that is, the distance from object *a* to object *b* is the same as the distance from object *b* to object *a*. Contrast with **metric** and **semimetric.** [12]

non-metric multidimensional scaling (NMDS) A method of ordination that preserves the rank ordering of the original distances or dissimilarities among observations. This general and robust method of ordination can use any distance or dissimilarity measure. [12]

non-parametric statistics A branch of statistical analysis that does not depend on the data being drawn from a defined random variable with known probability distribution. Largely superseded by Monte Carlo analysis. [5]

normal random variable A random variable whose probability distribution function is symmetrical around its mean, and which can be described by two parameters, its mean (μ) and its variance (σ^2). Also called a Gaussian random variable. [2]

nuisance parameter A parameter in a statistical model that is not of central concern but that still has to be estimated before we can estimate the parameter in which we are most interested. See also **parameter.** [14]

null hypothesis The simplest possible hypothesis. In ecology and environmental science, the null hypothesis is usually that any observed variability in the data can be attributed entirely to randomness or measurement error. Compare with **alternative hypothesis.** See also **statistical null hypothesis.** [1, 4]

oblique rotation A linear combination of the common factors in a factor analysis that results in common factors that may be correlated with each other. Contrast with **orthogonal rotation** and **varimax rotation.** [12]

occupancy The probability that a species is present at a site. Estimating it requires an estimate of **detection probability.** [14]

one-tailed test A test of the statistical alternative hypothesis that the observed test statistic is either greater than or less than (but not both) the expected value under the statistical null hypothesis. Because statistical tables normally give the results for two-tailed tests, the one-tailed probability value can be found by dividing the two-tailed probability value by 2. Contrast with **two-tailed test.** [5]

open interval An interval that does not include its endpoints. Thus, the open interval (0, 1) is the set of all numbers between

0 and 1, excluding the numbers 0 and 1. Contrast with **closed interval.** [2]

ordinate The vertical, or *y*-axis, on a plot. The opposite of **abscissa.** [7]

ordination Methods of reducing multivariate data by arranging the observations along a smaller number of axes than there are original variables. [12]

orthogonal In a multifactor analysis of variance (ANOVA), the property that all treatment combinations are represented. In a multiple regression design, the property that all values of one predictor variable are found in combination with each value of another predictor variable. In matrix algebra, two vectors whose product = 0. [7, 9, A]

orthogonal rotation A linear combination of the new common factors in a factor analysis that results in common factors that are uncorrelated with each other. Contrast with **oblique rotation** and **varimax rotation.** [12]

orthonormal A transformation applied to a matrix such that the Euclidean distances of the rows, columns, or both equal 1. [12]

outcome A result, usually of an observation or trial. [1]

outliers Unexpectedly large or small data points. [3, 8]

P-value See **probability value.** [4]

paradigm A research framework about which there is general agreement. Thomas Kuhn used the term paradigm in describing "ordinary" science, which is the activity scientists engage in to fit observations into paradigms. When too many observations don't fit, it's time for a new paradigm. Kuhn called changes in paradigms scientific revolutions. [4]

parameter A constant in statistical distributions and equations that needs to be estimated from the data. [3, 4]

parametric The assumption that statistical quantities can be estimated using probability distributions with defined and fixed constants. The requirement of many statistical tests that the data conform to a known and definable probability distribution. [3]

parametric analysis One of three major frameworks for statistical analysis. It relies on the requirement that the data are drawn from a random variable with probability distributions defined by fixed parameters. Also called frequentist analysis or frequentist statistics. Contrast with **Bayesian inference** and **Monte Carlo analysis.** [5]

partial regression parameters Parameters in a multiple regression model whose values reflect the contribution of all the other parameters in the model. [9]

path analysis A method for attributing multiple and overlapping causes and effects in a special type of multiple regression model. [9]

path coefficient A type of partial regression parameter calculated in path analysis that indicates the magnitude and sign of the effect of one variable on another. [9]

PCA See **principal component analysis.** [12]

PDF See **probability density function.** [2]

percentage Literally, "of 100" (Latin *per centum*), hence the relative frequency of a set of observations multiplied by 100. Percentages can be calculated from a contingency table, but the statistical analysis of such a table is always based on the original frequency counts. [11]

percentile The 100th part of something. Usually qualified, as in 5th percentile (a twentieth), 10th percentile (decile), etc. [3]

permutation An arrangement of discrete objects. Unlike a combination, a permutation is any rearrangement of objects. Thus, for the set {1,2,3}, there are six possible permutations: {1,2,3}, {1,3,2}, {2,1,3}, {2,3,1}, {3,2,1}, {3,1,2}. However, all of these represent the same combination of the three objects 1, 2, and 3. There are $n!$ permutations of n objects, but only

$$\frac{n!}{X!(n-X)!}$$

combinations of n objects taken X at a time. The symbol "!" is the mathematical operation **factorial.** See also **binomial coefficient, combination.** [2]

PEV See **proportion of explained variance.** [10]

PIE See **probability of an interspecific encounter.** [11, 13]

platykurtic A probability distribution is said to be platykurtic if it has relatively many values in the center of the distribution and relatively skinny tails. The kurtosis of a platykurtic distribution is greater than 0. Contrast with **leptokurtic.** [3]

plot An illustration of a variable or dataset. Technically, plots are elements of graphs. [8]

Poisson random variable The discrete result from a given experiment conducted in either a finite area or a finite amount of time. Unlike a binomial random variable, a Poisson random variable can take on any integer value. [2]

posterior odds ratio In a Bayesian analysis, the ratio of the posterior probability distributions of two alternative hypotheses. [9]

posterior probability The probability of a hypothesis, given the results of an experiment and any prior knowledge. The result of Bayes' Theorem. Contrast with **prior probability.** [1]

posterior probability distribution The distribution resulting from an application of Bayes' Theorem to a given dataset. This distribution gives the probability for the value of any hypothesis of interest given the observed data. [5, 9]

postmultiplication In matrix algebra, the product **BA** of two square matrices **A** and **B.** Also called the **postproduct.** [A]

postproduct See **postmultiplication.** [A]

power The probability of correctly rejecting a false statistical null hypothesis. It is calculated as $1 - \beta$, where β is the probability of committing a Type II error. [4]

precision The level of agreement among a set of measurements on the same individual. It is also used in Bayesian inference to mean the quantity equal to 1/ variance. Distributions with high precision have low variance. Contrast with **accuracy.** [2, 5]

predictor variable The hypothesized causal factor. See also **independent variable** and contrast with **response variable.** [7]

premultiplication In matrix algebra, the product **AB** of two square matrices **A** and **B.** Also called the **preproduct.** [A]

preproduct See **premultiplication.** [A]

press experiment A manipulative experiment in which the treatments are applied, and reapplied throughout the duration of the experiment in order to maintain the strength of the treatment. Contrast with **pulse experiment.** [6]

principal axis A new variable created by an ordination along which the original observations are ordered. [12]

principal components The principal axes ordered by the amount of variability that they account for in the original multivariate dataset. [12]

principal component analysis (PCA) A method of ordination that reduces multivariate data to a smaller number of variables by creating linear combinations of the original variables. It was originally developed to assign racial identity to individuals based on multiple biometric measurements. [12]

principal component score The value of each observation that is a linear combination of the original variables measured for that observation. It is determined by a principal component analysis. [12]

principal coordinates analysis A generic ordination method that can use any measure of distance. Principal components analysis and factor analysis (for example) are special cases of principal coordinates analysis. [12]

prior odds ratio The ratio of the prior probability distributions of two alternative hypotheses. [9]

prior probability The probability of a hypothesis before the experiment has been conducted. This can be determined from previous experience, intuition, expert opinion, or literature reviews. Contrast with **posterior probability**. [1]

prior probability distribution One of two distributions used in the numerator of Bayes' Theorem to compute the posterior probability distribution (the other is the likelihood). It defines the probability for the value of any hypothesis of interest before the data are collected. See also **prior probability** and **likelihood**. [5, 9]

probability The proportion of particular results obtained from a given number of trials, or the proportion expected for an infinite number of trials. [1]

probability calculus The mathematics used to manipulate probabilities. [1]

probability density function (PDF) The mathematical function that assigns the probability of observing outcome X_i to each value of a random variable X. [2]

probability distribution A distribution of outcomes X_i of a random variable X. For continuous random variables, the probability distribution is a smooth curve, and a histogram only approximates the probability distribution. [2]

probability of an interspecific encounter (PIE) A diversity index that measures the probability that two randomly selected individuals from an assemblage represent two different species. *PIE* is also a measure of the slope of the individual-based rarefaction curve measured at its base and is closely related to the Gini coefficient used in economic analysis. [11, 13]

product-moment correlation coefficient (*r*) A measure of how well two variables are related to one another. It can be calculated as the square root of the coefficient of determination, or as the sum of cross products divided by the square root of the product of the sums of squares of the X and Y variables. [9]

proportion Relative frequency. The number of observations in a given category divided by the total number of observations. Proportions can be calculated from a contingency table, but the statistical analysis of such a table is always based on the original frequency counts. [11]

proportion of variance explained (PEV) The amount of variability in the data that is accounted for by each factor in an analysis of variance. It is equivalent to the coefficient of determination in a regression model. [10]

probability value (*P*-value) The probability of rejecting a true statistical null hypothesis. Also called the probability of a Type I error. [4]

pulse experiment A manipulative experiment in which the treatments are applied only once, and then the treated replicate is allowed to recover from the manipulation. Contrast with **press experiment.** [6]

quadrat A sampling plot, usually square or circular of a fixed and known area that is used for standardized ecological sampling in both terrestrial and aquatic habitats. [2]

quadratic form A 1×1 matrix equal to the product of $\mathbf{x}^T\mathbf{A}\mathbf{x}$, where $\mathbf{x}$ is a $n \times 1$ column vector and $\mathbf{A}$ is an $n \times n$ matrix. [A]

quality assurance and quality control (QA/QC) The process by which data (or any activity) are monitored to determine that they have the expected accuracy and precision. [8]

quantile An *n*th part of something. Specific types of quantiles include deciles, percentiles, and quartiles. [3]

quantile regression A regression model on a subset (an analyst-defined quantile) of the data. [9]

quartile One-quarter of something. The lower and upper quartiles of a dataset are the bottom 25% and top 25% of the data. [3]

random effects In an analysis of variance design, the set of treatment levels that represents a random subset of all possible treatment levels. Also known as a random factor. Contrast with **fixed effects.** [7, 10]

random variable The mathematical function that assigns a numerical value to each experimental result. [2]

randomization The assignment of experimental treatments to populations based on random number tables or algorithms. Contrast with **haphazard.** [6]

randomization tests Statistical tests that rely on reshuffling the data using bootstrap, jackknife, or other re-sampling strategies. The reshuffling simulates the data that would have been collected if the statistical null hypothesis were true. See also **Monte Carlo methods, bootstrap,** and **jackknife.** [5]

randomized block design An analysis of variance design in which sets of replicates of all treatment levels are placed into a fixed area in space or a fixed point in time (a block) in order to reduce the variability of unmanipulated conditions. [7, 10]

randomized intervention analysis A Monte Carlo technique for the analysis of BACI designs. [7]

rarefaction The statistical procedure of randomly subsampling individuals, plots, or other sampling units to estimate diversity indices at smaller sample sizes. Rarefaction allows for the standardization of biodiversity sample data to a common sampling effort. [13]

rate parameter The constant that is entered into the formula for a Poisson random variable. Also, the average, or expected value, and the variance of a Poisson random variable. [2]

RDA See **redundancy analysis.** [12]

reciprocal transformation The function $Y^* = 1/Y$. It is often used for hyperbolic data. [8]

redundancy analysis (RDA) A type of multiple regression used when both predictor and response variables are multivariate. Unlike the related canonical correspondence analysis, redundancy analysis makes no assumptions about the shape of either the predictor or response axes. [12]

reference sample A random or representative collection of individuals or of replicated sampling units for the purposes of estimating biodiversity. [13]

regression The method of analysis used to explore cause-and-effect relationships between two variables in which the predictor variable is continuous. Contrast with **correlation.** [7, 9]

regression design The class of experimental layouts used to explore relationships between continuous predictor variables and continuous or categorical response variables. [7]

regression tree A model or visual display of continuous data that is constructed by splitting the dataset into successively smaller groups. Splitting continue until no further improvement in model fit is obtained. A regression tree is analogous to a taxonomic key, except that divisions in a regression tree reflect probabilities, not identities. See also **classification tree; cluster diagram**. [11]

reification Converting an abstract concept into a material object. [8]

relative frequency See **proportion.** [11]

repeated measures design A type of ANOVA in which multiple observations are taken on the same individual replicate at different points in time. [7, 10]

replicate An individual observation or trial. Most experiments have multiple replicates in each treatment category or group. [1, 7]

replication The establishment of multiple plots, observations, or groups within the same experimental or observational treatment. Ideally, replication is achieved by randomization. [6]

residual The difference between an observed value and the value predicted by a statistical model. [8, 9]

residual plot A diagnostic graph used to determine if the assumptions of regression or ANOVA have been met. Most commonly, the residuals are plotted against the predicted values. [9]

residual sum of squares What's left over (unexplained) by a regression or ANOVA model. It is calculated as the sum of the squared deviations of each observation from the mean of all observations in its treatment groups, and then these sums are summed over all treatment groups. Also called the **error variation** or the **residual variation.** [9, 10]

residual variation What's left over (unexplained) by a regression or ANOVA model. Also called the **error variation** or the **residual sum of squares.** [9, 10]

response surface design An experimental design used to study competition between a target or focal species of interest and its competitors. In a response surface design, the density of both the target species and its competitors are systematically varied. Response surface designs can also be used in lieu of ANOVA when the factors are better represented as continuous variables. Contrast with **additive design** and **substitutive design.** [7]

response variable The effect of an hypothesized causal factor. See also **dependent variable** and contrast with **predictor variable.** [7]

right-skewed A distribution that has most of its observations smaller than the arithmetic mean but a long tail of observations larger than the arithmetic mean. The skewness of a right-skewed distribution is greater than 0. Compare with **left-skewed.** [3]

robust regression A regression model that is relatively insensitive to outliers or extreme values. Examples include least-trimmed regression and *M*-estimators. [9]

rotation In a factor analysis, rotation is a linear combination of common factors that are easy to interpret. See **oblique rotation**, **orthogonal rotation**, and **varimax rotation.** [12]

row total The marginal sum of the entries in a single row of a contingency table. See also **column total** and **grand total.** [11]

row vector A matrix with many columns but only one row. Contrast with **column vector.** [12]

Rule of 10 The guiding principle that says you should have a minimum of 10 replicates per observational category or treatment group. It is based on experience, not on any mathematical axiom or theorem. [6]

---

sample A subset of the population of interest. Because it is rarely possible to observe or manipulate all the individuals in a population, analyses are based on a sample of the population. Probability and statistics are used to extrapolate results from the smaller sample to the larger population. [1]

sample-based rarefaction Rarefaction by random sub-sampling of plots, traps, or other replicated sampling units followed by calculation of a diversity index for the smaller set of samples. The reference sample consists of a matrix with rows representing species, columns representing replicate samples, and entries in the matrix representing either the abundances or the incidence (occurrence) of species. Sample-based rarefaction should always be used when the independently replicated observations are the samples, not the individuals within a sample. See also **individual-based rarefaction.** [13]

sample space The set of all possible results or outcomes of an event or experiment. [1]

sample standard deviation An unbiased estimator of the standard deviation of a sample. Equal to the square root of the sample variance. [3]

sample variance An unbiased estimator of the variance of a sample. Equal to the sums of squares divided by the sample size – 1. [3]

sampling effect The observation that nearly all indices of biodiversity are sensitive to the number of individuals and samples that are collected. Therefore, comparisons of biodiversity among treatments or habitats must be standardized to control for the sampling effect. [13]

scalar A single number, not a matrix. [A]

scalar product In matrix algebra, the product of a $1 \times n$ row vector and a $n \times 1$ column vector. [A]

scatterplot A two-dimensional plot or graph used to illustrate bivariate data. The predictor or independent variable is usually placed on the x-axis and the response or dependent variable is usually placed on the y-axis. Each point represents the measurements of these two variables for each observation. [8]

scatterplot matrix A graph made up of many scatterplots, and used to illustrate multivariate data. Each plot is a simple scatterplot, and they are grouped together by rows (which have common Y variables) and columns (which have common X variables). The arraying of these plots into a matrix of plots illustrates all possible bivariate relationships between all the measured variables. [8]

scientific hypothesis A statement of cause and effect. The accuracy of a scientific hypothesis is inferred from the results of testing statistical null and alternative hypotheses. [4]

scree plot A diagnostic plot used to decide how many components, factors, or axes to retain in an ordination. It is shaped like a collapsed mountainside (or scree slope); the meaningful components are the slope and the meaningless components are the rubble at the bottom. [12]

SD See **standard deviation.** [3]

seasons In occupancy or mark-recapture models, the time periods between which local colonization and extinction could occur. See also **closure assumption.** [14]

semi-metric A type of a distance (or dissimilarity) measure that satisfies three properties: (1) the minimum distance equals 0 and the distance between identical objects equals 0; (2) the distance between two non-identical objects is positive; (3) distances are symmetric—that is, the distance from object a to object b is the same as the distance from object b to object a. Contrast with **metric** and **non-metric.** [12]

set A collection of discrete objects, outcomes, or results. Sets are manipulated using the operations **union, intersection**, and **complement.** [1]

shared events In set theory, shared events are collections of simple events. If A and B are two independent events, then $C = A$ *and* B, written as $C = A \cap B$ and read as "A intersection B," is a shared event. If A and B are independent events, each with associated probabilities P_A and P_B, then the probability of C (P_C) equals $P_A \times P_B$. Contrast with **complex events.** [1]

shift operation The transformation of a normal random variable brought about by adding a constant to each observation. See also **change of scale.** [2]

simultaneous prediction interval An adjustment to a confidence interval in regression analysis that is necessary if multiple values are to be estimated. [9]

single factor design An analysis of variance design that manipulates only one variable. Contrast with **multi-factor design.** [7]

singletons The number of species represented by exactly one individual in an individual-based reference sample. See also **doubletons, duplicates**, and **uniques**. [13]

singular values The values along the diagonal of a square matrix **W** such that an $m \times n$ matrix **A** can be re-expressed as the product $\mathbf{VWU}^T$, where **V** is an $m \times n$ matrix of the same dimension as **A**; **U** and **W** are square matrices (dimension $n \times n$); and **V** and **U** are column-orthonormal matrices. [A]

site scores One result of a redundancy analysis. The site scores are the matrix **F** resulting from an ordination of the original data relative to the environmental characteristics: $\mathbf{F} = \mathbf{YA}$, where **Y** is the multivariate response variable and **A** is the matrix of eigenvectors resulting from a principal component analysis of the expected (fitted) values of the response variable. Contrast with **fitted site scores** and **species scores.** [12]

skewness A description of how far off a distribution is from being symmetrical. Abbreviated as g_1 and calculated from the third central moment. See also **right-skewed** and **left-skewed.** [3]

slope Rise over run. The change in the expected value of a response variable for a given change in the predicted variable. The parameter β_1 in a regression model. [9]

snapshot experiment A type of natural experiment in which all the replicates are sampled at one point in time. Replication is based on spatial variability. Contrast with **trajectory experiment.** [6]

species accumulation curve A graph of the relationship between the number of individuals sampled (x-axis) and the observed number of species (y-axis). As more individuals are accumulated, the curve rises relatively steeply from its origin (1,1), and then less steeply as the curve approaches an asymptote, beyond which no additional species will be found. [13]

species density The number of species represented in a collection, but not properly standardized to account for sampling effort. [13]

species evenness The extent to which the relative abundances of species in an assemblage or reference sample approach a uniform distribution. [13]

species richness The number of species in an assemblage or reference sample, ideally standardized by rarefaction or extrapolation to a common number of individuals, samples, or coverage for the purposes of comparison. [13]

species scores One result of a redundancy analysis. The species scores are the matrix of eigenvectors **A** resulting from a principal component analysis of the standardized sample variance-covariance matrix of the predicted values of the response variable ($\hat{\mathbf{Y}}$). It is used to generate both site scores and fitted site scores. [12]

sphericity The assumption of multivariate hypothesis testing methods that the covariances of each group are equal to one another. [12]

split-plot design A multi-factor analysis of variance design in which each experimental plot is subdivided into subplots, each of which receives a different treatment. [7, 10]

spread A measure of the variation within a set. [3]

spreadsheet A computer file in which each row represents a single observation and each column represents a single measured or observed variable. The standard form in which ecological and environmental data are placed into electronic form. See also **flat files.** [8]

square matrix A matrix with the same number of rows as columns. [12]

square-root transformation The function $Y^* = \sqrt{Y}$. It is used most frequently for count data that are Poisson random variables. [8]

SSCP See **sums of squares and cross products.** [12]

standard deviation (SD) A measure of spread. Equal to the square-root of the variance. [3]

standard error of the mean An asymptotic estimate of the population standard deviation. Often presented in publications in lieu of the sample standard deviation. The standard error of the mean is smaller than the sample standard deviation, and may give the misleading impression of less variability in the data. [3]

standard error of regression For a regression model, the square root of the residual sums of squares divided by the degrees of freedom of the model. [9]

standard normal distribution A normal probability distribution with mean = 0 and variance = 1. [2]

standard normal random variable A probability distribution function that yields a normal random variable with mean = 0 and variance = 1. If a variable X is standard normal, it is written as $X \sim N(0,1)$. Often indicated by Z. [2]

standardized residual A residual transformed to a Z-score. [9]

statistical null hypothesis The formal hypothesis that there is no mathematical relationship between variables, or the statistical form of the scientific hypothesis that any observed variability in the data can be attributed entirely to randomness or measurement error. See also **null hypothesis.** [4]

stem-and-leaf plot A data-rich histogram in which the actual value of each datum is illustrated. [8]

stepwise regression Any type of regression procedure that involves assessing multiple models with shared parameters. The goal is to decide which of the parameters are useful in the final regression model. Examples of methods used in stepwise regression include **backward elimination** and **forward selection.** [9]

stress A measure of how well the results of a non-metric multidimensional scaling predict the observed values. Also, the result of spending too much time on statistical analysis and not enough time in the field. [12]

studentized residuals See **standardized residuals.** [9]

subplot factor In a split-plot design, the treatment that is applied within a single block or plot. Contrast with **whole-plot factor.** [7]

subset A collection of objects that are also part of a larger group or set of objects. See also **set.** [1]

substitutive design An experimental design used to study competition between a target or focal species of interest and its competitors. In a substitutive design, the total density of both the target species and its competitors is maintained at a constant level, but the relative proportion of the target species and its competitors are systematically varied. Contrast with **additive design** and **response surface design.** [7]

sum of cross products (SCP) The sum of the product of the differences between each observation and its expected value. The sum of cross products divided by the sample size, is the **covariance.** [9]

sums of squares (SS) The sum of the squared differences between the value of each observation and the arithmetic mean of all the values. Used extensively in ANOVA calculations. [3]

sums of squares and cross products (SSCP) A square matrix with the number of rows and columns equal to the number of variables in a multivariate dataset. The diagonal elements of this matrix are the sums of squares of each variable, and the off-diagonal elements are the sums of the cross products of each pair of variables. SSCP matrices are used to compute test statistics in MANOVA. [12]

summary statistics Numbers that describe the location and spread of the data. [3]

syllogism A logical deduction, or conclusion, (e.g., this is a pig) is derived from two premises (e.g., 1, all pigs are yellow; 2, this object is yellow). Syllogisms were first described by Aristotle. Although syllogisms are logical, they may lead to erroneous conclusions (e.g., what if the yellow object is actually a banana?). [4]

symmetric An object whose two sides are mirror images of each other. A matrix is said to be symmetric if a mirror were to be aligned along its diagonal; the (*row*, *column*) values above the diagonal would be equal to the (*column*, *row*) values below the diagonal. [2,12]

---

t-distribution A modification of the standard normal probability distribution. For small sample sizes, the *t*-distribution is leptokurtic, but as sample size increases, the *t*-distribution approximates the standard normal distribution. The tail probabilities of the *t*-distribution are used to compute confidence intervals. [3]

tail probability The probability of obtaining not only the observed data, but also all more extreme data that were possible, but not observed. Also called a *P*-value (probability value) and calculated using the cumulative distribution function. See also **probability value.** [2, 5]

test statistic The numerical result of manipulating the data that is used to examine a statistical hypothesis. Test statistics are compared to values that would be obtained from identical calculations if the statistical null hypothesis were in fact true. [4]

theory In science, a theory is a collection of accepted knowledge that has been built up through repeated observations and statistical testing of hypotheses. Theories form the basis for **paradigms.** [4]

time series model A statistical model for which time is included explicitly as a predictor variable. Many time-series models are also **autoregressive models.** [6]

tolerance A criterion for deciding whether or not to include variables or parameters in a stepwise regression procedure. Tolerance reduces multicollinearity among candidate variables. [9]

trace The sum of the main diagonal elements of a square matrix. [A]

trajectory experiment A type of natural experiment in which all the replicates are sampled repeatedly at many points in time. Replication is based on temporal variability. Trajectory experiments often are analyzed using time series or autoregressive models. Contrast with **snapshot experiment**. [6]

transpose A rearrangement of a matrix such that the rows of the original matrix equal the columns of the new (transposed matrix), and the columns of the original matrix equal the rows of the transposed matrix. Symbolically, for a matrix **A** with elements a_{ij}, the transposed matrix $\mathbf{A}^T$ has elements $a'_{ij} = a_{ji}$. Transposition of matrices is indicated either by a superscript T or by the prime (′) symbol. [12]

treatment A set of replicated environmental conditions imposed and maintained by an investigator. In the context of statistical analysis, the set of categories of predictor variables used in an analysis of variance design. Treatments are equivalent to factors, and are comprised of levels. [7]

trial An individual replicate or observation. Many trials make for a statistical experiment. [1]

triangle inequality For three multivariate objects *a*, *b*, and *c*, the distance between *a* and *b* plus the distance between *b* and *c* is always greater than or equal to the distance between *a* and *c*. [12]

triangular matrix A matrix in which all of the elements on one side of the diagonal = 0 [A].

two-tailed test A test of the statistical alternative hypothesis that the observed test statistic is not equal to the expected value under the statistical null hypothesis. Statistical tables and software packages normally provide critical values and associated *P*-values for two-tailed tests. Contrast with **one-tailed test**. [5]

two-way design An ANOVA design with two main effects, each of which has two or more treatment levels. [7, 10]

Type I error Falsely rejecting a true statistical null hypothesis. Usually indicated by the Greek letter alpha (α). [4]

Type II error Incorrectly accepting a false statistical null hypothesis. Usually indicated by the Greek letter beta (β). Statistical power equals $1 - \beta$. [4]

unbiased The statistical property of measurements, observations, or values that are neither consistently too large nor consistently too small. See also **accuracy** and **precision**. [2]

unbiased estimator A calculated statistical parameter that is neither consistently larger nor smaller than the true value. [3, 9]

uniform random variable A random variable for which any result is equally likely. [2]

union In set theory, all the unique elements obtained by combining two or more sets. In probability theory, the probability of two independent events occurring. In both cases, union is written with the symbol $\cup$. Contrast with **intersection**. [1]

uniques The number of species represented in exactly one sample in a sample-based reference sample. See also **doubletons**, **duplicates**, and **singletons**. [13]

univariate data Data consisting of a single variable for each observation. Often illustrated using box plots, histograms, or stem-and-leaf plots. [8, 12]

upper triangular matrix A matrix in which all of the elements below the diagonal = 0. [A]

variance A measure of how far the observed values differ from the expected values. [2]

variance-covariance matrix A square matrix **C** (n rows $\times$ n columns, where n is the number of measurements or individual variables for each multivariate observation) for which the diagonal elements $c_{i,i}$ are the variances of each variable and the off-diagonal elements $c_{i,i}$ are the covariances between variable i and variable j. [9, 12]

variation Uncertainty; difference. [1]

variation among groups In an analysis of variance, the squared deviations of the means of the treatment groups from the grand mean, summed over all treatment groups. Also called the sum of squares among groups. [10]

variation within groups See **residual sum of squares.** [10]

varimax rotation A linear combination of the common factors in a factor analysis that maximizes the total variance of the common factors. Contrast with **oblique rotation** and **orthogonal rotation.** [12]

Venn diagram A graphical display used to illustrate the operations union, intersection, and complement on sets. [1]

---

white noise An error distribution for which the errors are independent and uncorrelated (see **IID**). It is called white noise by analogy with light: just as white light is a mixture of all wavelengths, white noise is a mixture of all error distributions. [6]

whole-plot factor In a split-plot design, the treatment that is applied to an entire block or plot. Contrast with **subplot factor.** [7]

within-subjects factor In ANOVA, the term that accounts for variance within treatment groups. In a repeated measures design, the within-subjects factor is equivalent to the subplot factor of a split-plot design. [10]

---

x-axis The horizontal line on a graph, also called the abscissa. The x-axis usually represents the causal, independent, or predictor, variable(s) in a statistical model. [1]

---

y-axis The vertical line on a graph, also called the ordinate. The y-axis usually represents the resulting, dependent, or response variable(s) in a statistical model. [1]

Z-score The result of transforming a variable by subtracting from it the sample mean and dividing the difference by the sample standard deviation. [12]

Literature Cited

Abramsky, Z., M. L. Rosenzweig and A. Subach. 1997. Gerbils under threat of owl predation: Isoclines and isodars. *Oikos* 78: 81–90. [7]

Albert, J. 1996. Bayesian selection of log-linear models. *Canadian Journal of Statistics* 24: 327–347. [11]

Albert, J. H. 1997. Bayesian testing and estimation of association in a two-way contingency table. *Journal of the American Statistical Association* 92: 685–693. [11]

Albert, J. 2007. *Bayesian computation with R*. Springer Science+Business Media, LLC, New York. [5]

Alexander, H. M., N. A. Slade and W. D. Kettle. 1997. Application of mark-recapture models to estimation of the population size of plants. *Ecology* 78: 1230–1237. [14]

Allison, T. and D. V. Cicchetti. 1976. Sleep in mammals: Ecological and constitutional correlates. *Science* 194: 732–734. [8]

Allran, J. W. and W. H. Karasov. 2001. Effects of atrazine on embryos, larvae, and adults of anuran amphibians. *Environnmental Toxicology and Chemistry* 20: 769–775. [7]

Alroy, J. 2010. The shifting balance of diversity among major marine animal groups. *Science* 329: 1191–1194. [13]

Anderson, M. J. et al. 2011. Navigating the multiple meanings of β diversity: a roadmap for the practicing ecologist. *Ecology Letters* 14: 19–28. [13]

Anscombe, F. J. 1948. The transformation of Poisson, binomial, and negative binomial data. *Biometrika* 35: 246–254. [8]

Arnould, A. 1895. *Les croyances fondamentales du bouddhisme; avec préf. et commentaries explicatifs.* Société théosophique, Paris. [1]

Arnqvist, G. and D. Wooster. 1995. Meta-analysis: Synthesizing research findings in ecology and evolution. *Trends in Ecology and Evolution* 10: 236–240. [10]

Bailey, L. L., J. E. Hines, J. D. Nichols and D. I. MacKenzie. 2007. Sampling design trade-offs in occupancy studies with imperfect detection: examples and software. *Ecological Applications* 17: 281–290. [14]

Baillargeon, S. and L.-P. Rivest. 2007. Rcapture: loglinear models for capture-recapture in R. *Journal of Statistical Software* 19: 5. [14]

Barker, S. F. 1989. *The elements of logic*, 5th ed. McGraw Hill, New York. [4]

Beltrami, E. 1999. *What is random? Chance and order in mathematics and life*. Springer-Verlag, New York. [1]

Bender, E. A., T. J. Case and M. E. Gilpin. 1984. Perturbation experiments in community ecology: Theory and practice. *Ecology* 65: 1–13. [6]

Berger, J. O. and D. A. Berry. 1988. Statistical analysis and the illusion of objectivity. *American Scientist* 76: 159–165. [5]

Berger, J. O. and R. Wolpert. 1984. *The likelihood principle*. Institute of Mathematical Statistics, Hayward, California. [5]

Bernardo, J., W. J. Resetarits and A. E. Dunham. 1995. Criteria for testing character displacement. *Science* 268: 1065–1066. [7]

Björkman, O. 1981. Responses to different quantum flux densities. Pp. 57–105 in O. L. Lange, P. S. Nobel, C. B. Osmond and H. Ziegler (eds.). *Encyclopedia of plant physiology*, new series, vol. 12A. Springer-Verlag, Berlin. [4]

Bloor, M. 2005. Population estimation without censuses or surveys: a discussion of mark-recapture methods illustrated by results from three studies. *Sociology* 39: 121–138. [14]

Blume, J. D. and R. M. Royall. 2003. Illustrating the Law of Large Numbers (and confidence intervals). *American Statistician* 57: 51–57. [3]

Boecklen, W. J. and N. J. Gotelli. 1984. Island biogeographic theory and conservation practice: Species–area or specious-area relationships? *Biological Conservation* 29: 63–80. [8]

Bolker, B. M. 2008. *Ecological models and data in R.* Princeton University Press, Princeton, NJ. [5]

Boone, R. D., D. F. Grigal, P. Sollins, R. J. Ahrens and D. E. Armstrong. 1999. Soil sampling, preparation, archiving, and quality control. Pp. 3–28 in G. P. Robertson, D. C. Coleman, C. S. Bledsoe and P. Sollins (eds.). *Standard soil methods for long-term ecological research.* Oxford University Press, New York. [8]

Box, G. E. P. and D. R. Cox. 1964. An analysis of transformations. *Journal of the Royal Statistical Society, Series B* 26: 211–243. [8]

Boyer, C. B. 1968. *A history of mathematics.* John Wiley & Sons, New York. [2]

Breiman, L., J. H. Friedman, R. A. Olshen and C. G. Stone. 1984. *Classification and regression trees.* Wadsworth, Belmont, CA. [11]

Brett, M. T. and C. R. Goldman. 1997. Consumer versus resource control in freshwater pelagic food webs. *Science* 275: 384–386. [7]

Brezonik, P. L. et al. 1986. Experimental acidification of Little Rock Lake, Wisconsin. *Water Air Soil Pollution* 31: 115–121. [7]

Brown, J. H. and G. A. Leiberman. 1973. Resource utilization and coexistence of seed-eating desert rodents in sand dune habitats. *Ecology* 54: 788–797. [7]

Burnham, K. P. and D. R. Anderson. 2010. *Model selection and multi-modal inference: A practical information-theoretic approach*, 2nd ed. Springer-Verlag, New York. [5, 6, 7, 9]

Butler, M. A. and J. B. Losos. 2002. Multivariate sexual dimorphism, sexual selection, and adaptation in Greater Antillean *Anolis* lizards. *Ecological Monographs* 72: 541–559. [7]

Cade, B. S. and B. R. Noon. 2003. A gentle introduction to quantile regression for ecologists. *Frontiers in Ecology and the Environment* 1: 412–420. [9]

Cade, B. S., J. W. Terrell and R. L. Schroeder. 1999. Estimating effects of limiting factors with regression quantiles. *Ecology* 80: 311–323. [9]

Caffey, H. M. 1982. No effect of naturally occurring rock types on settlement or survival in the intertidal barnacle, *Tesseropora rosea* (Krauss). *Journal of Experimental Marine Biology and Ecology* 63: 119–132. [7]

Caffey, H. M. 1985. Spatial and temporal variation in settlement and recruitment of intertidal barnacles. *Ecological Monographs* 55: 313–332. [7]

Cahill, J. F., Jr., J. P. Castelli and B. B. Casper. 2000. Separate effects of human visitation and touch on plant growth and herbivory in an old-field community. *American Journal of Botany* 89: 1401–1409. [6]

Carlin, B. P. and T. A. Louis. 2000. *Bayes and empirical Bayes methods for data analysis*, 2nd ed. Chapman & Hall/CRC, Boca Raton, FL. [5]

Carpenter, S. R. 1989. Replication and treatment strength in whole-lake experiments. *Ecology* 70: 453–463. [6]

Carpenter, S. R., T. M. Frost, D. Heisey and T. K. Kratz. 1989. Randomized intervention analysis and the interpretation of whole-ecosystem experiments. *Ecology* 70: 1142–1152. [6, 7]

Carpenter, S. R., J. F. Kitchell, K. L. Cottingham, D. E. Schindler, D. L. Christensen, D. M. Post and N. Voichick. 1996. Chlorophyll variability, nutrient input, and grazing: Evidence from whole lake experiments. *Ecology* 77: 725–735. [7]

Caswell, H. 1988. Theory and models in ecology: a different perspective. *Ecological Modelling* 43: 33–44. [4, 6]

Chao, A. 1984. Non-parametric estimation of the number of classes in a population. *Scandinavian Journal of Statistics* 11: 265–270. [13]

Chao, A. and L. Jost. 2012. Coverage-based rarefaction and extrapolation: standardizing samples by completeness rather than size. *Ecology* (in press). dx.doi.org/10.1890/11-1952.1 [13]

Chao, A. and T.-J. Shen. 2003. Nonparametric estimation of Shannon's index of diversity when there are unseen species in the sample. *Environmental and Ecological Statistics* 10: 429–443. [13]

Chao, A., R. L. Chazdon, R. K. Colwell and T.-J. Shen. 2005. A new statistical approach for assessing compositional similarity based on incidence and abundance data. *Ecology Letters* 8: 148–159. [13]

Chao, A., C.-H. Chiu and T. C. Hsieh. 2012. Proposing a resolution to debates on diversity partitioning. *Ecology* 93: 2037–2051. [13]

Chao, A., R. K. Colwell, C. W. Lin and N. J. Gotelli. 2009. Sufficient sampling for asymptotic minimum species richness estimators. *Ecology* 90: 1125–1133. [13]

Chapman, D. G. 1951. Some properties of the hypergeometric distribution with applications to zoological sample censuses. *University of California Publications in Statistics* 1: 131–160. [14]

Chiarucci, A., G. Bacaro, D. Rocchini and L. Fattorini. 2008. Discovering and rediscovering the sample-based rarefaction formula in the ecological literature. *Community Ecology* 9: 121–123. [13]

Chipman, H. A., E. I. George and R. E. McCulloch. 2010. Bart: Bayesian additive regression trees. *Annals of Applied Statistics* 4: 266–298. [11]

Clark, J. S. 2007. *Models for ecological data: an introduction*. Princeton University Press, Princeton, New Jersey. [5]

Clark, J. S., J. Mohan, J. Dietze and I. Ibanez. 2003. Coexistence: How to identify trophic trade-offs. *Ecology* 84: 17–31. [4]

Clarke, K. R. 1993. Non-parametric multivariate analysis of changes in community structure. *Australian Journal of Ecology* 18: 117–143. [12]

Clarke, K. R. and R. H. Green. 1988. Statistical design and analysis for a "biological effects" study. *Marine Ecology Progress Series* 46: 213–226. [12]

Cleveland, W. S. 1985. *The elements of graphing data*. Hobart Press, Summit, NJ. [8]

Cleveland, W. S. 1993. *Visualizing data*. Hobart Press, Summit, NJ. [8]

Cochran, W. G. and G. M. Cox. 1957. *Experimental designs*, 2nd ed. John Wiley & Sons, New York. [7]

Coddington, J. A., I. Agnarsson, J. A. Miller, M. Kuntner and G. Hormiga. 2009. Undersampling bias: the null hypothesis for singleton species in tropical arthropod surveys. *Journal of Animal Ecology* 78: 573–584. [13]

Cody, M. L. 1974. *Competition and the structure of bird communities*. Princeton University Press, Princeton, NJ. [6]

Colwell, R. K. 2011. *EstimateS, Version 8. 2: Statistical Estimation of Species Richness and Shared Species from Samples (Software and User's Guide)*. Freeware for Windows and Mac OS. www.purl.oclc.org/estimates [13]

Colwell, R. K. and J. A. Coddington. 1994. Estimating terrestrial biodiversity through extrapolation. *Philosophical Transactions of the Royal Society of London B* 345: 101–118. [12, 13]

Colwell, R. K., A. Chao, N. J. Gotelli, S -Y. Lin, C. X. Mao, R. L. Chazdon and J. T. Longino. 2012. Models and estimators linking individual-based and sample-based rarefaction, extrapolation, and comparison of assemblages. *Journal of Plant Ecology* 5: 3–21. [13]

Colwell, R. K., C. X. Mao and J. Chang. 2004. Interpolating, extrapolating, and comparing incidence-based species accumulation curves. *Ecology* 85: 2717–2727. [13]

Congdon, P. 2002. *Bayesian statistical modeling*. John Wiley & Sons, Chichester, UK. [9]

Connolly, S. R. and M. Dornelas. 2011. Fitting and empirical evaluation of models for species abundance distributions. Pp. 123–140 in A. E. Magurran and B. J. McGill (eds.). *Biological diversity: frontiers in measurement and assessment*. Oxford University Press, Oxford. [13]

Connor, E. F. and E. D. McCoy. 1979. The statistics and biology of the species–area relationship. *American Naturalist* 113: 791–833. [8]

Connor, E. F. and D. Simberloff. 1978. Species number and compositional similarity of the Galapágos flora and avifauna. *Ecological Monographs* 48: 219–248. [12]

Craine, S. J. 2002. *Rhexia mariana* L. (Maryland Meadow Beauty) New England Plant Conservation Program Conservation and Research Plan for New England. New England Wild Flower Society, Framingham, MA. www.newfs.org/ [2]

Creel, S., J. E. Fox, A. Hardy, J. Sands, B. Garrott and R. O. Peterson. 2002. Snowmobile activity and glucocorticoid stress responses in wolves and elk. *Conservation Biology* 16: 809–814. [4]

Crisp, D. J. 1979. Dispersal and re-aggregation in sessile marine invertebrates, particularly barnacles. *Systematics Association* 11: 319–327. [3]

Cutler, D. R., T. C. Edwards, Jr., K. H. Beard, A. Cutler, K. T. Hess, J. Gibson and J. J. Lawler. 2007. Random forests for classification in ecology. *Ecology* 88: 2783–2792. [11]

Darwin, C. 1859. *On the origin of species by means of natural selection, or the preservation of favoured races in the struggle for life.* John Murray, London. [13]

Darwin, C. 1875. *Insectivorous plants.* Appleton, New York. [1]

Davies, R. L. 1993. Aspects of robust linear regression. *Annals of Statistics* 21: 1843–1899. [9]

Day, R. W. and G. P. Quinn. 1989. Comparisons of treatments after an analysis of variance in ecology. *Ecological Monographs* 59: 433–463. [10]

De'ath, G. 2007. Boosted trees for ecological modeling and prediction. *Ecology* 88: 243–251. [11]

De'ath, G. and K. E. Fabricius. 2000. Classification and regression trees: A powerful yet simple technique for ecological data analysis. *Ecology* 81: 3178–3192. [11]

Denison, D. G. T., C. C. Holmes, B. K. Mallick and A. F. M. Smith. 2002. *Bayesian methods for nonlinear classification and regression.* John Wiley & Sons, Ltd, Chichester, UK. [11]

Dennis, B. 1996. Discussion: Should ecologists become Bayesians? *Ecological Applications* 6: 1095–1103. [4]

Diamond, J. 1986. Overview: Laboratory experiments, field experiments, and natural experiments. Pp. 3–22 in J. Diamond and T. J. Case (eds.). *Community ecology.* Harper & Row, Inc., New York. [6]

Doherty, P. F., G. C. White and K. P. Burnham. 2012. Comparison of model building and selection strategies. *Journal of Ornithology* 152 (Supplement 2): S317–S323. [9]

Doornik, J. A. and H. Hansen. 2008. An omnibus test for univariate and multivariate normality. *Oxford Bulletin of Economics and Statistics* 70 (Supplement 1): 927–939. [12]

Dorazio, R. M. and D. T. Rodríguez. 2012. A Gibbs sampler for Bayesian analysis of site-occupancy data. *Methods in Ecology and Evolution* dx.doi.org/10.1111/j.2041-210x.2012.00237.x [14]

Dorazio, R. M. and J. A. Royle. 2005. Estimating size and composition of biological communities by modeling the occurrence of species. *Journal of the American Statistical Association* 100: 389–398. [14]

Dorazio, R. M., N. J. Gotelli and A. M. Ellison. 2011. Modern methods of estimating biodiversity from presence-absence surveys. Pp. 277–302 in O. Grillo and G. Venora (eds.). Biodiversity loss in a changing planet. InTech Europe, Rijeka, Croatia. [14]

Dunne, J. A., J. Harte and Kevin J. Taylor. 2003. Subalpine meadow flowering phenology responses to climate change: integrating experimental and gradient methods. *Ecological Monographs* 73: 69–86. [10]

Ebert, C., F. Knauer, I. Storch and U. Hohmann. 2010. Individual heterogeneity as a pitfall in population estimates based on non-invasive genetic sampling: a review and recommendations. *Wildlife Biology* 16: 225–240. [14]

Edwards, A. W. F. 1992. *Likelihood: Expanded edition.* The Johns Hopkins University Press, Baltimore. [5]

Edwards, D. 2000. Data quality assurance. Pp. 70–91 in W. K. Michener and J. W. Brunt (eds.). *Ecological data: Design, management and processing.* Blackwell Science Ltd., Oxford. [8]

Efron, B. 1982. The jackknife, the bootstrap, and other resampling plans. *Monographs of the Society of Industrial and Applied Mathematics* 38: 1–92. [5]

Efron, B. 1986. Why isn't everyone a Bayesian (with discussion). *American Statistician* 40: 1–11. [5]

Ellison, A. M. 1996. An introduction to Bayesian inference for ecological research and environmental decision-making. *Ecological Applications* 6: 1036–1046. [3, 4]

Ellison, A. M. 2001. Exploratory data analysis and graphic display. Pp. 37–62 in S. M. Scheiner and J. Gurevitch (eds.). *Design and analysis of ecological experiments,* 2nd ed. Oxford University Press, New York. [8]

Ellison, A. M. 2004. Bayesian inference for ecologists. *Ecology Letters* 7: 509–520. [3, 4]

Ellison, A. M. and B. Dennis. 2010. Paths to statistical fluency for ecologist. *Frontiers in Ecology and the Environment* 8: 362–370. [1, 2]

Ellison, A. M. and E. J. Farnsworth. 2005. The cost of carnivory for *Darlingtonia californica* (Sarraceniaceae): evidence from relationships among leaf traits. *American Journal of Botany* 92: 1085–1093. [8]

Ellison, A. M. and N. J. Gotelli. 2001. Evolutionary ecology of carnivorous plants. *Trends in Ecology and Evolution* 16: 623–629. [1]

Ellison, A. M. et al. 2005. Loss of foundation species: consequences for the structure and dynamics of forested ecosystems. *Frontiers in Ecology and the Environment* 9: 479–486. [13]

Ellison, A. M., A. A. Barker-Plotkin, D. R. Foster and D. A. Orwig. 2010. Experimentally testing the role of foundation species in forests: the Harvard Forest Hemlock Removal Experiment. *Methods in Ecology and Evolution* 1: 168–179. [13]

Ellison, A. M., E. J. Farnsworth and R. R. Twilley. 1996. Facultative mutualism between red mangroves and root-fouling sponges in Belizean mangal. *Ecology* 77: 2431–2444. [10]

Ellison, A. M., N. J. Gotelli, J. S. Brewer, D. L. Cochran-Stafira, J. Kneitel, T. E. Miller, A. C. Worley and R. Zamora. 2003. The evolutionary ecology of carnivorous plants. *Advances in Ecological Research* 33: 1–74. [1]

Ellison, A. M., N. J. Gotelli, E. J. Farnsworth and G. D. Alpert. 2012. *A field guide to the ants of New England*. Yale University Press, New Haven. [13, 14]

Emerson, B. C., K. M. Ibrahim and G. M. Hewitt. 2001. Selection of evolutionary models for phylogenetic hypothesis testing using parametric methods. *Journal of Evolutionary Biology* 14: 620–631. [12]

Englund, G. and S. D. Cooper. 2003. Scale effects and extrapolation in ecological experiments. *Advances in Ecological Research* 33: 161–213. [6]

Farnsworth, E. J. 2004. Patterns of plant invasion at sites with rare plant species throughout New England. *Rhodora* 106: 97–117. [11]

Farnsworth, E. J. and A. M. Ellison. 1996a. Scale-dependent spatial and temporal variability in biogeography of mangrove-root epibiont communities. *Ecological Monographs* 66: 45–66. [6]

Farnsworth, E. J. and A. M. Ellison. 1996b. Sun–shade adaptability of the red mangrove, *Rhizophora mangle* (Rhizophoraceae): Changes through ontogeny at several levels of biological organization. *American Journal of Botany* 83: 1131–1143. [4]

Felsenstein, J. 1985. Confidence limits on phylogenies: An approach using the bootstrap. *Evolution* 39: 783–791. [12]

FGDC (Federal Geographic Data Committee). 1997. FGDC-STD-005: Vegetation Classification Standard. U.S. Geological Survey, Reston, VA. [8]

FGDC (Federal Geographic Data Committee). 1998. FGDC-STD-001–1998: Content Standard for Digital Geospatial Metadata (revised June 1998). U.S. Geological Survey, Reston, VA. [8]

Fienberg, S. E. 1970. The analysis of multidimensional contingency tables. *Ecology* 51: 419–433. [11]

Fienberg, S. E. 1980. *The analysis of cross-classified categorical data*, 2nd ed. MIT Press, Cambridge, MA. [11]

Fisher, N. I. 1993. *Statistical analysis of circular data*. Cambridge University Press, Cambridge. [10]

Fisher, R. A. 1925. *Statistical methods for research workers*. Oliver & Boyd, Edinburgh. [5]

Fisher, R. A., A. S. Corbet and C. B. Williams. 1943. The relation between the number of species and the number of individuals in a random sample of an animal population. *Journal of Animal Ecology* 12: 42–58. [13]

Fiske, I. J. and R. B. Chandler. 2011. unmarked: an R package for fitting hierarchical models of wildlife occurrence and abundance. *Journal of Statistical Software* 43: 10. [14]

Fitzpatrick, M. C., E. L. Preisser, A. M. Ellison and J. S. Elkinton. 2009. Observer bias and the detection of low-density populations. *Ecological Applications* 19: 1673–1679. [14]

Fitzpatrick, M. C., E. L. Preisser, A. Porter, J. S. Elkinton and A. M. Ellison. 2012. Modeling range dynamics in heterogeneous landscapes: invasion of the hemlock woolly adelgid in eastern North America. *Ecological Applications* 22: 472–486. [13]

Flecker, A. S. 1996. Ecosystem engineering by a dominant detritivore in a diverse tropical stream. *Ecology* 77: 1845–1854. [7]

Foster, D. R., D. Knight and J. Franklin. 1998. Landscape patterns and legacies resulting from large infrequent forest disturbance. *Ecosystems* 1: 497–510. [8]

Fox, D. R. 2001. Environmental power analysis: A new perspective. *Environmetrics* 12: 437–448. [7]

Franck, D. H. 1976. Comparative morphology and early leaf histogenesis of adult and juvenile leaves of *Darlingtonia californica* and their bearing on the concept of heterophylly. *Botanical Gazette* 137: 20–34. [8]

Fretwell, S. D. and H. L. Lucas, Jr. 1970. On territorial behavior and other factors influencing habitat distribution in birds. *Acta Biotheoretica* 19: 16–36. [4]

Friendly, M. 1994. Mosaic displays for multi-way contingency tables. *Journal of the American Statistical Association* 89: 190–200. [11]

Frost, T. M., D. L. DeAngelis, T. F. H. Allen, S. M. Bartell, D. J. Hall and S. H. Hurlbert. 1988. Scale in the design and interpretation of aquatic community research. Pp. 229–258 in S. R. Carpenter (ed.). *Complex interactions in lake communities.* Springer-Verlag, New York. [7]

Gabrielson, I. N. 1962. Obituary. *The Auk* 79: 495–499. [14]

Gaines, S. D. and M. W. Denny. 1993. The largest, smallest, highest, lowest, longest, and shortest: Extremes in ecology. *Ecology* 74: 1677–1692. [8]

Garvey, J. E., E. A. Marschall and R. A. Wright. 1998. From star charts to stoneflies: Detecting relationships in continuous bivariate data. *Ecology* 79: 442–447. [9]

Gauch, H. G., Jr. 1982. *Multivariate analysis in community ecology.* Cambridge University Press, Cambridge. [12]

Gelman, A. 2006. Prior distributions for variance parameters in hierarchical models. *Bayesian Analysis* 1: 515–534. [14]

Gelman, A., J. B. Carlin, H. S. Stern, and D. S. Rubin. 2004. *Bayesian data analysis, second edition.* Chapman & Hall, Boca Raton, FL. [5, 11, 14]

Gelman, A., A. Jakulin, M. G. Pittau and Y.-S. Su. (2008). A weakly informative default prior distribution for logistic and other regression models. *Annals of Applied Statistics* 2: 1360–1383. [14]

Gifi, A. 1990. *Nonlinear multivariate analysis.* John Wiley & Sons, Chichester, UK. [12]

Gilbreth, F. B., Jr. and E. G. Carey. 1949. *Cheaper by the dozen.* Thomas Crowell, New York. [6]

Gill, J. A., K. Norris, P. M. Potts, T. G. Gunnarsson, P. W. Atkinson and W. J. Sutherland. 2001. The buffer effect and large-scale population regulation in migratory birds. *Nature* 412: 436–438. [4]

Gnanadesikan, R. 1997. *Methods for statistical data analysis of multivariate observations,* 2nd ed. John Wiley and Sons, London. [12]

Goldberg, D. E. and S. M. Scheiner. 2001. ANOVA and ANCOVA: Field competition experiments. Pp. 77–98 in S. Scheiner and J. Gurevitch (eds.). *Design and analysis of ecological experiments,* 2nd ed. Oxford University Press, New York. [7]

Gotelli, N. J. 2008. *A Primer of Ecology,* 4th ed. Sinauer Associates, Sunderland, MA. [3]

Gotelli, N. J. 2004. A taxonomic wish-list for community ecology. *Transactions of the Royal Society of London B* 359: 585–597. [13]

Gotelli, N. J. and A. E. Arnett. 2000. Biogeographic effects of red fire ant invasion. *Ecology Letters* 3: 257–261. [6]

Gotelli, N. J. and R. K. Colwell. 2001. Quantifying biodiversity: procedures and pitfalls in the measurement and comparison of species richness. *Ecology Letters* 4: 379–391. [12, 13]

Gotelli, N. J. and R. K. Colwell. 2011. Estimating species richness. Pp. 39–54 in A. E. Magurran and B. J. McGill (eds.). *Biological diversity: frontiers in measurement and assessment.* Oxford University Press, Oxford. [13]

Gotelli, N. J. and A. M. Ellison. 2002a. Biogeography at a regional scale: determinants of ant species density in New England bogs and forest. *Ecology* 83: 1604–1609. [6, 9, 12]

Gotelli, N. J. and A. M. Ellison. 2002b. Assembly rules for New England ant assemblages. *Oikos* 99: 591–599. [6, 9, 12]

Gotelli, N. J. and A. M. Ellison. 2006. Food-web models predict abundance in response to habitat change. *PLoS Biology* 44: e324. [7, 9]

Gotelli, N. J. and G. L. Entsminger. 2003. *EcoSim: Null models software for ecology.* Version 7. Acquired Intelligence Inc. & Kesey-Bear. Burlington, VT. Available for free at www.uvm.edu/~ngotelli/EcoSim/EcoSim.html [5]

Gotelli, N. J. and G. R. Graves. 1996. *Null models in ecology.* Smithsonian Institution Press, Washington, DC. [5, 11, 12, 13, 14]

Gotelli, N. J., A. Chao, R. K. Colwell, W.-H. Hwang and G. R. Graves. 2012. Specimen-based modeling, stopping rules, and the extinction of the ivory-billed woodpecker. *Conservation Biology* 26: 47–56. [14]

Gotelli, N. J., A. M. Ellison and B. A. Ballif. 2012. Environmental proteomics, biodiversity statistics, and food-web structure. *Trends in Ecology and Evolution* 27: 436–442. [13]

Gotelli, N. J., A. M. Ellison, R. R. Dunn and N. J. Sanders. 2011. Counting ants (Hymenoptera: Formicidae): biodiversity sampling and statistical analysis for myrmecologists. *Myrmecological News* 15: 13–19. [13, 14]

Gould, S. J. 1977. *Ontogeny and phylogeny.* Harvard University Press, Cambridge, MA. [8]

Gould, S. J. 1981. *The mismeasure of man.* W. W. Norton & Company, New York. [12]

Graham, M. H. 2003. Confronting multicollinearity in ecological multiple regression. *Ecology* 84: 2809–2815. [7, 9]

Green, M. D., M. G. P. van Veller and D. R. Brooks. 2002. Assessing modes of speciation: Range asymmetry and biogeographical congruence. *Cladistics* 18: 112–124. [8]

Gurevitch, J. and S. T. Chester, Jr. 1986. Analysis of repeated measures experiments. *Ecology* 67: 251–255. [10]

Gurevitch, J., P. S. Curtis and M. H. Jones. 2001. Meta-analysis in ecology. *Advances in Ecological Research* 3232: 199–247. [10]

Gurland, J. and R. C. Tripathi. 1971. A simple approximation for unbiased estimation of the standard deviation. *American Statistician* 25: 30–32. [3]

Halley, J. M. 1996. Ecology, evolution, and 1/*f*-noise. *Trends in Ecology and Evolution* 11: 33–37. [6]

Hamaker, J. E. 2002. A probabilistic analysis of the "unfair" Euro coin. www.isip.pineconepress.com/publications/presentations_misc/2002/euro_coin/presentation_v0.pdf [11]

Hand, D. J. and C. C. Taylor. 1987. *Multivariate analysis of variance and repeated measures.* Chapman & Hall, London. [12]

Hanski, I. 1991. Single-species metapopulation dynamics: concepts, models and observations. *Biological Journal of the Linnean Society* 42: 17–38. [14]

Harris, R. J. 1985. *A primer of multivariate statistics.* Academic Press, New York. [12]

Harrison, P. J., I. Hanski and O. Ovaskainen. 2011. Bayesian state-space modeling of metapopulation dynamics in the Glanville fritillary butterfly. *Ecological Monographs* 81: 581–598. [14]

Harte, J., A. Kinzig and J. Green. 1999. Self-similarity in the distribution and abundance of species. *Science* 284: 334–336. [8]

Hartigan, J. A. 1975. *Clustering algorithms.* John Wiley & Sons, New York, New York. [12]

Harville, D. A. 1997. *Matrix algebra from a statistician's perspective.* Springer-Verlag, New York. [Appendix]

Heck, K. L., Jr., G. Van Holle and D. Simberloff. 1975. Explicit calculation of the rarefaction diversity measurement and the determination of sufficient sample size. *Ecology* 56: 1459–1461. [13]

Hilborn, R. and M. Mangel. 1997. *The ecological detective: Confronting models with data.* Princeton University Press, Princeton, NJ. [4, 5, 6, 11]

Hill, M. O. 1973a. Diversity and evenness: a unifying notation and its consequences. *Ecology* 54: 427–432. [13]

Hill, M. O. 1973b. Reciprocal averaging: An eigenvector method of ordination. *Journal of Ecology* 61: 237–249. [12]

Hoffmann-Jørgensen, J. 1994. *Probability with a view toward statistics.* Chapman & Hall, London. [2]

Holling, C. S. 1959. The components of predation as revealed by a study of small mammal predation of the European pine sawfly. *Canadian Entomologist* 91: 293–320. [4]

Horn, H. S. 1986. Notes on empirical ecology. *American Scientist* 74: 572–573. [4]

Horswell, R. 1990. *A Monte Carlo comparison of tests of multivariate normality based on multivariate skewness and kurtosis.* Ph.D. Dissertation, Louisiana State University, Baton Rouge, LA. [12]

Hotelling, H. 1931. The generalization of Student's ratio. *Annals of Mathematical Statistics* 2: 360–378. [12]

Hotelling, H. 1933. Analysis of a complex of statistical variables into principal components. *Jour-*

nal of Educational Psychology 24: 417–441, 498–520. [12]

Hotelling, H. 1936. Relations between two sets of variables. *Biometrika* 28: 321–377. [12]

Hubbard, R. and M. J. Bayarri. 2003. Confusion over measures of evidence (p's) versus errors (α's) in classical statistical testing. *American Statistician* 57: 171–182. [4]

Huber, P. J. 1981. *Robust statistics.* John Wiley & Sons, New York. [9]

Huelsenbeck, J. P., B. Larget, R. E. Miller and F. Ronquist. 2002. Potential applications and pitfalls of Bayesian inference of phylogeny. *Systematic Biology* 51: 673–688. [12]

Huggins, R. M. 1989. On the statistical analysis of capture experiments. *Biometrika* 76: 133–140. [14]

Humboldt, A. 1815. *Nova genera et species plantarum* (7 vols. folio, 1815–1825). [13]

Hurlbert, S. H. 1971. The nonconcept of species diversity: A critique and alternative parameters. *Ecology* 52: 577–585. [11, 13]

Hurlbert, S. H. 1984. Pseudoreplication and the design of ecological field experiments. *Ecological Monographs* 54: 187–211. [6, 7]

Hurlbert, S. H. 1990. Spatial distribution of the montane unicorn. *Oikos* 58: 257–271. [3]

Inouye, B. D. 2001. Response surface experimental designs for investigating interspecific competition. *Ecology* 82: 2696–2706. [7]

Ives, A. R., B. Dennis, K. L. Cottingham and S. R. Carpenter. 2003 Estimating community stability and ecological interactions from time-series data. *Ecological Monographs* 73: 301–330. [6]

Jaccard, P. 1901. Étude comparative de la distribution florale dans une portion des Alpes et du Jura. *Bulletin de la Société Vaudoise des Sciences naturalles* 37: 547–549. [12]

Jackson, C. H. N. 1933. On the true density of tsetse flies. *Journal of Animal Ecology* 2: 204–209. [14]

Jackson, D. A. 1993. Stopping rules in principal component analysis: a comparison of heuristical and statistical approaches. *Ecology* 74: 2204–2214. [12]

Jackson, D. A. and K. M. Somers. 1991. Putting things in order: The ups and downs of detrended correspondence analysis. *American Naturalist* 137: 704–712. [12]

Jackson, D. A., K. M. Somers and H. H. Harvey. 1989. Similarity coefficients: measures of co-occurrence and association or simply measures of occurrence? *American Naturalist* 133: 436–453. [12]

Jaffe, M. J. 1980. Morphogenetic responses of plants to mechanical stimuli or stress. *BioScience* 30: 239–243. [6]

James, F. C. and N. O. Warner. 1982. Relationships between temperate forest bird communities and vegetation structure. *Ecology* 63: 159–171. [13]

Järvinen, O. 1982. Species-to-genus ratios in biogeography: a historical note. *Journal of Biogeography* 9: 363–370. [13]

Jennions, M. D. and A. P. Møller. 2002. Publication bias in ecology and evolution: An empirical assessment using the "trim and fill" method. *Biological Reviews* 77: 211–222. [10]

Jolly, G. M. 1965. Explicit estimates from capture-recapture data with both dead and immigration-stochastic model. *Biometrika* 52: 225–247. [14]

Jost, L. 2006. Entropy and diversity. *Oikos* 113: 363–375. [13]

Jost, L. 2007. Partitioning diversity into independent alpha and beta components. *Ecology* 88: 2427–2439. [13]

Jost, L. 2010. The relation between evenness and diversity. *Diversity* 2: 207–232. [13]

Juliano, S. 2001. Non-linear curve fitting: Predation and functional response curves. Pp. 178–196 in S. M. Scheiner and J. Gurevitch (eds.). *Design and analysis of ecological experiments*, 2nd ed. Oxford University Press, New York. [9]

Kareiva, P. and M. Anderson. 1988. Spatial aspects of species interactions: the wedding of models and experiments. Pp. 38–54 in A. Hastings (ed.). *Community ecology*. Springer-Verlag, Berlin. [6]

Kass, R. E. and A. E. Raftery. 1995. Bayes factors. *Journal of the American Statistical Association* 90: 773–795. [9]

Kéry, M. 2010. *Introduction to WinBUGS for ecologists: a Bayesian approach to regression, ANOVA, mixed models, and related analyses.* Academic Press, Amsterdam. [5]

Kéry, M. and J. A. Royle. 2008. Hierarchical Bayes estimation of species richness and occupancy in spatially replicated surveys. *Journal of Applied Ecology* 45: 589–598. [13]

King, R. 2012. A review of Bayesian state-space modelling of capture-recapture-recovery data. *Interface Focus* 2: 190–204. [14]

Kingsolver, J. G. and D. W. Schemske. 1991. Path analyses of selection. *Trends in Ecology and Evolution* 6: 276–280. [9]

Knapp, R. A., K. R. Matthews and O. Sarnelle. 2001. Resistance and resilience of alpine lake fauna to fish introductions. *Ecological Monographs* 71: 401–421. [6]

Knüsel, L. 1998. On the accuracy of statistical distributions in Microsoft Excel 97. *Computational Statistics and Data Analysis* 26: 375–377. [8]

Koizol, J. A. 1986. Assessing multivariate normality: A compendium. *Communications in Statistics: Theory and Methods* 15: 2763–2783. [12]

Kramer, M. and J. Schmidhammer. 1992. The chi-squared statistic in ethology: Use and misuse. *Animal Behaviour* 44: 833–841. [7]

Kuhn, T. 1962. *The structure of scientific revolutions.* University of Chicago Press, Chicago. [4]

Lakatos, I. 1978. *The methodology of scientific research programmes.* 1978. Cambridge University Press, New York. [4]

Lakatos, I. and A. Musgrave (eds.). 1970. *Criticism and the growth of knowledge.* Cambridge University Press, London. [4]

Lambers, H., F. S. Chapin III and T. L. Pons. 1998. *Plant physiological ecology.* Springer-Verlag, New York. [4]

Laplace, P. S. 1786. Sur les naissances, les mariages et les morts. A Paris, depuis 1771 jusq'en 1784, et dans toute l'étendue de la France, pendant les années 1781 et 1782. *Histoire de L'Académie Royale des Sciences, année 1783*: 35–46. [14]

Larson, S., R. Jameson, M. Etnier, M. Fleming and B. Bentzen. 2002. Low genetic diversity in sea otters (*Enhydra lutris*) associated with the fur trade of the 18th and 19th centuries. *Molecular Ecology* 11: 1899–1903. [3]

Laska, M. S. and J. T. Wootton. 1998. Theoretical concepts and empirical approaches to measuring interaction strength. *Ecology* 79: 461–476. [9]

Lavine, M. 2010. Living dangerously with big fancy models. *Ecology* 91: 3487. [1]

Law, B. E., O. J. Sun, J. Campbell, S. Van Tuyl and P. E. Thornton. 2003. Changes in carbon storage and fluxes in a chronosequence of ponderosa pine. *Global Change Biology* 9: 510–524. [6]

Lawton, J. H. et al. (1998) Biodiversity inventories, indicator taxa and effects of habitat modification in tropical forest. *Nature* 391: 72–76. [13]

Lebreton, J.-D., K. P. Burnham, J. Clobert and D. R. Anderson. 1992. Modeling survival and testing biological hypotheses using marked animals: a unified approach with case studies. *Ecological Monographs* 62: 67–118. [14]

Legendre, P. and M. J. Anderson. 1999. Distance-based redundancy analysis: Testing multi-spe–cies responses in multi-factorial ecological ex–periments. *Ecological Monographs* 69: 1–24. [12]

Legendre, P. and L. Legendre. 1998. *Numerical ecology.* Second English edition. Elsevier Science BV, Amsterdam. [6, 11, 12, Appendix]

Leonard, T. 1975. Bayesian estimation methods for two-way contingency tables. *Journal of the Royal Statistical Society, Series B* (Methodological). 37: 23–37. [11]

Levings, S. C. and J. F. A. Traniello. 1981. Territoriality, nest dispersion, and community structure in ants. *Psyche* 88: 265–319. [3]

Levins, R. 1968. *Evolution in changing environments: Some theoretical explorations.* Princeton University Press, Princeton, NJ. [9]

Levins, R. 1969. Some demographic and genetic consequences of environmental heterogeneity for biological control. *Bulletin of the Entomological Society of America* 15: 237–240. [14]

Lichstein, J. W, T. R. Simons, S. A. Shriner and K. E. Franzreb. 2003. Spatial autocorrelation and autoregressive models in ecology. *Ecological Monographs* 72: 445–463. [6]

Lilliefors, H. W. 1967. On the Kolmogorov-Smirnov test for normality with mean and variance unknown. *Journal of the American Statistical Association* 62: 399–402. [11]

Lilliefors, H. W. 1969. On the Kolmogorov-Smirnov test for the exponential distribution with mean unknown. *Journal of the American Statistical Association* 64: 387–389. [11]

Lincoln, F. C. 1930. Calculating waterfowl abundance on the basis of banding returns. *USDA Circular* 118. [14]

Lindberg, M. S. 2012. A review of designs for capture-mark-recapture studies in discrete time. *Journal of Ornithology* 152: 355–370.

Lindley, D. V. 1964. The Bayesian analysis of contingency tables. *Annals of Mathematical Statistics* 35: 1622–1643. [11]

Link, W. A. and R. J. Barker. 2006. Model weights and the foundations of multimodal inference. *Ecology* 87: 2626–2635. [9]

Loehle, C. 1987. Hypothesis testing in ecology: Psychological aspects and the importance of theory maturation. *Quarterly Review of Biology* 62: 397–409. [4]

Lomolino, M. V. and M. D. Weiser. 2001. Towards a more general species–area relationship: Diversity on all islands, great and small. *Journal of Biogeography* 28: 431–445. [8]

Longino, J. T. and R. K. Colwell. 1997. Biodiversity assessment using structured inventory: Capturing the ant fauna of a lowland tropical rainforest. *Ecological Applications* 7: 1263–1277. [13]

Longino, J. T., J. Coddington and R. K. Colwell. 2002. The ant fauna of a tropical rain forest: estimating species richness three different ways. *Ecology* 83: 689–702. [13]

Looney, S. W. 1995. How to use tests for univariate normality to assess multivariate normality. *American Statistician* 49: 64–70. [12]

Lorenz, M. O. 1905. Methods of measuring concentration of wealth. *American Statistical Association* 70: 209–219. [13]

Lustenhouwer, M. N., L. Nicoll and A. M. Ellison. 2012. Microclimatic effects of the loss of a foundation species from New England forests. *Ecosphere* 3: 26. [13]

MacArthur, R. H. 1962. Growth and regulation of animal populations. *Ecology* 43: 579. [4]

MacArthur, R. H. 1965. Patterns of species diversity. *Biological Reviews* 40: 510–533. [13]

MacArthur, R. H. and E. O. Wilson. 1967. *The theory of island biogeography.* Princeton University Press, Princeton, NJ. [8, 14]

MacKay, D. J. C. 2002. 140 heads in 250 tosses—suspicious? www.inference.phy.cam.ac.uk/mackay/abstracts/euro.html [11]

MacKenzie, D. I. and J. A. Royle. 2005. Designing occupancy studies: general advice and allocating survey effort. *Journal of Applied Ecology* 42: 1105–1114. [14]

MacKenzie, D. I., J. D. Nichols, J. E. Hines, M. G. Knutson and A. B. Franklin. 2003. Estimating site occupancy, colonization, and local extinction when a species is detected imperfectly. *Ecology* 84: 2200–2207. [14]

MacKenzie, D. I., J. D. Nichols, G. B. Lachman, S. Droege, J. A. Royle and C. A. Langtimm. 2002. Estimating site occupancy rates when detection probabilities are less than one. *Ecology* 83: 2248–2255. [14]

MacKenzie, D. I., J. D. Nichols, N. Sutton, K. Kawanishi and L. L. Bailey. 2005. Improving inferences in population studies of rare species that are detected imperfectly. *Ecology* 86: 1101–1113. [14]

MacKenzie, D. I., J. D. Nichols, J. A. Royle, K. H. Pollock, L. L. Bailey and J. E. Hines. 2006. *Occupancy estimation and modeling.* Academic Press, Burlington, MA. [14]

MacNally, R. 2000a. Modelling confinement experiments in community ecology: Differential mobility among competitors. *Ecological Modelling* 129: 65–85. [6]

MacNally, R. 2000b. Regression and model-building in conservation biology, biogeograph, and ecology: The distinction between—and reconciliation of—"predictive" and "explanatory" models. *Biodiversity and Conservation* 9: 655–671. [7]

Madigan, D. and A. E. Raftery. 1994. Model selection and accounting for model uncertainty in graphical models using Occam's window. *Journal of the American Statistical Association* 89: 1535–1546. [11]

Magurran, A. E. 2011. Measuring biological diversity in time (and space). Pp. 85–94 in A. E. Magurran and B. J. McGill (eds.). *Biological diversity: frontiers in measurement and assessment.* Oxford University Press, Oxford. [13]

Magurran, A. E. and P. A. Henderson. 2003. Explaining the excess of rare species in natural species abundance distributions. *Nature* 422: 714–716. [2]

Magurran, A. E. and B. J. McGill (eds.). 2011. *Biological diversity: frontiers in measurement and assessment*. Oxford University Press, Oxford. [13]

Manly, B. F. J. 1991. *Multivariate statistical methods: A primer*. Chapman & Hall, London. [12]

Mardia, K. 1970. Measures of multivariate skewness and kurtosis with applications *Biometrika* 78: 355–363. [12]

Martin, H. G. and N. Goldenfeld. 2006. On the origin and robustness of power-law species-area relationships in ecology. *Proceedings of the National Academy of Sciences, USA* 103: 10310–10315. [8]

May, R. M. 1975. Patterns of species abundance and diversity. Pp. 81–120 in M. L. Cody and J. M. Diamond (eds.). *Ecology and evolution of communities*. Belknap, Cambridge, MA. [2]

Mayr, E. 1963. *Animal species and evolution*. Harvard University Press, Cambridge, MA. [8]

McArdle, B. H. and M. J. Anderson. 2001. Fitting multivariate models to community data: A comment on distance-based redundancy analysis. *Ecology* 82: 290–297. [12]

McArdle, B. H., K. J. Gaston and J. H. Lawton. 1990. Variation in the size of animal populations: Patterns, problems, and artifacts. *Journal of Animal Ecology* 59: 439–354. [8]

McCollough, M. 2011. Eastern puma (= cougar) (*Puma concolor couguar*) 5-year review: summary and evaluation. U. S. Fish and Wildlife Service, Maine Field Office, Orono, ME. [14]

McCullagh, P. and J. A. Nelder. 1989. *Generalized linear models*, 2nd ed. Chapman & Hall, London. [9, 10]

McCullough, B. D. and B. Wilson. 1999. On the accuracy of statistical procedures in Microsoft Excel 97. *Computational Statistics and Data Analysis* 31: 27–37. [8]

McGarvey, R. 2009. Methods of estimating mortality and movement rates from single-tag recovery data that are unbiased by tag non-reporting. *Reviews in Fisheries Science* 17: 291–304. [14]

McGill, B. J. 2011. Species abundance distributions. Pp. 105–122 in A. E. Magurran and B. J. McGill (eds.). *Biological diversity: frontiers in measurement and assessment*. Oxford University Press, Oxford. [13]

McKelvey, K. S., K. B. Aubry and M. K. Schwartz. 2008. Using anecdotal occurrence data for rare or elusive species: the illusion of reality and a call for evidentiary standards. *BioScience* 58: 549–555. [14]

Mead, R. 1988. *The design of experiments: Statistical principles for practical applications*. Cambridge University Press, Cambridge. [7]

Mecklin, C. J and D. J. Mundfrom. 2003. On using asymptotic critical values in testing for multivariate normality. www.interstat.statjournals.net/YEAR/2003/articles/0301001.pdf [12]

Merkt, R. E. and A. M. Ellison. 1998. Geographic and habitat-specific morphological variation of *Littoraria* (*Littorinopsis*) *angulifera* (Lamarck, 1822). *Malacologia* 40: 279–295. [12]

Michener, W. K and K. Haddad. 1992. Database administration. Pp. 4–14 in G. Lauff and J. Gorentz (eds.). *Data management at biological field stations and coastal marine laboratories*. Michigan State University Press, East Lansing, MI. [8]

Michener, W. K. 2000. Metadata. Pp. 92–116 in W. K. Michener and J. W. Brunt (eds.). *Ecological data: Design, management and processing*. Blackwell Science Ltd., Oxford. [8]

Michener, W. K., J. W. Brunt, J. Helly, T. B. Kirchner and S. G. Stafford. 1997. Non-geospatial metadata for the ecological sciences. *Ecological Applications* 7: 330–342. [8]

Miller, J. A. 2003. Assessing progress in systematics with continuous jackknifing function analysis. *Systematic Biology* 52: 55–65. [12]

Mitchell, R. J. 1992. Testing evolutionary and ecological hypotheses using path-analysis and structural equation modeling. *Functional Ecology* 6: 123–129. [9]

Mooers, A. Ø., H. D. Rundle and M. C. Whitlock. 1999. The effects of selection and bottlenecks on male mating success in peripheral isolates. *American Naturalist* 153: 437–444. [8]

Moore, J. 1984. Parasites that change the behavior of their host. *Scientific American* 250: 108–115. [7]

Moore, J. 2001. *Parasites and the behavior of animals.* Oxford Series in Ecology and Evolution. Oxford University Press, New York. [7]

Morgan, J. 1962. The anatomy of income distribution. *The Review of Economics and Statistics* 44: 270–283. [13]

Motzkin, G. and D. R. Foster. 2002. Grasslands, heathlands and shrublands in coastal New England: historical interpretations and approaches to conservation. *Journal of Biogeography* 29: 1569–1590. [13]

Murray, K. and M. M. Conner. 2009. Methods to quantify variable importance: implications for the analysis of noisy ecological data. *Ecology* 90: 348–355. [9]

Murtaugh, P. A. 2002a. Journal quality, effect size, and publication bias in meta-analysis. *Ecology* 83: 1162–1166. [4]

Murtaugh, P. A. 2002b. On rejection rates of paired intervention analysis. *Ecology* 83: 1752–1761. [6, 7]

Newell, S. J. and A. J. Nastase. 1998. Efficiency of insect capture by *Sarracenia purpurea* (Sarraceniaceae), the northern pitcher plant. *American Journal of Botany* 85: 88–91. [1]

Niklas, K. J. 1994. *Plant allometry: The scaling of form and process.* University of Chicago Press, Chicago. [8]

O'Hara, R. B. 2005. Species richness estimators: how many species can dance on the head of a pin. *Journal of Animal Ecology* 74: 375–386. [13]

Olszweski, T. D. 2004. A unified mathematical framework for the measurement of richness and evenness within and among multiple communities. *Oikos* 104: 377–387. [13]

Onken, B. and R. Reardon. 2011. *Implementation and status of biological control of the hemlock woolly adelgid.* U.S. Forest Service Publication FHTET-2011–04, Morgantown, WV. [13]

Orlóci, L. 1978. *Multivariate analysis in vegetation research*, 2nd ed. Dr. W. Junk B. V., The Hague, The Netherlands. [12]

Orwig, D., D. Foster and D. Mausel. 2002. Landscape patterns of hemlock decline in southern New England due to the introduced hemlock woolly adelgid. *Journal of Biogeography* 29: 1475–1487. [14]

Osenberg, C. W., O. Sarnelle, S. D. Cooper and R. D. Holt. 1999. Resolving ecological questions through meta-analysis: Goals, metrics, and models. *Ecology* 80: 1105–1117. [10]

Pacifici, K., R. M. Dorazio and M. J. Conroy. 2012. A two-phase sampling design for increasing detections of rare species in occupancy surveys. *Methods in Ecology and Evolution* 3: 721–730. [14]

Payton, M. E., M. H. Greenstone and N. Schenker. 2003. Overlapping confidence intervals or standard error intervals: What do they mean in terms of statistical significance? *Journal of Insect Science* 3: 34. [13]

Pearson, K. 1900. On the criterion that a given system of deviations from the probable in the case of a correlated system of variables is such that it can be reasonably supposed to have arisen from random sampling. *The London, Edinburgh and Dublin Philosophical Magazine and Journal of Science, Fifth Series* 50: 157–172. [11]

Pearson, K. 1901. On lines and planes of closest fit to a system of points in space. *The London, Edinburgh and Dublin Philosophical Magazine and Journal of Science, Sixth Series* 2: 557–572. [12]

Peres-Neto, P. R., D. A. Jackson and K. M. Somers. 2003. Giving meaningful interpretation to ordination axes: assessing loading significance in principal component analysis. *Ecology* 84: 2347–2363. [12]

Petersen, C. G. J. 1896. The yearly immigration of young plaice into the Limfjord from the German Sea. *Report of the Danish Biological Station to the Home Department*, 6 (1895): 5–84. [14]

Petersen, C. G. J. 1903. What is overfishing? *Journal of the Marine Biological Association* 6: 577–594. [14]

Petraitis, P. 1998. How can we compare the importance of ecological processes if we never ask, "compared to what?" Pp. 183–201 in W. J. Resetarits Jr. and J. Bernardo (eds.). *Experimental ecology: Issues and perspectives.* Oxford University Press, New York. [7, 10]

Petraitis, P. S., A. E. Dunham and P. H. Niewiarowski. 1996. Inferring multiple causality: The limitations of path analysis. *Functional Ecology* 10: 421–431. [9]

Pielou, E. C. 1981. The usefulness of ecological models: A stock-taking. *Quarterly Review of Biology* 56: 17–31. [4]

Pimm, S. L. and A. Redfearn. 1988. The variability of population densities. *Nature* 334: 613–614. [6]

Platt, J. R. 1964. Strong inference. *Science* 146: 347–353. [4]

Podani, J. and I. Miklós. 2002. Resemblance coefficients and the horseshoe effect in principal coordinates analysis. *Ecology*. 83: 3331–3343. [12]

Pollock, K. H. 1982. A capture-recapture design robust to unequal probability of capture. *Journal of Wildlife Management* 46: 752–757. [14]

Pollock, K. H. 1991. Modeling capture, recapture, and removal statistics for estimation of demographic parameters for fish and wildlife populations: past, present, and future. *Journal of the American Statistical Association* 86: 225–238. [14]

Popper, K. R. 1935. *Logik der Forschung: Zur Erkenntnistheorie der modernen Naturwissenschaft.* J. Springer, Vienna. [4]

Popper, K. R. 1945. *The open society and its enemies.* G. Routledge & Sons, London. [4]

Potvin, C. 2001. ANOVA: Experimental layout and analysis. Pp. 63–76 in S. M. Scheiner and J. Gurevitch (eds.). *Design and analysis of ecological experiments*, 2nd ed. Oxford University Press, New York. [6]

Potvin, C., M. J. Lechowicz and S. Tardif. 1990. The statistical analysis of ecophysiological response curves obtained from experiments involving repeated measures. *Ecology* 711: 1389–1400. [10]

Potvin, C., J. P. Simon and B. R. Strain. 1986. Effect of low temperature on the photosynthetic metabolism of the C_4 grass *Echinocloa crus-galli*. *Oecologia* 69: 499–506. [10]

Poulin, R. 2000. Manipulation of host behaviour by parasites: A weakening paradigm? *Proceedings of the Royal Society Series B* 267: 787–792. [7]

Preisser, E. L., A. G. Lodge, D. A. Orwig and J. S. Elkinton. 2008. Range expansion and population dynamics of co-occurring invasive herbivores. *Biological Invasions* 10: 201–213. [14]

Preisser, E. L., M. R. Miller-Pierce, J. L. Vansant and D. A. Orwig. 2011. Eastern hemlock (*Tsuga canadensis*) regeneration in the presence of hemlock woolly adelgid (*Adelges tsugae*) and elongate hemlock scale (*Fiorinia externa*). *Canadian Journal of Forest Research* 41: 2433–2439. [14]

Press, W. H., B. P. Flannery, S. A. Teukolsky and W. T. Vetterling. 1986. *Numerical recipes.* Cambridge University Press, Cambridge. [9, Appendix]

Preston, F. W. 1948. The commonness and rarity of species. *Ecology* 29: 254–283. [2]

Preston, F. W. 1962. The canonical distribution of commonness and rarity: Part I. *Ecology*. 43: 185–215. [8, 9]

Preston, F. W. 1981. Pseudo-lognormal distributions. *Ecology* 62: 355–364. [2]

Price, M. V. and N. M. Waser. 1998. Effects of experimental warming on plant reproductive phenology in a subalpine meadow. *Ecology* 79: 1261–1271. [10]

Pynchon, T. 1973. *Gravity's rainbow.* Random House, New York. [2]

Quinn, G. and M. Keough. 2002. *Experimental design and data analysis for biologists.* Cambridge University Press, Cambridge. [7, 10]

Rao, A. R. and W. Tirtotjondro. 1996. Investigation of changes in characteristics of hydrological time series by Bayesian methods. *Stochastic Hydrology and Hydraulics* 10: 295–317. [7]

Rasmussen, P. W., D. M. Heisey, E. V. Nordheim and T. M. Frost. 1993. Time-series intervention analysis: Unreplicated large-scale experiments. Pp. 158–177 in S. Scheiner and J. Gurevitch (eds.). *Design and analysis of ecological experiments.* Chapman & Hall, New York. [7]

Real, L. 1977. The kinetics of functional response. *American Naturalist* 111: 289–300. [4]

Reckhow, K. H. 1996. Improved estimation of ecological effects using an empirical Bayes method. *Water Resources Bulletin* 32: 929–935. [7]

Reese, W. L. 1980. *Dictionary of philosophy and religion: Eastern and western thought.* Humanities Press, New Jersey. page 572. [4]

Rice, J. and R. J. Belland. 1982. A simulation study of moss floras using Jaccard's coefficient of similarity. *Journal of Biogeography* 9: 411–419. [12]

Ripley, B. D. 1996. *Pattern recognition and neural networks.* Cambridge University Press, Cambridge. [11]

Robson, D. S. and H. A. Regier. 1964. Sample size in Petersen mark-recapture experiments. *Transactions of the American Fisheries Society* 93: 215–226. [14]

Rogers, D. J. 1972. Random search and insect population models. *Journal of Animal Ecology* 41: 369–383. [9]

Rohlf, F. J. and R. R. Sokal. 1995. *Statistical tables,* 3rd ed. W. H. Freeman & Company, New York. [3, 11]

Rosenzweig, M. L. and Z. Abramsky. 1997. Two gerbils of the Negev: A long-term investigation of optimal habitat selection and its consequences. *Evolutionary Ecology* 11: 733–756. [10]

Royle, J. A. and R. M. Dorazio. 2008. *Hierarchical modeling and inference in ecology: The analysis of data from populations, metapopulations, and communities.* Academic Press, London. [13, 14]

Royle, J. A. and R. M. Dorazio. 2012. Parameter-expanded data augmentation for Bayesian analysis of capture-recapture models. *Journal of Ornithology* 152 (Supplement 2): S521–S537. [14]

Royston, J. P. Some techniques for assessing multivariate normality based on the Shapiro-Wilk *W. Applied Statistics* 32: 121–133. [12]

Sackett, T. E., S. Record, S. Bewick, B. Baiser, N. J. Sanders and A. M. Ellison. 2011. Response of macroarthropod assemblages to the loss of hemlock (*Tsuga canadensis*), a foundation species. *Ecosphere* 2: art74. [13]

Sale, P. F. 1984. The structure of communities of fish on coral reefs and the merit of a hypothesis-testing, manipulative approach to ecology. Pp. 478–490 in D. R. Strong, Jr., D. Simberloff, L. G. Abele and A. B. Thistle (eds.). *Ecological communities: Conceptual issues and the evidence.* Princeton University Press, Princeton, NJ. [4]

Salisbury, F. B. 1963. *The flowering process.* Pergamon Press, Oxford. [6]

Sanders, H. 1968. Marine benthic diversity: a comparative study. *The American Naturalist* 102: 243–282. [13]

Sanderson, M. J. and A. C. Driskell. 2003. The challenge of constructing large phylogenetic trees. *Trends in Plant Science* 8: 374–379. [12]

Sanderson, M. J. and H. B. Shaffer. 2002. Troubleshooting molecular phylogenetic analyses. *Annual Review of Ecology and Systematics* 33: 49–72. [12]

Scharf, F. S., F. Juanes and M. Sutherland. 1998. Inferring ecological relationships from the edges of scatter diagrams. *Ecology* 79: 448–460. [9]

Scheiner, S. M. 2001. MANOVA: Multiple response variables and multispecies interactions. Pp. 99–115 in S. M. Scheiner and J. Gurevitch (eds.). *Design and analysis of ecological experiments,* 2nd ed. Oxford University Press, Oxford. [12]

Schick, R. S. et al. 2008. Understanding movement data and movement processes: current and emerging directions. *Ecology Letters* 11: 1338–1350. [14]

Schindler, D. W., K. H. Mills, D. F. Malley, D. L. Findlay, J. A. Shearer, I. J . Davies, M. A. Turner, G. A. Lindsey and D. R. Cruikshank. 1985. Long-term ecosystem stress: The effects of years of experimental acidification on a small lake. *Science* 228: 1395–1401. [7]

Schluter, D. 1990. Species-for-species matching. *American Naturalist* 136: 560–568. [5]

Schluter, D. 1995. Criteria for testing character displacement response. *Science* 268: 1066–1067. [7]

Schluter, D. 1996. Ecological causes of adaptive radiation. *American Naturalist* 148: S40–S64. [8]

Schnabel, Z. E. 1938. The estimation of the total fish population of a lake. *American Mathematical Monthly* 45: 348–352. [14]

Schoener, T. W. 1991. Extinction and the nature of the metapopulation: A case system. *Acta Oecologia* 12: 53–75. [6]

Schofield, M. R. and R. J. Barker. 2008. A unified capture-recapture framework. *Journal of Agricultural, Biological and Environmental Statistics* 13: 458–477. [14]

Schroeder, R. L. and L. D. Vangilder. 1997. Tests of wildlife habitat models to evaluate oak mast production. *Wildlife Society Bulletin* 25: 639–646. [9]

Schroeter, S. C., J. D. Dixon, J. Kastendiek, J. R. Bence and R. O. Smith. 1993. Detecting the ecological effects of environmental impacts: A case study of kelp forest invertebrates. *Ecological Applications* 3: 331–350. [7]

Seber, G. A. F. 1965. A note on the multiple-recapture census. *Biometrika* 52: 249–259. [14]

Shipley, B. 1997. Exploratory path analysis with applications in ecology and evolution. *American Naturalist* 149: 1113–1138. [9]

Shrader-Frechette, K. S. and E. D. McCoy. 1992. Statistics, costs and rationality in ecological inference. *Trends in Ecology and Evolution* 7: 96–99. [4]

Shurin, J. B., E. T. Borer, E. W. Seabloom, K. Anderson, C. A. Blanchette, B. Broitman, S. D. Cooper and B. S. Halpern. 2002. A cross-ecosystem comparison of the strength of trophic cascades. *Ecology Letters* 5: 785–791. [10]

Silvertown, J. 1987. *Introduction to plant ecology*, 2nd ed. Longman, Harlow, U.K. [7]

Simberloff, D. 1978. Entropy, information, and life: Biophysics in the novels of Thomas Pynchon. *Perspectives in Biology and Medicine* 21: 617–625. [2]

Simberloff, D. and L. G. Abele. 1984. Conservation and obfuscation: Subdivision of reserves. *Oikos* 42: 399–401. [8]

Simpson, E. H. 1949. Measurement of diversity. *Nature* 163: 688. [13]

Sjögren-Gulve, P. and T. Ebenhard. 2000. *The use of population viability analyses in conservation planning.* Munksgaard, Copenhagen. [6]

Smith, G. D. and S. Ebrahim. 2002. Data dredging, bias, or confounding. *British Medical Journal* 325: 1437–1438. [8]

Sneath, P. H. A. and R. R. Sokal. 1973. *Numerical taxonomy: The principles and practice of numerical classification.* W. H. Freeman & Company, San Francisco. [12]

Soberón, J. and J. Llorente. 1993. The use of species accumulation functions for the prediction of species richness. *Conservation Biology* 7: 480–488. [13]

Sokal, R. R. and F. J. Rohlf. 1995. *Biometry*, 3rd ed. W. H. Freeman & Company, New York. [3, 4, 5, 8, 9, 10, 11]

Somerfield, P. J., K. R. Clarke and F. Olsgard. 2002. A comparison of the power of categorical and correlational tests applied to community ecology data from gradient studies. *Journal of Animal Ecology* 71: 581–593. [12]

Sousa, W. P. 1979. Disturbance in marine intertidal boulder fields: the nonequilibrium maintenance of species diversity. *Ecology* 60: 1225–1239. [4]

Spearman, C. 1904. "General intelligence," objectively determined and measured. *American Journal of Psychology* 15: 201–293. [12]

Spiller, D. A. and T. W. Schoener. 1995. Long-term variation in the effect of lizards on spider density is linked to rainfall. *Oecologia* 103: 133–139. [6]

Spiller, D. A. and T. W. Schoener. 1998. Lizards reduce spider species richness by excluding rare species. *Ecology* 79: 503–516. [6]

Stewart-Oaten, A. and J. R. Bence. 2001. Temporal and spatial variation in environmental impact assessment. *Ecological Monographs* 71: 305–339. [6, 7]

Stewart-Oaten, A., J. R. Bence and C. W. Osenberg. 1992. Assessing effects of unreplicated perturbations: No simple solutions. *Ecology* 73: 1396–1404. [7]

Strauss, R. E. 1982. Statistical significance of species clusters in association analysis. *Ecology* 63: 634–639. [12]

Strong, D. R. 2010. Evidence and inference: Shapes of species richness-productivity curves. *Ecology* 91: 2534–2535. [10]

Sugar, C. A. and G. M. James. 2003. Finding the number of clusters in a dataset: An information-theoretic approach. *Journal of the American Statistical Association* 98: 750–763. [12]

Sugihara, G. 1980. Minimal community structure: an explanation of species abundance patterns. *American Naturalist* 116: 770–787. [2]

Taper, M. L. and S. R. Lele (eds.). 2004. *The nature of scientific evidence: statistical, philosophical, and empirical considerations.* University of Chicago Press, Chicago, IL. [4]

Thomas, D. L., E. G. Cooch and M. J. Conroy (eds.). 2009. *Modeling demographic processes in marked populations*. Springer Science+Business Media, LLC, New York. [14]

Thomas, L., S. T. Buckland, E. A. Rexstad, J. L. Laake, S. Strindberg, S. L. Hedley, J. R. B. Bishop, T. A. Marques and K. P. Burnham. 2010. Distance software: design and analysis of distance sampling surveys for estimating population size. *Journal of Applied Ecology* 47: 5–14. [14]

Thompson, J. D., G. Weiblen, B. A. Thomson, S. Alfaro and P. Legendre. 1996. Untangling multiple factors in spatial distributions: Lilies, gophers, and rocks. *Ecology* 77: 1698–1715. [9]

Thompson, S. K. 2012. *Sampling*, 3rd ed. Wiley-Blackwell, New York. [14]

Tipper, J. C. 1979. Rarefaction and rarefiction-the use and abuse of a method in paleoecology. *Paleobiology* 5: 423–434. [13]

Tjørve, E. 2003. Shapes and functions of species-area curves: a review of possible models. *Journal of Biogeography* 30: 827–835. [8]

Tjørve, E. 2009. Shapes and functions of species-area curves (II): a review of new models and parameterizations. *Journal of Biogeography* 36: 1435–1445. [8]

Trexler, J. C. and J. Travis. 1993. Nontraditional regression analysis. *Ecology* 74: 1629–1637. [9]

Tuatin, J. 2005. Frederick C. Lincoln and the formation of the North American bird banding program. *USDA Forest Service General Technical Report* PSW-GTR-191: 813–814. [14]

Tufte, E. R 1986. *The visual display of quantitative information*. Graphics Press, Cheshire, CT. [8]

Tufte, E. R. 1990. *Envisioning information*. Graphics Press, Cheshire, CT. [8]

Tukey, J. W. 1977. *Exploratory data analysis*. Addison-Wesley, Reading, MA. [8]

Tuomisto, H. 2010. A diversity of beta diversities: straightening up a concept gone awry. Part 1. Defining beta diversity as a function of alpha and gamma diversity. *Ecography* 33: 2–22. [13]

Turchin, P. 2003. *Complex population dynamics: A theoretical/empirical synthesis*. Princeton University Press, Princeton, NJ. [6]

Ugland, K. I. and J. S. Gray. 1982. Lognormal distributions and the concept of community equilibrium. *Oikos* 39: 171–178. [2]

Underwood, A. J. 1986. The analysis of competition by field experiments. Pp. 240–268 in J. Kikkawa and D. J. Anderson (eds.). *Community ecology: Pattern and process*. Blackwell, Melbourne. [7]

Underwood, A. J. 1994. On beyond BACI: sampling designs that might reliably detect environmental disturbances. *Ecological Applications* 4: 3–15. [6, 7]

Underwood, A. J. 1997. *Experiments in ecology: Their logical design and interpretation using analysis of variance*. Cambridge University Press, Cambridge. [7, 10]

Underwood, A. J. and P. S. Petraitis. 1993. Structure of intertidal assemblages in different locations: How can local processes be compared? Pp. 39–51 in R. E. Ricklefs and D. Schluter (eds.). *Species diversity in ecological communities: Historical and geographical perspectives*. University of Chicago Press, Chicago. [10]

Varis, O. and S. Kuikka. 1997. BeNe-EIA: A Bayesian approach to expert judgement elicitation with case studies on climate change impacts on surface waters. *Climatic Change* 37: 539–563. [7]

Velland, M., W. K. Cornwell, K. Magnuson-Ford and A. Ø. Mooers. 2011. Measuring phylogenetic diversity. Pp. 194–207 in A. E. Magurran and B. J. McGill (eds.). *Biological diversity: frontiers in measurement and assessment*. Oxford University Press, Oxford. [13]

Venables, W. N. and B. D. Ripley. 2002. *Modern applied statistics with S*, 4th ed. Springer-Verlag, New York. [9, 11]

Wachsmuth, A., L. Wilkinson and G. E. Dallal. 2003. Galton's bend: A previously undiscovered nonlinearity in Galton's family stature regression data. *The American Statistician* 57: 190–192. [9]

Wallace, A. R. 1878. *Tropical nature, and other essays*. Macmillan. [13]

Watson, J. D. and F. H. Crick. 1953. Molecular structure of nucleic acids. *Nature* 171: 737–738. [4]

Watts, L. P. and D. Paetkau. 2005. Noninvasive genetic sampling tools for wildlife biologists: a review of applications and recommendations for accurate data collection. *Journal of Wildlife Management* 69: 1419–1433. [14]

Weiher, E. 2011. A primer of trait and functional diversity. Pp. 175–193 in A. E. Magurran and B. J. McGill (eds.). *Biological diversity: frontiers in measurement and assessment.* Oxford University Press, Oxford. [13]

Weiner, J. and O. T. Solbrig. 1984. The meaning and measurement of size hierarchies in plant populations. *Oecologia* 61: 334–336. [3, 13]

Weisberg, S. 1980. *Applied linear regression.* John Wiley & Sons, New York. [9]

Werner, E. E. 1998. *Ecological experiments and a research program in community ecology.* Pp. 3–26 in W. J. Resetarits Jr. and J. Bernardo (eds.). *Experimental ecology: Issues and perspectives.* Oxford University Press, New York. [7]

White, G. C. and K. P. Burnham. 1999. Program MARK: survival estimation from populations of marked animals. *Bird Study* 46 (Supplement): 120–138. [14]

Whittaker, R. J. 2010. Meta-analyses and mega-mistakes: Calling time on meta-analysis of the species richness-productivity relationship. *Ecology* 91: 2522–2533. [10]

Whittaker, R. H. 1967. Vegetation of the Great Smoky Mountains. *Ecological Monographs* 26: 1–80. [12]

Wiens, J. A. 1989. Spatial scaling in ecology. *Functional Ecology* 3: 385–397. [6]

Williams, C. B. 1964. *Patterns in the balance of nature and related problems in quantitative ecology.* Academic Press, London. [13]

Williams, D. A. 1976. Improved likelihood ratio tests for complete contingency tables. *Biometrika* 39: 274–289. [11]

Williams, M. R. 1996. Species–area curves: The need to include zeroes. *Global Ecology and Biogeography Letters* 5: 91–93. [9]

Williamson, M., K. J. Gaston and W. M. Lonsdale. 2001. The species–area relationship does not have an asymptote! *Journal of Biogeography* 28: 827–830. [8]

Willis, E. O. 1984. Conservation, subdivision of reserves, and the anti-dismemberment hypothesis. *Oikos* 42: 396–398. [8]

Wilson, J. B. 1993. Would we recognise a broken-stick community if we found one? *Oikos* 67: 181–183. [2]

Winer, B. J., D. R. Brown and K. M. Michels. 1991. *Statistical principles in experimental design*, 3rd ed. McGraw-Hill, New York. [7, 10]

Wolda, H. 1981. Similarity indices, sample size, and diversity. *Oecologia* 50: 296–302. [12]

Index

Entries printed in *italic* refer to information in a table or illustration. The letter "n" refers to information in a footnote.

About the book

This book was designed by Jefferson Johnson, with illustrations and page layout by The Format Group LLC, Austin, Texas.

The main text, tables, figures, and figure legends are set in Minion, designed in 1989 by Robert Slimbach. The footnote text is set in Myriad, designed in 1991 by Robert Slimbach and Carol Twombly. Fonts were first issued in digital form by Adobe Systems, Mountain View, California.

The cover was designed by Joan Gemme, with original artwork by Elizabeth Farnsworth.

The book and cover were manufactured by The Courier Companies, Inc., Westford, Massachusetts.

Editor: Andrew D. Sinauer

Project Editor: Azelie Aquadro

Review Editor: Susan McGlew

Copy Editor: Randy Burgess

Production Manager: Christopher Small

Project Production Staff: Joan Gemme